Critical Approaches to Science & Philosophy

with a new introduction

Critical Approaches to Science & Philosophy

MARIO BUNGE

EDITOR

Routledge
Taylor & Francis Group

LONDON AND NEW YORK

Originally published in 1964 by The Free Press.

Published 1999 by Transaction Publishers

Published 2017 by Routledge
2 Park Square, Milton Park, Abingdon, Oxon OX14 4RN
711 Third Avenue, New York, NY 10017, USA

Routledge is an imprint of the Taylor & Francis Group, an informa business

Library of Congress Catalog Number: 98-4673

Library of Congress Cataloging-in-Publication Data

Critical approach to science and philosophy.
    Critical approaches to science and philosophy / edited by Mario Bunge, with a new introduction by the editor.
        p. cm. — (Science and technology studies)
    Originally published: The critical approach to science and philosophy. Glencoe, London : The Free Press, 1964.
    Includes bibliographical references.
    ISBN 0-7658-0427-1 (paper : alk. paper)
    1. Philosophy. 2. Science—Methodology. 3. Logic, Symbolic and mathematical. I. Bunge, Mario Augusto. II. Title. III. Series.
B29.C695   1998
100—dc21   98-4673
                                                                              CIP

ISBN 13: 978-0-7658-0427-3 (pbk)

CONTENTS

v

Part III
The Critical Approach to Science

Part IV
The Critical Approach to Society and History

*Deceased, 1961.

Introduction to the Transaction Edition

WHEN this book first appeared in 1964, Popper was little known out-side a small circle of philosophers and scientists. The perception of his stature has greatly changed since then. He was knighted in 1965, and shortly thereafter a large international conference was held in London in his honor. That was when the memorable Popper-Carnap and Popper-Kuhn controversies took place. Since then Popper's views have been discussed by hundreds of scholars and quoted by thousands of scientists. For example, the quarterly *Philosophy of the Social Sciences* has been publishing regularly essays on Popper for over more than a quarter of a century.

Whether Popper's ideas have always been correctly understood is another matter. I doubt that they have, not only because he was always swimming against the current, but also because sometimes he changed his views without saying so. Moreover, he was seldom overly precise. Nor was he the prisoner of the *esprit de système*. Imprecision and unsystematicity always breed multiple interpretations, which in turn breed unending disputes. No wonder that his followers are divided on many important points of exegesis.

Undoubtedly, Popper has been the most influential philosopher of science and social philosopher of the second half of the twentieth century. His *Logic of Scientific Discovery* (1935, or 1934 if you belong to Popper's inner circle) and *The Open Society and Its Enemies* are modern classics. The former all but destroyed positivism, and the latter was the first scholarly study of the philosophical roots of totalitarianism left and right.

One of the reasons for Popper's popularity is that he wrote in a simple and beautiful style, always avoiding unnecessary jargon. (Joseph Conrad comes to mind.) Another reason is that he tackled interesting problems, usually topical ones: he abhorred mental acrobatics as much as pomposity. A third reason is that, even when he was confused, Popper did no pussy-footing: he had strong opinions and expressed them bluntly—to the annoyance of experts in subtlety without substance.

For all these reasons, it is always a delight to read or re-read Popper, even when some of his views—or the lack of them in cases that call for definite views—may be irritating. For myself, I have been reading and re-reading him for over four decades.

I have also had the privilege of enjoying his friendship for a quarter of a century—until we split over the mind-body problem and the question of Popper's "world 3." I visited him and his wife Hennie several times, and we met at various scholarly meetings. Karl and Hennie Popper lived in the small town of Penn, south of London. They had a large and beautiful house, in the middle of a well-tended garden. The house was full of books, and Karl's manuscripts cluttered tables, chairs, and couches. (Karl had made much of the furniture himself, when his father insisted that everyone should have a manual trade.)

Hennie was not only Karl's life companion: she was also his muse and secretary. She was also a gracious hostess. She would not only feed us, but also serve me grape juice, knowing that it was my main addiction at the time. And, when the conversation slackened, she would prompt us, saying: "Go on. You may not have this opportunity again." Unlike Karl, who was a perfectionist, hence a bit of a procrastinator, she knew that life is short. Alas, hers turned out to be shorter than his.

Through the Poppers I got to know a number of members of their circle. Among them were Peter Medawar, Ernest Gombrich, Frederick von Hayek, Joseph Agassi, Imre Lakatos, and John Watkins. Karl and Hennie gave me and my wife Marta the first glimpse of the magnificent English theatre: they took us to Stratford upon Avon to watch *Hamlet.* But enough of reminiscences.

Because of the continuing interest of Popper's philosophy, it has been decided to reissue this volume, which was Popper's first Festschrift—and which pleased him greatly. This book helps one understand Popper's ideas, and it is full of interesting thoughts independent of them. It is also a window into the minds of the members of an extraordinary cast of characters, several of whom have earned the Nobel prize.

Mario Bunge
Foundations & Philosophy
of Science Unit
McGill University
January 1998

Preface

THIS collection of philosophical essays, written on four continents by scientists, philosophers, and humanists, is presented to Karl Raimund Popper on his sixtieth anniversary as a token of critical admiration and in recognition of his work, as well as in expectancy for his future contributions.

The author of *The Logic of Scientific Discovery, The Open Society and its Enemies, The Poverty of Historicism, Conjectures and Refutations,* and other important and controversial works, is acknowledged as one of the most original, penetrating, rigorous, and versatile thinkers of our time. He is also a very influential author, although most of the influence of his thought has been either indirect—namely, through the writings of his students and students of his work—or by way of reaction.

The depth of Popper's thought is related to its amazing range—an indication that nothing human is alien to him. Not unlike the philosophers of old, Popper has worked out and published his thoughts on the most varied subjects: logic, epistemology, metaphysics, physics, sociology, political science, and history. What is most remarkable is that he has not done it in a cursory and discursive way, but with the utmost rigor and an admirable clarity.

The astonishing clarity, depth, and range of Popper's thought, along with his caustic style, may not have helped him find followers prepared to agree with him on all counts. Perhaps many of those who tend to agree with Popper's philosophy of science will tend to disagree with Popper's social philosophy and vice versa. The important thing, however, is not that one would reject this or that idea of Popper's but that it is both instructive and stimulating to rethink what he has thought, whether agreeing with him or not. It is interesting and fruitful, though slightly dangerous, to argue with Popper, because he deals with deep problems in a way that is both careful and ingenious—a rare combination indeed. This is precisely one reason both Popper's followers and his intellectual opponents feel such a great respect for him: because he cannot help but stimulate his readers and listeners. Sometimes this is true to the point of shock, particularly when he sets out to criticize prestigious doctrines or authors.

Criticism is one of Popper's favorite occupations. It is not just that Popper can *afford* to be a merciless critic because he is a creative thinker, but that his *attitude* to problems and their proposed solutions is critical. He knows that we never start from scratch but always from a certain cultural heritage which we must enrich by way of invention, discovery, and criticism. And he knows that invention and discovery, if they are to add to our rational understanding of the world and ourselves, must be criticizable, that is, testable—and criticized, that is, tested.

Of all intellectual endeavors, criticism is the most widespread and the least respected. Criticism can be applied to fact-gathering, to description, and to theory-building. It must be so applied if these activities are to be rational. Criticism, regarded as a minor or even shameful activity in some highly production-minded and success-oriented quarters, is the supervisor of creativity. Henri Poincaré believed that "criticism is the life of science," and Popper feels this is true of all intellectual work aiming at truth, as well as of all social endeavor aiming at freedom. Reason proper is constantly testing its own creations as well as those of intuition; and testing is criticizing.

The progress of knowledge—the perfecting of truths—is not a linear accumulation of definitive acquisitions but a zigzagging process in which counter-examples and unfavorable evidence ruin generalizations and prompt the invention of more comprehensive and sometimes deeper generalizations, to be criticized in their turn. A *critical approach* to problems, procedures, and results in every field of inquiry is, therefore, a necessary condition for the continuance of progress. One would even speak of a *critical method* for advancing knowledge, on any occasion but the present one. (Indeed, Popper's dislike of the word 'method', with the silly connotation of infallible recipe, is so intense that the writer heard him begin one of his lecture courses on scientific method by declaring that his subject did not exist.)

The title of this volume, then, is, in a sense, a homage to Popper's critical rationalism, critical empiricism, and critical realism. It is a tribute to his unceasing and uncompromising quest, not for final certainty, since he is a fallibilist, but for closer truth and increased clarity. Needless to say, ours is a critical homage.

No tribute to Popper could fail to be, at the same time, a homage to his wife Hennie, the closest of his friends and the most efficient of his collaborators. It is the writer's conviction that, were it not for Mrs. Popper, her husband would hardly cease answering letters and retouching—that is, criticizing—his manuscripts and galley proofs. In an important sense she is the coauthor of all of Popper's works and therefore deserves our gratitude.

The general recognition of the importance of Popper's work made it quite easy to compile this volume. In this task the editor had the valuable and friendly advice of Joseph Agassi, William W. Bartley III, Paul K. Feyerabend, Ernst Gombrich, Alan Musgrave, George Nadel, and John Watkins. Dr. Bartley is to be thanked for having compiled the biographical notes on the contributors to this volume, and Mr. Musgrave for having looked after Popper's bibliography. The editor's thanks are also due to the editorial staff of The Free Press of Glencoe.

MARIO BUNGE

Biographical Notes
on the Contributors

AGASSI, JOSEPH : Associate Professor of Philosophy, University of Illinois, and former chairman of the Department of Philosophy, University of Hong Kong. Author of *Towards an Historiography of Science,* and of papers published in *British Journal for the Philosophy of Science, British Journal of Sociology, Economica, Isis,* and *Mind.*

ALBERT, HANS : Professor of General Sociology and Methodology, Wirtschaftshochschule, Mannheim, Germany. Author of *Oekonomische Ideologie und politische Theorie,* and of numerous papers on problems in the social sciences and philosophy of science.

BARTLEY, WILLIAM WARREN III : Lecturer in the History of the Philosophy of Science, Warburg Institute, University of London; Tutor in Philosophy at the London School of Economics. Author of *The Retreat to Commitment,* and of papers in *British Journal for the Philosophy of Science, Commentary, Harper's, The New Republic, Papers of the Architectural Association, Philosophical Studies,* and other journals. Associate Editor of *History and Theory.*

BERNAYS, PAUL : Professor emeritus of Mathematics, Eidgenössiche Technische Hochschule, Zürich. Author of *Grundlagen der Mathematik* (with D. Hilbert) and *Axiomatic Set Theory* (with A. Fraenkel), and of papers in *Abhandlungen der Fries'schen Schule, Dialectica, Journal of Symbolic Logic, Mathematische Annalen, Mathematische Zeitschrift, Ratio, Synthèse,* and other journals. Membre titulaire of the Académie internationale de philosophie des sciences, Brussels; and Member of the Norwegian Academy of Science.

BOHM, DAVID : Professor of Theoretical Physics, Birkbeck College, University of London. Author of *Quantum Theory* and *Causality and Chance in Modern Physics,* and of papers in *British Journal for the Philosophy of Science, Physical Review,* and other journals.

BRIDGMAN, PERCY W. : Late Higgins University P   ʃ at Harvard University. Author of *Logic of Modern Physics;   .ysics of High Pressure;*

Thermodynamics of Electrical Phenomena in Metals; Nature of Physical Theory; Reflections of a Physicist; The Nature of Some of Our Physical Concepts, and numerous other books and papers. Foreign Member, The Royal Society, London. Nobel Laureate in Physics, 1946.

BUNGE, MARIO : Professor of Philosophy and former professor of Theoretical Physics, University of Buenos Aires, Argentina. Author of *Causality; Metascientific Queries; Intuition and Science; The Myth of Simplicity; Cinemática del electrón relativista; Etica y ciencia,* and other works, and of papers in *Nature, American Journal of Physics, Nuovo Cimento, Austria Physica Acta, Minerva, British Journal for the Philosophy of Science, Methodos, Philosophy of Science, Journal of Philosophy, Mind,* and other journals. President of the Agrupación Rioplatense de Lógica y Filosofía Científica.

CURRY, HASKELL BROOKS : Evan Pugh Research Professor, Pennsylvania State University. Author of *Outlines of a Formalist Philosophy of Mathematics; A Theory of Formal Deducibility; Leçons de logique algébrique; Combinatory Logic* (with Robert Feys), and of *Foundations of Mathematical Logic,* and of numerous papers. Member, Académie internationale de philosophie des sciences, Brussels. Recipient of medal and title of Honorary Visiting Professor, University of Louvain.

DAVIES, J. T. : Head of Department and Professor of Chemical Engineering, University of Birmingham, England. Author of *Interfacial Phenomena* (with Sir Eric Rideal), and of numerous papers in chemical engineering, physical chemistry, and philosophy of science.

ECCLES, JOHN C. : Professor of Physiology, Australian National University. Author of *Neurophysiological Basis of Mind: The Principles of Neurophysiology,* and *The Physiology of Nerve Cells,* and of papers in *Journal of Physiology, Journal of Neurophysiology, Proceedings of the Royal Society,* and *Ergebnisse der Physiologie.* Fellow of the Royal Society, Member of the Pontifical Academy of Sciences, Fellow of the Australian Academy of Science, Foreign Honorary Member of the American Academy of Arts and Sciences. Nobel Laureate 1963.

FEIGL, HERBERT : Professor of Philosophy, University of Minnesota, and Director of the Minnesota Center for Philosophy of Science. Co-editor of *Readings in Philosophical Analysis; Readings in the Philosophy of Science; Minnesota Studies in the Philosophy of Science,* Vols. I, II, and III; and *Current Issues in the Philosophy of Science;* and author of numerous papers in *Erkenntnis, Journal of Philosophy, Philosophy of Science, Journal of Unified Science, Journal of Symbolic*

Logic, and other journals. President, Western Division, American Philosophical Association.

FEYERABEND, PAUL K. : Professor of Philosophy, University of California at Berkeley. Author of papers in *British Journal for the Philosophy of Science, Proceedings of the Aristotelian Society, American Journal of Physics, Zeitschrift für Physik, Minnesota Studies in the Philosophy of Science, Philosophical Studies, Philosophical Review, Kantstudien, Daedalus,* and other journals.

FRISCH, O. R. : Jacksonian Professor of Natural Philosophy, University of Cambridge. Author of *Meet the Atoms* and *Atomic Physics Today,* and of numerous scientific papers. Order of the British Empire. Fellow of the Royal Society.

GALLIE, W. B. : Professor of Logic and Metaphysics, Queen's University of Belfast. Author of *An English School; Peirce and Pragmatism* and *A New University: A. D. Lindsay and the Keele Experiment;* and of papers in *Mind, Philosophy, French Studies, History and Theory, British Journal for the Philosophy of Science, Proceedings of the Aristotelian Society,* and other journals.

GEYL, PIETER : Professor emeritus of Modern History, University of Utrecht. Author of *The Revolt of the Netherlands; The Netherlands in the 17th Century; Napoleon, For and Against; Debates with Historians; Encounters in History;* and numerous other books and papers. Honorary doctor of Harvard, Oxford, and St. Andrews Universities. Fellow of the Royal Netherlands Academy. Fellow of the Royal Flemish Academy. Fellow of the British Academy. Commander of the British Empire.

GIEDYMIN, JERZY : Lecturer in Logic and Philosophy of Science, University of Poznań, Poland. Author of *Logical Problems of Historical Enquiry,* and of papers in *Studia Logica, British Journal for the Philosophy of Science, Studia Filozoficzne,* and *Studia Zrodoznaweze.*

GOMBRICH, E. H. : Director of the Warburg Institute and Professor of the History of the Classical Tradition, University of London; and Slade Professor of Fine Art, University of Cambridge. Author of *Weltgeschichte für Kinder; Caricature* (with E. Kris); *The Story of Art;* and *Art and Illusion;* and of papers in the *Journal of the Warburg and Courtauld Institutes, The Burlington Magazine, Proceedings of the British Academy, Daedalus, Proceedings of the Aristotelian Society, International Journal of Psycho-Analysis, Gazette des Beaux Arts,* and other journals. Fellow of the British Academy. Fellow of the Society of Antiquaries.

GRÜNBAUM, ADOLF : Andrew Mellon Professor of Philosophy, University of Pittsburgh. Author of *Philosophical Problems of Space and Time,* and of numerous papers in *Philosophy of Science, Review of Metaphysics, Analysis, American Scientist, Journal of Philosophy, Scripta Mathematica, British Journal for the Philosophy of Science, The Scientific Monthly, Philosophical Review, Nature,* and other journals.

HARE, R. M. : Fellow of Balliol College, Oxford. Author of *The Language of Morals,* and of numerous papers in *Mind, Proceedings of the Aristotelian Society, Journal of Philosophy, Analysis,* and other journals.

HAYEK, FRIEDRICH A. VON : Professor of Economics, University of Freiburg i. B., Germany. Author of *Prices and Production; Monetary Theory and the Trade Cycle; Monetary Nationalism and International Stability; The Pure Theory of Capital; The Road to Serfdom; Individualism and Economic Order; John Stuart Mill and Harriet Taylor; The Counter-Revolution of Science; The Sensory Order; The Constitution of Liberty;* and other works. Fellow of the British Academy.

KNEALE, WILLIAM : White's Professor of Moral Philosophy, University of Oxford. Author of *Probability and Induction; The Development of Logic* (with Martha Kneale); and of papers in *Mind, Proceedings of the Aristotelian Society, Philosophical Quarterly, Philosophy, Analysis,* and *British Journal for the Philosophy of Science.* Fellow of the British Academy. Honorary Doctor of the University of Aberdeen.

MUNZ, PETER : Associate Professor of History, The Victoria University of Wellington, New Zealand. Author of *The Place of Hooker in the History of Thought; Problems of Religious Knowledge;* and *The Origin of the Carolingian Empire;* and of papers in *Hibbert Journal, Philosophical Quarterly, Cambridge Journal; Journal of Ecclesiastical History, Dialectica, Historical Studies,* and other journals.

NADEL, GEORGE H. : Honorary Nuffield Fellow, Warburg Institute. Author of *Australia's Colonial Culture,* and of papers in *Journal of the History of Ideas, Comparative Studies, Church History, Historical Studies,* and other journals. Editor of *History and Theory.*

RESCHER, NICHOLAS : Professor of Philosophy, University of Pittsburgh. Author of *Epistemology of the Inexact Sciences* (with Olaf Helmer); *Al-Farabi: An Annotated Bibliography; Al-Farabi's Short Commentary on Aristotle's Prior Analytics; Introduction to Logic;* and numerous papers in *Journal of Symbolic Logic, Journal of the History of Ideas, Philosophical Review, Archiv für Geschichte der Philosophie, Journal of Philosophy,* and other journals.

STAHL, GEROLD : Professor of Logic, The University of Chile, Santiago. Author of *Introducción a la lógica simbólica; Enfoque moderno de la lógica clásica,* and of papers in *Zeitschrift für mathematische Logik und Grundlagen der Mathematik, Revista de Filosofía de la Universidad de Chile,* and *Revue Philosophique.* President of the Sociedad Chilena de Lógica, Metodología y Filosofía de la Ciencia.

ULLMO, JEAN : Maître de Conférences on Mathematics at the École Polytechnique, Paris. Author of papers on mathematics and on the philosophy of physics, and of *La pensée scientifique moderne.*

WATKINS, J. W. N. : Reader in the History of Philosophy, London School of Economics and Political Science, University of London. Author of papers in *British Journal for the Philosophy of Science, Mind, Philosophy, Ratio, Philosophical Studies, Philosophical Quarterly, The Listener,* and other journals. Holder of Distinguished Service Cross.

WISDOM, J. O. : Reader in Logic and Scientific Method, London School of Economics and Political Science, University of London. Author of *The Metamorphosis of Philosophy; Foundations of Inference in Natural Science; The Unconscious Origin of Berkeley's Philosophy;* and of numerous papers. Editor of the *British Journal for the Philosophy of Science.* Past President of the Society for Psychosomatic Research.

YOURGRAU, WOLFGANG : Professor of Philosophy of Science, University of Denver; member of the Denver Research Institute. Author of *Variational Principles in Dynamics and Quantum Theory* (with S. Mandelstam), and of papers and reviews in *British Journal for the Philosophy of Science, Physics Today, Methodos, Philosophy, Nature, Journal of Symbolic Logic, Philosophy and Phenomenological Research, Nuclear Physics, Scientia, Theoria,* and other journals.

I

THE CRITICAL APPROACH
TO PHILOSOPHY

CHAPTER 1

Rationality versus the Theory of Rationality

BY WILLIAM W. BARTLEY III

I

THE THREE PRINCIPAL PROBLEMS of philosophy are the problem of knowledge, the problem of rationality, and the problem of reconciling knowledge and rationality.[1]

This essay will be concerned primarily with the second problem, to which it will present a solution. I will assume as correct most of the epoch-making contributions Karl Popper has made to the first problem, that of knowledge and its growth, which has been one of his principal philosophical interests. And I will use many of his remarks which bear more or less directly on the other problems. Thus the essay is in part an application of his philosophical thought; but it is also an attempt to interpret and generalize it, and to test and illustrate its power by extending it to further fundamental problems.[2]

---

[1] I shall use "intellectualism" to designate the movement of seventeenth-century rationalism, represented by philosophers such as Descartes and Spinoza, and reserve the words "reason," "rationality," and "rationalism" to refer broadly to that tradition whose members are dedicated to learning more through critical discussion. One of my main hopes will be to make this latter sense more precise. Needless to say, this view of rationalism is not at all dependent on the false notion that men can act rationally most of the time.

[2] Almost all other philosophical problems are directly related yet subordinate to the three fundamental ones. Of those not so related the most important are certain cosmological problems that may be deemed philosophical; and even these very often connect at least indirectly with the main three. For example, Descartes's most important cosmological theory—his identification of matter with extension—is directly sponsored by his views about the growth of knowledge.

The third problem, that of conflict between knowledge and rationality, typically arises when it is found that according to one's theory of rationality, knowledge is impossible; or that according to one's theory of knowledge, rationality is impossible. This conflict is usually occasioned by the existence of another conflict within the theory of rationality itself, a conflict that appears when it is discovered that according to one's theory of rationality, rationality is impossible. The problem of resolving the latter conflict I take to be the *problem of rationality* or of the *limits of rationality*.

My attempt to solve this problem will proceed in the following manner. First, in Section II, I shall indicate the kind of limitation of rationality I have in mind, specify the most important philosophical positions this limitation engenders, and summarize the claim and argument on which these positions rest. Then in Sections III and IV, I shall explain briefly why it is important to refute these positions by solving the problem, and shall outline my strategy for dealing with it. I shall proceed to consider, in Sections V–VII, the historical background behind the construction of theories of rationality, and will discuss critically two influential but unsuccessful specimen theories of rationality. In Sections VIII–X, as an historical hypothesis, I shall specify several important but unrecorded philosophical dogmas I believe are responsible for the perennial failure of theories of rationality. Finally in Sections XI and XII, I shall present an alternative theory of rationality which, freed of these dogmas, is able successfully to solve the problem of rationality and its limits.

II

There are, in the history of philosophy, a number of different senses in which rationality has been viewed as limited. On the one hand, some philosophers have maintained, as did Kant in some of his moods, that there are certain factual limitations connected with the nature of thought, or with the psychological and biological structure of the human mind. Or the limits of human reason may be thought to coincide with the limits of sense experience; or it may be argued that they coincide with the limits of science, which cannot provide answers to all questions that may be posed to it. Other philosophers have stressed certain physical conditions in nature, largely independent of man's psyche and its limitations, which make scientific investigation exceedingly difficult if not impossible in certain areas. The existence of radiation chaos,[3] or the velocity of light, might be cited as limiting our research somewhat in certain parts of the universe.

Still other limitations are said to be connected with our historical existence and with the impossibility of predicting the historical future

---

[3] See K. R. Popper, "The Aim of Science," *Ratio*, 1 (1957), p. 35.

because—among other reasons—we cannot predict the future growth of human knowledge.[4] This limitation Popper himself has particularly stressed : we learn by refuting our present theories, by deriving predictions from them and trying to falsify those predictions; but we cannot derive or predict a refutation of these theories *from* these theories (provided they are consistent). Related limitations appear in physics and economics due to the existence of indeterminacy and "feedback." Yet another limitation is connected with the necessarily selective character of description. Then there are the practical limitations of rationality such as those explored by Freud—due to human weakness, physical frailty, humanity.

In what follows, I shall not be at all concerned with any of these limitations. Rather, I shall deal with a so-called *logical* limitation of rationality, which is sufficient, without aid from other limitations,[5] to perpetuate the aforementioned conflict between rationality and the theory of rationality.

The logical problem of rationality can be considered as that of defeating the two main philosophical positions that have grown from the claim that rationality is logically limited : *skepticism* and *fideism*. Paradoxically, although both positions are, I believe, inimical to philosophy, both are sheltered by similar philosophical arguments. To defeat these self-appointed guardians of the boundaries of reason it is necessary (1) to deal with a philosophical claim both make and (2) to refute the argument on which both rest their claim.

The *claim* is that *from a rational point of view, the choice between competing beliefs and positions and ways of life, whether scientific, mathematical, moral, religious, metaphysical, political, or other, is arbitrary.* In short, it is claimed to be demonstrable by rational argument that it is logically impossible to act and decide on rational grounds when it comes to such choices—even though the making of such choices in a nonarbitrary way can be considered to be the main task of rationality.

The core of the *argument* used by skeptics and fideists to back their claim consists in a simple analysis of what is commonly regarded as the rational way to defend ideas.[6] The argument based on this analysis might

---

[4] K. R. Popper, *The Poverty of Historicism*, Preface, pp. ix-xi.

[5] This is not, of course, to deny that in individual philosophical discussions, the previously stated limitations—some of which have more to do with the limitation of knowledge than the limitation of rationality—have often been fused and even confused with the particular limitation that concerns me. Indeed, some writers who have advocated fideistic irrationalism, Pascal for example, have blended together many different kinds of rationality limitations in their writings—in order, for instance, to stress the weakness and misery of any man who insists on relying on his pitiful reason "without God."

[6] For a discussion of some related issues, see my article, "Achilles, the Tortoise, and Explanation in Science and History," *British Journal for the Philosophy of Science*, May 1962.

be called the argument about the limits of rationality, the *tu quoque* or boomerang argument,[7] the dilemma of ultimate commitment, the problem of ultimate presuppositions, or any number of other names. The argument is a commentary on the fact that any view may be .challenged with such questions as "How do you know?" "Give me a reason," or even "Prove it!" When such challenges are accepted by the citation of further reasons which entail those under challenge, these may be questioned in turn. And so on forever. Yet if the burden of proof or rational justification can be perpetually shifted back to a higher-order premise or reason, the belief originally questioned is never effectively defended. In order to justify the original conclusion, it appears that one must eventually stop at something not open to question, for which one need not provide reasons when demanded. Usually, but not necessarily, the role of halting point of rational discussion is played by such things as standards, criteria, ends, or goals which one accepts irrationally in order to avoid infinite regress. But if to defend a position rationally is to give good reasons in justification of it, it would appear that this stopping point is not rationally defensible. If men conclude their rational justifying at different points, moreover, "ultimate relativism" arises; for some way of choosing rationally among competing ultimate stopping points by appeal to a common standard is excluded *in principle* by the way the problem is set. Even if all men did subjectively stop at the same place, the problem would remain of determining rationally whether this universal subjective stopping point led to objectively true statements about the world.

Skepticism and fideism, then, share both the claim and the powerful argument which underlies it. They differ in their opposed practical attitudes toward the claim. Whereas the skeptic suspends his judgment about competing positions, or so he says, the fideist makes an irrational commitment to one or another of them, or to some authority or tradition claiming to possess the competence or the right to make such decisions for him. The claim and the argument are, however, far more basic features of the two positions than are the different concluding attitudes they choose : a position that cannot escape skepticism cannot refute fideism.

III

Philosophical incapacity to answer the fideistic and skeptical claim and argument has some serious consequences. In the first place, the argument, if correct, implies that it is pointless from a rational point of view for men to argue rationally about their extremely different "ultimate presuppositions" or commitments. T. S. Eliot, who found a way to become a

---

[7] The name "boomerang argument" was suggested to me by Professor Hans Albert.

fideist Christian when he had learnt why his teacher Bertrand Russell could not refute skepticism, embraced such an implication in *After Strange Gods*, in which he wrote :

> I am not arguing or reasoning or engaging in controversy with those whose views are radically opposed to such as mine. In our time, controversy seems to me, on really fundamental matters, to be futile . . . . It requires common assumptions.

So it would seem that the limits of rational argument within any ultimate philosophical position are defined by reference to that object or belief in respect to which commitment is made or imposed, in regard to which argument is brought to a close. And if, since the limitation is a logical one, all men share it, if *no one* can escape irrational commitment, then no one can be criticized rationally for having made such a commitment, no matter how idiosyncratic. Bertrand Russell chose to commit himself to the so-called scientific principle of induction, and T. S. Eliot chose Anglo-Catholicism; and that is all there is to it, so it is often argued.

This means that any irrationalist has a *rational excuse for irrationalism* and a secure refuge from any criticism of this irrational commitment. The irrationalist can reply *"tu quoque"* to any critic, and remind him that people whose own rationality is similarly limited should not berate others for admitting to and acting on the limitation.

On the other hand, it should be noted that however useful his rational excuse for irrationalism may seem, the irrationalist pays a high price for using it. The strength and survival of the *tu quoque* argument does indeed make rational criticism of irrationalist commitment boomerang. But the *tu quoque* argument itself also has a boomerang effect on those who use it. I have discussed this point at some length elsewhere, with examples, and wish only to suggest the character of this backfire here by quoting a brief passage from my discussion : [8]

> To the extent that anyone employing [the *tu quoque*] strengthens his own position by insuring that it is parallel to his opponent's, to that extent he increases the invulnerability of the *opponent* to criticism. For the opponent, if criticized, may also use the *tu quoque*. Those who gain a refuge of safety for themselves through appeal to the limits of rationality thereby provide a similar refuge for all others whose commitments differ from theirs. Thus,

---

[8] See W. W. Bartley III, *The Retreat to Commitment* (New York: Alfred A. Knopf, Inc., 1962), Chap. IV, section 2, "How Shoes Can Be Danced Into Holes," pp. 95–104. In this book I have applied and developed some aspects of the present arguments in a less general way in terms of a specific case study of the development and stresses of American Protestant religious thought in the past century, particularly in its turn from "Protestant liberalism" or "religious modernism" to "neo-orthodoxy." See also my discussions in my dissertation for the Ph.D., "Limits of Rationality: A Study of Some Logical Problems in Contemporary Pragmatism and Related Movements," University of London Library.

the many criticisms which [irrationalists] have leveled at rationalism and liberalism become as pointless as those the liberals have directed at [irrationalism]. Ultimately, the use of the *tu quoque* makes nonsense of the idea of the historical development and change of ideas in the face of criticism. . . . In sum, the belief that rationality is ultimately limited, by providing an excuse for irrational commitment, enables . . . any . . . irrationalist to make an irrational commitment without losing intellectual integrity. But at the same time, anyone who makes use of this excuse may not, in integrity, *criticize* the holder of a different commitment. One gains the right to be irrational at the expense of losing the right to criticize. One gains immunity from criticism for one's own commitment by making any criticism of commitments impossible.[9]

IV

If the cost of his excuse makes it hard for the irrationalist to live comfortably with it, this does not reduce the argumentative effectiveness of the excuse on his opponent. Is the rational excuse for irrationalism really flawless from a rational point of view? Or can the excuse be defeated with rational arguments?

Before attempting to answer these questions, I would like to make the task more manageable; and without distorting any issues or minimizing any difficulties. At the same time I want to focus the skeptico-fideistic argument as sharply as possible in a single example by considering some specific tradition, philosophical viewpoint, or way of life of our culture to see whether irrational commitment is really required of its adherents. Some necessary care in the choice of a tradition to be examined will, I think, satisfy both these desires. The tradition I propose to study briefly is the rationalist tradition itself. Although this is very much a way of life in its own right, with its own history, its proponents rarely consider it historically and critically. Since, in any case, our main problem is about the limits of rationality in the choice of competing ways of life, we might turn the problem on itself by asking *what are the limits of rationality when it comes to choosing between the rationalist and some other way of life?* Must the rationalist, for instance, make an irrational commitment to reason? If so, would this not preclude him from rationalism, a tradition which has taught that everything one accepts must be rationally defensible? Is a real rationalist, then, even *possible* theoretically?

In dealing with the development of the rationalist tradition, its triumphs and vicissitudes, and with the growth and occasional decay of rationality within it, I have found helpful the two concepts of "crisis of

---

[9] *The Retreat to Commitment*, pp. 96–97, 103.

identity" and "crisis of integrity."[10] In their technical psychological sense, these terms are meant to refer to two turning points in the lives of most people. The problem of identity, usually encountered in late adolescence and early adulthood, is to mold out of one's heritage, one's conception of oneself, and one's conjectures about the image one presents to others, some satisfactory self-image, identity, or personal identification. The problem of integrity, which typically occurs in later life, although it is often fused with the problem of identity in the life of the *homo religiosus* or *philosophicus*, is this : Given one's identity, purpose, and claims about oneself, how can one live up to them, and cope with the fact that one can rarely if ever be entirely successful in such aims?

My thesis is that the perpetual crisis of integrity into which rationalists are continually falling or being forced is due to a neglected crisis of identity in the rationalist tradition; neglected partly because of a general failure on the part of philosophers to make deliberate efforts to develop a theory of rationality as well as a theory of knowledge. Because of these crises, the valuable handy man in the house of irrationalism—the *tu quoque*—is the skeleton in the cupboard of rationalism. Rationalists are overcommitted to a notion of rationality, or rationalist identity, that is impossible to attain; and the inevitable frustration of the effort to satisfy this overcommitment prevents them from achieving integrity. At the same time, the failure of the rationalist tradition to resolve its crisis of integrity enables many irrationalists, whatever their affiliations, to preserve their own identities without loss of integrity.

My attitude toward this situation is suggested by an anecdote about Gottlob Frege that Popper enjoys telling occasionally in his lectures. When Frege heard about Russell's discovery of paradoxes in his own and Frege's theories, the latter cried out : *"Die Arithmetik ist in's Schwanken geraten!"* (Roughly, "Arithmetic has been set spinning !") In fact, it was not arithmetic, but Frege's theory of arithmetic, that went spinning in vicious circles. Such mistakes Popper has dubbed (doubtless a bit unfairly to Frege) examples of "Frege's Mistake." A similar mistake is made about rationality and the possibility of being a rationalist. The blame for continued failure by rationalists to answer skeptical and fideistic arguments should, in fact, be placed on the inadequacy and primitive character of our *theories of rationality* or on our conception of rationalist identity, rather than on our rationality or reasoning capacity itself, where Pascal, Kant, and many others have put it. It is not the miserable state of the human creature that sponsors irrationalistic fideism and skepticism, but

---

[10] I use these terms in a sense I borrow from the noted sociologist and psychoanalyst Erik H. Erikson, of Harvard University. See his *Childhood and Society* (London: Imago Publishing Company, 1950); *Young Man Luther* (New York: W. W. Norton & Company, 1958), p. 14 *passim*; and *Identity and the Life Cycle, Psychological Issues*, Monograph 1, 1959. See also my discussion of Erikson in *The Retreat to Commitment*, pp. 3, 10, 59, 218.

the miserable state of the philosophical theory of rationality that that human creature has accepted.

To illustrate this thesis I shall single out three different answers to the problem of rationality, three stages in the quests for identity and integrity in the rationalist tradition. Needless to say, these stages are not to be interpreted as exhaustive. They are *comprehensive rationalism, critical rationalism,* and *comprehensively critical rationalism.*

The history of modern philosophy is largely that of the failure of *comprehensive rationalism* to defeat skepticism and fideism. Most contemporary philosophies are forms of *critical rationalism. Comprehensively critical rationalism,* the third stage of this triad, is my attempt to weave together the comprehensive aims of comprehensive rationalism and the critical spirit of critical rationalism—and thereby to preserve the good intentions of the two earlier concepts while avoiding their fatal difficulties. Although my theory about rationalist identity rejects most traditional and contemporary characterizations, I believe it can both solve the main problem in response to which theories of rationalist identity arose and avoid those aspects that engendered the crisis of integrity in rationalism.

V

How did theories of rationalist identity originate? And which of their tenets or assumptions brought them to defeat before the argument about the limits of rationality?

The problem of rationality, and of rationalist identity, has been the standard and controlling predicament of philosophy since the late sixteenth century, although its origins are much earlier and its classical developments considerably later. It had already been discussed extensively in antiquity; and during the Hellenistic period it became for a short time dominant in philosophical discussion, its most detailed statement at that period having been made by Sextus Empiricus. For understandable reasons, the problem became centrally important again in modern intellectual history during the Renaissance and Reformation, when for the first time since the secular decline of Rome men were forced to choose among a great number of radically competing views. Identical arguments about the limits of rationality could be used skeptically by Protestants, Humanists, and scientists to attack the traditional authority and intellectual pretensions of the Roman Catholic Church; and fideistically by the Roman Catholics to attack the Protestant and scientific rejection of the authority of the Pope and of Aristotle in the name of reason; and again fideistically by the Protestants to answer Roman Catholic allegations that Protestantism led inevitably to skepticism. Similar arguments about the limits of rationality were widely quoted in support of fideism and skepticism during

Descartes's lifetime, and importantly influenced the emphases and themes of his philosophical program. Indeed, his thought can hardly be understood without some acquaintance with the currents of skeptical thought in European philosophy flowing not only from the religious and political upheavals and the scientific revolution, but also from such events as the Latin translation and publication in 1562 of the *Hypotyposes* of Sextus Empiricus, whose ideas were widely disseminated during the following century.[11]

At their modern origins, theories of rationality served a strong utilitarian purpose. It was important for those who challenged traditional ecclesiastical and political authorities to be able to show that disputes could nevertheless still be settled in an orderly way; that traditional political, religious, and intellectual authorities, having come to seem rather arbitrary and irrational, could be displaced without producing social anarchy and intellectual chaos since they would be replaced by the authority of reason. The various schools of modern philosophy arose in an attempt to adjudicate among competing views by providing *rational* authorities to substitute for unwanted forms of traditional and hereditary authority. Since then, almost every "philosophical revolution," each being a phase in the search for rationalist identity, has disclosed that the previous candidate for intellectual authority was unsatisfactory and has proposed a new, supposedly more satisfactory, rational authority. The church was to be replaced by intellectual intuition, intellectual intuition by sense experience, sense experience by a certain language system, and so on. The story is almost always the same: past philosophical error is to be given a positive explanation by attributing it to the acceptance of the guidance of a false rational authority. Where these successive revolutions are fundamentally defective—apart from such assumptions as that error rather than knowledge needs to be explained—is in a depressingly similar structural pattern they share, which predestines them to failure.

VI

The traditionally dominant, and perhaps still most common, conception of rationalist identity—comprehensive rationalism—can be traced back at least to Epictetus, who wrote in his *Discourses* (Chapter II) that "To be a reasonable creature, that alone is insupportable which is unreasonable; but everything reasonable may be supported." Such conceptions combine two main requirements for rationalist identity.

---

[11] Sextus Empiricus, *Works* (Cambridge, Mass.: Harvard University Press), 4 vols. See also Richard H. Popkin: *The History of Scepticism from Erasmus to Descartes* (Assen: Van Gorcum; 1960).

1. In the first place, a rationalist accepts any position that can be justified or established or supported by appeal to the rational criteria or authorities.
2. In the second place, a rationalist accepts *only* those positions that can be justified in this way.

In their highly trite and stylized way, histories of modern philosophy usually say nothing about theories of rationality taken as a whole, but instead focus their attention on several basically subordinate questions which arise only when comprehensive rationalism is assumed as correct. Of these subordinate questions, the most important has probably been : "What is the nature of the rational authority or criterion to which a rationalist appeals to justify all his opinions ?" The well-known modern theories of knowledge are usually functions of the answers philosophers have given to the question, and fall into two main categories :

1. According to the intellectualists (or Rationalists in the seventeenth century sense), the rational authority is to be found in the intellect or Reason. A rationalist justifies his beliefs by appealing to intellectual intuition.
2. According to empiricists, the rational authority lies in sense experience. An empiricist justifies his beliefs by appealing to sense observation.

There is no need to tell the familiar story of the failure of both these approaches. To use Popper's phrase, they were "too narrow and too wide."[12] Both intellectualism and empiricism endorsed as rational some obviously irrational theories (they were too wide), and excluded as irrational some theories that seemed highly rational (they were too narrow). Kant, with his antinomies, showed that intellectualism was too narrow; and Hume and others brought to light the narrowness of sense experience, which could not even provide a rational endorsement for so great an achievement as Newton's dynamics.

Although a recalcitrant comprehensive rationalist might diffidently admit his failure to meet such objections and thus far to have provided an adequate version of rationalism in which all rational beliefs could be justified and all irrational ones excluded, some such person might insist that this indicated no more than temporary bad luck, to be followed by the eventual triumph of some kind of comprehensive rationalism. That such optimism would be misguided can be seen in the fact that the two requirements for comprehensive rationalism cannot be held simultaneously. If we take the second requirement seriously, we will have to justify the first; but this cannot be done by sense experience, intellectual intuition, or any other rational authority ever proposed. Moreover, any such justification

---

[12] See Popper, "The Demarcation Between Science and Metaphysics," in *Conjectures and Refutations* (New York: Basic Books; 1963).

of the practice of accepting the results of argument, even if it could be achieved, would be pointless unless it were accepted in advance that a justification should be accepted at least here—which is just what is in part at issue. The argument would be acceptable and convincing only to those who had already adopted the principle that arguments should count.

As a consequence, the first and second requirements cannot be maintained simultaneously, for the second forbids the holding of any unjustifiable principles. This means that at least one of the requirements must be rejected. Surely this must be the second; for it can be shown by independent argument to be self-contradictory since it too cannot be justified by appeal to the rational criteria or authorities. Therefore the requirement asserts its own unacceptability : if it is true, it must, according to its own stipulations, be rejected.[13]

Such a collapse of comprehensive rationalism would seem to strengthen the position of the fideist, and has in fact done so historically. If the so-called rationalist not only cannot justify his own presuppositions rationally, if his position is basically self-contradictory, an irrationalist need hardly worry if his commitments are unjustifiable. He may really seem better off, since he did not claim to be able to justify them rationally, and he therefore surpasses the rationalist in integrity. Hence, when a rationalist accuses another of irrationalism, the irrationalist can reply that what is impossible cannot be morally demanded, and that the sort of comprehensive rationalism which the rationalist demands is, *rationally speaking,* impossible.

VII

That such difficulties have battered rationalists severely, from antiquity to the present, and have even driven some, if reluctantly, into outright irrationalism, can hardly be doubted. Many rationalists, however, have attempted to face and dispose of their difficulties without abandoning rationalism. Their responses to the problem, crude and sophisticated alike, amount to quests for some alternative theory of rationalist identity that might escape the quandaries of a primitive comprehensive rationalism.

These attempts vary widely in substance as well as sophistication. Here I propose to examine one prominent representative of what I regard as the most influential contemporary approach to the theory of rationality, which

---

[13] Some of the components of my argument here are taken from Popper, *The Open Society and Its Enemies* (Princeton: Princeton University Press, 1950), Chap. XXIV, p. 416 *passim.* I believe that Popper's formulation, which shows that the position—by asserting its own falsity—is self-contradictory in a way analogous to the original statement of the liar paradox (which was, strictly speaking, nonparadoxical), is an important improvement on previous attempts to refute comprehensive rationalism.

I shall give the name of "critical rationalism."[14] In applying this name to a number of philosophers who differ widely among themselves on important philosophical issues, I suggest only that their approaches to skepticism and fideism, and to the problem of the limits of rationality, share some common fundamental characteristics. The representative I have chosen to illustrate such an approach is A. J. Ayer, who, in 1956—having earlier abandoned the comprehensive rationalism of sense experience he championed in *Language, Truth and Logic* (1936)—presented a variety of critical rationalism in *The Problem of Knowledge*. Along with Morton White's *Toward Reunion in Philosophy,* published in the same year, Ayer's statement provides a most perceptive and self-conscious development of the general approach—and one which is lucid and precise enough to make the implications of the position clear. Ayer's account has the added merit of being explicitly devoted to considering and answering the skeptic's claims.

Ayer's strategy is to minimize the importance of the skeptical arguments while granting their cogency, to make the skeptic's *victory* "bloodless" or even "fictitious."[15] He flatly concedes to any critic of skeptical or fideist bent that it is impossible to provide a rational justification of one's basic philosophical standards, principles, procedures—among which Ayer seems to regard "inductive inference" as one of the most important. It is, he says, impossible to give a proof "that what we regard as rational procedure really is so; that our conception of what constitutes good evidence is right."[16]

Whatever its faults, Ayer's approach does possess the initial advantage of honesty about the unjustifiability of ultimate standards and principles. By candidly discarding the claim that all legitimate positions must be rationally justifiable, he seems to avoid the problem of integrity, of claiming the ability to do more than he logically can. This is the main reason for calling his position a *critical* rationalism. If one's theory of rationality does not *require* one's standards of rationality to be rationally justifiable, then the skeptic's argument that it is *impossible* to do so loses its force even when it is admitted as correct.

It is not, of course, sufficient for Ayer simply to abandon the demand

---

[14] Although the critical rationalism discussed here resembles terminologically the view Popper discusses under the same name in *The Open Society,* Chap. XXIV, and elsewhere, the problem he was discussing there is somewhat different from my own; his argument was concerned less to defeat the rational excuse for irrational commitment than to deal with more radical and—to my mind—less intellectually serious forms of irrationalism which were prevalent during the pre-war and wartime years when *The Open Society* was being conceived and written. I believe Popper would largely agree with the view I call "comprehensively critical rationalism," a view that I could hardly have stated without the help of his views.

[15] A. J. Ayer, *The Problem of Knowledge,* pp. 75 and 80.

[16] *Ibid.,* p. 74.

that his standards of rationality be justified. He must go on to show why his approach, *as a theory of rationality,* can dispense with the justifiability requirement in respect to standards. If Ayer could do this he would indeed thereby trivialize his admission that the skeptic succeeds in showing the impossibility of any such justification. The skeptic's argument about impossibility would "draw blood" from the rationalist only if the latter needed such justification.

As might be expected, Ayer's discussion conveys the impression that he is fully alert to the scope of his task. He writes of the importance of showing, "in a way that satisfactorily disposes of the skeptic's disproof," that the procedures "which sustain our claim to knowledge . . . do not *require* a proof of their legitimacy."[17]

But Ayer's response to his task, as contrasted with his appraisal of it, is curious and disappointing. Why do our standards of rationality not need rational justification? Simply, according to Ayer, because any such standard "could be irrational only if there were a standard of rationality which it failed to meet; whereas *in fact* it goes to set the standard : arguments are judged to be rational or irrational by reference to it."[18] "When it is understood," Ayer explains, "that there logically could be no court of superior jurisdiction, it hardly seems troubling that inductive reasoning should be left, as it were, to act as judge in its own cause. The skeptic's merit is that he forces us to see that this must be so."[19] Since there can be no proof that what we take to be good evidence really is so," then "it is not sensible to demand one."[20]

There are several reasons why Ayer's position is unsatisfactory. Here I shall consider only the question of whether his viewpoint, even if assumed to be internally coherent, is successful *as a theory of rationality.* Unfortunately, the answer to this question is negative. The nub of the skeptical and fideistic objection was not, after all, simply the argument that comprehensive justification is impossible. It was, rather, that *since* comprehensive justification is impossible, the choice between competing ultimate positions is arbitrary. It is this argument, in its strongest form, that a successful theory of rationality would have to meet.

Viewed in this light, Ayer's argument falls to pieces. For (1) it begs the question; and (2) it can be shown to be itself a variety of fideism and hence no answer to it.

First, how does it beg the question? His argument is that our standards of rationality enjoy a special immunity from the justification demand since it would be impossible to judge these standards to be irrational.

---

[17] *Ibid.* (my italics).
[18] *Ibid.*, p. 75.
[19] *Ibid.*
[20] *Ibid.*, p. 81.

These standards, Ayer argues, *set the standards on which any such judgment of their own irrationality would have to be based.*

Ingenious as it might seem, Ayer's argument could not be relevant to the present discussion, let alone valid, *unless* some particular standards and procedures of rationality, such as Ayer's own, which include "scientific induction," are *assumed* to be the correct ones. If some particular standards of rationality *are* the correct ones, then there can exist no other rational standards which are also correct but which can nevertheless invalidate the former as irrational. Yet this "if" marks a very big assumption; for *this is precisely what is at issue in the problem of rationality.* Whether regarded historically or theoretically, the most important criticisms of putative standards of rationality have questioned *whether* they were correct. Alternative conceptions of scientific argument and method, such as Popper's, which denies the *existence* of inductive procedure, let alone its rational legitimacy, do claim that there are standards of rationality which positions like Ayer's fail to meet.

Not surprisingly, a position that begs the question in this way turns out to be fideistic itself. Indeed, the "ultimate irrational commitment" of the fideist might alternatively be described as a self-conscious deliberate begging of the question. The main doubt about whether Ayer's position is fideistic would be whether in this case the question-begging is conscious and deliberate.

Ayer's position is fideistic in other respects. Apart from suggesting that any critic's demands would not be sensible, Ayer says rather little about how an unjustifiable rationalist position might be defended against a critic —whether a fellow philosopher-critic like Popper, or the most hyperbolical irrationalist—who simply does not "understand" why logically there can be no court of superior jurisdiction to arbitrate among standards like the principle of induction, whose rationality he does not *in fact* accept.

On this point Morton White has been more explicit. Although it might be asked whether Ayer fully shares White's position here, White has made it clear, in a review of *The Problem of Knowledge,* that he shares Ayer's. In the course of his review White warmly endorses Ayer's book and discusses some of its similarities to various philosophical movements in America, particularly pragmatism and its successors. When we want to defend something like the "general practice of basing our knowledge claims on experience," White suggests, "we can do no more than appeal to the *accepted code* for the transmissibility of the right to be sure . . . to the accepted way of speaking."[21] To test the adequacy of his analyses, the philosopher can do no more than check them "against the moral convic-

---

[21] Morton White, *Religion, Politics and the Higher Learning* (Cambridge: Harvard University Press, 1959), p. 47.

tions which he and others share."[22] Consequently the rationalist position, unable to be rationally based or justified, is finally based on irrational moral commitment. And the choice of this commitment is throughout dominated by *conservative* attitudes toward the best entrenched standards already accepted by one's own philosophical community—indeed, just as conservative as those of the sixteenth- or seventeenth-century fideists who argued on similar grounds for adherence to the Church of Rome. How apt then—and how ironic—that in a different connection, in his book *Toward Reunion in Philosophy,* White should express his general preference of Erasmus to Luther. He writes :

Sometimes . . . the sanitation [i.e., removal of intellectual "refuse"] can be carried on with crusading enthusiasm, with a sense of deep antipathy to a philosophy which one opposes on every major issue, and which one would like to see extirpated. But at other times criticism can be conducted in the spirit of Erasmus rather than Luther, by one who admires a philosophical framework so much that he wishes to cleanse it of its shortcomings rather than demolish it. Much of the negative part of this book is intended in the second spirit, since I believe that the analytic, the empiricist, and the pragmatic movements . . . have been the most important and enlightening tendencies in twentieth-century philosophy.[23]

How can one tell when to adopt the spirit of Erasmus and when that of Luther? If the final appeal is to the code and convictions of one's colleagues, and if the view under consideration is one shared with them— indeed admired by them—then are the cards not stacked in favor of Erasmus? If White had lived during the Reformation and had practiced then the method he advocates in the twentieth century, would he not, like Erasmus, have remained within the fold of the Roman Catholic Church, fully critical of its shortcomings, but admiring its framework too much to want to see it extirpated?

One further, rather interesting point brings out more clearly the fideistic character of positions like Ayer's and White's. However radically their substantive opinions may differ, from a structural standpoint the positions of arch rationalists and antitheologians like Ayer and White are closely parallel not only, as might be expected, to fideistic positions like contemporary "Oxford theology," but also that of the arch theologian and belligerently fideistic irrationalist Karl Barth, the leading Protestant Christian exponent of modern-day Calvinist "neo-orthodoxy." In order to appreciate this striking parallel, it is necessary to note that as an alternative to the task of justifying rational standards Ayer sees the task of the philosopher as that of *describing* them—an important enough task if critical judgments are to be made by reference to the accepted procedures

---

[22] *Ibid.,* p. 8.
[23] Morton White, *Toward Reunion in Philosophy* (Cambridge: Harvard University Press, 1956), pp. 289–90.

of rationality; to avoid errors we would need accurate descriptions of these procedures.[24] White agrees with Ayer that the business of the philosophically minded person is to analyse and to express in principles the patterns of the accepted ways of speaking or reasoning.

Interestingly enough, Barth argued that although the content of the "Word of God" might be expounded or described, it is neither possible nor necessary for the theologian to apologize for it, to defend it against skeptical criticism by trying to justify it. To do so is not only logically impossible anyway, but more strongly—since logical impossibility does not necessarily worry a theologian—it is unnecessary and irreverent for anyone genuinely committed to the "Word of God." There can be no assessment of the Word of God or of the Christian's ultimate commitment to it *since* the Word of God is itself *assumed* to be the standard or criterion which any such assessment would have to use.[25]

In a similar vein, the theologian Paul Tillich blended fideism with a little "theological positivism" to write that "The assertion that something has sacred character is meaningful only for the asserting faith . . . . There is no criterion by which faith can be judged from outside the correlation of faith."[26] Ayer did not put the point much differently: "Inductive reasoning should be left, as it were, to act as judge in its own cause."[27] In Ayer's case as in Barth's, when we confront that to which commitment has been made—just as when we read *Wind in the Willows*—"it is we who are being judged."[28]

Indeed, the shift made by Ayer and others from comprehensive rationalism to critical rationalism can be described quite accurately in the traditional theological terminology. Both Ayer and the Christian fideists abandon *apologetic theology,* the kind of theology which tries to give rational justification for commitment, and replace it with *kerygmatic theology,* the theology devoted exclusively to the exposition and description of the fundamental message. Unable to justify his basic position, the logical empiricist, just like the neo-orthodox theologian, begins to describe the position, to preach it without regard to the critical situation within

---

[24] At this point Ayer's position links closely with many themes of the so-called Oxford philosophy of descriptive analysis, and also with the positions of many other leading American philosophers, such as W. V. Quine and Nelson Goodman. Goodman's unsuccessful but instructive attempts to describe the principles of inductive reasoning are chronicled in his *Fact, Fiction and Forecast.* For a discussion and criticism of Quine's holistic view, which radically differs from my own despite misleading similarities, see my book *The Retreat to Commitment,* where I discuss and reject his theory about the revisability of logic, and propose a "revisability criterion."

[25] See Barth's *Church Dogmatics,* particularly Vol. I, Part I, and my discussion of it in *The Retreat to Commitment,* Chap. III.

[26] Paul Tillich, *Dynamics of Faith* (New York: Harper & Brothers, 1957), pp. 58 ff.

[27] *The Problem of Knowledge,* p. 75.

[28] See the introduction by A. A. Milne to Kenneth Grahame's well-known book.

which it is being considered. As these basic similarities in their positions indicate, Ayer and like-minded rationalist philosophers beg the main question as flagrantly as do Barth and those who may share his commitments. While Barth begs the question of the existence and righteousness of God and his Word, Ayer begs the question of the existence and legitimacy of induction. And in common with various fideistic stances, Ayer's position bars in advance criticism of his fundamental standards by laying down what might be called a "persuasive definition" of "rational" in terms of which his own standards are automatically rational.[29]

Presented with such a defensive definition, a critic can perhaps refuse to be persuaded by the prestigious title and ask whether, if what is being suggested is "being rational," it is right to be rational. He might also ask whether someone with a genuine concern to criticize and test his standards should not—instead of comparing them with the "accepted code" and the convictions of the group of which he is a member—seriously explore the views of those who are convinced the code is wrong but who may, for all that, be able to defend their ideas, and perhaps even be right. But more of this latter point in the following sections.

In sum, Ayer's position, his critical rationalism—like thousands of similar stands taken by contemporary philosophers—may indeed be one that can be held with considerable intellectual integrity, or with at least as much as that of any other fideistic position. But it is not a satisfactory theory of rationality. Indeed, it can only be held with integrity because it *fails* to solve the problem of rationality. To be satisfactory, a theory of rationality which, like Ayer's, began with the admission that the standards of rationality were unjustifiable, would have to go on to show—without begging the question—that the arbitrary irrationalism of skepticism and fideism could be escaped without comprehensive justification. This task Ayer's theory fails even to attempt.

VIII

Critical rationalism, like comprehensive rationalism, was doomed to failure by its structure; no future theory of rationality will succeed until this structure is, to use White's word, "extirpated." By the word "structure," I have in mind nothing more mysterious than certain features of these theories of rationality which predetermine the kinds of questions asked in philosophical discussion and limit the range of answers deemed appropriate. My attempt to bring this structure to light in its historical

---

[29] Herbert Feigl had already made the last point, though not in reference to Ayer, in a fascinating paper published some four years before Ayer's book. See Feigl's "Validation and Vindication: An Analysis of the Nature and the Limits of Ethical Arguments," in *Readings in Ethical Theory*, p. 676.

context, to break it, and to put forward an alternative begins firstly from an important historical observation or discovery about our traditional philosophical questions: namely, that *our philosophical tradition is authoritarian in structure, even in its most liberal forms.* My second step will be to show how two unfamiliar and unnoticed philosophical dogmas, assumed throughout our philosophical tradition without question, are responsible for this structure. These two dogmas are : (1) The assumption that criticism is necessarily fused with justification; and (2) The assumption that the quality and degree of rationality pass through the relationship of logical deducibility from justifying premises to justified conclusions.

The authoritarian structure of our philosophical questioning has been concealed from the attention of philosophers, and might still be doubted by some of them, because of the standard oversimplified way in which the rise of modern philosophy is described as part of a rebellion against authority. So it was—but a rebellion against *particular* traditional authorities, not against authority as such. Far from repudiating the very appeal to authority, modern philosophy has entertained seriously only one alternative to the practice of basing opinions on irrational and traditional authorities; namely, that of basing them on *rational authorities.*

The traditional questions of philosophy are authoritarian in structure in the sense that *they all beg authoritarian answers.* Questions like "How do you know?" "How do you justify your beliefs?" or "With what do you guarantee your opinions?" demand authoritarian answers, whether those authorities in particular cases be the Bible, the leader, the social class, the nation, the fortune teller, the Word of God, the intellectual intuition, or sense experience. A demand for a justification or a guarantee cannot be answered except by providing something authoritative in the sense that it is unquestionable, does not itself need justification, and hence can guarantee the correctness of a conclusion.

This historical observation I owe to an address Popper read to the British Academy in 1960 : "On the Sources of Knowledge and of Ignorance."[30] His simple but brilliant observation has, I believe, an almost revelatory character when the history and problems of philosophy are reconsidered in its light. I shall try to build on this observation, first by putting it in some philosophical context, then by explaining it, and finally by suggesting the principal outlines of my new theory of rationality—comprehensively critical rationalism—which can be erected within the new, roomier structure Popper's observation makes possible.

But first we might ask whether Popper himself had an alternative

---

[30] Printed in *The Proceedings of the British Academy,* 1960, and published separately in 1961 by Oxford University Press (Henriette Hertz Trust monographs). The address is reprinted as the Introduction to *Conjectures and Refutations.* A preliminary statement of the view appears in Popper, "On the Sources of Our Knowledge," *Indian Journal of Philosophy,* August 1959.

approach to suggest. What new questions would be asked in philosophy if the old ones were abandoned because of their authoritarian character? Indeed, is it really possible to forsake the old questions without abandoning philosophy? Is a nonauthoritarian theory of rationality possible?

I think the importance of Popper's observation should have become apparent already, before it has even been applied, in that the questions just posed, and most interestingly the last one, could not even *arise* until the observation about the authoritarian character of traditional philosophy had been made. Yet, as we shall see below, these new questions lead directly to the solution of the problem of rationality.

In explaining the implications of his observation, Popper not only rejected, as had Ayer, the demand for rational proofs of our rational standards; he went further, to suggest that the demand that *everything else except* the standards be proved by appealing to infallible intellectual authorities (which do not exist anyway) should also be abandoned. Instead of following Ayer in replacing philosophical justification by philosophical description, Popper urged the *criticism* of standards as the main task of the philosopher. Philosophers, he argued, should not demand and search for infallible intellectual authorities, but should instead try to build a philosophical program for counteracting intellectual error. Rather than search for an infallible intellectual authority to guarantee the truth of his views, the philosopher should seek a way of eliminating error from them. Within such a program, the traditional "How do you know?" question would not legitimately arise. And if it arose in fact, the philosopher would have to reply : *"I do not know; I have no guarantees."* To be more precise, such a philosopher might elaborate : "Some of the theories I hold may in fact be true; but since there are no criteria of truth, I can never know for sure whether what I believe to be true is in fact so." For such a philosopher a different question would become important : "How can our intellectual life and institutions be arranged so as to expose our beliefs, conjectures, policies, sources of ideas, traditional practices, and the like— whether justifiable or not—to maximum criticism, in order to counteract and eliminate as much intellectual error as possible?

Coming from Popper, of course, a program of criticism is no surprise. In this context it is a very general extension of the familiar "criticize and try to falsify" approach he uses in the philosophy of science. Specific elements of his solution, including a discussion of the way philosophical questions acquire an authoritarian structure, are already to be found in his discussion of political leadership in *The Open Society,* Chapter VII. But the explicit link he establishes in the academy address between the more familiar parts of his theory of science on one hand, and the general structuring of philosophical inquiry on the other, is remarkably fresh and revealing, not only in regard to the solution of philosophical problems, but also in assessing Popper's own contributions, in setting his position in its proper philosophical context, and in helping to explain why it is so often misunderstood.

I believe that the general shift from the demand for authoritative justification to the demand for criticism is a genuine innovation in philosophy —by no means limited to the "falsification versus verification" issue— whose importance can hardly be overemphasized. All the same, it might be objected that the "innovation" I have been heralding so confidently is imaginary; that the emphasis on the critical evaluation of competing views, far from being novel, is just another rather off-key refrain of the "hymn to criticism" which has been in the philosophical psalter since the pre-Socratics. Talk of the critical attitude tends to become boring when almost everybody fervently approves of it—or says he does. Indeed, I have several times heard philosophers deplore Popper's "tiresome repetitiveness" about the critical attitude. "One hears it in every lecture. As if such things need to be said—particularly to philosophers!"

Although such reactions are often understandable, they are to be deplored. Such closing of the ears at the mention of the word "criticism" suggests that neither the words nor the music of Popper's particular song have been heard. And this is a pity, for the notion of criticism, far from being trite, is one of the most unexplored, puzzling, and perhaps exciting and rewarding areas of philosophy. To show why such reactions are mistaken, and to explain why this particular kind of tone deafness is so prevalent, I wish to try to bring out as clearly as possible the crucial difference between the *new* idea of criticism I believe is being advocated here and the old familiar themes of traditional critical philosophies.

IX

This can be done fairly directly by asking for an explanation of Popper's observation. *Why has an authoritarian structure been retained— and even gone unnoticed—in modern philosophies that have been intentionally and often explicitly antiauthoritarian and critical in spirit?*

The following assumption provides the most important element of the answer to this question : *The task of solving the problem of critical rational arbitration among competing positions has been frustrated by the fact that in almost all traditional and modern philosophies—those that have called themselves critical as well as those that have not—the idea of criticism has been fused with the idea of justification.*[31] Since demands for justification are, of course, satisfied by the appeal to authority, the dilemma of

---

[31] A number of historical positions doubtless approached toward a separation of justification and criticism. That Kant may have been doing so is suggested by his ambivalent use and description of the categorical imperative. However, I know of no philosophy before Popper's that succeeded in separating the two.

ultimate commitment arises *in respect to criticism* in such philosophies. As a group, the philosophies in which this fusion of justification and criticism occurs may be called justificational philosophies of criticism.

The main originality of Popper's position lies in the fact that it is the first *nonjustificational philosophy of criticism* in the history of philosophy. This aspect of his thought may help explain why some misunderstandings of his position have been so persistent. Many of his critics, not surprisingly failing to see this point, have tried to deal with his thought within a philosophical structure and system of classification foreign to it, and have consequently led themselves to conclude that Popper's theory is subject to the same difficulties as its predecessors. I suspect that the fusion of justification and criticism in Ayer's thought may explain why he turned to description when justification broke down; for criticism would only appear as an *alternative* to justification after the two notions had been separated.

The purpose of the view advocated here is to help make future hymns to the critical attitude valuable by making *explicit* the nonjustificational character of Popper's critical thinking—an aspect which it may be especially useful to bring to attention since Popper has throughout his writings *practiced* nonjustificational criticism without explicitly discussing it in general terms.

But what does it mean to talk of the fusion of justification and criticism? Until this is explained, the importance of a separation of the two notions can hardly be assessed. They have been fused in a number of different ways. One of these, historically probably the most important, dominates most kinds of comprehensive rationalism. In this view, the way to criticize a view is to see whether it can be logically derived from, i.e., "justified by," the rational criterion or authority. In Hume's empiricist view, for instance, the strongest way to criticize a particular theory was to show that it could not be justified or established properly—in his case by an appeal to sense experience. On examining Hume's philosophical writings, one finds him making fairly consistent use of this basic strategy of criticism. He takes one idea after another—the idea of God, of the soul, of memory, of other minds—and asks whether it can be justified by being derived from sense experience, which he regards as man's only source of knowledge, or rational authority. If it can be justified as required, he accepts it; if not, he either rejects it or implies that from a rational point of view it should be rejected.

The analytic and synthetic procedures of Descartes's method "for conducting the reason well and for searching for truth in the sciences," however different in other respects, are closely parallel to Hume's in this. Ideas that cannot be reduced to clear and distinct ideas, and thus rationally justified, Descartes thinks should be rejected, just as everything that is to be accepted should be so justified. For both men, the rational way to criticize an idea is to see whether or not it can be rationally justified.

Another popular strategy of criticism, although weaker than the first in its demands, also fuses justification and criticism. It is weaker in that it employs a kind of "elastic clause" similar to that in the United States Constitution. What matters is not whether a belief can be derived from the rational authority but whether it conflicts with it. In other words, it is not irrational to hold a belief that cannot be derived from, i.e., justified by, the rational authority unless its denial can be derived from the rational authority. The second strategy can be varied in many quite subtle ways. Indeed, a typology of theories of authority, developed in terms of the different possible critical moves consistent with the general strategy, might illuminate some of the particular twists taken in historical controversies.[32] Yet all varieties fuse justification and criticism in one way or another; to criticize a position, one must either show that it cannot be derived from, or else that it conflicts with, the rational authority, which is itself not open to criticism.

X

Since such an account of justificational strategies of criticism leaves undecided the character, or even the possibility, of a nonjustificational type of criticism, we need at least a rough picture of the latter type in operation, as well as some further clarification of the difference between justificational and nonjustificational theories. I think we can get both from trying to answer yet another question: *Why have justification and criticism repeatedly been fused in the ways described?*

In effect, this is a request for an explanation of the assumption just discussed, and may be obtained by bringing to light a further hidden philosophical dogma: *Most philosophical views have tacitly taken for granted that rational character and degree of rationality are properties that pass from premises to conclusion in the same manner as the property of truth through the relationship of logical deducibility.*[33] That is, most

---

[32] For example, it might illuminate a number of controversies in the history of jurisprudence, such as the question of the province of, and the demarcation between, positive law and morality. See H. L. A. Hart, "Positivism and the Separation of Law and Morals," *Harvard Law Review*, February 1958, p. 599: "What both Bentham and Austin were anxious to assert were the following two simple things: first, in the absence of an expressed constitutional or legal provision, it could not follow from the mere fact that a rule violated standards of morality that it was not a rule of law; and conversely, it could not follow from the mere fact that a rule was morally desirable that it was a rule of law."

[33] Several philosophers of science, including Carnap, Hempel, Goodman, Popper, and Watkins, have in recent years—without discussing its general historical background—considered the place of a similar assumption in theories of "confirmation" in the natural sciences, referring to this assumption by names like "consequence con-

theories about the rational evaluation and criticism of competing views include (1) some more or less well-defined notion about the character of whatever standard of rationality or intellectual respectability is to be used in making evaluations and criticisms, and (2) the assumption that this intellectual respectability, whatever its nature, must be fully transmissible — like truth and unlike falsity (which is retransmitted)—from premises to conclusion through the deducibility relationship. According to the assumption, the logical derivatives of a theory inherit its quality and degree of intellectual respectability.

The historical origins of this assumption, as well as its unusual tenacity, are probably to be explained by the fact that the earliest attempted criteria of respectability were regarded as *criteria of truth*. In such views, the demarcation between the respectable and the disreputable coincides with the demarcation between the true and the false. These early attempts met many difficulties. Criteria of truth proved to be either unattainable or practically inapplicable to the issues for which they were needed; and the search for criteria of truth was gradually displaced—although never completely abandoned—by a search for some weaker but more attainable measures. The property most often used for this task, particularly by scientists and mathematicians, was that of probability. Prior to the development of modern probability theory several different senses of "probability" came to be used for this purpose.[34] With the development of the probability calculus, however, probability *in the sense of the probability calculus* has increasingly—but not universally!—been applied in attempts to provide the measure of rational respectability which would be the best substitute for a criterion of truth.

Truth and *probability* (to be used henceforth in the sense of the probability calculus) happen, however, to be two of the very few characteristics which are indeed transmissible from premises to conclusion through

---

dition," "entailment condition," and "content condition." Although, as will be apparent, I agree with Popper's position, I do not intend my remarks to apply only to scientific subject matters; the problem of rational evaluation, and the conditions under which it has been discussed, extend through the various subject matters of philosophy. For some earlier remarks about this assumption see my article, "A Note on Barker's Discussion of Popper's Theory of Corroboration," *Philosophical Studies*, January-February 1961, p. 8. It is more accurate to read there "deducibility relationship" for "implication sign" since we are speaking metalinguistically. The idea of "condition sign" (Quine) or "implication sign" (Carnap) would coincide with this only in demonstrable relationships.

[34] There is, for example, the very interesting sense of probability Sextus Empiricus attributed, during the Hellenistic period, to Carneades. Sextus reports that Carneades was compelled to adopt a criterion for the conduct of life which made use of both "probable presentations" and those which are "at once probable and irreversible and tested." Although it is hard to reconstruct from the texts exactly how this conception of probability operated, it could hardly be a type of probability in the sense of the probability calculus. See Sextus Empiricus, *Against the Logicians*, Book I, pp. 166 ff.

the deducibility relationship.[35] I suggest, as an historical conjecture, that these two concepts exerted such a determining influence over the early developments of the theory of rationality that it has been unquestioningly assumed that other putative measures and standards of respectability, however they might differ from truth and probability, would nevertheless automatically share their logical transmissibility. Indeed, the demand for justification made undesirable any measure unable to justify its derivatives by lending them its respectability.

The whole self-reinforcing structure of traditional philosophy so endorses and protects this assumption that even today, when criteria of truth are as unavailable as ever, and when probability measures have been found to be quite incapable of arbitrating rationally among competing scientific hypotheses, not to mention less precise ideas, the property of logical transmissibility is still expected of other evaluatory measures without regard to their real logical capabilities.

These remarks may help explain the persistence of some empiricists in attempting to retain probability as a measure despite devastating criticism. The transmissibility capacity of truth and probability led, historically, to the general assumption that intellectual respectability is transmissible from premises to conclusion. But historical origins are easily forgotten, so that the situation now seems ironically reversed; probability measures are retained *because* they are logically transmissible. The transmissibility requirement itself is taken for granted.[36]

It is a matter of interest, if not surprise, that many theories that have relied on probability as a rationality measure have also included "empirical character" as a further requirement for rationality. In such a view legitimate statements would have to be truth-functionally reducible to something like individual "basic statements" reporting sense experience; and the logical derivatives of a legitimate statement would inherit not only its degree of probability but also its empirical character. Empirical character, however, turns out *not* to be transmissible through the deducibility relationship. From every basic empirical statement both nonempirical metaphysical statements and all tautologies follow logically. Adding to the

---

[35] See Popper, "The Demarcation Between Science and Metaphysics," *op. cit.*, footnote 63. Another such property would be "the right to be sure," as discussed and used by both Ayer and White. In saying that probability is transmissible, I mean of course that a consequence will be *at least as probable* as the premise from which it is derived. It may of course be *more* probable.

[36] This account does not, of course, fully explain such apparent anomalies as Carnap's simultaneous *rejection*, in *Logical Foundations of Probability*, p. 474, of the applicability of the transmissibility assumption or "consequence condition" to the theory of confirmation, and *acceptance* of the view that degree of confirmation coincides with degree of probability. Since probability is itself transmissible from premises to conclusion, Carnap's position would appear to be inconsistent.

difficulty, universal scientific hypotheses cannot be reduced to truth functions of a finite class of basic empirical observation statements—which denies empirical character to scientific hypotheses themselves. Such unwanted results hound empiricists with the well-known "paradoxes" of induction and confirmation in any case; but when transmissible probability is mixed with the nontransmissible property of "empirical character," the results may be bizarre. Nonempirical consequences of empirical statements inherit the probability of the original empirical statement, and consequently become respectable from the point of view of the probability standard, but rather disreputable from the empirical standpoint. When "meaning criteria" are added as still further criteria of respectability, still more anomalies appear.[37]

If it were really fundamentally similar to other critical philosophies, Popper's theory of critical evaluation might be expected to share some of these perennial characteristics and difficulties. Actually, when we turn to the theory of testability he devised for evaluating scientific theories, we find an entirely different approach. Although the theory of testability is a genuine approach to the familiar problem of assessing and criticizing competing theories rationally, it does not contain the deducibility assumption and is, in fact, incompatible with it; and the criticism involved is entirely nonjustificational.

His theory of testability is an attempt to assess as one (but not the only) measure of rationality not the degree to which a theory is probabilified or confirmed or justified by evidence, but the degree to which it is testable or falsifiable and has been subjected to deliberate criticisms, tests, or attempts to falsify it. The measures of degree of testability or degree of corroborability, unlike truth and probability, are not logically transmissible from premises to conclusion. Quite the contrary, these properties share with *falsity* the property of *retransmission* from conclusion to premises. Consequently, the difference between Popper's theory and the others referred to can be stated—and even defined—in the following terms. Whereas evaluational properties like probability and degree of confirmation are transmitted in the same direction as truth, degree of testability is retransmitted in the *opposite* direction like falsity. Thus the difference between the testability theory and various probabilistic theories of confirmation can be defined by reference to the irreducible difference in the ways the two concepts are logically transmitted.

If Popper's theory of testability did share with rival theories the assumption that the measure of intellectual respectability flowed from premises to conclusions which they justified, then any logical consequence of an hypothesis would have to be as highly testable as the original hypo-

---

[37] Popper has of course discussed these many times. See for examples "The Demarcation Between Science and Metaphysics," *op. cit.*

thesis : testability would be transmissible from premises to conclusion. But in fact the opposite happens : since a hypothesis is testable or falsifiable by the falsification of any of its consequents, the hypothesis must possess at least as high a degree of testability as any of its consequents. But it *may* possess a higher degree of testability—if, for example, it possesses other consequents independent of the first. And if the hypothesis can possess a higher degree of testability than its consequents, then the consequent does not inherit this particular evaluational property through the deducibility relationship. A high level, highly testable scientific theory does not, then, bequeath its degree of falsifiability or testability to those lower level theories it entails, its necessary conditions, which traditional accounts would represent it as justifying.

This point might be illustrated by examining the testability relations of these three hypotheses :

1. All who dwell in London are English.
2. All who dwell in Hampstead are English.
3. All who dwell in Bloomsbury are English.

Assuming, correctly, that Bloomsbury and Hampstead are both in London and that both the second and the third statements follow from the first, let us suppose that the second statement is falsified. By *modus tollens,* the rule of retransmission of falsity, the first statement is falsified too.

But suppose that the second statement has not been falsified, and that another of the first statement's consequents, the third statement, has never been tested. Clearly, the first statement will be falsified by the third just in case the third is subsequently tested and found false. But the second statement will not be falsified thereby, since it is logically unrelated to the third. Thus the first statement is more testable or falsifiable than the second, since the first is falsifiable by something that does not falsify the second. And the second cannot be more falsifiable than the first, since anything that falsifies the second falsifies the first.

This example, and similar ones, bring into relief not only the absence of any assumption that a measure of intellectual responsibility or rationality (in this case "degree of testability") is logically transmissible from premises to conclusion, but also the nonjustificational character of Popper's theory of criticism.

To avoid confusion of this approach with the second strategy of *justificational* criticism discussed in the previous section, it should be noted that in the theory of corroboration the falsity of a view is not *established* in a refutation thereof. Rather, the view is provisionally rejected because it conflicts with some other better tested, less problematic view. But the view that occasions the refutation is itself open to criticism by the testing of its own consequences. And these in turn are criticizable; and so on forever. This *process* of testing is, of course, in principle infinite; but there is no infinite *regress*, because the aim of justifying or establishing has been abandoned.[38]

If all measures of degree of intellectual respectability resembled truth and probability in passing from premise to conclusion through the relationship of logical deducibility, all criticism would have to be justificational. The very fact that degree of corroborability and testability are not such qualities establishes not only that not all criticism need be justificational, but also that this type of criticism *cannot* be justificational. Hence a nonjustificational, nonauthoritarian theory of knowledge and rationality is certainly possible.[39]

The features of traditional philosophy discussed in the last three sections—the authoritarian structuring of its questions, and the two dogmas that are largely responsible for this structuring—each deserve further detailed exploration and explanation, as does the important, more or less technical question of how to construct critical institutions in philosophical inquiry. I hope only to have suggested where some profitable investigations may lie. But even in their present form I believe the matters just discussed are sufficient to suggest immediately the character of a theory of rationalist identity that can escape the dilemma of ultimate commitment and thereby avoid both skepticism and fideism.

XI

The new concept of rationalist identity I have in mind is already implicit in the nonjustificational approach just sketched. I suggest that within

---

[38] See in this connection Popper, *Logic of Scientific Discovery*, p. 105. It might be objected that this passage from the *Logic of Scientific Discovery* shows that Popper did not at the time he wrote that book hold a nonjustificational theory of criticism. For he writes on page 105: "The basic statements at which we stop . . . have admittedly the character of *dogmas* . . . in so far as we may desist from justifying them by further arguments (or by further tests)." The word "justifying" here does not, however, support this objection. It is not a remnant in Popper's writing of justificational *philosophy* (which is inconsistent with the position he expounds in the book—see for example Appendix *9, *Logic of Scientific Discovery*); rather it is a remnant of the *terminology* of justificational philosophy. This could hardly have been otherwise at a time when the basic contrast between justification and criticism had not yet been made explicit. I tried to do this when writing my book; and during Popper's seminar at the London School of Economics in October 1961 I suggested the distinction to him in an attempt to explain why certain aspects of his thought had so often been misunderstood, and in part to excuse and defend some of his more persistent critics. By pointing up the incorrect but understandable assumptions on which the critics had based their interpretations of and objections to his thinking, I suggested that some of their arguments, although indeed incorrect, deserved more credit for their intelligence and seriousness than is sometimes accorded them.

[39] I am indebted to Professor Popper, and to Dr. Jerzy Giedymin, for correspondence and conversation which helped me to correct and improve this section—although I am not sure that they would agree with my final formulation.

this new framework the rationalist identity might be characterized as that of one who holds *all* his beliefs, including his standards and his basic philosophical position itself, open to criticism; one who protects nothing from criticism by justifying it irrationally. The position, which I call comprehensively critical rationalism, has several affinities with comprehensive rationalism : it shares its comprehensive aims, and it follows from, or is implied by it. That is, any comprehensive rationalist who succeeded *per impossible* in justifying *all* his opinions rationally would clearly not have to justify any of them irrationally.

Nevertheless the two concepts differ considerably. For instance, if they were equivalent positions, the traditional requirement would also be implied by the new one—which it is not. Significantly, it does not follow that a man who justifies none of his beliefs irrationally will be able to justify all of them rationally. If the aim of comprehensive justification has been abandoned he need not justify some, or even most, of them at all.

The new concept of rationalist identity also differs from critical rationalism, wherein the rationalist accepts the *rational* unjustifiability of his position but goes on to justify it in an *irrational* commitment to principles that are not themselves open to criticism.

If rationality lay in justification, such eventual irrational justification or commitment might seem inescapable. However, if rationality lies in criticism, and if we can subject everything to criticism and continuing test, including the rationalist way of life itself, then rationality is in this important sense unlimited. If all justification—rational as well as irrational— is truly abandoned, there is no need to justify irrationally a position that is rationally unjustifiable. The position may be held rationally regardless of justification, provided it can be, and is, held open to criticism and survives severe testing. The question of how well a position is justified is quite different from the question of how well a position is criticized.

The general separation of justification and criticism *can,* of course, be extended to the examination of the rationalist position itself. Whereas under traditional conceptions of rationalism the rationalist position was itself not rational and thus perpetuated crises of integrity, under the new conception the crisis disappears. The new rationalist identity satisfies its own requirements; without any contradiction or other difficulty the practice of critical argument can be criticized. Just as it is possible for a democracy, during the democratic process of voting, to commit suicide, so a comprehensively critical rationalist, who is not committed to the view that his position is the correct one, could be argued out of rationalism by himself or someone else. Fideistic irrationalists have, by contrast, argued that, even without actually abandoning allegiance to the object of commitment, merely subjecting the allegiance to criticism is to forsake the allegiance. For a comprehensively critical rationalist, continued subjection to criticism of his allegiance to rationality is explicitly part of his rationalism.

This spells defeat for the *tu quoque* argument that supports the skeptical and fideistic claim. The case for irrational commitment, as well as the rational excuse for it, rests on the argument that rationality is so limited logically that such commitment is inescapable. For the comprehensively critical rationalist there is no such limitation. If he accuses his opponent of protecting some belief from criticism through irrational commitment to it, he is not open to the charge that he is similarly committed. Consequently, the *tu quoque* argument cannot be used against comprehensively critical rationalism, and criticism of commitments no longer boomerangs.

XII

Since antiquity, irrationalists have embraced as fact—indeed as one of the few facts established in philosophy—the philosophical contention that rationality is logically limited; that at least insofar as his standards are concerned, every man must make some irrational commitment. They have often used this contention to excuse on rational grounds their own particular irrational commitments, however bizarre in character. Thereby they have been able in principle, though not in practice, to avoid loss of intellectual integrity. I have tried in the preceding pages to refute the philosophical theory about the limits of rationality by turning the tables, by showing how we can shift the emphasis in rational discussion from justification to nonjustificational criticism. I hope in this way to have contributed to a clarification of the issues of identity and integrity in the rationalist tradition. If my argument is sound, irrationalists lose the most formidable weapon in their intellectual armory, their rational excuse for irrational commitment. Those who continue as irrationalists will really be so, in the sense that they can no longer retain their various irrationalist identities while preserving the intellectual integrity the *tu quoque* has so long afforded them.

CHAPTER 2

Reflections on Karl Popper's Epistemology

BY PAUL BERNAYS

Introduction

MY CONTRIBUTION TO THIS VOLUME in honor of Karl Popper will be related to his paper "On the Sources of Knowledge and of Ignorance" (1960). This article establishes a connection between Popper's inquiries on the methods of science, in particular his book *The Logic of Scientific Discovery,* to which I shall also refer, and his contributions to the history of ideas in *The Open Society,* wherein he defends the tendencies for rationality in such a decided way, and with such power and warmth, as is hardly to be found elsewhere in the philosophical literature of our times.

Popper is well aware of the objections to which traditional rationalism is exposed, but he is not misled by these objections to pass to the other extreme, as are so many philosophers today. The said objections were provoked by the oversimplifying theories of knowledge that had been suggested by the confidence in reason. Popper characterizes such a simplifying doctrine as one which holds that "truth is manifest." Such a doctrine, as Popper observes, has been developed in an empiricist as well as in an intellectualistic way. The empiricist doctrine of Bacon and his followers is expressed by the image that "the book of Nature" lies before our eyes and that we need only to spell it out, not influenced by anticipations, in order to get the truth about nature. The intellectualistic doctrine, as it originated with Descartes, says that we have only to rely on our clear and distinct intuition, without being diverted by traditional views, in order to obtain true knowledge.

Such philosophical claims of certain knowledge, untainted by error, were promoted in opposition to the claims of authority of the traditional scholastic doctrines, and became "the major inspiration of an intellectual and moral revolution without parallel in history."

Popper highly values this very wholesome influence of the doctrine that truth is manifest. Nevertheless, he rejects the doctrine. His epistemological theses have a strongly restricting character. Some of them are quoted here:

Knowledge cannot start from nothing . . . nor yet from observation. The advance of knowledge consists, mainly, in the modification of earlier knowledge.

Clarity and distinctness are not criteria of truth, but . . . obscurity and confusion indicate error.

. . . most of the sources of our knowledge are traditional . . . . But . . . every bit of our traditional knowledge (and even our inborn knowledge) is open to critical examination and may be overthrown.

The most important function of observation and reasoning, and even of intuition and imagination, is to help us in the critical examination of those bold conjectures which are the means by which we probe into the unknown.

There are no ultimate sources of knowledge. Every source, every suggestion, is welcome; and every source, every suggestion, is open to critical examination.

Popper calls his point of view "critical rationalism." Its essential difference from the rationalism of the Enlightenment, described above, is that it strongly restricts our claims to knowledge. This means that one does not assume that at the basis of our investigations there must be something absolutely certain; rather, one admits that knowledge proceeds by trials, guesses, and hypothetical mental constructs which are subjected to combination, comparison, and criticism.

In accordance with his quoted theses, Popper stresses in particular that the question of the ultimate source of our knowledge is "misconceived." He compares this question to the political question "Who should rule?" which "begs for an authoritarian answer" and "should be replaced by a completely different question such as 'How can we organize our political institutions so that bad or incompetent rulers . . . cannot do too much damage?'" In a similar way he proposes to replace the question of the sources of our knowledge by the entirely different one: "How can we hope to detect and eliminate error?"

The idea of error, as Popper points out, "involves . . . the idea of an objective truth. . . . Thus the doctrine of fallibility should not be regarded as part of a pessimistic epistemology." It "implies that we may seek . . . for objective truth, even though we may miss it by a wide margin."

I

As we see, Popper confronts his critical rationalism with the kind of rationalist epistemology which maintains that truth is manifest. The two philosophic doctrines of this kind considered by Popper, that of Bacon and that of Descartes, both amounted to an attack upon authority and tradition; nevertheless, Popper thinks that they did not succeed in freeing epistemology from authority : "They could only replace one authority— that of Aristotle and the Bible—by another." The first one appealed "to the authority of the senses, and the other to the authority of the intellect."

Here one might wonder. In fact, what we usually call authoritarian is the appeal to submit to something superior to us. Reference to the testimony of our senses or to our clear and distinct comprehension can hardly restrict our autonomy.[1]

I think that Popper's objection to these two doctrines is justified— although not with respect to their authoritarian character. These doctrines strongly contributed to liberation, as indeed Popper himself states (as mentioned above). But there is the epistemological point that we should not give an absolute competence to some particular organ of knowledge. For example : It was not the lack of clearness or distinctness of the geocentric view, and even less the contradictory testimony of sense-perception, which induced science to abandon it. Rather the decisive circumstance was the discovery of a theoretical aspect which proved to be much superior to the old view in its explanatory power.

These remarks concern Popper's opposition to the doctrine that truth is manifest, as this doctrine is embodied in the philosophies of Bacon and Descartes. However, as we have stated, Popper more generally opposes those epistemologies which concentrate on the question of the source of our knowledge. While considering the case of individual assertions in daily life, he readily shows that in fact we do not inquire very far into the sources of such assertions, and that a pursuing of the question of sources would lead to infinite regress. Similar considerations can be applied with regard to single empirical scientific assertions, such as the statement, for instance, that some concretely given experimental instrument has certain prescribed properties.

But Popper goes much further in denying that the search for sources of knowledge is a task of epistemology. In his *Logic of Scientific Discovery*,[2] one purpose of which is to give a suitable definition of empirical

---

[1] That Popper himself admits that *inner* authority does not indeed contradict autonomy may be seen from passages like this: ". . . the open society is one in which men have learned . . . to base decisions on the authority of their own intelligence." (*The Open Society*, note to the Introduction.)

[2] Referred to in subsequent discussion as *L.S.D.*

science, he expressly refrains from any reference to perception—at least he does not include it as a constitutive element in that definition. Similarly, with respect to the theoretical aspect of science, Popper refuses to acknowledge any epistemological relevance in the search for the leading heuristic ideas by which the main scientific theories were motivated in the stage of their invention. In both cases he describes the views he opposes as psychologistic.

Popper regards a science as a system of statements. For him, justification of statements means logical justification, and statements, as he stresses and as one must certainly concede, can be logically justified only by statements. Hence it follows that those statements from which the logical derivations start have no justification at all, insofar as they are not analytic. They are either hypotheses or such "basic statements" as "are accepted as the result of a decision or agreement."

This aspect is not changed, as one could think, by the essential role that testing and falsifying has in Popper's epistemology. For falsifying is here described as finding a refuting hypothesis which has undergone many tests and is not found in any one of them to be contradicted by accepted basic statements. And by this procedure no new element of knowledge is introduced. The issue of empirical science in this way appears fully agnostic.

Tending to give to this criticism a more positive turn, I think that I can get support from Popper's own philosophical aims, especially his stress on the need for modesty in our pretensions to knowledge. Indeed, modesty not only precludes the claim to kinds of knowledge which in fact we do not possess but also keeps us from despising restricted forms and means of knowledge which are essential ingredients of our way of learning and of gaining understanding. This point of view enables us to overcome various difficulties by way of reconciling theses which otherwise seem to be incompatible. Let us explain this for the case of dealing, in the theory of knowledge, with sense-perception.

We have to admit that sense-perception is not "immediate knowledge," and also that there is no purely logical passage from perceptions to statements. But this does not prevent us from learning from experience and learning to use language; and we do succeed in communicating what we observe and in getting mutual understanding about our observations.

How does observation occur in science? When science begins, we already have a rich set of concepts embodied in ordinary language, and a great amount of experience at our disposal. On the basis of this previous stage of knowledge, with its suppositions (its "practical *a priori*," as we might say), the descriptive sciences proceed in a relatively unproblematic way, i.e. without being obliged to introduce a new world of objects and new conceptual categories. Here observation is used in a rough sense: we do not separate perception proper from our instinctive interpretations, at least in so far as these interpretations are established in the experience of our daily life.[3] And, in fact, investigation is collecting experiences, though certainly not only collecting.

In theoretical science, especially in recent theoretical physics, the situation is very different. Here we have to deal with objects that do not belong to our familiar perceptual world. In order that such a theory may have empirical content, it must have either a phenomenological part, or there must be joined to it an experimental semantics, by means of which it becomes possible to test the theory. Let us see how this can work. By means of the experimental semantics we can draw from the theory, in conjunction with known facts about experimenting, consequences amounting to assertions—like the assertion that, given a certain experimental arrangement, we can observe, by looking through the eyepiece of some apparatus, a certain definite phenomenon (for instance a yellow line between two marked strokes). Such an assertion can certainly be tested. The possibility of testing holds, of course, for any theoretical hypothesis from which experimental consequences can be drawn, not only for a whole theory.

In both descriptive and theoretical science the passage from observations to statements raises no difficulty in principle. For in descriptive science we have, so to speak, a standard form of expression in our usual language. And in theoretical science, observations in testing can, by suitable experimental arrangements (as indicated above), be reduced to examination of : (1) Are the experimental conditions really satisfied? and (2) Is the predicted phenomenon present? The answer to both 1 and 2 can be stated as "yes" or "no." For 2 this is indeed sufficient; but if the answer to 1 is "no," it will be necessary to describe the deficiency of the experimental procedure. However, this can be done in the experimental language, which belongs to that which is taken to be accepted at the stage of arranging the experiment.

Popper's description of the procedure of testing, in the form which he regards as the correct one, contains some additional elements. According to his view, the only proper test effect is refutation or, as he calls it, falsification, whereas corroboration of an hypothesis means only that it has not been falsified, even by strong tests.

Now the process of falsifying an hypothesis or a theory is conceived by him in a kind of normal form (already briefly mentioned above, on p. 35): A falsifying hypothesis which contradicts some consequence of the theory (or hypothesis) to be refuted has to be formulated; and this falsifying hypothesis must be corroborated by accepted basic statements.

How this is to be understood, may be illustrated, for the case of an experiment realizing some conditions C, upon which, according to the

---

[3] There occur, of course, statements like that about a crystal, that it looks as differently colored according to the circumstances of lighting. But such an assertion does not transgress the means of our usual language.

theory, a phenomenon P is to be expected; whereas as a matter of fact a phenomenon P' has been found. Here the falsifying hypothesis would be the statement that upon the conditions C a phenomenon of the kind P', differing from P, is to be observed. And the basic statements are singular existential assertions designed to express the results of the various single observations. Moreover they have to be accepted, which means that eventually some of the observations might be canceled as unreliable.

In this description of refutation Popper adds two things to the usual way of regarding an experimental test : the falsifying hypothesis and the accepted basic statements.

Popper's argument in favor of the need of a falsifying hypothesis for a refutation is very persuasive : the falsification must be reproducible. However for this purpose it seems to be sufficient to concentrate on that special prognosis of an experimental effect (derived from the theory or hypothesis to be refuted) which has been found to disagree with the observations. The refutation is strengthened when we succeed in corroborating a falsifying hypothesis, though this hypothesis might be of only a low generality.

But the possibility of refutation without a falsifying hypothesis appears just as advantageous according to Popper's general point of view. Indeed, he stresses the superiority which refutation has over corroboration as to the determination of the result. This superiority is certainly diminished by maintaining that every refutation includes the corroboration of a falsifying hypothesis.

Regarding the basic statements we have first to remember that they are introduced by Popper in order to replace the protocol sentences of the Vienna school. They are also meant to amount to a kind of protocol, not of personal experiences, to be sure, but of the results of observations. The role ascribed to the basic statements is that they "are accepted as the result of a decision or agreement" (*L.S.D.*, section 30) and "accepted basic statements are the basis for the corroboration of hypotheses" (*L.S.D.*, section 22) and thus indirectly for falsification. "Every test of a theory, whether resulting in its corroboration or falsification, must stop at some basic statement or other which we *decide to accept.*" (*L.S.D.*, section 29).

The element of decision is here stressed by Popper, and he argues against the view that basic sentences can be justified by reference to perceptions, which he regards as a kind of psychologism. "I admit," he says (*L.S.D.*, section 29), "that the decision to accept a basic statement . . . is causally connected with our experiences—especially with our *perceptual experiences*. But we do not attempt to *justify* basic statements by these experiences. Experiences can *motivate a decision,* and hence an acceptance or a rejection of a statement, but a basic statement cannot be *justified* by them—no more than by thumping the table."

The following may be observed about this pregnant argumentation. In the first sentence Popper speaks of the causal connection between the experiences and the decision, but then he says "experiences can motivate a decision." Now motivation in general means more than mere causality. That a circumstance motivates an action of a person means that the person adapts his action to that circumstance *with respect to a purpose,* and, in particular, motivation occurs in the case of some tactics. Yet decisions about accepting basic statements are just a kind of tactics, and motivating these tactics amounts to at least a partial justification, and even, when the motivation is sufficiently strong, to a practically full justification.[4]

Also, another point is to be considered. Popper's arguments are directed against reference to sense data or to convictions. But reference to perception can be made in the sense of referring to possible repetition of the perception by the reader—as a reference to a procedure of derivation (with indication of the rules and axioms to be used and of the successive steps to be performed) refers to the possibility of the reader's repeating, and thereby checking, the derivation.

There are also cases where the use of perception as a means of knowledge is so familiar and unproblematic (upon premises which can be taken as previous for the investigation) that a special reference to perception is not even needed. In zoology, botany, and mineralogy it will not be necessary, in general, to refer expressly to perception. (What is to be indicated is, for instance, the places where certain animals or plants or minerals are to be found, or the measures to be taken in order that certain described processes occur, and the like.)

An essential circumstance here is that our dealing with general concepts (concepts of kinds of things and of beings, concepts of materials, of behaviors) is established already in our daily experience, which thereby provides the practical solution of what Popper calls the "problem of universals," residing in the circumstance that, by the use of universal names, even simple statements like "here is a glass of water" have "the character of a theory, of a hypothesis" (*L.S.D.,* section 25).

If the solution of this problem were not effected, to a great extent at least, by prescientific experience, due in the first place to properties of external nature—Kant spoke indeed in this respect of a *"transzendentale Affinität"*—and in the second place to the instinctive faculties (involving elements of more or less unconscious rationality), then science would not be able to start, and decisions could not help.

Generally, on principle, there seems to be hardly a need to make decisions about accepting basic statements.

Popper gives an illustration of his view by drawing an analogy with a jury's verdict in some kinds of actions. "By its decision, the jury accepts,

---

[4] Popper's argument that perceptual experiences cannot justify a statement amounts to the thesis "What can be said, cannot be shown," which is an equivalent transformation of Wittgenstein's thesis "What can be shown, cannot be said." It means denying that language can fully serve its purpose.

by agreement, a statement about a factual occurrence—a basic statement, as it were" (*L.S.D.*, section 30). However in this case of criminal law there is a peculiar necessity : here a sentence has to be obtained. In science we are not obliged to give our results a greater certainty than they have according to our standard of knowledge.

It is true that in experimental science it is a familiar device to reject from a series of observations some sporadic members, which seem to be unreliable. Proceeding in this way amounts to an assumption, which however can be mentioned explicitly in the publication of the experiment.

There are indeed also other kinds of assumptions, implicitly contained in most experimenting. Thus a physicist who uses an instrument has not generally produced it himself, but has to rely on the work of the producing firm.

Yet I do not think that empirical science by its nature proclaims a freedom in deciding on statements about single facts, in the way that for instance, mathematics, proclaims its freedom in the formation of conceptions. There is something like decision which belongs to empirical science : the adopting of previous knowledge at any stage. But this adoption is to a great extent implicit and not especially concerned with single existential statements. Certainly there are also cases of explicit adoption, where, for instance, a scientist explicitly bases a theory on some laws which he regards as sufficiently secured. (Through the theory such a law may then appear in a new light, and if the theory is very successful, then the law may be regarded more and more only under the aspect of a consequence of the new concepts.)

This view about the role of perception in empirical science may be summarized as follows.

1. We need not deny that empirical science starts (in a harmless sense) from observations and begins with collecting and arranging experiences.
2. An essential epistemological function of perception for empirical science has to be acknowledged, and we can acknowledge it'without falling into psychologism, since the reference to perception is not to the *fact,* but to the *use* of perception.
3. In accordance with Popper's criticism we state that the starting point of empirical science is not epistemologically absolute : At any stage empirical science is relying on a store of *previous* knowledge which contains many unanalyzed constituents, which however, may become accessible to analysis at further stages.

II

Until now we have dealt with the question of the empirical basis of natural science.

Concerning the question of the rational element in the formation of scientific theories, a main point in Popper's considerations is his opposition to the view that induction affords a universal method for obtaining general laws from a greater quantity of observations.

In fact it is very pretentious to speak in this sense of induction as a general method. Most of what has been offered in the name of induction consists either of devices for proceeding cautiously and circumspectly in exploring regularities, or of information about typical forms of value distributions of quantities, or pairs (or triplets) of quantities, which may help in detecting lawful connections.[5] All this, however, is concerned with the investigation of what are called empirical laws, whereas Popper is mainly interested in those scientific advances by which new ideas and new conceptions are introduced.

There is certainly no general recipe for conceiving such new ideas. But Popper goes much beyond stating this. From considering that "there is no such thing as a logical method of having new ideas" and that "every discovery contains 'an irrational element' "[6] he goes on to deny altogether that a rational reconstruction of the steps leading to the conception of a new theory can be a task of epistemology. It is true that Popper does not formulate it quite this way; he speaks of "the steps that have led the scientist to a discovery," and he argues that analyzing these steps is a matter of psychology, whereas what interests epistemology are "the *subsequent tests* whereby the inspiration may be discovered to be a discovery, or become known to be knowledge." However, this way of arguing bypasses the essential possibility of a rational heuristics : I mean the possibility of extracting a discovery's *objective heuristic contents* subsequently, that is, after the discovery is made. Often this is done by the author himself (either in the first publication of his discovery, or in a later presentation intended to make the scientific idea of the discovery accessible to an interested public).

According to Popper all the rationality contained in the development of science is to be found in testing and criticism. He describes the process of critically testing (in section 3 of *L.S.D.*) as follows : "From a new idea, put up tentatively, and not yet justified in any way . . . conclusions are drawn by means of logical deduction. These conclusions are then com-

---

[5] Quite different is the penetrating analysis given by Nelson Goodman in his book *Fact, Fiction and Forecast* (Chapters III and IV). Goodman conceives the question of induction as a special case of the problem of "projection," i.e., of conceptual generalization, which is mainly the same as that which Popper calls the "problem of universals." Goodman's "projections" correspond to Popper's "guesses." Also Goodman's way of questioning conforms to that of Popper. "We ask not," he says, "how predictions come to be made, but how—granting they are made—they come to be sorted out as valid and invalid."

[6] By the way, an irrational element might amount to an unconscious rationality.

pared with one another and with other relevant statements, so as to find what logical relations . . . exist between them."

And about the dominant role of critical testing there is, among many other passages, the pregnant one in Popper's "Three Views concerning Human Knowledge" (section V) : ". . . it is only in searching for refutations that science can hope to learn and to advance. It is only in considering how its various theories stand up to tests that it can distinguish between better and worse theories and so find a criterion of progress."

Here two things are certainly to be acknowledged :

1. In testing a hypothesis we need not care about the motivation of the hypothesis,
2. Without the possibility, on principle, of being refuted, a theory or an hypothesis has no empirical content.

Nevertheless, the picture which Popper gives of the advance of science seems to me in some respects not satisfying. It reminds me of the view given by extreme Darwinism of the advance in organic formations, according to which the whole development comes about by irregular variations and by the selecting effect of the struggle for life; no positively formative element is admitted.

If we mistrust such a view, at least with respect to the advance of empirical science, then we have to look for those factors by which progress may possess a more positive character than that of depending exclusively on test and criticism. In this respect the following circumstances deserve attention.

First of all, in spite of the great changes which the development of empirical science brings about in our views of the constitution of matter, the nature of light, the kinds of physical reality, the character of causality, and even about space and time, there are many laws of lower generality, which, once they have been discovered, remain in force (though they may be differently interpreted in different stages)—for instance, the mechanical laws of equilibrium, the laws on electric circuits, those on reflection and refraction of light, laws of thermodynamics, laws of spectroscopy, and so on. Laws of this kind perform the role of the facts to be explained by the theories of higher generality. And as more and more such laws are established, the task of theoretical science becomes successively more rationally circumscribed.

Moreover, testing does not generally have the character of searching for refutations. Certainly, as Popper states, for the testing of predictions which result as consequences of a new theory, "those are selected which are not derivable from the current theory, and more especially those which the current theory contradicts" (*L.S.D.*, section 3). But the most obvious way of regarding this choice is by considering that the founder of a new theory is opposing the current theory and tends to show that this theory is wrong. Indeed, refutation of one theory is often the triumph of a new theory. Experimental testing of consequences of a theory is mostly done, not in the somewhat pessimistic spirit that we "try as hard as we can to overthrow" the theory, but rather in a sanguine spirit, which often succeeds in fruitful discoveries and inventions. Indeed important advances are not only possible by recognizing theories to be defective, but also by becoming confident in assumptions which at first were only timidly adopted. A theory can not only "clash with reality";[7] there can also be a surprising agreement of a theory or a hypothesis with reality, which gives us a strong confidence. Thus, for instance, the impressive agreement between the results of the various, quite different methods (some astronomical, some terrestrial) of experimentally determining the velocity of light gives us a strong confidence in our conception of the velocity of light.

Finally, rational discussion does not exclusively consist of criticism. Popper himself, in his 1958 preface to *L.S.D.*, says about rational discussion : "The method I have in mind is that of stating one's problem clearly and of examining its various proposed solutions *critically*." And in section 3, in considering "different lines along which the testing of a theory could be carried out" he mentions "the comparison with other theories, chiefly with the aim of determining whether the theory would constitute a scientific advance should it survive our various tests."

Here one thing especially to be observed is that reasoning having these objectives and purposes can serve as well for preparing the formation of a theory as for criticizing a theory. In fact, stating a problem clearly, regarding proposed solutions, comparing their merits and their deficiencies, reflecting in what directions an advance should be attempted—activities of this kind can be well suited to developing a discovery in a rational way (and, on behalf of the scientific public, to developing the understanding of a new theory).

This remark suggests the idea that it is, so to speak, the same kind of rationality which directs the inventions of theories as that which criticizes them.

We also find an utterance of Popper's which leads in this direction. I refer to the summarizing passage in his *L.S.D.* (section 85) :

We do not know : we can only guess. And our guesses are guided by the unscientific, the metaphysical . . . faith in law, in regularities which we can uncover—discover. . . . But these marvellously imaginative and bold conjectures or "anticipations" of ours are carefully and soberly controlled by systematic tests. . . . Even the careful and sober testing of our ideas by experience is in its turn inspired by ideas : experiment is planned action in which every step is governed by theory.

Thus Popper in fact states that our testing is "inspired by ideas," and, as he surely here means, by ideas of the same kind as those which direct

---

[7] Cf. "Three Views concerning Human Knowledge," Section VI.

our guesses. Of course in the personal realization the roles are mostly separated, and a theory which is the work of a genius may be tested by a quite modest physicist. But both appeal to a common super-individual standard of reasoning.

And indeed this standard is a kind of rationality, though not merely of logical and mathematical rationality. Popper speaks of faith in law, faith in regularities. But it might be possible to analyze it further: all our instinctive ways of arguing about causality, matter, material, action, interaction, dispositions, states, faculties, probability, are here included.

As is well known, David Hume denied any rational character to causal thinking. Indeed, he tended to show that our conclusions from experience "are *not* founded on reasoning, or any process of the understanding."[8] His argument was based on the implicit premise that all rationality must have the character of logic or mathematics. And the embarrassment that his arguments brought to the philosophy of his time is certainly in the main due to the circumstance that the said premise was mostly also inherent in this philosophy.

The doctrine of Kant expressly abandoned this premise. Here the methods of theoretical physics were acknowledged as a kind of rationality, as a way of understanding depassing the scope of pure mathematics. But at the same time this philosophy pretended to establish Newtonian mechanics and its conceptuality as *a priori* valid and as a definitive frame of empirical science. This claim could not stand against the great revolutions that theoretical physics has had to undergo since the nineteenth century.

In reflecting on how Kant's opposition to Hume can be maintained without the Kantian claim, we find that there is again a tacit premise which has to be abandoned: namely, that rationality must include absolute certainty and hence must provide fully definitive results. It seems that this assumption was first on principle rejected in Ferdinand Gonseth's *Philosophie ouverte*. Yet the assumption has nothing cogent. Rationality and understanding can be present not only in full evidence, but in reflecting theoretical speculation as well.

We may recall here our devise of modesty in the pretensions to knowledge. It is surely in the spirit of this devise and of critical rationalism if we acknowledge the elements of rationality—our restricted human rationality—in theoretical speculation. This does not disagree with our admission that the theories and hypotheses of empirical science are "guesses," insofar as they are not determined as consequences of our relevant knowledge. Indeed, we have to distinguish between more and less reasonable guesses, and we nevertheless can recognize that adoption or rejection of theoretical proposals is directed by rational, though not exclusively mathematical ways of arguing.

---

[8] See Hume's *Enquiry Concerning Human Understanding*, section IV, § 28.

Our learning by empirical science is indeed—as Popper also points out in his opposition to instrumentalism—not only a learning of facts and of techniques, but a development of our methods of thinking and talking about nature with the intention of a successively better understanding of the processes in nature. And if there really exists a growth of scientific empirical knowledge along these lines, then this growth should, in the retrospective view, be describable for any science as an epistemological process, with carefully prepared and motivated guesses and including the rational elaboration of those eminently successful theories which for a considerable domain of experience have a permanent significance.

Such descriptions, if given in the spirit of critical rationalism, could provide a more positive color to this philosophical attitude than it gains from the assumption that rationality consists only in criticism and in detecting and eliminating error. Believing in a more positive character of our scientific rationality however does not exclude the vivid consciousness that "we do not know."

A final remark : I hope that the reader will recognize that I am in full agreement with Popper regarding the main tendencies. It was against my original intention that I became engaged in various criticisms by tending to oppose an overestimation of criticism.

CHAPTER 3

What Hume Might Have Said to Kant

(And a few questions about induction and meaning)

BY HERBERT FEIGL

THE AIM OF THIS BRIEF ESSAY is to present some conjectures as to how Hume might have criticized certain central epistemological doctrines of Kant.[1] By a natural extension of this theme I shall also raise a few questions concerning Karl Popper's views on induction and his critique of the empiricist meaning criterion. I do this in the spirit of my very great admiration for Popper's work and his friendship, which I cherish sincerely. If I have misunderstood his basic outlook, I hope that he will set me right in future discussions.

There can be little doubt that one of the main purposes of Kant's theory of knowledge was to overcome Hume's skepticism, and to establish the apodictic validity of a number of basic principles of factual knowledge. It has been said that though Kant admitted to have been "awakened from his dogmatic slumbers" by Hume, he nevertheless went quickly to sleep again by blandly accepting the synthetic *a priori*. It is true that Kant hardly questions the existence of such knowledge, and that his primary aim was not to demonstrate its existence, but rather to *account* for it. Hume, had he lived to read the first *Critique* and the *Prolegomena*, would hence have had to assess Kant's reasoning in the transcendental deduction.

Following current custom we may distinguish the "anthropological" from the "presuppositional" strands in Kant's epistemology. Kant's own explicit programmatic remarks in the preface to the *Critique of Pure*

---

[1] I have attempted a similarly anachronistic exercise in my article "Matter Still Largely Material," *Philosophy of Science*, January 1962, in which a "Locke redivivus" criticizes Berkeley.

Reason indicate that he was very clear about the distinction. *Quid facti* and *quid juris* are the terms Kant himself used in this connection. But, as is evident also in many other great thinkers, it is not easy to avoid the psychologism which results from mistaking questions regarding the validation of knowledge-claims for questions regarding their origin. In the "anthropological" strands of the *Critique* Kant attempted to explain the existence of synthetic *a priori* knowledge of space, time, substance, and causality on the basis of the structure of the human mind. There we are told that the forms of pure reason, i.e., the forms of intuition and the categories of the understanding, are impressed upon the raw materials of sensory experience; and that knowledge, being the product of the combination of these two factors or components, could not conceivably fail to exhibit the *a priori* forms which entered into it in the first place and inevitably. If, for example, we interpret with Kant the principle of causality as the assertion of the strict lawfulness of nature, then Kant's transcendental deduction in the anthropological version boils down to the well-known claim that it is the human mind which imposes this order upon experience.

How would Hume have reacted to the anthropological version? I think it is fairly obvious that he might well have confronted Kant with the following dilemma. If the order of nature is to be a product of the organization of the human mind, then Kant must have assumed that this organization is fixed and unchanging. And if this assumption itself is to be justified, it can be done *either by induction or by deduction*. In the first alternative, our confidence in the order of nature cannot rest on *a priori* grounds, although it remains synthetic in that it consists in an assertion about matters of fact. Hume's skepticism would apply with its full force to the extrapolative assumption that the human mind will essentially retain the same forms for all time to come. Such an assumption is then clearly synthetic *a posteriori,* and could not possibly warrant apodictically true knowledge of the uniformity of nature. If Kant were to avail himself of the second alternative, i.e., to justify the constancy of reason by deduction, he (according to both Kant and Hume) could do this only by incorporating such constancy in the very *concept of reason*. But then we are merely explicating what we have put by definition into this concept, and we arrive at a statement which in Kant's own terms would be *analytic*. It is obvious that the synthetic principle of causality could not be derived from an analytic statement regarding the constancy of human reason.

Hume would certainly not have allowed Kant to escape between the horns of this dilemma by declaring the constancy assumption itself to be synthetic *a priori,* for in that case Kant would be faced with an infinite regress in transcendental deductions. Hence, Hume might well rest his case against the anthropological version by contending that in the light of the criticisms just presented, Kant could arrive at either synthetic *a pos-*

teriori or analytic *a priori* formulations of the principle of causality, but never at a synthetic *a priori* formulation. Paraphrasing the well-known saying of Einstein's, Hume could maintain that, inasmuch as the principle of causality makes an assertion concerning the world, it is in principle always open to doubt; and inasmuch as it is not open to doubt (viz., as an analytic proposition), it provides no information about matters of fact.

What about the "presuppositional" version of Kant's *Critique*? Here we are told that the validity of the principle of causality is a necessary condition for the very possibility of knowledge concerning matters of fact. It is easy to understand how tempting this interpretation has been to Kant and many of his followers. Knowledge at its best, as in the great theories of the physical sciences, is knowledge in terms of *laws*. Scientific explanation, explicitly or implicitly, always utilizes lawlike premises. Science would be reduced to mere descriptions and narratives if there were no regularities of nature. Science as we understand it, at least since the Renaissance, is essentially a body of knowledge claims which formulate the regularities of nature in lawlike statements, be they deterministic or statistical, "low-level empirical" or "high-level theoretical."

It may well be granted that this is the current common understanding of the significance of the term "scientific knowledge." It certainly was also Kant's understanding. For him Newtonian physics was the paradigm of factual knowledge. Now, all that the presuppositional version of Kant's epistemology warrants is the claim : "If there are no regularities there can be no scientific knowledge." However, from what we have said above, this must be recognized as an *analytic* statement, i.e., as a simple consequence of the meaning of the term "scientific knowledge." We have no assurance that scientific knowledge in this sense will always be possible in all domains of experience. The optimistic confidence of scientists that such knowledge will always be attainable is defensible only on inductive grounds, and hence, if formulated in a proposition, would be synthetic *a posteriori*— making it again open to Hume's skeptical doubts. Hume's reply to Kant's presuppositional version would once more confront him with an ineluctable dilemma.

(I realize that I have—for brevity's sake—presented only the bare bones of the issue. But I don't think that it is necessary for my purposes to enter into the finer details of exegesis of either Hume or Kant. In what follows I shall, in the same spirit, concentrate on fundamentals only, and discuss some questions concerning Popper's critique of inductivism, and related issues. This is pertinent since Popper's point of view appears as a *tertium quid* between Hume's and Kant's epistemologies. I expect he would agree with the criticisms advanced by Hume *redivivus*—as I construe them.)

In his brilliant and powerful analysis of scientific knowledge Popper has emphasized again and again that there can be no justification of inductive inference. By explicating the logic of science in terms of the hypothetico-deductive model, he arrived at the position that laws and theories can be refuted (and that, indeed, if they were not conceivably refutable, they would not be scientific), but that they can never be verified. Of course, Popper allows for corroboration in the sense that we may say that a theory which has withstood very severe tests is, at least until further notice, acceptable as part of the justified corpus of scientific knowledge-claims.

It is precisely on this point that I wish to ask Popper whether he does not in effect fall back on the much maligned inductivism. Here is a simple illustrative example : The law of the conservation of energy is one of the best and most severely tested principles of physics. Although there have been occasional suspicions (as by Bohr, Kramers, and Slater in 1924) that the law might not hold strictly on the micro-level of atomic physics, such suspicions have been quite effectively removed by further experimental testing. I should think that the law deserves to be considered extremely well corroborated. To any adherent of inductive logic this would provide excellent grounds for the assumption that the energy law will not be refuted in the future. If "corroboration" does not entail this sort of justified expectation, I don't know what reasons we could give to would-be inventors of perpetuum mobiles (of the "first kind") to abandon their efforts. If Popper feels entitled to give such (negative) advice to inventors, he is trading on the *inductive* implications which the word "corroboration" has in common language. (These inductive implications are the same as those connected with such words as "confirmation," "evidential support," "substantiation," etc.) In the light of his critique of inductivism, is Popper really justified in utilizing these customary connotations of "corroboration"? All he is entitled to say on the grounds of his own analysis is that the energy principle is thus far unrefuted though it has been subjected to very incisive and multifarious tests.

My point can also be made by an opposite type of example. Some experiments are generally considered to be extremely decisive refutations. For instance, the Michelson-Morley experiment and some of its more recent replications are considered as highly conclusive refutations of the stationary ether hypothesis. (Actually, the situation is more complex, but I am sure there is no disagreement between Popper and me on the significance of other hypotheses, experiments, and observations in this context. For example, I agree with Popper that even observation statements are revisable. Still, there would be no "corpus of science" if we did not rely on some observation statements in a given context and "until further notice.") Of course, according to the simple model of *modus tollens* reasoning, a universal statement can be definitively refuted, once and for all. In its purely logical form this is unassailable. However, consider the situation in the empirical sciences. Popper's policy of the critical, rational approach must (and does) leave open the possibility that the Michelson-Morley type of experiment might give positive results beginning tomorrow and forever after. It is only by *induction* that we can assume that a well-refuted theory will *stay* refuted. After all, it is logically conceivable that such a "knocked-out" theory might begin to "stand up" at any time; and from there on out for all *future* concerns this would be just as good as a theory which had never been refuted.

Perhaps I cannot have what I am asking for : an inductive logic which provides an *objective* justification for the confidence which goes with corroboration as well as with refutation. Popper and I surely agree that Hume's psychological account—even if modernized in the light of recent psychological theory—does not give us a justification. The element of "animal faith" remains. Moreover, any psychological account of belief, expectation—in short, of induction—in relying on psychological laws makes use of inductive extrapolation. Hume knew this very well, and the Kantians who took him to task on this point misunderstood or distorted Hume's intentions. The various attempts in the direction of an inductive logic by Zilsel, Keynes, Nicod, Jeffreys, Reichenbach, Williams, Kneale, Braithwaite, Carnap, and others are, in their different ways, all subject to serious philosophical scruples. The sort of vindication of inductive inference that Reichenbach, Kneale, Salmon, and I[2] have been proposing is a very weak one, it must be admitted. Perhaps nothing stronger can be attained than such a pragmatic justification of the adoption of inductive procedures. It rests on the analytic truth that if any method of extrapolation succeeds, then inductive extrapolation will, too. This yields the "degenerate" justification : If there are regularities of nature, not too complex and not too deeply hidden, they are discoverable by the hypothetico-deductive method. Hence, it is reasonable to employ this method and tentatively, i.e., "until further notice," to rely extrapolatively on its results, be they corroborations or refutations. Despite its "weakness," such a vindication appears to me philosophically more enlightening than the shortcut of the Wittgensteinians (Strawson, Edwards, Hanson, Toulmin, and others), who essentially fall back on the common usage of such words or phrases as "good reasons," "strong evidence," etc. The problem is precisely to show what entitles us to use these honorific descriptions.

My final questions concern the issue of criterion of demarcation versus criterion of meaning. The logical empiricists, very much in the spirit of Hume (as well as of Peirce), and more radically than Kant, have opposed a certain type of metaphysics by declaring its questions as well as its answers devoid of factual meaning. Popper has been at pains to separate himself from this radicalism. He proposed a criterion of demarcation, drawing a

---

[2] See my article, "On the Vindication of Induction" in *Philosophy of Science*, 1961, pp. 212–16, as well as the references listed at the end of that article.

line between questions that pertain to empirical enquiry, and those that fall outside this domain. He considers the proposal of a meaning criterion fruitless and even harmful. His arguments deserve our fullest attention. Popper is surely right in repudiating the narrow meaning criterion of the early logical positivists. With painful clarity he pointed out that *that* criterion not only eliminated metaphysics but, alas, science as well. This was not intended by the positivists, of course. As I see it, the shift to logical empiricism involved a welcome and indispensable liberalization of the criterion. I wonder if Popper fully appreciates why the logical empiricists, while agreeable to the demarcation criterion, wish to go further and insist on a criterion of factual meaningfulness. If we make it impossible "in principle" to test a given hypothesis, i.e., if the manner in which an hypothesis is construed logically excludes refutation or corroboration, we have made it impossible to give empirical reasons for either the rejection or the acceptance of such an hypothesis. This may be achieved by making it immune to tests either by its outright formulation, or by hedging it about with special additional *ad hoc* assumptions. Such was clearly the case, for example, in the final stage of the ether hypothesis of H. A. Lorentz. While the earlier formulations of the ether hypothesis were clearly testable, the negative results of such tests motivated Lorentz to safeguard the hypothesis absolutely and unconditionally by his peculiar assumptions. It became "in principle," that is, logically, impossible to provide evidence for or against any assertion concerning the velocity of a body with respect to the stationary ether. Paraphrasing Peirce and James, there is not—and there could not be—any difference that makes a difference between the assertion and the denial of a given proposition concerning velocities relative to the ether so conceived. To be sure, the differences in question pertain to observational evidence; and since I am in full agreement with Popper's realistic epistemology, I would insist on distinguishing sharply the truth conditions of statements from their confirming (or corroborating) evidence. Nevertheless, if we isolate statements radically from whatever evidence might speak for or against them, do we still know what the statement is about? I believe that the appearance of meaning in this case is engendered by the pictorial appeals of the statement—but its factual meaning seems to me to have been "emptied out" just as decisively as in the case of tautologies or contradictions.

I admit that many statements which the logical empiricists diagnosed as factually meaningless are open to more charitable interpretations. Often it is the vagueness so characteristic of theological and metaphysical assertions which makes it possible to interpret them as having at least a modicum of factual meaning, or as presenting some sort of "promissory note" of a specification of meaning to be "made good" later.

In its least offensive form, the meaning criterion, very much like Popper's criterion of demarcation, may be construed as a *proposal*.

(Carnap settled for this many years ago.) Personally, I think a somewhat stronger case can be made. If we distinguish the cognitive functions of language from the noncognitive (pictorial, emotional, motivational) functions, the acceptance of the meaning criterion amounts to preventing confusions of the latter with the former. This still seems to me an eminently clarifying procedure, and thus a contribution of logical empiricism which has enduring value. I admit that Popper's demarcation criterion is adequate for the distinction of scientific from nonscientific enterprises. But the illusions and the confusions of transcendent metaphysics can be more effectively exposed by attention to the functions of language and the meanings of "meaning." In this way, I think we can steer clear of Hume's reductive fallacies, i.e., his skepticism and his phenomenalism, and nevertheless retain what is valuable in his empiricism and his critique of metaphysics. Popper's position seems in this respect closer to Kant than to Hume. With Kant he considers problems of transcendent metaphysics meaningful but undecidable. Logical empiricists admit that many questions will forever remain unanswered. But they insist that questions which can be shown to be unanswerable because of immunization to tests of any kind are not merely transempirical but are devoid of precisely the sort of meaning which certain metaphysicians impute to them.

CHAPTER 4

Strength, Confirmation, Compatibility

BY JERZY GIEDYMIN

IN THIS ESSAY I will discuss two concepts of compatibility of a hypothesis with the evidence, associated with different concepts of confirmation of hypotheses. One of the concepts is logical, the other pragmatic (in the sense given to this term by Charles W. Morris and Rudolf Carnap). I shall discuss the matter by analyzing two arguments advanced in the recent symposium on simplicity of hypotheses by Nelson Goodman and by Stephen Barker. This will give me an opportunity to make a few comments on the pragmatic character of some of the basic concepts of K. R. Popper's methodology.

I shall first give a summary of Goodman's and Barker's arguments and then proceed to an analysis of these arguments.

I. Goodman's and Barker's Arguments

The object of N. Goodman's critical argument[1] is the following methodological rule attributed by him to K. R. Popper:

R : The strongest hypothesis not falsified by the evidence should be chosen.

To show that strength is not sufficient as a criterion of choice of a hypothesis, Goodman presents the following two hypothesis and an evidence statement:

H_1 : Every examined A is a B and every unexamined A is a B.

H_2 : Every examined A is a B and every unexamined A is not a B.

E : Every examined A is a B.

Goodman maintains that : (a) both H_1 and H_2 conform to the evidence E; (b) H_1 and H_2 are equally strong, therefore (c) neither the incompati-

---

[1] Nelson Goodman, "Safety, Strength, Simplicity," *Philosophy of Science*, April, 1961.

bility relation nor strength are competent criteria of choice in this case; (d) for every hypothesis strong enough to go beyond the evidence, there is an equally strong conflicting hypothesis based on the same evidence, therefore (e) though desirable, strength is by itself not a competent criterion of choice; (f) simplicity cannot be identified with strength; neither can acceptibility of a hypothesis.

To establish the last contention, (e), Goodman produces the following example:

E_1 : All maple trees in a sample, consisting of many and widely distributed specimens, are deciduous.

H_3 : All maples are deciduous.

H_4 : All maples whatsoever, and all sassafras trees in Eagleville, are deciduous.

H_4 in the above example is stronger than H_3 and yet less simple and acceptable.

S. Barker argues similarly,[2] producing examples analogous to H_3 and H_4. Suppose, Barker says, that we have to account for the fact that "the maid has been acting strangely and the silver spoons are missing from their cabinet." This fact may be explained, for example, by the hypothesis: "The maid has stolen the spoons," or by any more detailed hypothesis such as "the maid has stolen the spoons and pawned them to raise money which she has used in order to elope with the butler on the 10:30 plane to Acapulco." The latter is stronger and yet less acceptable and less simple.

In my criticism of the above arguments I shall raise some doubts as to Goodman's contention (b). Further, I shall maintain that R is misleading when isolated from Popper's views on the role of tests and corroboration and that both Goodman and Barker have chosen a wrong interpretation of R, based on the logical concept of compatibility (conformity to facts, or the relation of "being based on the evidence"), where the pragmatical concept was necessary to formulate Popper's position.

II. The Concept of Strength Involved

Contentions (a) and (b) in Goodman's argument must be true to establish contention (c). However, (b) seems to me far from obvious.

In what sense are H_1 and H_2 said to be equally strong? Surely, in a criticism of Popper's methodology, and Goodman explicitly refers to Popper's book,[3] strength must be identified with testability (empirical content, simplicity).

Popper proposed two methods of comparing the testability (strength) of statements: (1) comparison by means of the subclass relation; (2)

---

[2] S. Barker, "On Simplicity in Empirical Hypotheses," *Philosophy of Science*, April 1961, pp. 169–70.

[3] K. R. Popper, *The Logic of Scientific Discovery* (London: Hutchinson, 1959).

comparison by reference to dimensions.[4] The former method is not applicable to H_1 and H_2 since neither of the classes of potential falsifiers of the statements includes the other as a proper subclass. This can be shown in the following way:

Let H_1 and H_2 be written as infinite sequences of statements of the form $a_i \epsilon B$ or $a_i \epsilon B'$, the first n elements of the sequence referring to examined cases:

$$H_1': a_1 \epsilon B, \ldots a_n \epsilon B, a_{n+1} \epsilon B, \ldots$$

$$H_2': a_1 \epsilon B, \ldots a_n \epsilon B, a_{n+1} \epsilon B', \ldots$$

The classes of potential falsifiers of the statements may be written as follows:

$$a_1 \epsilon B', \ldots a_n \epsilon B', a_{n+1} \epsilon B', \ldots$$

$$a_1 \epsilon B', \ldots a_n \epsilon B', a_{n+1} \epsilon B, \ldots$$

This shows that the classes of potential falsifiers of the two hypotheses H_1 and H_2 have the first n elements in common and differ with respect to the remaining elements. Therefore comparison by means of the subclass relation is impossible.

I do not think that the second method, i.e., comparison by reference to dimensions is applicable either. Ax, Bx E(xamined)x, are not "homotypic events," and statements obtained from these functions do not form a field in the sense of Popper's definition.[5]

Perhaps, however, the strength (simplicity) of H_1 and H_2 might be compared as follows:

H_1 is equivalent to the universal statement:

(1) $$(x) \, (Ax \supset Bx)$$

in which the predicates Examined, Unexamined do not occur. The information whether or not an a_i has been examined is not necessary to form a potential falsifier of H_1. It is necessary, however, to form a potential falsifier of H_2. Not that we cannot rewrite H_2 in a form similar to (1) above. We can—by replacing, for example, the predicates BE and B'E' (where E stands for "examined" and E' for "unexamined"), in

(2) $$(x) \, [(AEx \supset Bx) \cdot (AE'x \supset B'x)]$$

by, say, a predicate K in accordance with the definition

(3) $$K \overset{df}{=} EB \text{ or } E'B'$$

to obtain

(4) $$(x) \, (Ax \supset Kx)$$

whose antecedent is identical with that of (1). This exercise in translation, however, will not alter the fact that to form a potential falsifier of (4) we need more information (observation statements) than we need to form a

---

[4] *Ibid.*, pp. 115 ff., 126 ff.
[5] *Ibid.*, Appendix 1.

potential falsifier of (1)—in the former case $Ax.Ex.B'x$ or $Ax.E'x.Bx$; in the latter just $Ax.B'x$. This would suggest that (1) is simpler than (4).

I believe that in *The Logic of Scientific Discovery* Popper makes use of a pragmatic concept of strength (simplicity, testability)[6] along with the logical (syntactical or semantical) one, both as classificatory and relative concepts. The pragmatic concept, like the logical, is in terms of potential falsifiers, but while to determine the class of potential falsifiers in the latter case we need only the rules of language and logic, in the former case we have to know the rules of behavior of the experimenter testing the hypothesis in question, *viz.* when he is ready to reject the hypothesis.[7] To determine the degree of testability (strength) of a hypothesis in the pragmatic sense it is essential to know whether or not the class of the potential falsifiers of the hypothesis includes statements referring to objects which the experimenter can (practically) examine by observation and whether or not the class of potential falsifiers includes statements not yet accepted by the experimenter (i.e., whether or not the content of the hypothesis includes "free" or "unbound" components). A hypothesis fitted *ex post*[8] to all known facts in the given domain and leaving no practical possibility of further testing, is less testable from the pragmatic point of view than another one always capable of independent testing. A hypothesis which in given circumstances makes severe testing possible is more testable (stronger) than another one not susceptible to severe tests in these circumstances.

---

[6] In his contribution to the symposium on simplicity Mario Bunge distinguishes, besides other concepts of simplicity, the pragmatic notion. Cf. his "The Weight of Simplicity in the Construction and Assaying of Scientific Theories," *Philosophy of Science*, April 1961.

[7] For the classificatory concept of strength (testability), see *The Logic of Scientific Discovery*, chapter IV; for the relative concept, chapter VI. The pragmatic concept of strength (testability) is involved, I believe, in the following passages: ". . . I admit that my criterion of falsifiability does not lead to an unambiguous classification. Indeed, it is impossible to decide, by analyzing its logical form, whether a system of statements is a conventional system of irrefutable implicit definitions, or whether it is a system which is empirical in my sense; that is a refutable system. This, however, only goes to show that my criterion of demarcation cannot be applied immediately to a system of statements—a fact I have already pointed out in sections 9 and 11. . . . Only with reference to the method applied to a theoretical system is it at all possible to ask whether we are dealing with a conventionalist or an empirical theory. The only way to avoid conventionalism is by making a decision: the decision not to apply its method . . ." (p. 82). ". . . From my point of view, a system must be described as complex in the highest degree if, in accordance with conventionalist practice, one holds fast to it as a system established forever which one is determined to rescue, whenever it is in danger, by the introduction of auxiliary hypotheses. For the degree of falsifiability of a system thus protected is equal to zero . . ." (p. 145).

[8] Goodman's case involving H_1, H_2 and E, is clearly a case of hypothesis fitting rather than hypothesis testing, in Popper's sense of this term. The distinction is, of course, pragmatic in character.

It is not clear how Goodman's hypotheses H_1 and H_2 would compare if strength were taken in this pragmatic sense.

The objections and doubts raised in this section do not affect Goodman's second example, involving H_3 and H_4. They do not affect Barker's argument either. The strength of hypotheses in this case may be compared with the help of the derivability relation. H_3 follows from H_4, but not vice versa; therefore H_4 is stronger although less acceptable on the evidence.

We now turn to other objections which apply, I believe, to all arguments under discussion. These objections concern the interpretation of R and of the concept of compatibility of a hypothesis with the evidence.

III. The Concept of Compatibility with Facts

Goodman and Barker attribute to Popper and criticize the following view on the role of strength (testability):

T_1: Given the evidence (or assuming the evidence to be constant) scientists choose the strongest hypothesis compatible with the evidence.

T_1 is the descriptive correlate of R and is explicitly formulated by Barker[9] while R is attributed to Popper by Goodman.[10]

Now, when Popper writes that testability " . . . actually dominates the progress of science . . ."[11] or that ". . . theoretical science aims, precisely, at obtaining theories which are easily falsifiable . . . ,"[12] he obviously does not assert T_1. For T_1 explicitly assumes the evidence to be constant, which would be absurd in any account of "the growth of science."[13] What Popper does maintain is something like the following:

T_2: The history of theoretical science is a series of theories of increasing testability (strength).[14]

Neither of the critics has distinguished T_1, with its static assumption, from T_2. It should also be noted that, while Popper discusses the simplicity of theories, both Goodman and Barker in their arguments make use of statements which do not qualify as theories according to Popper's criteria. However, I am not going to consider the relevance of this difference.

In the remaining part of my analysis I want to concentrate on the concept of compatibility of the hypothesis with the evidence (or conformity to the evidence, or the relation of being based on the same evidence). It seems to me that both in Goodman's and in Barker's arguments

[9] *Op. cit.*, p. 169.

[10] *Op. cit.*, p. 150.

[11] "Some Comments on Truth and the Growth of Knowledge," lecture delivered at the Philosophy of Science Conference, Stanford, California, 1960; mimeographed, p .6.

[12] *Logic of Scientific Discovery*, p. 113.

[13] *Ibid.*, Preface.

[14] Cf. "Some Comments on Truth . . .," p. 6: ". . . and in each of these cases the progress made was towards a stronger . . . theory. . . ."

there occurs a concept of compatibility, appearing also in R and in T_1, which is inappropriate for the formulation of Popper's views on the role of testability.[15] I shall refer to this concept of compatibility as logical compatibility. There is a weaker and a stronger concept of logical compatibility as distinguished by the following definitions:

$Df1$: Two statements e and h are compatible if and only if neither of them logically implies the negation of the other.

$Df2$: A statement e is compatible with another, h, if and only if e is logically implied by h and h is not contradictory.

Compatibility in the sense of $Df2$ is associated with the conception of confirmation according to which any asserted observation statement implied by the hypothesis confirms this hypothesis. Incompatibility in the sense of $Df1$ occurs in the formulation of the sufficient condition for the falsification.

H_1 and H_2 are compatible with E, and H_3 and H_4 with E_1, both in the sense of $Df1$ and of $Df2$. The same applies to the statements in Barker's argument.

In Popper's methodology simplicity (testability, strength) as a criterion of choice of hypotheses is combined with corroboration (confirmation), defined however not in terms of logical compatibility in the sense either of $Df1$ or $Df2$ but in pragmatic terms of severe criticism, or ingenious and honest attempts at falsification. The identification of strength and simplicity with testability is significant: strong (simple) hypotheses are preferred, so Popper maintains, both because they are rich in content and because they make severe testing possible. Now, how severe the testing is in a given case depends, of course, not only on the strength of the hypothesis but also on other factors such as the ingenuity of the experimenter, his sincerity, his technical possibilities, the accessible evidence, etc. To make a choice of a hypothesis in accordance with the testability-corroboration criterion it is not sufficient to know that the hypotheses under consideration are compatible with the evidence in the logical sense: it is necessary to know how the evidence was obtained in order to appraise the severity of testing. That corroboration is not identical with logical compatibility was explicitly stated by Popper.[16]

---

[15] See R. Ackerman, "Inductive Simplicity," *Philosophy of Science*, April 1961, p. 153: ". . . The notion of Compatibility is not sufficiently clarified. . . ."

[16] *Logic of Scientific Discovery*, p. 266: ". . . It might perhaps be suggested that a theory should be accorded some positive degree of corroboration if it is compatible with the system of accepted basic statements, and if, in addition, part of this system can be derived from the theory . . ."; further, on p. 267: ". . . this last formulation seems to be insufficient for an adequate characterization of the positive degree of corroboration of a theory . . . it is not so much the number of corroborating instances which determine the degree of corroboration as the severity of the various tests to which the hypothesis in question can be, and has been, subjected. . . ." In the footnote on the same page: "If . . . we restrict the class of the derived accepted basic

As corroboration in Popper's methodology is restricted to certain test statements, satisfying certain pragmatical conditions, compatibility with evidence is relevant only if it is compatible with the results of severe tests and not with any observation statements. It seems to me that this concept of compatibility is widely accepted in the procedure of many branches of science, but it is most conspicuous in those in which statistical methods are applied. It is only the results of "good" tests that confirm a hypothesis and therefore only such compatibility matters in the choice. In statistical procedure, random sampling is one of the conditions of good tests, and it seems to me that it shares at least some of the intuitions of Popper's severe testing. In what follows I shall discuss Goodman's hypotheses and evidence statement on the additional assumption that the evidence statement is the result of the examination of a (sufficiently large) random sample chosen from the population A.

Let us denote by A_s the sample from the population A and assume that it consists of n elements. We may write then:

$$E' : \frac{N(A_s B)}{N(A_s)} = \frac{n}{n} = 1$$

$$H_1'': \frac{N(AB)}{N(A)} = 1$$

$$H_2'': \frac{N(AB)}{N(A)} = \frac{n}{N(A)}$$

H_1'' and H_2'' are consequences of H_1 and H_2 in Goodman's argument. H_1'' asserts that the proportion of the elements of A having the property B equals 1. H_2'' states that the number of the elements of A having the property B equals exactly n and that, therefore, n over the number of all elements of A gives the proportion we are interested in.

Now, assuming that the population A is infinite or very great in comparison to the sample (and this is the usual situation in scientific research), the probability of obtaining the result E' given H_1'' equals or is very close to 1, while the probability of obtaining E' given H_2'' equals or is very close to zero. Obviously, we would choose H_1'' and reject H_2'' given E'. We may say that E' is statistically compatible with H_1'' and is statistically incompatible with H_2''. Compatibility is meant here in the following sense:

$Df3$: The result of an experiment e is (statistically) compatible with the hypothesis h if and only if the probability of e given h is great.

Although E does not contradict H_2 or H_2'' and is implied by H_2, nevertheless it is statistically incompatible with one of the consequences of H_2. We may also say that, granted the assumption of random sampling, H_2 is almost self-contradictory, and would by no means be a "good fit."

statements further, by demanding that they should be accepted as the result of sincere attempts to refute the theory, then our definition becomes an adequate definition of 'positively corroborated'. . . ."

Can we assume, however, that the evidence statement E in Goodman's argument was the result of the examination of a random sample? Two objections may be raised against this assumption, one of a general character and the other especially concerning Goodman's argument.

Firstly, we may doubt the possibility of choosing a random sample from a "dynamic" population, i.e. from one which practically does not exist at the moment of sampling as it includes future elements besides the past and present ones. It is with such populations that we are usually concerned in science. The assumption of equiprobability of choosing any element from such a population seems never to be satisfied. Dynamic populations are not models of sets of events with which the probability theory is dealing.

However interesting theoretically the above objection may seem, we shall reject it on pragmatic grounds : it questions the possibility of applying a theory which has proved useful in practice. Moreover, it may be argued that the conditions of sampling are at least approximately satisfied if we know that the "structure" of the population in question will not change (considerably) in future, where "structure" is defined in terms of some parameters in which we are interested.

Ignoring, therefore, the general objection we now turn to the special one. It may be formulated as follows :

If random sampling conditions require the stability of the structure of the population, then the stability assumption must not be made for the purpose of testing H_1 and H_2. This assumption would be compatible with H_1, but H_2 may be interpreted as the prediction that the population A will change in some future moment t, following the moment of sampling. H_2 would, therefore, contradict this assumption. It would be inappropriate to assume the stability of A when trying to decide between H_1 and H_2, therefore E is not the result of the examination of a random sample. If the sample is not random, then E is not incompatible (statistically) with H_2, or at least we have no ground to believe that it is and cannot appeal to this relaxation as to our criterion of choice.

I think that by accepting the interpretation of H_2 on which the above objection was based, we do not have to give up confirmation as our criterion of choice, although we no longer can explicate confirmation in terms of random sampling of the population A. A hypothesis predicting the change of the structure of the population may be tested against the consequences of this prediction referring to the behavior of the factors on which the structure of the population depends. If no such observable consequences are available, the hypothesis is untestable under these circumstances. If applied to H_1 and H_2 in Goodman's argument, this would mean, however, that there is no problem of strength as a criterion of choice : we are interested in choosing one of two well-tested hypotheses but not in choosing any strong hypothesis.

To sum up our analysis of Goodman's argument : Popper's criterion of choice of hypotheses is testability–corroboration. To see how it worked when applied to H_1, H_2, and E, more information would be necessary on E. If E is the result of severe testing, e.g. the result of an examination of a random sample from A, its relation to either hypothesis may be different, thus providing a criterion of choice. This is clear in the other of the two sets of hypotheses discussed by Goodman : the evidence statement E_1 is not a good test of H_4 as the sample was not chosen from both populations referred to in H_4.

Similar remarks may be directed toward Barker's argument. He also disregards the fact that, according to Popper's methodology, an increase in testability is desirable to better test the hypothesis. Of two hypotheses considered by him, the one "embroidered with details" could be interesting (even as a gossip to a detective looking for "clues"). However, it would have to be tested, and the assumed evidence is insufficient for this purpose.

I conclude, therefore, that both Goodman and Barker mistakenly attribute to Popper T_1 or R, while he in fact asserts either T_2 or the following thesis :

T_3 : If two universal hypotheses, h_1 and h_2, have both successfully passed severe tests and h_2 is logically implied by h_1, but not vice versa, then scientists prefer h_1 to h_2, both because h_1 is more informative and stimulating and because it may be better tested in future.

I do not think that Popper would advise anybody always to choose a stronger, even if well-tested, hypothesis when a practical, single decision is to be based upon it, and when, moreover, evidence cannot be increased and the eventual gain in content (information) is inapplicable to any future cases. This is just why I think Barker's argument a failure as a criticism of Popper.[17]

Many authors have emphasized the inadequacy of logical (syntactical) concepts for the methodology of empirical science. Among them was Nelson Goodman. What a pity that he should disregard Popper's pragmatical concept of "positively corroborated"[18] and of testability or strength,[19] both proposed in 1934, and Popper's pragmatical characterization of corroboration, given in the 1954–58 notes on confirmation.[20]

---

[17] It should be noted, however, that Barker refers in his article to "the manner in which he [Popper] presents his doctrine," which "makes it seem as though . . . ," and that "Popper leaves his readers with the impression . . .," etc. I would agree with Barker, if he just meant the criticism of single, isolated formulations of Popper's views. If, however, one considers Popper's whole methodology, I do not think that Barker's is a good account of the impression it leaves.

[18] See note 16, *supra.*

[19] See note 7, *supra.*

[20] *Logic of Scientific Discovery*, New Appendix IX.

CHAPTER 5

A Question about Plato's Theory of Ideas

BY R. M. HARE

A VOLUME PUBLISHED in honor of Professor Popper would be incomplete without some contribution specifically intended to honor him as the writer of one of the very few recent books about Plato worth reading. Since Platonic scholarship, seriously pursued, can easily become a fulltime occupation, not many philosophers have been prepared to leave their main field of interest, even temporarily, to raise historical questions about the beginnings of their subject. Yet unless they do so, Platonic studies will lose contact with Plato the philosopher. Unless he is studied, at least sometimes, by those whose principal interest is in the philosophical questions which he and his master Socrates first raised, scholars will be left in a position not unlike that of historians trying to piece together Hannibal's route from ancient sources without taking a look at the Alps. I make no apology, therefore, for following Popper's example, but only for my lack of equipment for the task.

The most useful contribution philosophers can make to Platonic studies is to ask questions in the hope that specialist scholars may find the answers. I wish in this paper to ask one such question. It is a question which is extremely hard to make clear, and to distinguish from other apparently similar questions. I shall therefore go further, and suggest a tentative answer to it, and even consider what light is shed on the question by the text of the dialogues; but I do this only to clarify the question as that question to which this answer is an appropriate (though possibly a false) one. The major task remains of attempting to test this hypothesis by an exhaustive study of the textual and other evidence. This enormous task I have barely begun; one of the purposes of publishing this paper is to enlist help in it. All that can be said is that I have failed so far to find much evidence which seems to tell against the hypothesis, and that even this evidence may admit of alternative explanations. I have, however, found quite compelling evidence, as it seems to me, against some alternative hypotheses. With the object, therefore, of presenting a clear-cut proposal for discussion, I shall argue the case for my suggested answer with greater apparent confidence than I in fact feel.

My question, it must be emphasized, is not a philosophical one, and no answer to it should contain any philosophical terminology—though a good deal of philosophy will be required in order to distinguish this question from others. Nor is it a question of exegesis. It is a historical question of a psychological sort. That is to say, I am not going to ask either what Plato may have meant by certain things that he said or whether he was right. To answer either of these questions would require at the outset a statement in philosophical terms of certain propositions to which, in my view, Plato would have assented.

I shall not discuss (except in passing) what Plato's philosophical views were or were not. What I want to know (because it has a crucial bearing on many philosophical and exegetical questions about Plato) is the answer to a plain question of psychological fact : What, in certain circumstances and on certain occasions which I shall specify, was actually going on in Plato's experience? The circumstances and occasions in question are, roughly, those in which Plato himself would have said "I am seeing (or apprehending) an Idea."[1] The answer to my question must be couched in terms which would enable us, if the same thing happened in our own experience, to recognize it and to say "That was the sort of thing that Plato was referring to when he said . . . etc." It is therefore necessary that the answer be given in words which we can all understand—in particular, not in philosophical language. Since my question is not a philosophical one, I can demand an answer to it in plain English without incurring any of Popper's strictures against "ordinary-language philosophy."

I

I will now illustrate the character of my question by asking another question which is of the same kind, but to which the answer is fairly obvious. Suppose we were to ask, in the context of *Republic VI* and *VII,* what was actually happening in Plato's experience when he was engaging in what he calls *eikasiâ*. This word is one of the four key terms which Plato uses in expounding his theory of knowledge in these books, in the simile known as "The Line." The other three are called in 511*d pistis,*

---

[1] The difference between "seeing" Ideas and "apprehending" them is no doubt important, since to see a bird is very different from catching it; but I shall have no room to deal with this distinction.

dianoiâ, and *noêsis*—though *noêsis* is later used generically to include *dianoiâ* (534*a*).

In 511*d* it is said that these four "experiences in the mind" (*pathêmata en têi psŷchêi*) have as their objects four kinds of thing, which have been described earlier. *Eikasiâ* has as its objects "shadows, and reflections in water and in anything that is fine-textured and smooth and shiny, and so forth"; and *pistis* has as its objects "the animals in our environment and the entire classes of plants and of man-made objects" (510*a*). From this it is clear that when Plato was engaging in *eikasiâ* he was looking at, for example, a reflection in water; and that when he was engaging in *pistis* he was looking at, for example, a horse or a table.

It is this kind of answer that we require to the question, "What was Plato doing when he was engaging in *dianoiâ* or *noêsis*?" We all sometimes look at horses, tables, and reflections in water; therefore, when Plato mentions these activities, or experiences, we all know to what he is alluding. We need to know in just the same way to what sort of experience he is alluding when he talks about seeing or apprehending Ideas (which, as we learn from 511*c* and elsewhere, are the objects of *noêsis*). We need to be able to catch ourselves doing (if we ever do) what Plato describes in these terms; for if we cannot do this, we shall never be able to understand all the further things which he says about this activity. We do not, however, need to accept all these further things in order to be able to identify, in our own experience, that activity about which he is saying them.

The distinction between identifying the experience and understanding the philosophical things which are said about the experience will perhaps become clearer if we consider what we should say to a person who answered, to our question about *eikasiâ*, that when Plato was doing this he was "having sense data." If no importance is being attached, in this answer, to the difference, if any, between the experience (whatever it is) called "having sense data" and the experience called "looking at a shadow or reflection," then, indeed, this answer becomes a somewhat affected way of expressing the correct answer, namely, that Plato was looking at a shadow or reflection. If, on the other hand, some importance *is* being attached to the alleged difference between having sense data and looking at a shadow or reflection, the answer becomes either inapposite or wrong. If the difference is an empirical one, then since the correct answer to the question "What experience was Plato having?" is "Plato was looking at a shadow or reflection," "Plato was having sense data" must be an incorrect answer. The two experiences are, on this supposition, different. If the difference is not an empirical one, but a difference between two philosophical theories about the same experience, the answer is inapposite —an answer to the wrong question. The question asked was not what philosophical theories Plato had about this experience, but what the experience was.

We want, therefore, an answer to the question, "What was happening in Plato's experience when he was, as he would have put it, 'seeing or apprehending an Idea'?" which is like the answer "Plato was looking at a shadow or reflection" and not like the answer "Plato was having sense data." We must, accordingly, rule out from the start, as not being answers (even wrong answers) to our question, such suggestions as "Plato was intuiting a transcendental universal." While it may be true that this is what Plato, if we could ask him, and if he knew the English of the schools, would say as a philosopher about the experience in question, such an "answer" does absolutely nothing to help us identify the experience which is the subject of this philosophical pronouncement. How are we to know when we are "intuiting a transcendental universal" and when we are not? To answer that "Plato was intuiting a transcendental universal" is therefore no more illuminating than to answer that "Plato was *noôn, or pros tên ideân blepôn*" (or one of the other ways in which he puts it). We want to know what was happening.

II

A possible, and indeed plausible, answer to our question would be the disappointing one that no single thing was happening on all the occasions on which Plato would have said he was seeing or apprehending an Idea. It might be that a great many different things were happening on different occasions, and that he failed to distinguish them. However, it may not be necessary to be so pessimistic. Even though it would be surprising if it could be shown that on *every* occasion of which Plato would have spoken in these terms the same thing was going on in his experience, we ought perhaps to be content with the lesser satisfaction of discovering that *in paradigm or typical cases* a certain kind of thing was happening, and that Plato's theory was an attempt to assimilate the less elementary and more interesting cases to these paradigm cases. This would be a temptation into which many philosophers have fallen.

We have, however, at the outset to answer a possible objection to our procedure of looking for something in our own experience which might be similar to what was going on in Plato's when he would have said he was looking at an Idea. This would be a vain search if Plato, when he was "looking at an Idea," was having a very high-grade mystical experience such as we ordinary mortals cannot aspire to. Fortunately, this suggestion runs counter to some quite strong evidence in the dialogues. For, while it is true that Plato does sometimes speak of apprehending certain of the Ideas as a matter of great difficulty requiring the highest powers of intellect, he speaks elsewhere as if the simpler Ideas were open to inspection by the meanest minds. We learn from *Rep.* 596b and *Crat.* 389a that a carpenter (who would have been in the lowest class in Plato's city) looks at the relevant Idea every time he makes a table or a bed or a shuttle. It is also implied by the Plato-Diogenes story (see below) that Plato thought that anybody who cannot see the Ideas of Cup and Table has, to say the least, a low I.Q.

As we shall see, Plato thought that some Ideas were more difficult to apprehend than others. It may also be true that until the highest Idea, that of the Good, was apprehended, he thought there was something lacking in our *knowledge* even of the lower ones which depended upon it (*Rep.* 511c, d). And we must not assume that Plato thought that ordinary carpenters (any more than ordinary mathematicians) could give a *logos* or definition of the Ideas with which they were concerned. To this extent they lacked knowledge or art (*epistêmê, technê*), for which the ability to give a *logos* is a necessary condition (*Rep.* 510c, 533c; *Meno* 98a; *Gorg.* 465a).

However, it remains true that, according to Plato, ordinary men could look at and see the simpler Ideas, even if it took a philosopher to make them attend to what they were seeing. Anybody, therefore, who has ever made a table can properly ask himself which, of all the things he did when making it, corresponds to what Plato said the carpenter was doing when he was "looking at the Idea."

III

It will be helpful if we start our search, not among the carpenters, about whom Plato tells us too little, but among the mathematicians, about whom he tells us a lot, though obscurely. In *Rep.* 510d he says of them, among other things, that they

. . . use, in addition, visible forms (*eidê*) [as diagrams], and talk about them, though it is not about them that they are thinking (*dianooumenoi*), but rather about those other [Forms] whose likenesses they are. Thus, the object of their discussion is the Square itself, and the Diagonal itself, not this diagonal which they scratch on the sand; and so with the rest of the figures. These scratched or handmade things which have as we saw *their* likenesses—shadows and reflections in water—are themselves, in their turn, used by the mathematicians as likenesses (*eikones*) when they are trying to see those things-themselves, which are not to be seen with anything but the mind.

It is not certain that here "the Square itself" is an Idea; but those who have argued that it is have, besides good general grounds, support in the text of this passage : there is the familiar wording, "the Square itself," echoed later in the phrase "those other things-themselves"; and it is surely correct, as it certainly is natural, to supply the noun *tôn eidôn,* as understood, after "those other" in the first sentence of the quotation. Indeed, the opportunity for this veiled allusion to his technical term may be the reason for Plato's use of the otherwise rather odd expression "visible *eidê*" for mathematical diagrams.

However this may be, it is hard to deny that the language which Plato uses in speaking of the mathematicians' dealings with "the Square itself" is entirely typical of the sort of language which he generally uses in speaking of our dealings with Ideas. We may notice in particular that the Square itself and the Diagonal itself are the objects of their discussion (*logoi*) and that they are trying to "see [them] . . . with . . . the mind." From this language it is at any rate clear that "the Square itself" is not any kind of talk, mental or otherwise, but rather something that is talked about; and it seems that "making it the object of their discussion" (literally "making their discourse for the sake of it") is a different activity from actually seeing it, since the mathematicians are already doing the former, but only trying to do the latter. They are trying to see it with an organ called "mind" (*dianoiâ*); and this visual language is repeated later (511*c*), where it is applied equally to the higher activity called *noêsis,* which is indisputably concerned with Ideas.

From the use of words for seeing in connection with both *dianoiâ* and *noêsis* throughout the two "Line" passages, it is clear that, whatever the differences between the two upper segments of the Line may be,[2] the activities which they represent both consist, in part, in some sort of *seeing* which is not the seeing of ordinary visible objects with the eyes, but the seeing of something else by the use of some other faculty. Indeed, this entire part of the *Republic,* and especially the simile of the Sun (508ff.), is full of this metaphor (if Plato thought of it as a metaphor) of seeing; and he constantly seems to envisage that faculty with which we do our thinking—or at any rate the best and proper kind of thinking—as "the eye of the mind" (533*d*).

We are reminded of the story told by Diogenes Laertius about an encounter between his namesake the Cynic and Plato (6, 2, 53; for another version, with Antisthenes instead of Diogenes, see Simplicius *ad Arist. Cat.* 66*b*, 67*b*). Diogenes, poking fun at the Theory of Ideas, said, "I see the table and the cup; but *tableness* and *cupness*—I can't see *them* at all"; to which Plato replied, "Of course; because the cup and the table are seen with the eyes, which you have; but tableness and cupness are seen with the *nous,* which you lack."

This assimilation of thinking to a kind of mental seeing had a past and was to have a future. In Homer the verb *noein* (with which *nous, dianoiâ,* and *noêsis* are all cognate) is used of visual perception (*Iliad* 15, 422), of "noticing" consequent upon such perception (*Iliad* 11, 599), and of pure thought (*Iliad* 19, 112). The future of the comparison is testified to by the

---

[2] I hope to examine these differences in another paper, to be included in a collection of essays on topics in Greek philosophy, edited by J. R. Bambrough.

passage into philosophical terminology of the words 'theory' and 'intuition' (both derived from words for looking); but we must be careful not to be misled, through our familiarity with the very watered-down meaning which these terms now bear, into thinking that the comparison was not conceived of by Plato as a very exact one.

IV

Let us now revert to our original problem. If we make the working assumption that Plato's mental processes were not entirely unlike our own, the evidence which we have so far noticed suggests a possible answer. Speaking very roughly and crudely, what we do when we are thinking may be classified into, first, talking to ourselves, and second, forming "mental images."[3] In the latter we must include, not only quasi-visual images, but also imagined sounds, smells, etc.; and it is perhaps necessary to include also the feelings which may be associated with these images— as when someone imagines a pretty girl and experiences the feeling called "lust." This classification is, I repeat, crude; but it may be exact enough to deal with Plato's first attempts to give an account of thinking.

It has already become apparent, and will shortly become even more so, that Plato is not equating the "seeing" of Ideas with talking to oneself—although in the *Sophist* he does describe thinking (*dianoiâ*) as "the discourse of the mind with itself" (263*e*). He generally speaks as if the talking that we do, to ourselves and to others, is not itself the seeing of the Ideas but something *about* the objects of this seeing, in order to assist the talker to "see" them, or to show that he had "seen" them. It seems to follow, by elimination, that what was happening when Plato was "seeing" an Idea was something falling within the vague class of "the forming of mental images."

Consider the famous passage in the Seventh Letter (342*a*) in which "the necessary conditions of knowledge" are discussed. Even if Plato did not write this letter himself, it is probably near enough to him in time to be of some use as evidence; and it contains the fullest list available of the things that are, in the Platonic view, involved in knowledge. Three things are listed: the name of the thing known, e.g., "circle"; its definition (*logos*), e.g., "the figure whose circumference is everywhere equidistant from the center"; and what is called the "image" (*eidôlon*). By this last, it is said, is meant "that which is drawn and rubbed out and turned on a lathe and broken"—what we should call a physical object or diagram.

Plato often speaks of physical objects as "images" or "likenesses" of

---

[3] Aristotle held that "there is no thought without a mental image" (*De An.* 431*a* 16, cf. 432*a* 8); the interpretation of this is, however, doubtful.

Ideas. This is, no doubt, the main reason why the answer which I have proposed to our problem seems at first sight scandalous. However, once the answer is properly understood, it is seen to be no cause for scandal at all. For I am not suggesting that Plato would have said "The Ideas are mental images"; but that this would be *our* way of describing what was happening, stripped of philosophical interpretations, when Plato would have said "I am seeing an Idea." It goes without saying that *he* would have dissented from this identification. For to call something a mental image is, in our way of speaking, to deny that it is a real thing (in the ordinary sense of those words), whereas Plato wanted to attribute the highest kind of reality to what he said he was seeing. This, however, is part of the philosophical interpretation of the experience, and therefore has no bearing on our problem. What is being suggested is that the experience which Plato, interpreting it one way, would have called "seeing an Idea," we, interpreting it another way, would call "forming a mental image." We shall later return to this distinction.

Once this misunderstanding is cleared away, the use of the words 'image' and 'likeness' of physical objects and diagrams actually supports the suggested answer. It is of the essence of an image or likeness to be like the object imaged; and since similarity is a symmetrical relation (cf. *Parm.* 132*d*) the object must also be like the image. Therefore, if Plato attributed the highest sort of reality to what we would call "a mental image," it was very natural for him to call the physical object an "image" or "likeness" of it, instead of the other way around. Since there is nothing so like a physical object, in one way, as the mental image of it, it is obvious that "forming a mental image" is a much better candidate for the position of what Plato was doing when he was, in his terms "seeing an Idea" than, for example, is "holding a mental discourse with himself."

The passage in the Seventh Letter affords even more positive evidence than this. After listing the three things just mentioned (name, definition, image), the Letter adds two other things : the knowledge itself, and "that which is the object of the knowledge, and which is in the true sense real"— *viz.,* the Idea. Since the Idea appears separately on this list, it cannot be identified with any other member of the list. Therefore it is not the name for the *logos* or definition; nor is it, needless to say, the physical object; nor even the mental activity, knowledge, which consists in contemplating the Idea, though this activity, being mental (*en psychais*), is said to have the closest affinity with the Idea itself. We must ask, then, what else is left for the Idea to be but the object of this activity, i.e., that which is before the mind when one is engaged in the activity. And what could this be but a mental image, to which reality is ascribed by Plato? With his (perhaps mistaken) conception of knowledge as, at least in part, an occurrent— that is, not merely dispositional—activity analogous to sight, it is hard to see what other candidates there could be.

V

Further light may perhaps be shed by a passage in Aristotle's *Metaphysics* (990*b* 8), in which Aristotle is attacking the Platonic version of the theory of Ideas. "None of the ways," he says, "in which we show that there are Ideas is adequate to show this; for some of them do not necessarily lead to the desired conclusion, and some of them lead to Ideas of objects of which we do not think there are Ideas." "We" here means the Platonists, whose view Aristotle is in process of abandoning. He then lists a number of arguments, the last of which falls into the class of "ways which lead to Ideas of objects of which we do not think there are Ideas" : "According to the argument, that we think of an object when it has been destroyed, there will be Ideas of destructible objects; for there is a sort of mental image (*phantasma*) of these."

The argument referred to seems to have been somewhat similar to one recently revived by Professor Price, that "thinking in absence" involves knowledge by acquaintance of universals. The fact that we can think about an object after it has been destroyed means that what is before our mind cannot be the object itself; it must therefore be, say the Platonists, the Idea of the object. But, says Aristotle, this entails there being Ideas of destructible objects—which, for some reason, he thinks would be repugnant to the position he is attacking. The fact that Aristotle here regards the existence of a mental image as requiring, for the Platonists, the existence of an Idea is support, of a kind, for our suggested answer to our problem, although, as we shall see, Aristotle was wrong if he thought that *whenever* Plato was having a mental image he would have said that he was seeing an Idea.

Consider next a passage which might be thought to tell against the suggested answer : *Parmenides* 132*b*. Here Socrates, to avoid a difficulty which does not concern us, suggests that the Idea of each thing might be a thought (*noêma*), which cannot properly exist anywhere but "in minds" (*en psŷchais*). Since this suggestion is firmly and summarily rejected by Parmenides, and does not recur, it is safe to assume that Plato did not think it a good one. It might therefore be argued that whatever else the Ideas were, they cannot have been mental images, which are also, in some sense, "in minds." But to argue thus is to misunderstand, in a way against which I have been at great pains to warn the reader, both the question which we are asking and the suggested answer. Plato's language here is perfectly consistent with the theory that what was happening when Plato said "I am seeing an Idea" was what *we* should describe as "Plato having a mental image," though it is inconsistent with the theory, if anyone held it, that Plato would *himself* have said "The Ideas are mental images." That is to say, the passage tells against the view that Plato would have accepted an utterly un-Platonic philosophical interpretation of what was happening, but not against our suggestion about what *was* happening.

Let us, however, examine the argument a little more closely. The word *noêma* clearly means, as Cornford says, not "object of thought," but "act of thinking." The first argument of Parmenides against the suggestion that Ideas are acts of thinking (the only argument that concerns us) is that there cannot be an act of thought without an object of thought, and this is the Idea. Here the same distinction is made as in the Seventh Letter between a mental activity and its object. Thus this passage supports rather than conflicts with our suggested answer. When Plato was having a mental image, he thought of what was happening as follows. He, Plato, was, as it were, seeing something, and that which he was seeing might be an Idea—that is to say, a real universal object of thought (though it might equally well not be an Idea—hence the need for the *elenchus* and all the other apparatus for sifting genuine from spurious objects of thought.) This view entailed that, at least, one could take another look at this Idea on another occasion, and it would be the same. But the fact that Plato thought of his experience in this way does not mean that what was actually happening was anything other than that for which the correct English description is "Plato having a mental image."

A parallel from modern philosophy (already briefly alluded to) may perhaps make the situation clearer. There is no experiential difference, it is sometimes said, between seeing a table and having table-ish visual sense-data. But there are two ways of talking about this experience— "sense-datum language" and "material-object language"—of which the former carries much weaker implications than the latter. If I say "I see a table," I imply, for example, that others can have similar or related experiences, and that I can at a future time myself "see the same table again." But (according to sense-datum theorists) if I say "I am having table-ish sense-data," I imply nothing about any other experience that I or somebody else might have. Plato's proposal, on the suggestion that I have made, could be put in the following way : he was proposing that we should start using material-object language about some of our mental images (for all of which we now use a kind of sense-datum language) and, perhaps, that we should start using a sense-datum type of language in speaking of what we now call material objects (though Plato is far from actually carrying out the latter proposal, for all that it is strongly suggested by his talk about *doxa* that this is what he really wanted to do). Thus material objects are no longer to be called "real" (because one cannot step into the same river twice); and those mental images which pass muster are to be treated as independently existing objects, available to repeated inspection by different people, and called "truly real." There seems to be, at any rate, no *greater* logical difficulty in talking like this than there would be in adopting a sense-datum language when speaking about tables (though both proposals are probably misguided). The adoption of a new language would not in either case make any difference to the experiences that were being had. The sense-datum theorists and Plato are both making suggestions about how we should speak, but the experiences about which they are speaking are the same experiences as we have learned to recognize and to describe in quite different terms.

VI

We must next ask whether the suggested answer to our question will fit the whole of "the population of the world of Ideas."[4] If there were any class of things of which it is impossible to form mental images, but of which Plato did admit Ideas, that would be evidence against the suggestion that I have made, unless it could be shown that, though it is in fact impossible to form mental images of these things, Plato did not recognize this impossibility. It will not be sufficient, in order to refute the suggestion, to show that we *actually* cannot form mental images of a class of things of which Plato admitted Ideas. For he may have thought that, by strenuous philosophical endeavor, he might come to be able to do what we would call "having a mental image of these things," and he would call "seeing the Ideas of them." In any case, even if the answer which I have proposed is the right one, it cannot be expected to fit all kinds of Ideas equally well. I suggested earlier that Plato might have been assimilating the more difficult cases to his paradigm cases. If this were so, we should expect to find some cases which fit in quite easily and others which present greater difficulty; we should also expect to find Plato giving evidence that in these latter cases he finds difficulties which he does not find in the easier ones.

The cases which fit in most easily are the simpler geometrical ones. Do we not all quite frequently, when thinking about the properties of the circle or the square, form to ourselves a mental image of a circle or a square, and think about the properties of the image? It will, of course, be said that the mental image we form is still, like a physical diagram, particular, and therefore can have nothing to do with what Plato was referring to. But is this so obvious? What would be the *experiential* difference between forming a particular image of a square and mentally seeing the universal, the Square, on a particular occasion? Or between forming a number of particular images of particular squares at different times, and seeing, on a number of particular occasions, the one universal called "the Square"? If, as I think—as it is easy to think—that there is no experiential difference, it is not difficult to see how Plato can have interpreted

---

[4] This phrase is borrowed from Sir David Ross's book, *Plato's Theory of Ideas*, to which, in spite of disagreements, my debt is obvious.

the experience which *we* call "having a particular mental image of a square" as "having, on a particular occasion, a mental look at the Square." It is true that one can look *simultaneously* at two imagined squares, whereas there could not be two Ideas of the Square; but this difficulty, which led to Plato's invention of the "mathematicals," probably did not occur to him until quite late in life. That it was, when noticed, felt as a difficulty is some evidence in favor of our suggestion.

The mental image, considered in this way, has many of the qualifications that Plato was obviously looking for in his search for reality. It is always available for inspection; it is independent of the chances and changes of the world, not being, as physical images are, drawn and then rubbed out. The mental image, it is true, vanishes when we turn our attention to something else; but Plato may have thought that, although we were no longer looking at it, it was still there to be looked at. Moreover, once one has come to look upon the experience which *we* call "having mental images" as looking at a transcendental reality, it is extremely tempting to think of physical squares and circles as images of the mental ones and not vice versa, since, as we have already seen, and as Plato himself pointed out, the relation of similarity is a symmetrical one. That is why mathematical diagrams are said to be "images" or "likenesses" of the Square and the Circle. On any other theory, how can these words be called in the least appropriate?

It is not necessary, therefore, to look for any more complicated explanations of *Republic*, 510d ff. (the passage first discussed) than that suggested by the text. What the mathematicians are doing is, in our own terms, drawing diagrams on the sand, and talking (apparently) about them, but trying to form a mental image of a square with a diagonal and to discover their properties (a process which Plato describes not as we have, but as "trying to see those things-themselves, which are not to be seen with anything but the mind").

It cannot be denied that Plato was starting a great deal more here than he was aware of. No doubt, once one has realized that geometry discovers universal truths and not truths about particular objects, one has started on a journey which does not end until one arrives at analytical geometry and dispenses with figures of all sorts, both physical and mental. Geometry then becomes a branch of algebra, and thus, according to some logicians, of logic. There are indications in Plato's account, in his talk about dialectic in the *Republic* and about *logoi* there and elsewhere, of the direction which the journey was to take. All I am denying is that Plato, when he wrote the *Republic,* had got as far on this journey as some people have supposed. Philosophical advances are much more difficult to make than often appears after they have been made. It is one of the commonest sources of error in interpreting Plato to suppose that, being a great philosopher, he must have seen what is obvious to us. These things are obvious to us only because pioneer philosophers—of whom, though Plato was one of the greatest, there have been many since—have discovered them and made them part of the common body of knowledge which is the despised inheritance of every schoolboy. No doubt it is obvious to us that the easiest way of conveying heavy loads is on wheels; but that does not mean that the man who first thought of conveying them on rollers was stupid.

There is one particular logical objection to thinking as, on my view, Plato did; and this might also be advanced as an objection to thinking that he thought in this way—for no admirer of Plato likes to suppose him guilty of not having discovered everything. It is the objection raised by Berkeley (*Principles,* Introduction, 13) against Locke (*Essay* IV, 7, 9.) Locke had said that one must form an idea of a triangle which is "neither oblique nor rectangle, neither equilateral nor equicrural nor scalenon, but all and none of these at once"; and Berkeley said that this is manifestly impossible—as indeed it is, if one means by "idea of a triangle" what Berkeley, and perhaps also Locke, meant by it. Berkeley's objection could be raised against Plato too, if Plato was thinking as I have suggested. One cannot form a mental image of a triangle which is neither equilateral nor equicrural nor scalenon; therefore, when Plato was "seeing the Idea of the Triangle," he would have, on my view, to have been doing the impossible.

There is, in fact, a passage in the *Meno* (74*d* ff.) in which Plato uses words which are very reminiscent of Locke. Socrates is discussing the definition of Shape (*schêma*). Quoting the words of an imaginary questioner, with whom he is in sympathy, he says, "Since you call these many things by one name, and say that every one of them is a shape, though they are, even, opposite to one another [he has been discussing the shapes called "curved" and "straight"], will you not say that there is this thing that comprises the curved no less than the straight, which you call 'shape'?" The Idea of Color, which is dealt with in the same passage, presents a similar difficulty.

It is possible that Locke meant by "idea" what Berkeley probably thought he meant, *viz.,* "mental image." In this case Berkeley's criticism of Locke is justified; but we can then argue that if Locke, writing with more than 2000 years of philosophical history behind him, could make the mistake of thinking that one can form mental images having incompatible characteristics, Plato may have done the same. If so, the possibility of this objection being raised does not make it impossible that Plato should have failed to notice it. Therefore, the possibility of the objection is not conclusive evidence against the answer to our problem which I have suggested.

On the other hand, it is possible that Berkeley misunderstood Locke, and that Locke did not mean "mental image" when he wrote of an "idea." In this case, Locke may have been right in thinking that it is possible, though difficult, to form an idea of a triangle having these incompatible characteristics—though in that case the difficult question remains : "In what sense *was* Locke using the word 'idea' ?" If, however, we let Locke escape in this way, we can argue that, since *Berkeley* could have been confused on this point, Plato too could have been confused. This is, indeed, one of the most confusing topics in philosophy, and it would be surprising if Plato, Locke, or Berkeley were entirely clear about it. I certainly am not.

In any case, even if the argument of the preceding paragraph be not accepted, there is a shorter way of dealing with this objection. For it would apply to any view which held that, for Plato, the Idea of the Triangle was a triangle, whether or not my "mental image" suggestion is a sound one. And (though this has been disputed) it seems certain in the light of passages like *Phaedo* 74 that Plato did accept the thesis of "self-predication"—*viz.*, that the Ideas themselves possessed the qualities of which they were Ideas.[5] If so, the Idea of the Triangle will have to be some sort of triangle; and the question of whether it is equilateral, etc., or all, or none of these things will arise. Since, therefore, the Berkeley objection arises for Plato whatever the truth of my contentions, it is easiest to suppose that he had not noticed that he was open to it.

There is a further possibility to be considered, which I suggest with great diffidence, since I have not yet had time to check it carefully against the text (which would require the close examination, with this specific question in mind, of all the dialogues). This is that the way of thinking that I have suggested was Plato's was only developed during the period of the middle dialogues, from which the most favorable examples seem to come. This was the period during which Plato appears often (though not always) to have taken "imitation" as his model of the relation between Ideas and particular things; and this model fits in very well with the "mental image" suggestion.

In the earlier dialogues, perhaps, he was rather following Socrates in his search for definitions in words, and had not developed any "Theory of Ideas" proper, in the sense that requires an answer to the question posed in this paper. If this were so, we should not be unduly disturbed to find, for example, in the *Laches* (192b), an early dialogue, a definition being given of Speed, of which it would be hard to form an image. The *Meno* might mark the transition from this stage to the fully developed theory. There is no reason to suppose that *Socrates* was trying to form mental images when he asked his "What is . . . ?" questions. Therefore, when we find him, in the early dialogues, asking and answering them about things which are scarcely imageable, this might be because the paradigm had not yet crystallized for Plato.

---

[5] See the illuminating series of articles in *Philosophical Review*, 1954–1960, initiated by Prof. G. Vlastos.

Another group of Ideas, already referred to, which fits our suggested answer very well is that of Ideas of manufactured objects. There is some evidence that at a later period Plato may have thought that there could not be such Ideas; but when he wrote the *Republic* and the *Cratylus* he certainly admitted them.[6] In general, Ideas of all concrete objects like horses, cups, and tables present no difficulty; but in the case of manufactured objects in particular our suggestion is especially attractive. If my experience is any guide, it is extremely common, when making something, to form first a mental image of the thing that one is going to make, and then to copy the image. This, translated into Platonic terms, is what Plato says the carpenter does when he makes a bed or a shuttle. "Where does the carpenter look when he makes the shuttle? Is it not at the sort of thing that has as its nature to fulfill the function of a shuttle? . . . If the shuttle breaks as he is making it, will he, in making a replacement, look at the broken one, or will he look at that idea (*eidos*) by reference to which he made also the one which he broke?" (*Crat.* 389*a, b*; *cf. Rep.* 596*b*).

It is significant that the carpenter is said to look at the Idea *and not* at the broken shuttle. Presumably, therefore (since there is no hint of zeugma), the word "look" is being used twice in very much the same sense. This means that the way in which the carpenter looks at the Idea must be very like the way he might (though he does not) look at the broken shuttle. And it seems plausible to suppose that the reason Plato speaks in this way is that he thought that when the carpenter visualizes in his mind's eye the shuttle or table that he is going to make, as most carpenters do, he is actually seeing the Idea of the Shuttle or the Table. This explains both the attractiveness of the whole "imitation" story in the *Republic* and elsewhere and Plato's failure to distinguish between the ideal table (the standard example) and the property of being a table. If you think you can tell whether a thing is an X by seeing whether it resembles the standard X which you can look at with your mind's eye, then knowing what an X is will be hard to distinguish from being able to summon up before your mind's eye this ideal or standard X.

VII

When, however, we come to Ideas of more abstract notions, especially value terms, the task of finding a mental image to correspond to them gets much more difficult. This is just where Plato himself apparently began to find difficulty. In *Phaedrus* 263*a* he says that when anyone says the word "iron" or "silver," we all think (*dianoein*) of the same thing; but when anyone says the word "right" or "good," one man goes one way and one another. And among value words some are easier than others.

---

[6] See Ross, *op. cit.*, pp. 171 ff.

Righteousness and Good Behavior and all the other admirable qualities of the mind have no earthly likenesses whose brightness picks them out; with our dull faculties we can see them only with difficulty—those few who can see them at all—by going to their images (*eikones*) and discerning the nature of that of which they are images. But Beauty, on the contrary, is bright to see, when in a happy dance the worshippers . . . , initiates in what may rightly be called the most blessed of mysteries, look on and contemplate the blessed sight. We who partake in such worship, being ourselves made perfect . . . , witness in a pure light the mystery of those perfect, single, steady visions, filling us with joy. . . . (250*b*)

We are told in the succeeding passage that the reason Beauty is easier to grasp is that its images in the world are grasped by sight, the clearest sense we have. That is why the man whose power of "recollection" is strong, "the man who in heaven has looked well about him, when on earth he sees a divine-looking face, or the shape of someone's body, which is a good likeness of Beauty . . . then the efflux of Beauty comes in through his eyes, and he is put in a fever . . . ." (251*a*)

It is a commonplace that visual words abound in passages like these. But why should Plato find Beauty so much easier to grasp than Righteousness or Good Behavior? Simply because, if one is trying to *visualize* an abstract quality, it is much easier to persuade oneself that one has succeeded in the case of Beauty than in the other cases. There goes the beautiful boy, and from him comes this "efflux" which is associated with a sort of fever in the blood. Is it not easy to imagine Plato thinking that it was possible to extract from this situation a pure image of Beauty itself, and thus to be able to grasp and identify the Idea which this image resembled? This becomes very easy to understand if Plato was actually treasuring a very idealized and transfigured image of the boy, forgetting about his boyishness and other irrelevant qualities, but retaining the "efflux" and the fever in the blood. Though, as sober philosophers, we have to admit that this "pure light" never was on sea or land,[7] it is perfectly easy to understand how Plato persuaded himself that he had done the impossible.

In the *Republic* we find him attempting something similar, but even more difficult. We can imagine him searching assiduously with his mind's eye for something that he could call "the Righteous," and finding it by finding its replica in large, the righteous city. To this example we shall recur shortly. And we can imagine him looking for something that he could call "the Good," and thinking that he had descried it when he had found something that was very bright, like the sun, and associated with pronounced feelings of reverence and admiration. It is because the Idea of the Good is so difficult to *visualize* that the simile or image (*eikôn*) of the sun is introduced (cf. *Republic* 368*d*, 433*a*, 509*a*). These conjectures have no bearing whatever on the philosophical importance of Plato's work in

---

[7] Wordsworth, *Peele Castle*.

putting philosophy on its way to the study of the nature of universal concepts; but they may perhaps shed some light on what it was like to be Plato.

The same kind of difficulty is described—more prosaically than in the *Phaedrus*—in *Politicus* 285*d* :

> Some things (*onta*—real things, i.e., Ideas) have by nature sensible images, which are easily learned about, and which can be indicated without difficulty, whenever anybody wants to point them out in answer to someone who is asking for an account (*logos*) of one of them; this can be done easily, without trouble and without saying anything (*chôris logou*). But in the case of the greatest things and the most precious there is no image (*eidôlon*) clearly wrought for men to look at, which one can show and fit to one of the senses if one wants to satisfy the mind of the questioner. That is why we have to learn by practice to be able to give an account of each thing in words, and understand it when given. For immaterial things, which are the fairest and greatest of things, can be clearly pointed out only by speech, and not by any other means.

Here we can see that Plato finds a difficulty just where, if my suggestion is right, we should expect him to find one. The things of which one can fairly easily persuade oneself that one has formed a mental image (which are the same cases as those in which, in Plato's terminology, there exists a clear *physical* image of the Idea) present no difficulty. The task for the philosopher is to do this same thing in cases in which this possibility does not exist. As an aid in this task we have to use words (*logoi*).

The *logos* is, in fact, regarded by Plato as a description in words of the Idea—a description of something seen by the mind. The words fulfill several important functions in the enquiry. They put on record what we have seen, or claim to have seen; to be able to give a *logos* which will stand up to questioning is the only way to substantiate such a claim. When someone has seen an Idea, he can help someone else to see it, even if this is an Idea which has no sensible image, by means of the *logos*. There is a good example of this process in the discovery of Righteousness in the *Republic* (432*b* ff.). Socrates is represented as looking at the ideal city (itself, we might say, an Idea or paradigm) and searching in it for the fourth virtue, having already discerned the other three. Do we not all, as we read this passage, picture the city to ourselves? And when Socrates finally locates what he is looking for, and gives the *logos*, do we not all direct our mind's eye to that aspect of the life of the city to which he draws our attention, namely, the fact that all the people in it are doing their own jobs? I am sure that at any rate I do; and if the objection is raised that this is because I have an abnormally active visual imagination, I can retort that this is a help to the understanding of Plato, who was certainly similarly endowed.

A passage that might be thought to tell against the view which I have been putting forward is *Parmenides* 130*b*, ff. There the young Socrates is represented as being quite happy about there being Ideas corresponding to value words, but doubtful about some Ideas of objects which are easily imageable (man, fire, and water), and still more doubtful about hair, mud (or clay), and filth, which are also imageable. But what is in question here is not, as in the passages referred to earlier, a difficulty in apprehending Ideas which are admitted to exist, but a difficulty in admitting the very existence of certain Ideas.

The reason for the latter difficulty probably lies in something not very relevant to our present question, viz., the fact that Plato thought of the Ideas as paradigms or perfectly good examples of the classes of things of which they were Ideas. He naturally found it easier, approaching the matter in this way, to suppose that there were Ideas of things which were in themselves good; possible though not so obviously necessary to admit Ideas of things which could be good or bad; and *prima facie* absurd to allow Ideas which would be "perfect mud" or "perfectly good filth"— and indeed the oddness of these expressions gives him some excuse for hesitation. But Parmenides, who speaks here with the voice of Plato, encourages Socrates to admit the existence of Ideas of any universal properties. Mud (or clay) is given in a later dialogue as a paradigm case of a *definiendum* (*Theat.* 147*a*), though the word "Idea" is not used.

In any case, whatever the explanation of the *Parmenides* passage, it has nothing to do with the difficulty or ease of forming images, and has therefore little bearing on our problem. It would constitute an objection only if I were claiming (which I am not) that Plato thought that *whenever* he was (as we should put it) having a mental image he was seeing an Idea.

VIII

The suggestion which I have been making would be put out of court if we ever found Plato talking in unambiguous terms about mental images and contrasting them with his Ideas. It will not do as counterevidence if we find him contrasting *part* of the class of mental images with Ideas. As I have just said, I do not wish to suggest that any mental image which Plato had would have been classified by him as an Idea. There were, no doubt, many mental images in his fertile imagination which he would not have called Ideas, and one of the tasks of philosophy was to separate the wheat from the chaff. I have looked for passages in which Plato is clearly talking about what we would call mental images. We can perhaps ignore as irrelevant his references to dream images, which are fairly frequent. And we must not think that whenever the word *phantasma* or one of its cognates occurs, Plato is speaking of mental images. This seems to have been Aristotle's word for a mental image, though he does not always so use it (see Bonitz, *Index Aristotelicus* s.v.). But Plato used words of this group more widely, to mean any kind of "appearance" whether mental or perceptual. An example is *Sophist* 263-4.

I have, in fact, discovered two passages in which Plato is clearly talking about what we should call mental images. In neither of them does he have a *word* for "mental image"; he borrows other words, and in one passage his language is highly metaphorical. The first passage is *Phaedo* 73*d*, in which he is talking about the "recollection" of Ideas from an earlier existence. "Lovers," he says, "when they see a lyre or a coat or something else that their loved one is accustomed to use, recognize the lyre and grasp in their mind the form (*eidos*) of the boy whose lyre it is—and this is recollection." It is significant that the words "grasp in their mind the *eidos*" might, in another context, have been used of the intellectual grasping of an Idea. The words are not being used here in this technical sense, for the individual Cebes, for example, does not have an Idea corresponding to him. But it is not fortuitous (any more than it is in *Republic* 510*d,* to which I referred earlier) that Plato uses the word *eidos*. His point here is that, just as we can see with our minds *eidê* of individuals, so we can see with our minds the *eidos* of, for example, the Equal (as he goes on to explain).

If the objection is made that it is impossible to form a mental image of the Equal, it must be admitted that in fact it is; but there is ample evidence in the text that Plato *thought* it possible. He thought, no doubt, first of two equal sticks or stones, and then of two equal nonphysical lines in the mind's eye (just as we do when we are doing geometry in a naïve way), and then said "The former have reminded me of the latter." And the latter he confusedly regarded as the *eidos* of Equality.

This explains how on the next page, in the course of three lines, he can alternate so readily between the three expressions 'the equals themselves' (plural), 'equality' (singular), and 'the equal itself' (singular). If one is confusedly equating the property of being equal with a transcendant perfectly equal pair of things, it will indeed be difficult to decide whether to use the singular or the plural. And what easier way of getting into this confusion than to think that to become acquainted with a property is to do what we should call "form a mental image of a thing perfectly exemplifying the property"?

Plato describes the following as examples of the very same thing: seeing a lyre and being reminded of its owner; seeing Simmias and being reminded of Cebes; seeing a picture of a horse or lyre and being reminded of a man (presumably its owner); seeing a picture of Simmias and being reminded of Cebes; seeing a picture of Simmias and being reminded of Simmias (here the analogy to being reminded of an Idea is obviously getting very close); and lastly, seeing a pair of equal sticks or stones and being reminded of "the equals themselves." There is absolutely no suggestion in the passage of a transition from a literal to a metaphorical sense of any of the words involved. We may therefore conclude that when Plato says of the last case "seeing equal sticks or stones or something, our mind passes from them to that [viz., the Idea] which is different from them," the experience to which he is referring is of exactly the same sort as in the earlier cases, namely, the experience which we should describe as seeing with our eyes one thing, and then seeing with our mind's eye a mental image of that thing or of something else. This passage, then, far from affording counter-evidence to my suggestion, supports it.

IX

The other passage (Philebus 38-9) is not so helpful, because in it Plato is not immediately concerned with the Theory of Ideas. He is trying to explain the origin of, and difference between, true and false belief. Lack of space precludes a detailed discussion of the passage. In it, Plato is considering a case in which a man sees an object, forms an opinion about it, and mentally records this opinion, first in words, and then with an image of the object. The image is variously called eikôn, zôgraphêma (picture), and phantasma ezôgraphêmenon (appearance pictured). Plato says that both the verbal record and the pictorial one can be recalled, and that when this happens, those images can be said to be true which are images "of" true opinions and words (or propositions).

Here Plato is clearly speaking of what we should call mental images — images, moreover, of particular objects which have been seen. It might be argued, therefore, that since Plato mentions mental images which are not Ideas, he cannot, when he speaks of Ideas, be referring to what we should call mental images. But it does not follow, from the fact that Plato sometimes mentions mental images which are not Ideas, that what he calls Ideas were not what we should call mental images. And Ideas do not in fact come in at all in this context; I have tried, but failed, to find in the passage more positive evidence than this relevant to the present question. With its "representational" model of thought, it is somewhat different from the usual Platonic approach, which regards thought as a direct apprehension of an object. But to discuss the difficulties raised by the passage would take us too far from our topic.

To sum up, we may say that the answer which I have suggested to our question has encountered some difficulties, but none which seem to put it altogether out of court. To this extent it remains tenable, and may have to hold the field until someone produces a more plausible answer. Or it may be shown that the question itself was illegitimate — though this would surprise me.

I wish to emphasize in conclusion that I am not trying to lower Plato's reputation as a philosopher. What he did in thinking out his Theory of Ideas was of lasting importance and value for logic. We have here the beginning of the philosophical study of universals, which was the genesis of Aristotle's work in this field; and of course it was soon apparent that mental imagery had little to do with them, and propositions (*logoi*) a great deal. But Plato may have put his thoughts to himself in this unsatisfactory way, because that way came easiest to him, and he was after all, struggling (without a philosophical terminology or any but the most rudimentary philosophical tradition behind him) with difficulties which would have defeated a lesser thinker. Columbus's discovery of America is not rated any less an achievement because he thought he was paying a visit to China.

CHAPTER 6

Popper and Wittgenstein

BY PETER MUNZ

THE PROBLEM OF INDUCTION has been the most persistently perplexing problem of philosophy. By induction I do not mean here only the technical process of deriving from a limited number of data a more or less general law as to their behavior. The problem of induction in this special narrow sense has from the days of Mill's famous canons of induction caused many difficulties to philosophers.

In the wider sense the problem of induction is the question of how the words we use to describe the world are related to the world we experience. Basically, the wider question is not at all unlike the narrower problem. For the narrower problem consists in the oddity that we discover general laws about the behavior of matter, for example, when we have no more than limited instances of experience or observation to support them or base them on. The wider problem is very similar : we have certain experiences—but when we want to describe them we must necessarily make use of words; and these words and the propositions they form must by the nature of the case have a meaning of a more general nature than the single and individual experiences they refer to. When we are talking about the world we are making use of words or concepts or propositions which are of a universal character; if they were not, nobody would understand their meaning.

Innumerable views have been put forward about the relationship that exists between our individual experiences and the words and propositions we use to describe them. Plato sought to account for this relationship by saying that our words describe not the world we experience with its individual facts or events, but ideal universal entities. And that the single events we know of by experience are simply shadow-like reflections of the universal entities. We have little taste today for the conception of ideal, universal entities. But we cannot deny that Plato saw the problem very clearly when he tried to solve it by postulating that, since our words and propositions are universal, they must also *refer* to universal facts; and that since these universal facts could not be discovered by ordinary experience, they must be postulated to exist ideally. In short, he sought to show that what really exists are universals. Our language, he held, consisting as it does of words which are not private, i.e., which are understood by more than one person, is a true reflection of reality, albeit of an ideal universal reality. And the individual facts or events which we know in our experience are not real at all. They are merely shadow-like appearances.

At the other extreme, there stands, of course, the view that the individual fact or event alone is real and that as soon as we choose to describe it by a word we are classing it in an unwarranted fashion with other facts or events which are really not at all like it, but which happen to be sufficiently similar to allow us to describe them by one and the same word or proposition. Plato's extremism had the weakness that it invited us to believe in the existence of ideal universal entities, and this opposite extreme has the weakness of making it almost impossible to talk about any experience for fear of lumping it together (through describing it by word or proposition) with experiences with which it has few similarities. In a sense the whole history of philosophy could be written as so many attempts to find acceptable compromises between these two extreme views.

Several decades ago there emerged logical positivism, the most striking and radical attempt ever made to solve the question. This view simply said that the meaning of a proposition is the method of its verification. If a proposition (or for that matter a concept) can be shown to be directly grounded in experience, it is a meaningful proposition or concept. If not, not. Sentences that do not report direct experience have no meaning, and, conversely, whether a sentence has meaning or not depends on whether it reports an experience.

This was startlingly simple. But it did not take long for careful observers to find out that this view of the matter begged the question. To begin with, there was the obvious rejoinder that if all propositions that were not grounded in experience were meaningless, then the proposition that this was so must itself be meaningless—for it could not demonstrably be grounded in experience. But this objection might seem trifling in comparison with two weightier ones. Firstly, no universal proposition can be completely verified : universal propositions cannot be said to be grounded in a limited number of experiences, and therefore they cannot be held to be meaningful. And secondly, not even a particular proposition or concept can ever be completely verified (i.e., can be said to be really grounded in experience) because by its very nature (being part of some intelligible language) it must have a wider reference than just one single experienced fact or event.

There were many varieties of logical positivism, with some quite elaborate attempts to meet these and other objections. But the more closely the matter came to be examined, the more clearly it appeared that in the long run logical positivism in its pure form, or in any other form, could not really be a valid account of the relations that exist between our experiences and the way in which we talk about them.

From the growing dissatisfaction with logical positivism there emerged two important efforts to find an alternative account: the philosophy of Wittgenstein's later period and Popper's theory of falsification. I think it is of vital importance to understand that *both* these views were a reaction against logical positivism.

It is, of course, very difficult to say precisely what Wittgenstein believed. But for the purpose of this argument I propose to assume that the main idea in Wittgenstein's later thought was the belief that there is no private language. There are two arguments to that effect that recurred in one form or another in his lectures in Cambridge after the Second World War. They can also be found in the *Philosophical Investigations*.

The first argument is that there is no such thing as an inner ostensive definition. If you try to record the occurrence of a certain sensation, for instance, by associating it with a sign *S,* the sign *S* will never be able to acquire meaning by your concentration on the sensation. A sign, Wittgenstein argued, can acquire meaning only if one learns the rules according to which it is used, not by concentration upon the sensation it is associated with. Hence, he insisted that one cannot learn a language by pointing to the objects to which the words refer or which they name.

The second argument is similar. Wittgenstein always maintained that one cannot *know* that one has a certain sensation. The concept of knowledge was not applicable to the sensations one had, because one could never be wrong about them. Our sensations are a matter of feeling, not objects of knowledge. And he thought that one could therefore never be wrong in saying "I am sad" in the way in which one could be wrong in saying "I see two chairs in this room." He concluded that the meaning of "I am sad" could only be learnt by learning the rules according to which the words in this expression are used—not by introspection and by watching the sensation in question. This means that he did not believe that our ability to use a language was grounded in experience (which may well be private). He believed that it was a public activity, something like learning the rules of a game.

It is not difficult to see that this way of looking at language might have been a perfect solution to the age-old problem. If a language is learnt by a knowledge of the rules and if one's ability to use that language does therefore not depend on experience, the old problem of induction simply disappears. It certainly disappears in its general form. (It is not so easy to say whether it would equally simply disappear in its narrow technical sense.) Thus the theory that there is no private language and that language is like a game, the rules of which have to be learned, appears at first sight to be a radical and powerful alternative to the views of logical positivism as well as to those of Wittgenstein's early work, the *Tractatus*. Logical positivism had made first language and then science grow upward from singular and *private* experiences. Wittgenstein's insistence that there is no private language must be seen as a reaction against, for example, Carnap's thesis that meaningful concepts are, so to speak, constituted out of private experiences—as well as a reaction against his own earlier view that meaningful propositions are truth-functions of atomic propositions.

But there are three comments to be made. Firstly, the view is false. Secondly, it is of such an unusual character that it is impossible to *prove* that it is false. Thirdly, I suspect very strongly that Wittgenstein was very close to realizing, or had actually realized, that it was false.

The view seems certainly false. No matter how similar a language and the way we learn it are to a game and the way we learn its rules, in the last instance a language is *not* a game. When we talk, we talk *about* something and no matter how difficult it may be to account for the relationship between the words we use to describe an experience and the experience itself, such a relationship *must* exist. Experience can therefore by no means be irrelevant to the way in which we use language and there is often complete justification for supporting something we say by stating that we feel it to be thus, or see it to be thus, or possibly even by an ostensive definition. To allow an appeal to experience is not to deny that a language cannot be *completely* private. In fact, the denial that a language cannot be completely private is just the problem : How can a language which is obviously *not* private give satisfactory descriptions of experiences that *are* private? Wittgenstein certainly tried to cut the Gordian knot with his argument that since language is not private, our feelings or experiences cannot pronounce on whether we are right or wrong when we say that we are seeing a red spot or that we feel a pain.

There is one consideration in particular which must throw considerable doubt upon Wittgenstein's arguments. When I feel sad, I can say "I am sad" and one could imagine that the formulation of this statement is due to my having learned the rules that govern the uses of the words involved. Everybody else can learn these rules too, and on certain occasions can say to me "you are sad." (Given certain rules, these statements are equivalent.) But it seems an incontrovertible fact that there is a difference between my feeling sad and the mere occasion which entitles an observer to state that I am sad. It follows, therefore, one would think, that when I say "I am sad" and when another person says to me "you are sad" we mean very different things. The other person's statement depends on his knowledge of the rules of the language. My statement depends on my knowledge of these rules *plus* something else. And that something else is precisely my own inner knowledge or awareness of feeling sad. Moreover, contrary to Wittgenstein's belief, one can very well be mistaken about the identification of one's inner sensations and feelings. Perhaps few people will erroneously identify sadness as joy; but one often makes mistakes in identifying jealousy as anger or aggressiveness as frustration. And since one can be mistaken about them, there is no good reason why one should not call them objects of knowledge. One could produce a great many such and similar considerations against Wittgenstein's views, but I will not pursue them.[1]

The next point is much more complicated. Once Wittgenstein and his partisans of various hues are convinced that there is no private language and that we learn a language by learning the rules, it becomes impossible for them to understand that Wittgenstein's theory is false. For if one believes that language is like a game and that all depends on the knowledge of the right rules, one will not be able to investigate the question of whether this view is correct. Suppose someone stated that there *was* a private language. Wittgenstein or one of his followers would not reply by adducing reasons to back their view that there is no private language. They would instead, true to the belief that all use of words, and hence the present argument, is a matter of certain rules about the use of words, retaliate with a question. They would ask: "When you say there is a private language, do you mean that there is such a language in the way in which you say that there is a chair in this room? Or do you mean to assert that there is a private language in the way in which you might say that there is a pain in your head? Or do you mean to assert the existence of such a language in the way someone might assert the existence of God?" In other words, they are forced, by their belief that the right use of words is merely a matter of rules, to start a discussion about the rules according to which the words in the initial question have been used. This discussion will proceed endlessly; but it can never come to the point. For to reach the point at which one is prepared to question the truth of the assertion that there is no private language, one would have to be willing to assume, for a moment at least, that the truth or falsity of statements which are clearly nonempirical, in the sense in which the assertion that there is a private language is nonempirical, depends on factors other than the right use of words.

Wittgenstein's theory, originally meant to be an alternative to logical positivism in general and to his own *Tractatus* in particular, has thus proved a powerful impetus to the habit of linguistic analysis. There is no point in inquiring into the precise origins of the habit. But it is worth recalling that Professor Ryle, one of the leading practitioners of linguistic analysis, recently justified the habit in terms that are completely Wittgensteinian: "The learning of the meaning of an expression is more like learning a piece of drill than like coming across a previously unencoun-

---

[1] I have presented them at greater length in the third chapter of my forthcoming book, *Relationship and Solitude*.

tered object."[2] If Wittgenstein is right, the practice of linguistic analysis is not only to be highly commended but is obviously the only proper philosophical procedure. But if he is not, it is at best a subsidiary, and at worst a futile undertaking which can have no results other than odd improvements on the Oxford English Dictionary.[3] The habit itself consists in the belief that all puzzles that are not clearly of an empirical kind (such as, for example, the question of whether a certain metal will dissolve in a certain acid) can be solved by a linguistic analysis, i.e., by an investigation of how words are used. If Wittgenstein's contention that we learn to speak a language correctly by learning its rules is sound, there can be no objection to this belief. But if it is not sound, there can be no possible justification for it.[4] Be this as it may; from our point of view the most tantalizing aspect of the whole matter is that Wittgenstein's view is not just a view like all other views, the truth or falsity of which one can discuss and assess and differ on. It is a view to end all views. For once it is accepted it is like a philosophical *perpetuum mobile* : it invites endless linguistic analyses without allowing anyone to advance arguments that would bring them to a stop. There is nothing very surprising in this. Any theory, of course, can be compared with and weighed against any other theory. But since all theories are linguistic expressions it is very difficult, if not impossible, to weigh a theory *A*, for example, about induction (i.e., about the relationship that exists between experiences and propositions about experiences) against theory *B* that all linguistic expressions are to be judged by the rules governing the particular language of which they are a part. The person who holds theory *A* can entertain any kind of reasonable criticism. But the person who holds theory *B* precludes all those criticisms which consist of considerations other than linguistic ones, since they too cannot be anything except linguistic formulations. Or perhaps one should not say "preclude." Criticism is allowed; but by the nature of the case, such criticism will be subject to a linguistic analysis and thus treated in the manner which is itself the subject of the criticism. The upholder of theory *B* is therefore proof against rational criticism of the theory he holds and we must therefore consider that he is not playing the game of philosophical discussion—no matter how strange this accusation may appear when it is raised against the contemporary masters of philosophical gamesmanship. They refuse to obey the rules, but in a very unobtrusive manner. They do not proclaim that they intend to defy the rules; they do not need to do this, for their theory enables them to defy the rules tacitly.

---

[2] In *British Philosophy in the Mid-Century*, C. A. Mace (ed.) (London, 1957), pp. 262–63.

[3] For an explanation of this view see my "Investigations of Philosophy," *Dialectica*, 49 (1959).

[4] It is certainly true that once we know one language, we can learn another by learning its rules. But it does not follow that we can learn the first language only by learning its rules.

The matter, however, is even more complicated. Wittgenstein always argued that a linguistic analysis should show the fly the way out of the bottle—that is, should resolve an intellectual cramp. He undoubtedly considered logical positivism to be such a cramp. He may well have been right, but the way in which he went about resolving it defeated his own purposes. To resolve an intellectual cramp caused by a belief in a false notion, one must criticize the notion or the idea. But this is precisely what Wittgenstein did *not* do. He practised linguistic analysis instead. And since he believed that we learn to speak a language correctly by learning the rules, he sought to resolve the cramp caused by a belief in logical positivism by analyzing the formulation of logical positivism. I remember an occasion when he was told of the problem of induction. He would not listen to an explanation of what it was supposed to be, but insisted on showing that there could be no such problem since the word "problem" was used according to certain rules and that the use of the expression "problem" in "problem of induction" corresponded to none of these rules. In this way his criticism was always directed against formulations or linguistic expressions, never against ideas or theories. And if criticism takes that form, it fails in its therapeutic effect. For such criticism only shows whether a certain expression can be used, whether a certain thing can or cannot be said *within* the framework of a certain language game. It fails by its very nature to express an opinion about the truth or viability of the idea or view expressed. Wittgenstein's teaching had therefore a bad pedagogic effect, for it taught young men to criticize formulations rather than ideas. When they found that certain things could be said within the framework of a certain language game, they went away happily, never bothering to inquire whether these things were true, good, or useful. It encouraged them to lull themselves into a false security. They had straightened out none of their problems, but appeared to have some good reason for holding all kinds of beliefs because they had convinced themselves that, given a certain framework of linguistic rules, what they happened to be saying did indeed make sense.

My third comment is concerned with the belief that Wittgenstein himself either realized or was very close to realizing that his view was an unworkable alternative to logical positivism. When he lectured or spoke in discussion he gave the impression of a man who was groping for something he could not really find. Many of his students believed that this was a sign of his great profundity. He himself said often enough that he was by no means profound and that he felt more and more incapable of coping with the problem he was discussing. I think we owe it to the memory of the serious thinker he was to take his words literally. Wittgenstein thought, I imagine, that he had made a discovery that would lead out of the impasse created by logical positivism. He saw fairly clearly that his view that there is no private language and that we learn to speak a language, not by basing ourselves on certain feelings or awarenesses or insights, but by learning the rules that govern its usage, was not a fully adequate view and that there was something missing. He felt that if he could only make a greater effort, he could find that missing factor. But he did in fact not find it because, since the view was mistaken, no amount of probing could lead him to the missing factor. I do not think that his followers and disciples are doing him a service by misunderstanding his tortuous doubts as an expression of a superhuman profundity, by mistaking the indefensibility of his views as they stand for a proof that they are too complicated to be understood. Wittgenstein hated all professions of profundity. It seems much more straightforward to admit that his views were false than to pretend that when they do not work or cannot be shared it is because they are not yet understood.

Popper, like Wittgenstein, does not believe that our knowledge is built up piecemeal and "inductively" from individual observations, experiments, and experiences. He sought to solve the problem of induction in an equally radical way. But unlike Wittgenstein, who tried to avoid the difficulties inherent in the verification criterion of meaning by insisting that the meaning of propositions depends on linguistic rules, Popper overcomes those difficulties by insisting that the verification criterion must be replaced by the falsification criterion. He argues that the importance of a proposition (i.e., whether it is empirical or not) depends on the possibility of its falsification. A proposition is empirical, in the sense that it says something about the world, if, and only if, there is a way in which it could be falsified. According to this view, whether a proposition (or a concept) is empirical in this sense does not depend on its being grounded in experience or derived inductively from experiences or the result of so many observations. A proposition is not inferred at all from experienced evidence. There is neither a logical nor a psychological induction. But the *falsity* of a proposition can be inferred from empirical evidence, and this inference is a purely deductive one. In the case of a particular proposition, its falsity can usually be established at once; in the case of a universal proposition, it may not be possible to establish it at once. Until it is established, the proposition is to be taken as provisionally true. This should, however, not be taken to mean that a number of satisfactory tests make the proposition probable or lend authority to it or stamp it as "almost verified." Popper insists that all universal propositions remain forever conjectural speculations.

The cogency of this argument depends upon the use it makes of a well-known logical phenomenon, of the asymmetry between verification and falsification. No finite number of positive instances (and there can never be more than a finite number) will ever conclusively verify a universal proposition; and similarly, no finite number of observations will ever entitle us to use a general concept or a word which means more than one thing, and one thing only. But a single negative instance will conclusively refute a general law. Using this logical phenomenon, Popper is able to say that whether a proposition is empirical or not does not depend on whether it can be verified, but on whether it can be falsified. With this seemingly simple observation, he shows the way out of the impasse created by logical positivism.

With this theory about the relationship between the world we experience and the way in which we talk about it or describe it, one is almost forced to the conclusion that knowledge precedes experience. But lest one gain the impression that Popper's ideas are a new form of idealism, it is necessary to add that his theory distinguishes very clearly between real knowledge (knowledge about the world) and sham knowledge. Real knowledge is formulated in such a way that the possibility of having certain experiences is denied. When I know that all metals dissolve in sulphuric acid or that Napoleon lost the battle of Waterloo, the possibility of certain experiences is denied by these formulations. When I find a metal that does *not* dissolve in sulphuric acid or when I find good evidence that obliges me to believe that Napoleon *won* the battle of Waterloo, the propositions are falsified. If observations are made which our statements have excluded as impossible, then our statements must be considered false. Even so they are genuine pieces of knowledge, for they assert something about the world. And even though knowledge in Popper's theory precedes experience, experience is by no means irrelevant to it. The meaning of knowledge depends on the possibility of its falsification. If a proposition is such that it does not exclude some experiences, it is either a tautology or asserts nothing at all.[5] In this way it is possible to distinguish genuine knowledge from sham knowledge without having to resort to the attempt to show that the genuineness of knowledge depends somehow or other on the fact that it is grounded in or derived from experience. As long as one does not realize that knowledge thus precedes experience and that its genuineness depends on the possibility of its falsification, not on the way in which it can be shown to be *derived* from experiences, one will never be able to avoid the tantalizing realization that all linguistic expressions (even particular propositions) are more universal than any experience or any finite number of experiences and therefore somehow unjustifiable. Popper's solution is as radical as Plato's. But it is sounder than Plato's in that he sees in the criterion of falsifiability a clear way of distinguishing empirical from nonempirical knowledge.

---

[5] Nevertheless, unfalsifiable, existential, and "all and some" statements do assert something about the world. Cf. the important papers by J. W. N. Watkins on this matter: "Between Analytical and Empirical," *Philosophy*, 32 (1957); "Confirmable and Influential Metaphysics," *Mind*, 67 (1958). My description of Popper here is merely designed to distinguish his views from idealism; it should not be taken to be incompatible with Watkins's arguments.

Popper's radical solution of the difficulties created by logical positivism may, of course, be false. If, for instance, it could be shown that,
contrary to general belief, there is no asymmetry between falsification and
verification, the solution would turn out to have been a wrong one. But in
any case, it is a solution that conforms to the rules of rational discussion.
Popper has made a proposal and it is now for others to see whether they
can accept it or find reasons for rejecting it. The proposal itself—and this
is where it differs vitally from Wittgenstein's proposal—in no way precludes the possibility of a discussion of its soundness. Popper's view is not
a view to end all views, but an invitation to consider the truth of one
particular view of the nature of knowledge. As such it is a serious move in
philosophical discussion and might well provoke a counter-move. It is not
self-validating like Wittgenstein's proposal and is entitled to demand,
therefore, to be treated like a genuine piece of knowledge about our knowledge.

CHAPTER 7

Confirmation, the Paradoxes, and Positivism*

BY JOHN W. N. WATKINS

I. Introduction

KARL POPPER'S philosophy of science has done what it requires a good scientific hypothesis to do—it has proved its mettle. This success creates a difficulty for a contributor to a *Festschrift* for him. The birthday present which he would enjoy most would be some effective criticism. Alas, I have none. So I fall back on a second-best policy.

Popper has pointed out that his falsificationist approach to the corroboration of scientific systems (according to which a system is corroborated if it entails falsifiable predictions and has withstood searching attempts to falsify them) leads to appraisals of scientific hypotheses strikingly different from those unhappily implied by a verificationist approach (according to which confirmation occurs when there is some sort of *upward*

* During 1960 I had a correspondence with Professor Israel Scheffler, arising out of his critical "A Note on Confirmation," (*Phil. Studies*, Jan.-Feb. 1960). He convinced me that I had made a mistake in discussing the problem of the avoidability of those "paradoxes of confirmation" which were pointed out by Professor C. G. Hempel. I acknowledged my mistake (*Phil. Studies*, Jan.-Feb. 1961, p. 17; the reference in footnote 3 is to an early version of the present paper), and set about clearing up that problem from a new angle. The present paper is the outcome. As first submitted, it contained a lengthy confession of past error. The editor required me to cut my contribution by forty per cent, which gave me an excuse for dropping my self-criticism. As my acknowledgments to Scheffler were also dropped, I should like to thank him here.

An early draft of its central sections was read and searchingly discussed at the Philosophy Seminar at the University of Chicago on March 1, 1961. I have been helped by criticisms and suggestions from Dr. J. Agassi, Professor P. K. Feyerabend, and Dr. I. Lakatos.

implication *from* the evidence *to* the hypothesis, so that confirmation is regarded as a sort of credibility-enhancing juice which seeps upwards, as it were by capillary attraction).

For verificationists, the ideal case of confirmation is that where the evidence actually entails (conclusively confirms) the hypothesis; and their problem has been to define some less strict relation between confirming evidence and confirmed hypothesis of which the entailment relation will be a limiting case. I know of two views about the direction in which this ideal relation should be relaxed and generalized to accommodate inconclusive confirmation—a majority view (associated with Keynes, Jeffreys, Reichenbach, Carnap, and others) weakens the strictness of the implication from evidence to hypothesis and makes it a probabilistic implication : confirming evidence raises the probability of the hypothesis, and in the ideal case raises it to one. Naturally enough, it was mainly with this majority view that Popper contrasted his own theory of corroboration.[1] The author of the minority view is C. G. Hempel,[2] who retains strict implications or entailments from the confirming evidence, but requires the evidence to entail at least *a weakened version of the hypothesis,* a surrogate or stand-in for it, the ideal case being that where the evidence entails the hypothesis itself.

My birthday present will consist primarily of a tidying-up operation. I shall show in Section II that Hempel's view has unfortunate implications similar to those which Popper has extracted from the majority view. In Section III, I shall show that it makes metaphysical hypotheses "confirmable," and in Section IV that it makes the exact hypotheses of mathematical physics "unconfirmable." In Section V, I shall show that Popper's falsificationist theory of corroboration does not involve the "paradoxes of confirmation" which Hempel's verificationist theory involves. And in Section VI, I shall use the previous results against positivism.

According to Hempel, an observation report *e* confirms an hypothesis *h* if *e entails,* not necessarily *h* itself, but at least the "development," as Hempel called it, of *h* for *e* : what *h* would state if what is mentioned in *e* were all that existed.

To formulate my thesis I will introduce the idea of a degenerating series of hypotheses, the first and strongest entailing, but not being entailed by, the second, the second entailing but not being entailed by the third, and so on.

Suppose we have three predicates of increasing determinateness, B_1,

---

[1] See his *The Logic of Scientific Discovery*—referred to as *L.S.D.* henceforth—especially Ch. X. and appendix *ix. Added in proof: see also his *Conjectures and Refutations*, espec. Chapters 10 and 11.

[2] See his two "Studies in the Logic of Confirmation," *Mind*, 1945—referred to as "Studies" henceforth—especially pp. 12–19.

B_2, and B_3, such that anything which is B_3 is necessarily B_2 and anything which is B_2 is necessarily B_1. Let our strongest hypothesis be :

A's exist and every A is B₃

A's exist and every A is B_3

let the next be :

Every A is B₂

Every A is B_2

let the next be :

Every A is B_1

I will refer to these as h_3, h_2, and h_1 respectively. We may conclude the series with h_0, where h_0 is any tautology. For by the generally accepted concept of logical consequence, a tautology is a consequence of every statement.

An artificial example : suppose that there exists a standard kind of apparatus and procedure for liquefying solid substances by electrolysis, known as "the usual electrolytic method." Suppose that this apparatus has a dial for controlling the strength I of the electric current, a lamp which lights when liquefaction is taking place, and a pointer which records the rate R of liquefaction.

Let h_3 be :

> Solid substances exist and all of them are always
> liquefiable by the usual electrolytic method at a
> rate R exactly proportional to the strength I of
> the electric current.

A feature of this hypothesis, which will prove crucial later, is that some of the predictions derivable from it are highly falsifiable but *too precise to be entailed* by experimental reports. We may call this hypothesis *transcendingly precise*. Since the time of Copernicus, Galileo, and Kepler, such hypotheses have been the glory of science. We shall find that they create a scandal for Hempel's theory of confirmation.

To generate a weaker hypothesis h_2 entailed by h_3 I will drop h_3's existence-clause and weaken its consequent clause. Let h_2 read :

> All solid substances are always liquefiable by the
> usual electrolytic method.

Clearly, h_2 is less falsifiable than h_3.

To generate a still weaker hypothesis I will weaken the consequent clause further so that h_1 reads :

> All solid substances are liquefiable.

Now h_1 is *not* falsifiable. Experimenters might have failed again and again to liquefy a particular solid substance, not only by electrolysis but by various other methods. That would falsify h_3 and h_2, but it would not falsify h_1 since it would not establish that this substance is liquefiable by no method whatever. However, its unfalsifiability does not render it analytic.[3] Its denial is not self-contradictory; and while it is compatible with all possible finite observation reports, it is not compatible with all scientific hypotheses. For instance, it is incompatible with the (unverifiable) hypothesis that manganese is not liquefiable. Hence h_1 is not factually empty. I would classify h_1 as a *metaphysical* hypothesis.[4] Additional justification for this classification will be given in Sections III and VI.

By Popper's falsificationist measure, h_3 is more corroborable than h_2 since it is exposed to, and might withstand, tests which are more severe than the severest tests to which h_2 is exposed, while h_1 and h_0 are uncorroborable, since untestable. Degrees of actual corroboration, as opposed to degrees of corroborability, would, of course, depend on how the tests went. If their outcomes were all favorable, and if h_3 had been tested by experiments some of which were not tests of h_2, then according to Popper's theory of corroboration h_3 would be better corroborated than h_2; h_1 and h_0 would both be uncorroborated.

To formulate my thesis I will introduce the following notation.

'$C(h, e) = 1$' MEANS : "The hypothesis h is conclusively confirmed, or verified, by the evidence e."

'$0 < C(h, e) < 1$' MEANS : "h is confirmed, though not conclusively, by e."

'$0 = C(h, e)$' MEANS : "h is neither confirmed nor disconfirmed by e" OR "e is neutral to h."

'$-1 < C(h, e) < 0$' MEANS : "h is discofirmed, though not conclusively, by e."

'$-1 = C(h, e)$' MEANS : "h is conclusively disconfirmed, or falsified, by e."

'$C(h_2, e) < C(h_1, e)$' MEANS : "h_1 is better confirmed than h_2 by e." NOTE : this does not mean that h_1 is positively confirmed by e. If h_2 were falsified and h_1 only disconfirmed by e, or if h_2 were disconfirmed by e and e were neutral to h_1, then h_1 would be "better confirmed" than h_2 by e.

I shall also use the sign '$\leqslant$'. Thus :

'$0 \leqslant C(h, e) \leqslant 1$' MEANS : "$h$ is unconfirmed or confirmed or verified by e"

'$-1 \leqslant C(h, e) \leqslant 0$' MEANS : "$e$ falsifies or disconfirms or is neutral to h"

'$C(h_2, e) \leqslant C(h_1, e)$' MEANS : "$h_1$ is at least as well confirmed as h_2 by e."

---

[3] See my "Between Analytic and Empirical," *Philosophy*, April 1957.
[4] See my "Influential and Confirmable Metaphysics," *Mind*, July 1958.

My thesis is this. Whatever e may be, Hempel's criterion yields the following ordering for our degenerating series of hypotheses :

$$- 1 \leqslant C(h_3, e) \leqslant C(h_2, e) \leqslant C(h_1, e) < C(h_0, e) = 1$$

where, moreover :

$$- 1 \leqslant C(h_3, e) \leqslant 0 \leqslant C(h_1, e) < 1$$

In words : whatever the evidence may be, our strongest hypothesis h_3 will never be "confirmed" though it may be disconfirmed, whereas our metaphysical hypothesis h_1 will never be disconfirmed though it may be "confirmed"; and while h_1 would never be so well "confirmed" as our tautology h_0, it will be at least as well "confirmed" as the scientific hypothesis h_2.

The original ground for my opposition to Hempel's conception of confirmation was that it turns what I regard as inferior metaphysical hypotheses into superior scientific hypotheses.

II.   Degrees of "Confirmation"

Hempel disavowed any intention of providing, in his "Studies," a measure of *degree* of confirmation. His aim there was only to lay down a criterion which would enable us only to answer "yes" or "no" to the question "Does this evidence confirm that hypothesis?" Nevertheless his criterion does (I shall argue) imply that, where we have a degenerating series of hypotheses, a weaker hypothesis is (in a sense I shall explain) more "confirmable" than a stronger one, and also that, whatever the evidence, the weaker hypothesis will be at least as well "confirmed" as those which entail it.

I will call any possible observation statement—whether it has been asserted or not and whether it is true or not—which would entail the development, for its domain, of an hypothesis, a *potential satisfier* of that hypothesis. We may say that if the class of all the potential satisfiers of the hypothesis h_2 is a sub-class of that of the hypothesis h_1, then h_1 is *more "confirmable"* than h_2.[5] For in that case more possible evidence could "confirm" h_1.

An accepted potential satisfier may be called an *actual satisfier*. If h_2's class of potential satisfiers is a sub-class of h_1's, and if evidence e "confirms" (is an actual satisfier of) h_2, then e also "confirms" (is also an actual satisfier of) h_1. But e may be an actual satisfier of h_1 without being an actual

---

[5] This would be the verificationist counterpart of Popper's idea that h_2 is more corroborable than h_1 if h_1's class of potential *falsifiers* is a sub-class of h_2's. See *L.S.D.* pp. 86, 115–26.

satisfier of h_2. Thus whatever e may be, the more "confirmable" h_1 will always be at least as well "confirmed" as h_2. We shall always get :

$$C(h_2, e) \leqslant C(h_1, e)$$

Now if h_2 entails (and is not entailed by) h_1, then h_2's class of potential satisfiers will indeed be a sub-class of h_1's : in the case of every possible observation statement, what h_2 would say if only those things mentioned in the observation statement existed will entail what h_1 would say if only those things existed. In other words, if h_2 entails h_1, then the development of h_2 for the domain of any observation statement will entail the development of h_1 for that domain. Thus every potential satisfier of h_2 is a potential satisfier of h_1; but some potential satisfiers of h_1 are not potential satisfiers of h_2. There are six possibilities :

1. e "confirms" h_2 and h_1
2. e is neutral to h_2 and "confirms" h_1
3. e is neutral to h_2 and to h_1
4. e disconfirms h_2 and "confirms" h_1
5. e disconfirms h_2 and is neutral to h_1
6. e disconfirms h_2 and h_1

These six possibilities hold good, *mutatis mutandis,* if h_3 entails (and is not entailed by) h_2.

With regard to 1, we can add that e "confirms" h_1 *no less strongly* than h_2; for by Hempel's consequence-condition ("Studies," p. 103 ff.), all the evidence which "confirms" or supports an hypothesis also supports all its consequences. Hypothesis h_1 is, so to speak, a less weighty hypothesis than h_2, and it has all the empirical support that h_2 has. Thus h_1 is at least as well supported or "confirmed" as h_2. The consequence-condition also implies that the tautology h_0 will be *conclusively* "confirmed" by e (see "Studies," p. 112).

The foregoing can be summarized as follows. If we have a degenerating series of hypotheses like the one given in Section I, then, whatever the evidence e may be,

$$- 1 \leqslant C(h_3, e) \leqslant C(h_2, e) \leqslant C(h_1, e) < C(h_0, e) = 1 \;^6$$

Now Hempel agrees that relative degrees of confirmation of hypo-

---

[6] A similar ordering is yielded by the definition of "degree of confirmation" of which Hempel is co-author. See Carl G. Hempel and Paul Oppenheim, "Definition of 'Degree of Confirmation'," *Philosophy of Science*, April 1945, where the degree of confirmation of an hypothesis is a function of its probability relative to the (statistically analyzed) evidence. Hempel and Oppenheim themselves point out that on their definition analytic hypotheses are maximumly confirmed (p. 10), though they do not add that this is only an extreme consequence and that in general a weaker hypothesis will, on their definition, be at least as well confirmed as a stronger one which entails it.

theses are a measure of their scientific acceptability.[7] In view of the fore-going, this would mean that a weaker hypothesis can never be less accept-able than a stronger one entailing it. For Popper, a stronger hypothesis is normally more acceptable than a weaker one entailed by it *provided* that its superior content has not brought it into conflict with observa-tions.

The result we have just obtained is implicit in Hempel's consequence-condition and provides a sufficient reason for rejecting that condition together with the idea behind it. But I have found that that idea is still very prevalent, despite Carnap's ingenious counter-examples (*Logical Foundations of Probability*, pp. 394 ff., 474 ff.); and it is often silently or explicitly presupposed in criticisms of Popper's theory of corroboration.[8] So I will add a short comment.

The consequence-condition is an expression of a verificationist view of confirmation, according to which, in the ideal case, e conclusively con-firms or verifies h, and in less than ideal cases, inconclusively confirms or quasi-verifies h. From this point of view it is natural to reason thus : If e verifies h, it verifies all h's consequences; so if e quasi-verifies or confirms h, it quasi-verifies or confirms all h's consequences.

But from a falsificationist point of view the situation is well-nigh re-versed. If e_1 is corroborating evidence obtained in testing h_1 and if h_2 entails h_1, then h_2 was also under test when e_1 was being obtained; thus if e_1 is congruous with h_2 it also corroborates h_2. On the other hand, if e_2 is evidence obtained from experiments which were tests on h_2 but not on h_1 (i.e., h_1 would not have been falsified or discorroborated whatever the outcome of these tests on h_2), then e_2 may corroborate h_2 but will not corroborate h_1.

I have no doubt that it is the latter, falsificationist view which accords with scientific practice. Suppose we have a powerful hypothesis about all heavenly satellites, and a much weaker hypothesis entailed by the first, about the moon only. It seems obvious that while certain observations of our moon might confirm both hypotheses, observations on Mars, say, or Jupiter's moons, might confirm the former hypothesis but would not normally be regarded as confirming the latter.

---

[7] "If an explicit definition of the degree of confirmation of a hypothesis were avail-able, then it might be possible to formulate criteria of acceptance in terms of the degree to which the accepted observation reports confirm the hypothesis in question" ("Studies," p. 115 note 1).

[8] In particular, it is presupposed by critics who employ a "tacking argument" against it. Their argument runs like this: let h be a testable hypothesis corroborated by e; and let m be a metaphysical sentence. Then the conjunction h *and* m will be corro-borated by e; therefore m alone will be corroborated by e because m is a consequence of h *and* m. Against this, see Popper, *L.S.D.*, p. 85 note *1, and W. W. Bartley, "A Note on Barker's Discussion of Popper's Theory of Corroboration," *Phil. Studies*, Jan.–Feb. 1961.

The hereditary principle does not work here. Each hypothesis must earn its own confirmation.

III.   "Confirmability" of Metaphysical Hypotheses

In this section I shall try to establish that the metaphysical h_1 in our original degenerating series is "confirmable" but not "disconfirmable" by Hempel's criterion; so that, whatever the evidence e may be:

$$0 \leqslant C(h_1, e) < 1$$

Hempel, on the other hand, would claim that, by his criterion, this hypothesis is confirmable *and disconfirmable*. For it certainly is "confirmable"; and he claims that "it is a theorem of [his] theory that *any* confirmable sentence is disconfirmable, namely by the evidence sentences confirming its negation."[9]

Let us approach the question of its disconfirmability via a consideration of a different hypothesis of a similar form, namely, the Hobbesian proposition :

Everyone is killable by someone.

This is "confirmable." An observation report which mentions just four individuals, namely the King and Queen of Denmark, Hamlet, and Laertes, and states that the Queen was killed by the King, Laertes by Hamlet, the King by Hamlet, and Hamlet by Laertes, implies that each of the individuals mentioned was killed, and therefore killable, by one of the individuals mentioned.

But is it disconfirmable : is its negation confirmable? Its negation is :

Someone is unkillable by everyone.

Hempel requires that an observation sentence "either asserts or denies that a given object has a certain observable property (such as 'a is a raven,' 'd is not black'), or that a given sequence of objects stand in certain observable relation (such as 'a is between b and c')" ("Studies," p. 23).

Relative to this requirement there is a crucial asymmetry between "a is killable by b" and "a is unkillable by b." The former would be entailed by an observation sentence which reported that a was killed by b; but the latter would not be entailed by a report that b tried hard but unsuccessfully to kill a.

Consider the observation report,

Nebuchadnezzer incarcerated Shadrach in a fiery
furnace; but he emerged unharmed.

---

[9] *Philosophy*, October 1958, p. 345 note 1; Hempel's italics.

This tells us that an unusually reliable method of disposing of people broke down when applied by Nebuchadnezzar to Shadrach. It does not tell us that every *alternative* method open to Nebuchadnezzar would have broken down. "Shadrach was unkillable by Nebuchadnezzar" would entail the development for these two individuals of "Someone is unkillable by everyone [else]," *but it is not an observation sentence.* "Unkillable by" is not an observable relation. To assert that *a* is unkillable by *b* is to assert a universal hypothesis to the effect that every possible method of killing will fail if applied by *b* to *a.* And for Hempel an observation sentence may not contain the universal quantifier ("Studies," p. 24 note 1).

Now a metaphysical hypothesis, as I conceive them, though neither entailed by, nor incompatible with, any finite observation report whatever, may very well be entailed by some, and incompatible with other, *unverifiable* scientific *hypotheses.* It is this which ensures that such metaphysical hypotheses have truth values and helps to explain their significance for science.[10] And there may be weaker relations between scientific hypotheses and a metaphysical hypothesis. For instance, an empirically supported scientific hypothesis may tell against, without contradicting, a metaphysical hypothesis. Thus the first-level hypothesis that Shadrach was unkillable by Nebuchadnezzar was supported by the zero-level report of Nebuchadnezzar's failure with the fiery-furnace method, and may in turn tell against the second-level hypothesis that Shadrach was killable by someone. But it would be a mistake to conclude that therefore that zero-level report indirectly disconfirms this second-level hypothesis. For that report would *also* have supported other *incompatible* first-level hypotheses—for instance, the hypothesis that Nebuchadnezzar could not kill Shadrach by the fiery-furnace method but could have killed him by the total-immersion-in-a-cask-of-malmsey-wine method; or the hypothesis that although Nebuchadnezzar could not kill Shadrach, Daniel could. And *these* two first-level hypotheses, far from telling against our second-level hypothesis, entail it as a metaphysical consequence.

Now Hempel's theory of confirmation rightly makes no provision for such indirect confirmation or disconfirmation. It does not allow us to interpose a first-level hypothesis to act as a broker between an observation report and a second-level hypothesis unless the mediating hypothesis actually *entails* the indirectly confirmed hypothesis. An observation report *e* "confirms" an hypothesis *h* if and only if it *entails* either *h* itself, or the development of *h* for *e*, or the development for *e* of some stronger hypothesis which *entails h.* And since, in our case, *h* ("Someone is unkillable by everyone"), and the developments for *e* of *h* and of any stronger hypothesis entailing *h*, will all contain the predicate "unkillable by," whereas *e* will at best contain the much weaker predicate "unkilled by," it follows that no *e* can

---

[10] See *Philosophy,* April 1957, pp. 129–30; *Mind,* July 1958, pp. 345–46, 356–57.

confirm our h or disconfirm the "confirmable" proposition "Everyone is killable by someone."

Similar considerations apply to our original h_1. The various developments for various domains of it and of stronger hypotheses entailing it will all contain the predicate "liquefiable" and may, therefore, be entailed by an observation report containing the predicate "liquefied." But the developments of its negation and of stronger hypotheses entailing its negation all contain the predicate "not liquefiable" and cannot, therefore, be entailed by an observation report containing the predicate "not liquefied." Hence h_1 is "confirmable" whereas its negation is not. Hence, whatever the evidence e may be,

$$0 \leqslant C(h_1, e) < 1$$

IV. "Unconfirmability" of Hypotheses in Mathematical Physics[11]

There is one difficulty which Hempel has in coping with exact hypotheses in mathematical physics which I do not wish to exploit : they cannot be formulated in the kind of language L to which Hempel confined his criterion of confirmation. L contains no sign of identity and its predicates are all observation-predicates ("Studies," p. 108, note 1); thus no mathematical statement, and no statement about theoretical entities, and hence no hypothesis in mathematical physics, could be formulated in L[12]. But Hempel looked forward to "the development of a definition of confirmation for languages of more complex logical structure than that incorporated in our model" ("Studies," p. 121); and in what follows I shall suppose that his definition of confirmation has indeed been extended to languages in which the transcendingly precise hypotheses of mathematical physics can be formulated.

Just as Hempel's satisfaction criterion generates a topsy-turvy ordering of degrees of "confirmation," which we considered in Section II, because of its conformity to the consequence-condition, so, I shall argue, it makes transcendingly precise hypotheses, even those which physicists would regard as well corroborated, altogether "unconfirmed" because of its conformity to another of his adequacy conditions—the consistency condition.

---

[11] The argument of this section occurred to me while reading in mimeograph "Explanation, Reduction, and Empiricism," by P. K. Feyerabend, since published in *Minnesota Studies in Philosophy of Science*, ed. H. Feigl and G. Maxwell, 1962, vol. III.

[12] It is curious that Hempel should throw stones from his glass house at what he calls the "prediction-criterion," objecting that *it* cannot cope with unobservables ("Studies," p. 99).

The consistency condition requires of any confirmation criterion that if, by it, an observation report (which is not self-contradictory) confirms two or more hypotheses, they and it must constitute a consistent set of statements ("Studies," pp. 102–6).

Hempel himself expressed misgivings about this "adequacy-condition." Carnap went further, bluntly stating that it "is not valid; it seems to me not even plausible . . . Hempel himself shows that a set of physical measurements may confirm several quantitative hypotheses which are incompatible with each other. He decides at the end to maintain it unchanged, without saying how he intends to overcome the difficulty which he has pointed out himself" (*Logical Foundations of Probability*, p. 476). Perplexed by this, Carnap asks : "What may be the reasons which have led Hempel to the consistency conditions . . . .?" (p. 477).

What answer can we suggest to Carnap's question ? To a verificationist philosopher who regards confirmation as quasi-verification it may seem incongruous that the same evidence should quasi-*verify* two inconsistent hypotheses at least one of which must be *false*. This would account for an initial bias in favor of the consistency condition. But it does not explain why Hempel stuck to it after acknowledging a serious shortcoming of it. Now Hempel's definition of the "development" of a hypothesis for the domain of an observation report e makes it impossible that e should entail the developments of two inconsistent hypotheses. Dare I voice the awful suspicion that, finding that his criterion had the awkward consequence that the same evidence cannot "confirm" two inconsistent hypotheses even in situations where working scientists would regard it as confirming both of them, Hempel boldly decided not to soft-pedal this consequence but to elevate it to the rank of an adequacy condition so that, instead of having to reproach his own criterion for having an unfortunate consequence, he could reproach rival criteria for *not* having this consequence?

Popper has pointed out that this condition implies that whatever Hempel "may mean by 'confirming evidence' of a theory, he clearly cannot mean the results of tests which corroborate the theory"; for a test result which corroborates one hypothesis is bound to corroborate other inconsistent hypotheses. Popper continues : "Let there be a hypothesis h, corroborated by the result e of severe tests, and let h_1 and h_2 be two incompatible theories (e.g., Einstein's and Newton's theories of gravitation) each of which entails h . . . . Then any test of h is one of both h_1 and h_2, since any successful refutation of h would refute both h_1 and h_2; and if e is the report of unsuccessful attempts to refute h, then e will corroborate both h_1 and h_2" (*L.S.D.*, p. 374 note 13). I shall now carry this criticism a stage further. I shall argue that the consistency condition means, not only that hypotheses are not "confirmed" by the results of tests, but that transcendingly precise hypotheses like our h_3 are not "confirmed" *at all*. My aim here is not so much to show that Hempel got his adequacy conditions wrong, but to clinch my argument against the satisfaction criterion itself (which, as we saw, necessarily conforms to the consistency condition).

For every transcendingly precise hypothesis there will be other hypotheses which are logically incompatible with it but also, because of experimental error, empirically indistinguishable from it—at least for the time being. For example, Newton's hypothesis that:

$$F = \frac{Mm}{d^2}$$

is, I imagine, likely to remain empirically indistinguishable for many years from the hypothesis that:

$$F = \frac{Mm}{d^{2.000\ 000\ 000\ 003}}$$

Or consider Galileo's and Newton's laws of gravitation. Hempel claimed that evidence which confirmed Newton's also confirmed Galileo's since Galileo's "is essentially a substitution instance of the stronger [Newtonian] one" ("Studies," p. 104). But far from being a substitution instance of it, Galileo's is *incompatible* with Newton's. Galileo's says that a body falling freely near the surface of the earth falls with *constant* acceleration. Newton's implies that such a body falls with an acceleration which *increases* as the distance between its and the earth's centers of gravity diminishes. True, it also implies that, if the proportionate reduction in distance is very small (as will be the case with bodies falling near the surface of the earth) the increase in acceleration will be very small—perhaps too small to be detected experimentally. But Newton's law remains *logically* incompatible with Galileo's (given that freely falling bodies exist). Given their incompatibility, and on the assumption that they give rise to empirically undistinguishable predictions, Hempel's consistency condition implies that Galileo's and Newton's laws have not been "confirmed" by any of the evidence which has corroborated them.

Let us discard the consistency condition and operate directly with Hempel's definition of "confirmation." In 1877, C. S. Peirce remarked that "whatever is held to be precisely true goes further than experience can possibly warrant" (*Collected Papers*, 5.382 n)—a remark which condemned in advance the idea that exact hypotheses are confirmed by observation reports entailing what the hypothesis would say if what the report mentions were all that existed.[13] Our original h_3 asserted that solid substances will liquefy at a rate R *exactly* proportional to I. The most that could be established experimentally would be that a substance

---

[13] Popper uses the impossibility of inferences from the inexact to the exact as one of his decisive arguments against induction ("On the Status of Science and Metaphysics," *Ratio*, Dec. 1958, p. 99).

liquefied at a rate which diverged from R by less than ε, where ε is an estimate of experimental error.[14]

Someone anxious to defend Hempel's criterion against this objection, especially someone with positivist tendencies, might try to do so by maintaining that logically incompatible but empirically indistinguishable hypotheses in mathematical physics should be regarded as making a common, slightly vague empirical statement to which each gives a different and merely metaphysical "precisification." There is some sanction for this defense in the writings of Bishop Berkeley, who held that it is meaningless to assert the existence of a physical quantity smaller than the *minimum sensibile*, and that, for example, "there is no such thing as the ten-thousandth part of an inch" (*Principles* § 127, § 132). Despite this episcopalian blessing there are decisive objections to this defense. For instance, it makes 1832 the date of the Copernican revolution. The pre-Copernican idea was that the earth's distance from the fixed stars is constant and the sun's distance from most of them varies as it moves round the earth. Copernicus claimed, on the contrary, that the sun's distance is constant and the earth's varies as it moves round the sun, adding that it is because the fixed stars are so far away that their annual parallax remains undetected. This stimulated efforts to improve the accuracy of astronomical observations, efforts which eventually led, in 1832, to the successful detection of an annual parallax.

Suppose that the positivist idea that incompatible but empirically indistinguishable hypotheses really say the same thing had prevailed, and that Copernicus had been regarded as essentially restating the Ptolemaic system in a different language. One or other of two things would have happened.

---

[14] I included the clause "A's exist" in h_3 to ensure that h_3 would be *dis*confirmed, rather than trivially "confirmed," by an observation report which stated of each thing it mentioned that it was non-A. But suppose we have an "ideal" law-statement with no existential clause which makes exact assertions about *non-existent* entities (ideal gases, absolutely elastic billiard balls, frictionless planes, etc.). Would not such an exact law-statement be "confirmable" by Hempel's definition?

A "law" which *merely* stated what would happen under conditions which never obtain, a "law" which did not make possible any predictions about what happens under actual conditions, would be untestable and would indeed be trivially satisfied and, by Hempel's criterion, "confirmed" by every situation.

But an ideal law in empirical science does more than this. It enables us, so to speak, to work back from what would happen under ideal conditions to what does happen under actual conditions. Typically, an ideal law states a functional dependence of the form: $y = f(x)$; and especially draws attention, so to speak, to the extreme case where the value of x is zero, though x always has a positive value in the world we know. Galileo's law of inertia does indeed say that a body moving solely under its own inertia along a frictionless horizontal plane would neither accelerate nor decelerate. But it also says, more generally, that acceleration and deceleration are a function of force. It applies to real bodies coming to rest through friction as well as to imaginary bodies gliding over ideal surfaces. Thus so-called "ideal" laws fare no better under Hempel's criterion than do other exact laws.

Either annual parallax would not have been discovered when it was owing to lack of theoretical stimulus, in which case science would have suffered a serious deprivation, or the annual parallax would nevertheless have been discovered when it was, in which case this positivist idea would oblige us to maintain that two theories had the same meaning until 1832, when they became incompatible.[15]

A further objection to this positivist idea is that scientists may have *cogent reasons* for asserting some perfectly *precise* relationship. For instance, there were decisive geometrical reasons for Newton's belief that force varies inversely with precisely the square of the distance.

I conclude that (in the absence of some *ad hoc* rescue operation) our h_3 is, for Hempel, falsifiable but not "confirmable."[16] Whatever the evidence e may be,

$$- 1 \leqslant C(h_3, e) \leqslant 0$$

I have established this for an hypothesis which transcends experience only in its universality and precision. Hypotheses in modern physics mostly transcend experience in their ontology also. Hempel's verificationist theory of confirmation can no more cope with this latter sort of transcendence than it can with transcending precision. But the fact that scientific hypotheses transcend experience in various ways creates no special difficulty for a falsificationist theory of corroboration.

I will now summarize the findings of the last three sections. Given a highly testable hypothesis in mathematical physics like our h_3, a less precise but still testable hypothesis entailed by it, a still vaguer and untestable metaphysical hypothesis entailed by it, and a tautology; and given that our "criteria of acceptance" for hypotheses in science are "in terms of the degree to which the accepted observation reports confirm the hypothesis in question";[17] then *whatever the accepted observation reports may be* we know that, by Hempel's criterion, the hypothesis in mathematical physics will be unacceptable, the tautology will be supremely acceptable, while the metaphysical hypothesis may be more, and cannot be less, acceptable than the less precise scientific hypothesis.

I find this result pretty devastating. In fairness I should add that much of the credit for it properly belongs to Hempel. He started with the idea that a hypothesis is confirmed if it is satisfied or borne out by the evidence. This idea seems intuitively plausible so long as it remains vague. Then

---

[15] Another example: this idea would mean that modern theories of the continuous creation of matter say the same, for the time being, as the older theory of the un-creatability of matter!

[16] We have here another counter-example to Hempel's claim (see p. 99, above) that by his criterion, "confirmability" and "disconfirmability" go hand in hand.

[17] See p. 97, above, and note 7.

Hempel gave it a desirable precision, and so made it possible to draw out its startling implications. Indeed, Hempel himself immediately pointed out that it has a very counter-intuitive or "paradoxical" implication. This we must now consider.

V.   The "Paradoxes of Confirmation"

Hempel went out of his way to emphasize that his satisfaction criterion has the counter-intuitive or "paradoxical" consequence that it makes every non-A constitute "confirming" evidence for a hypothesis of the form "All A's are B." For this hypothesis is "confirmed," according to that criterion, if it is satisfied or instantiated by the evidence; but it is equivalent to "Everything is B or non-A," and so will be satisfied or instantiated not only by any A which is B but also by anything which is B and by anything which is non-A. In short, it will be satisfied by everything except A's which are non-B. Hempel tried to remove the counter-intuitiveness of this consequence by pointing out that a hypothesis of the form "All A's are B" asserts something not only about A's, but about *every* thing (namely, that each thing is non-A or B). And since it asserts something about every thing we should not be surprised or perplexed to find that every thing may confirm it ("Studies," pp. 18–19).

What, if anything, is wrong with a criterion of confirmation which involves this "paradoxical" consequence? It is not, of course, paradoxical in the strict sense : it involves no logical inconsistency. It is only rather startlingly counter-intuitive. But there is not necessarily anything wrong with that. A counter-intuitive philosophical theory which shows up the mistakenness of our intuitions may well be superior to one which accords with them. And, indeed, Hempel claimed to have shown that the counter-intuitiveness of the idea that every non-A confirms "All A's are B" is a "psychological illusion" created by "a misleading intuition in the matter" ("Studies," p. 18).

But it *may* also happen that a counter-intuitive theory, by provoking us to examine the whole matter afresh, helps us to realize that the counter-intuitive theory is itself due to "a misleading intuition in the matter."

It is important to remember that the counter-intuitive conclusion which Hempel upheld was not just that a non-A may, in certain circumstances, provide confirmation for "All A's are B," but that every non-A in any circumstance automatically does so. A serious objection to this is that it makes the confirming of scientific hypotheses a terribly facile business. But we must remember that Hempel claimed that this conclusion is well-nigh unavoidable. (He conceded that there are certain *ad hoc* dodges for avoiding it, but claimed that to resort to any of these would have very unfortunate consequences—a claim I shall not dispute.) If it is unavoidable it would be no good complaining about facile "confirmation"; we should have to lump it.

I shall now argue, however, that it is not only avoidable but had already been avoided, easily and naturally, by Popper's theory of corroboration.

I shall also argue that Hempel arrived at his counter-intuitive conclusion because he stuck to a seemingly self-evident idea which is, however, mistaken. In the course of rejecting various ways in which one might try to avoid the "paradoxes," Hempel stated: "Nor does it seem possible to dispense with the stipulation that an object satisfying two conditions, C_1 and C_2, should be considered as confirming a general hypothesis to the effect that any object which satisfies C_1 also satisfies C_2." ("Studies," p. 14). I shall argue that Hempel's achievement here was to show, despite himself, that this seemingly obvious idea—the idea which governs his whole approach to confirmation—*does* have to be dispensed with.

In this problem area it is particularly desirable to have all the alternatives laid out before us. Call an A which is B an AB. With regard to the confirmation of "All A's are B," the seemingly obvious proposition which Hempel did not contemplate dispensing with was:

(1)   An AB always confirms it.

The counter-intuitive or "paradoxical" conclusion to which, Hempel showed, (1) leads is:

(1')   A non-A always confirms it.

To this pair of propositions we can add two alternative pairs:

(2)   An AB sometimes confirms it.

(2')   A non-A sometimes confirms it.

(3)   An AB never confirms it.

(3')   A non-A never confirms it.

It is not difficult to show that (3') leads to (3) and that (2) leads to (2') by reasoning similar to that whereby Hempel showed that (1) leads to (1').

Many people have felt that the ideal combination would be (1)-cum-(3')—is it not obvious that all white swans do, whereas no non-swan does, confirm "All swans are white"? Hempel showed that (1)-cum-(3') is an impossible combination, that *if* we accept (1) we must accept (1'); and he believed that there can be no question of dispensing with (1). But in fact we still have a *choice* here—between (1)-cum-(1') and (2)-cum-(2'). And if (1)-cum-(1') generates much too much "confirmation" while (3)-cum-(3') allows much too little (viz., none at all), we may conjecture that perhaps (2)-cum-(2') allows the right amount. Hempel's appraisal of (1') was revolutionary. But this was dictated by his *conservative* retention of (1).

Perhaps a moderately reformist approach to both yields better results. Popper's theory, I shall show in a moment, is less revolutionary than Hempel's in endorsing (2') rather than (1'), and less conservative in endorsing (2) rather than (1). I shall now try to show that (2)-cum-(2') is indeed the best combination. I shall do this by examining a possible defense of Hempel's (1)-cum-(1') against Popper's (2)-cum-(2').

It might be claimed that if Popper's position involves (2)-cum-(2') then it is even more satisfactory than Hempel's. First, it allows some non-A's to confirm "Every A is B" which is *nearly* as bad as making every non-A do so. Second, it has the shocking consequence that it *stops* some AB's doing so.

I shall take these two points in turn.

The first objection presupposes that (3' : a non-A never confirms) is the ideal from which (2' : a non-A sometimes confirms) lapses, though not so badly as does (1' : a non-A always confirms).

But there are various hypotheses which have surely been confirmed but for which we are just unable to obtain direct evidence about instances of their antecedent clauses. For instance, there are hypotheses about live dinosaurs for which there exists today a good deal of confirming evidence, though there are no live dinosaurs. Again, modern physics consists largely of hypotheses about unobservable entities for which confirming evidence is provided by things which are not unobservable entities.

Moreover, even in cases where instances of the antecedent clause of a hypothesis are open to observation it may very well happen that confirming evidence is provided by non-instances of it. Agassi has given a nice example of this. Before reproducing it I will indicate very briefly those features of Popper's theory of corroboration which are relevant to it.

Popper's fundamental idea is that a hypothesis is corroborated whenever it survives a test; the more severe the test the stronger the corroboration. The relative severity of tests is appraised in this way. Suppose we have background information k, an hypothesis h, and evidence e brought to light by an experimental test on h. Then the more improbable was e, given k alone, and the more probable was e, given the amalgamation of k and h, the more severe was the test and the better e corroborates h (*L.S.D.*, appendix *ix).

Here is Agassi's example.[18] He asks us to consider a situation where Galileo's hypothesis that freely falling bodies move with constant acceleration is being tested "by dropping steel balls off an electromagnet by switching off the electric current. These experiments can be carried out in deep mines to allow the steel balls to fall freely a long way." Let us suppose that the same test is being carried out in two mine-shafts.

---

[18] "Corroboration versus Induction," *Brit. Jour. Phil. of Science*, Feb. 1959, especially pp. 313–314.

Assume that background information by no means suggests that a freely falling body falls with constant acceleration, but does suggest that a ball dropped down a vertical shaft normally falls freely. Now introduce Galileo's hypothesis. This raises sharply the probability that a ball dropped down a vertical shaft will fall with constant acceleration. It *also* raises the probability that *if* such a ball does *not* fall with constant acceleration, it will be found *not to have fallen freely.* Suppose that observations in one of the mine-shafts confirm that the balls dropped there fell with constant acceleration. This evidence corroborates Galileo's hypothesis, since its probability was raised by that hypothesis. In this case a hypothesis of the form "All A's are B" has been corroborated by evidence of the form "These A's were not found to be non-B." Suppose that the balls dropped down the other mine-shaft are definitely observed to fall with inconstant acceleration; and suppose further that, as in Agassi's example, the cause of the aberration were found to be that the shaft passed through a magnetic rock. Then the evidence that these inconstantly accelerating balls were not falling freely *also* corroborates Galileo's hypothesis, since its probability was also raised by that hypothesis. In *this* case an hypothesis of the form "All non-B's are non-A" has been corroborated by evidence of the form "These non-B's were found to be non-A."

Thus it is wrong to suppose that, if Hempel's theory is pregnant with paradox in endorsing (1'), Popper's is "just a little pregnant" with paradox in endorsing (2').[19] With reference to Agassi's example the relative merits of propositions (1',) (2'), and (3') could be summarized thus:

Proposition (1') makes the steel balls which fell with inconstant acceleration because of the magnetic rock confirming evidence for Galileo's law; but it also makes the Tower of Pisa and all other things which are not freely falling bodies confirmatory.

Proposition (2') allows those steel balls to be confirmatory without requiring all other things which are not freely falling bodies to be confirmatory.

Proposition (3') makes those steel balls nonconfirmatory.

I conclude that (3') is as unsatisfactory, though in the opposite direction, as (1'), and that (2'), far from being a little "paradoxical," is entirely satisfactory.

I now turn to the relative merits of Hempel's (1): "An AB always confirms" and Popper's (2): "An AB sometimes confirms."

What I have to say here proceeds from two assumptions. First, confirmation theory bears on the appraisal of scientific hypotheses. Given two hypotheses and a body of evidence, the first hypothesis being more strongly confirmed than the second by that evidence according to a particular theory of confirmation; then my first assumption is that this theory

---

[19] I take the simile from I. Scheffler, *Philosophical Studies*, Jan.-Feb. 1961, p. 19.

of confirmation suggests that scientists should, other things being equal, *prefer* the first. Hempel accepts this.

My second assumption is that a predictively powerful hypothesis which has withstood severe testing *is* preferable (to put it mildly) to a hypothesis which is completely *ad hoc* relative to known evidence, a hypothesis from which no new predictions can be derived. This assumption is at the core of Popper's philosophy of science.

Verificationist theories of confirmation have an awkward tendency to imply that the *ad hoc* hypothesis is preferable, since it may be *verified,* whereas the predictively powerful hypothesis (especially if it is mathematically exact and postulates theoretical entities) may (as we have seen) turn out to be altogether "unconfirmable."

I shall now argue that proposition (1) is an expression of verificationism and leads precisely to a preference for *ad hoc* hypotheses over predictively powerful ones, whereas (2) enables us to avoid this. I will show this with the help of an artificial example.

Imagine a botanist living in Cornwall. Suppose that part of his background information k consists of the evidence e_1 that all the A's so far investigated in Cornwall have been found to be B, but that k does not include any information about A's in Sumatra, Greenland, or Peru. Now consider two hypotheses which state respectively :

All A's are B

All A's so far investigated in Cornwall are B

I will call these h_2 and h_1 respectively. Hypothesis h_1 is, of course, completely *ad hoc* relative to e_1. Now suppose that our botanist visits Sumatra, Greenland, and Peru in order to investigate A's under varying conditions, and that he finds many A's in each of these countries and that each of them turns out to be B. I will call the evidence he collects in these countries e_2.

By Popper's criterion (see p. 108, above) neither e_1 nor e_2 corroborates h_1. The evidence e_1 is a part of, and therefore entailed by, k so that the introduction of the *ad hoc* hypothesis h_1 cannot *raise* its probability. As for e_2, its probability was not raised by h_1, which predicts nothing about A's in Sumatra, Greenland and Peru.

Hypothesis h_2, on the other hand, is corroborated by e_2. Thus for Popper, h_2 is preferable to h_1.

But this appraisal is not possible for a theory of confirmation which involves (1) rather than (2). For by (1), an AB is always confirmatory. Thus the AB's investigated so far in Cornwall confirm h_1. Indeed, they *conclusively* confirm it. The AB's observed in Cornwall, Sumatra, Greenland, and Peru confirm h_2, but only *inconclusively*. So h_1 is more strongly confirmed than, and therefore preferable to, h_2.

My last remark to anyone who still hankers after (1) is a reminder that the choice is between (1)-cum-(1') and (2)-cum-(2'). If he still feels uneasy about (2): "An AB sometimes confirms," I ask him whether he does not feel even more uneasy about (1'): "A non-A always confirms."

Thus Popper's theory, I contend, avoids the "paradoxes" by saying that "Every A is B" is (strongly) corroborated by evidence (whether of AB's or non-A's) if and only if such evidence is rendered (considerably) more probable by the hypothesis than it was by background knowledge alone.

The appeal to background knowledge is essential. Now Hempel deliberately excluded background knowledge, citing in his support "the 'methodological fiction,' characteristic of every case of confirmation, that we have no relevant evidence for H other than that included in E" ("Studies," p. 20). He did not say where this strange fiction comes from or why we should submit to it.

Now genuine testing may still be possible even with very impoverished and sketchy background knowledge. But if we suppose *no background knowledge at all* (a completely unrealistic supposition, of course) the notion of testing becomes empty. Without *any* indications about where to look for likely counter-evidence, *all* undirected looking equally "tests" a scientific hypothesis. Anything, anywhere may be an A which is not B, and the discovery that something is, after all, a non-A will always constitute the favorable outcome of a pseudo-test.

Thus even a testability theory will generate the "paradoxes" (viz., the (1)-cum-(1') combination) if background knowledge is willfully excluded. But why in heaven's name (for I would rather appeal to heaven than to a methodological fiction) *should* it be excluded? Hempel claimed that the "paradoxes" are well-nigh unavoidable. Yet we have only to re-introduce background knowledge and replace a satisfaction theory involving (1)-cum-(1') by a testability theory involving (2)-cum-(2'), and they disappear.

VI. Positivism

In conclusion, I wish to draw out an implication of the foregoing for positivism.

The verification principle has suffered various setbacks in the course of its career, emerging from each in a more attenuated form. Eventually it came to this, that a non-analytic statement is cognitively meaningful if and only if it is at least confirmable or disconfirmable. I shall now show that even this modest version of the verification principle is untenable.

For expository reasons I will begin with a somewhat stronger version of the principle, according to which a nonanalytic statement is cognitively meaningful if and only if it is verifiable or falsifiable.[20] This version would

---

[20] Professor D. Rynin subscribed to this version of the Principle in his "Vindication of L*G*C*L P*S*T*V**M," (*Proc. Am. Phil. Ass.* vol. xxx, November 1957). How-

surely have to be abandoned if we could find a statement p and a statement q such that:

(i) p is "meaningful" and q is "meaningless" according to this criterion of meaningfulness;

(ii)   the conjunction p and q has more empirical content than has p alone.

Now the degenerating series of hypotheses described in Section I suggests an easy way to construct such a p and q.

Let p be a somewhat weakened version of our h_2, namely:

> *If* a solid substance is liquefiable at all, *then* it is always liquefiable by the usual electrolytic method.

Let q be our old h_1, namely:

> All solid substances are liquefiable.

It is easy to see that this pair of statements satisfies conditions (i) and (ii) above.

(i) p is "meaningful" because falsifiable. q is "meaningless" because neither verifiable nor falsifiable.

(ii) The conjunction p *and* q (which is actually equivalent to our old h_2) has more empirical content than has p alone. For p *and* q entitles us to predict of any solid substance that it will always liquefy if subjected to the usual electrolytic method, whereas p entitles us to predict this only of solid substances which are already known to be liquefiable by some other method.

Confronted by this counter-example to the verifiable-or-falsifiable version of the verification principle, a defender of the latter is likely to concede that that version was somewhat too strong and that it should be replaced by a confirmable-or-disconfirmable requirement. And he is also likely (I speak from experience) to add that the q in this counter-example (viz., our old h_1), though neither verifiable nor falsifiable, is surely confirmable, so that, with this modification, the difficulty vanishes.

Now there surely are definitions of "confirmation" by which h_1 is "confirmable." But they are *verificationist* definitions. Hempel's is one. Now a criterion of empirical meaningfulness based on the ideas of confirmation and disconfirmation requires a *materially adequate* definition which provides "a reasonably close approximation to that conception of confirmation which is implicit in scientific procedure and methodological discussion."[21]

---

ever, after considering unverifiable and unfalsifiable statements like our h_1, he wavered and indicated that he *might* "be willing to weaken the requirement to that of even a kind of confirmability not reducible to [verifiability-or-falsifiability]" (p. 62).

[21] Hempel, "Studies," p. 107; and see pp. 7–8.

I take it for granted that a theory of confirmation which involves trivial "confirmation" by all non-instances and also the sort of topsy-turvy ranking of hypotheses which we found Hempel's to involve is materially inadequate. Now Popper has shown that the same sort of topsy-turvy ranking is engendered by that other verificationist approach to confirmation where the evidence is required at least to raise the probability of the hypothesis it confirms.[22] Indeed, with post-*L.S.D.* hindsight one can see that any verificationist approach to confirmation is likely to produce such a ranking (provided it is able to endow a universal hypothesis with above-zero confirmation). For if e quasi-verifies h_2, and if h_1 is a weaker consequence of h_2, then e must surely quasi-verify h_1 no less strongly; and if h_1 "goes beyond" e less than h_2 does, or does not "go beyond" e at all and is actually verified by e, then h_1 will almost certainly be more strongly "confirmed" than h_2 according to any verificationist view of confirmation.

At any rate, writing on the occasion of Popper's birthday, I am now going to take it for granted that if we are to avoid that sort of topsy-turvy ranking and appraisal of scientific hypotheses, a verificationist approach must give way to a falsificationist approach. And I want now to show that if the verification principle is weakened from a verifiable-or-falsifiable requirement into a confirmable-or-disconfirmable requirement, and if "confirmable" is given a *falsificationist* rather than a verificationist interpretation, then our old h_1, and other unfalsifiable hypotheses of a similar character, will again prove its undoing.

A verifiable-or-falsifiable criterion is a two-pronged instrument which picks up some purely existential statements on one prong, some strictly universal statements on the other, some numerically universal and singular statements on both, and some "all-and-some" statements[23] on neither.

If 'confirmable' is given a verificationist interpretation — if 'confirmed' is thought of as "weakly verified" (and 'disconfirmed' as "weakly falsified") — then a confirmable-or-disconfirmable criterion is likely to be a wider but still a two-pronged instrument (though it will probably pick up more statements on both prongs simultaneously). For instance, if 'confirmable-or-disconfirmable' is interpreted according to Hempel's definitions, it is two-pronged.[24]

However, if "confirmable" is given a *falsificationist* interpretation, the confirmable-or-disconfirmable meaning criterion turns out to have one

---

[22] He has further shown that to equate confirmation with probabilification actually leads to contradictions. See *L.S.D.*, appendix *ix, especially p. 391; also §83; also Popper's "The Demarcation between Science and Metaphysics" in *The Philosophy of Rudolf Carnap*, P. A. Schilpp (ed.). Added in proof: this paper has now appeared as chapter 11 in *Conjectures and Refutations*.

[23] Such as our h_1. For the term 'all-and-some,' see *Philosophy*, April 1957, pp. 125 ff.

[24] We found that our h_1 is "confirmable" but not "disconfirmable" and that our h_3 is "conclusively disconfirmable" but not "confirmable"; moreover our tautological h_0 is "conclusively confirmable" but not "disconfirmable."

prong which will not pick up our untestable h_1. From a falsificationist point of view, *both* confirmability *and* disconfirmability are a function of *falsifiability* or at least of testability.

The point of the qualification "or at least of testability" is to allow for the confirmability of statistical or probabilistic hypotheses in science. It is often believed that these hypotheses, being strictly unfalsifiable, necessitate a radical revision of falsificationism.[25] And it is sometimes further supposed that a revision which renders probabilistic hypotheses confirmable will render unfalsifiable metaphysical hypotheses like our h_1 confirmable too. I wish now to indicate briefly why, by Popper's theory of corroboration, this is not so.

Suppose we have a non-probabilistic hypothesis h; a determinate prediction p which has a low probability relative to background knowledge k minus h and a probability of one relative to k plus h; and evidence e brought to light by putting p to the test. Suppose, first, that e matches p. Then h is, on Popper's view, corroborated. Suppose, second, that e conflicts quite glaringly with p. Then, relative to k and e, h is falsified.

Now suppose that we have a probabilistic hypothesis h', a statistical prediction p' which has a low probability relative to background knowledge k minus h' and a probability of almost one relative to k plus h'; and evidence e' brought to light by putting p' to the test. Suppose, first, that e' matches p'. Then h' is corroborated. Suppose, second, that e' conflicts quite glaringly with p'. Then, relative to k and e', h' will be strongly discorroborated.

No doubt it is easier for an experimental outcome e to conflict "quite glaringly" with a determinate prediction p—it may be sufficient if e diverges only a little from p—than it is for statistical evidence e' to conflict "quite glaringly" with a statistical prediction p'—e' may have to diverge widely from p'. (Though even this contrast is mitigated if, in the case of the determinate predictions of a nonprobabilistic hypothesis, the experimental measurements display a certain scatter to which the so-called "normal law of error" is applied.) But the point is that though no e' can strictly falsify a probabilistic h', e' may discorroborate h' very strongly or "almost falsify" h'.

One might say that (ignoring experimental error) a nonprobabilistic scientific system is absolutely intolerant of evidence inconsistent with its determinate predictions; a probabilistic scientific system grows increasingly intolerant of evidence the more it diverges from its statistical predictions;

---

[25] One way in which one might acquire this belief would be to read only the first page of Popper's chapter on "Probability" in *L.S.D.*, where he writes ". . . We shall discover what will at first seem an almost insuperable objection to my methodological views. For although probability statements play such a vitally important role in empirical science they turn out to be in principle *impervious to strict falsification*" (p. 146; his italics).

while a metaphysical hypothesis like our h_1 is tolerant of all evidence. Of any particular solid substance h_1 says that there is a way of liquefying it. Now there is an infinite number of *possible* ways of liquefying it, from saying "Abracadabra" to trying zymosis, and the empirical finding that a finite number of these ways have not worked goes no distance towards establishing that no possible way will work. It is much less important that such hypotheses share with probabilistic scientific hypotheses the feature of not being strictly falsifiable than that such metaphysical hypotheses do *not* share the feature of being highly discorroborable.

Now I revert to the destructive significance of our h_1 for the verification principle.

Suppose that an upholder of the confirmable-or-disconfirmable version of it has been converted to the view that confirmability has to be interpreted as testability. Then he may no longer say, "Ah, but your h_1, though admittedly unverifiable and unfalsifiable, is surely *confirmable.*" "Confirmable-or-disconfirmable" is no longer a genuine disjunction: it just says "testable" twice over; and our h_1 is neither confirmable by favorable outcomes nor disconfirmable by unfavorable outcomes of tests on it, because it is not testable.

Thus our h_1 is "meaningless" according to the most moderate version so far proposed of the verification principle, if the latter is interpreted (as I believe it has to be, if that topsy-turvy ranking is to be avoided) in a falsificationist spirit. Yet we know that our h_1 is meaningful, since conjoining it with a testable, meaningful statement results in a statement of greater empirical content. Any further relaxing of the verification principle would surely render it so tolerant as to be useless. I conclude that the only thing to do with it is to "commit it to the flames."

Verificationism is an expression of the wish for certainty. It leads naturally to positivism (for metaphysical speculations are inherently uncertain) and also to an inductivist view of confirmation whereby observation reports quasi-verify scientific hypotheses. Verificationism, positivism, and inductivism hang together—and they fall together too.

CHAPTER 8

Overlooked Aspects of Popper's Contributions to Philosophy, Logic, and Scientific Method

BY J. O. WISDOM

IN THIS PAPER I propose to draw attention to some contributions that Popper has made in the fields of philosophy, logic, and scientific method, the significance of which appears to have been overlooked. I will also add a brief description of a contribution which has not been published.

I

I will first outline a general view of logic that emerges from Popper's work.

Whether one looks at different treatments of modern logic or at textbooks of the subject since they first began to appear about thirty years ago, one is left without any explanation of the goal of the whole subject. There are various bits and pieces but their interconnectedness is not always evident, nor is it clear what are the central problems to which logic is devoted. Now intuitively it is reasonable to suppose that logic is about inference, and moreover that it is about valid inference. This is very simple and straightforward but not what immediately hits the eye in most treatments of the subject. It comes out very clearly, however, in Popper's treatment.

In the first place, one of his central concerns is the theory or definition of validity, and here he has added to Tarski's fundamental discovery, the upshot of which is a theory of validity defined by the concept of a counterexample. This is not simply a pleasant and elegant academic rounding-off of a loose end left by the history of the subject but it provides a powerful tool of demonstration. Incidentally, the additional material that I shall offer will be based upon it.

In the second place it is natural to inspect rules of inference, to sift out which of them are valid and which are invalid and then to attempt to prove the validity of those that are in fact valid.

So far, therefore, logic would consist of the theory of validity and the proof of the validity of rules of inference. And here we get an understandable view, at any rate, of elementary logic.

Around these conceptions, naturally, a number of other topics may cluster. It is not difficult to fit in truth tables, tautologies, the nature of proof itself, and so on. But special attention should be drawn to one thing, namely the role of the deduction theorem. Since it shows the equivalence between a demonstrated conditional tautology and a demonstrated rule of inference, it allows Popper to interpret logic in terms of rules of inference rather than in terms of tautologies. We may if we wish develop logic entirely in terms of tautologies if this is technically more convenient, while we are at the same time at liberty to interpret logic in terms of rules of inference. We would do this if we consider that after all what we are trying to do is to provide ourselves with rules that will apply to any inference that may be made, whether in ordinary life or in science or in mathematics, and whether the content is empirical or logical. The conception of logic as a set of rules of inference has universal application. It provides the more convincing interpretation of the subject, i.e., it is intuitively more satisfactory.

II

The general conception that Popper has put forward has a consequence for traditional logic that is of considerable interest. For we can see that the traditional treatment of the syllogism by reduction consists of making a collection of rules of inference which were called moods in certain figures; that these rules of inference may be divided into valid and invalid; and that Aristotle made a very serious, and in large measure successful, attempt to demonstrate a number of these rules of inference. (That is to say, he was successful insofar as he could demonstrate them in terms of a given basic rule of inference, namely Barbara; he was not able to take the further step of demonstrating this rule.)

Now with the introduction of Boolean algebra, Popper has been able to give proof of the basic rules of inference Barbara and Celarent in terms of which all the other Aristotelian rules of inference are proved.[1] This work of Popper's is little known. It has the charm of simplicity. Incidentally it provides an example of a very elementary kind to illustrate a rather sophis-

---

[1] K. R. Popper, "The Trivialization of Mathematical Logic," *Xth Inter. Congr. Philos.*, Amsterdam, 1948, 1, 722 ff.

ticated point about metamathematics. That is, there are results that cannot be obtained by mathematical methods but can be by metamathematics. The exemplification here is this: it is not possible to prove the rule of inference Barbara without transforming it metametalinguistically into terms of Boolean equations.

III

The problem I am now going to discuss concerns the classical rules of a syllogism, such as that the middle term of a valid syllogism must be distributed at least once. To avoid confusion with rules of inference, i.e., moods in certain figures, I will call those now to be discussed "rejection-rules." It is curious that in traditional logic, much of which is rigorous, the rejection-rules of the syllogism have been left for over twenty centuries without proof. The textbooks make discursive remarks upon them, which amount to saying that they are intuitively fairly clear.

What is the function of these rejection-rules? To answer this requires a word of contrast with the Aristotelian theory of reduction. This sought to prove those rules of inference that are in fact valid by reducing them to certain moods in the first figure. The rules of the syllogism now to be considered, however, fulfill the complementary role of disproving the invalid rules. Considering that a good deal of attention was devoted to the proof of the valid moods, it is strange that so little should have been done about these rules. But a little was done. It is known for instance that these rules can be reduced to three that are independent.[2] But there is a tendency to dismiss such efforts on the grounds that, even if the rules could be reduced to one, little would have been achieved. Naturally we should be left with something incomplete unless this one were also proved. But the interrelating of the rules would be an important step if it could be taken. In fact, however, it is likely that it cannot be taken, that in fact three of the rules are independent.

It occurred to me that it might be possible to prove these rejection-rules by having recourse to the Tarski-Popper method based on the notion of counter-example; and I found a rudimentary proof for one of them, the broad idea of which was to find suitable classes to replace the various subject, predicate, and middle terms in such a way that the premises would be necessarily true and the conclusion necessarily false if the proposed rejection-rules were denied. In other words, the procedure was to disprove by counter-example the contradictory of the rejection-rule. Popper then proceeded to find a proof for each of the independent rules, attaching the

---

[2] On the question of whether they can be reduced to two, see J. N. Keynes, *A Treatise on Formal Logic*, London, 1906, p. 292.

logical conditions to which they are subject, and prefixing appropriate definitions of key terms (e.g., "distribution"). For instance, he found it necessary to postulate a universe of discourse with classes containing at least one member—not, from a classical point of view, a very serious assumption to make. Thus he gave a general type of method of proof and rounded off the subject completely. Before sketching his proofs I may first remind the reader of the way in which the several rules can be reduced to three independent ones, with incidentally a shortcut or two that are not commonly mentioned in textbooks.

The three corollaries, as they are called, are rules of quantity, and proofs are available in any textbook. Given the first of these—that from two particular premises nothing follows—I wish only to mention that one of the corollaries—that if in a valid syllogism one premise is particular the conclusion is particular—can be proved very simply by the Aristotelian principle of indirect reduction, according to which one premise combined with the contradictory of the conclusion yields the contradictory of the other premise. For, supposing the conclusion is universal, then its contradictory is particular, which in combination with the particular premise constitutes a pair of particulars—from which no conclusion follows. Textbooks usually give a less simple proof, although J. N. Keynes drew attention to this "ingenious" method which he ascribes to de Morgan.

Turning to the two rules of quality: given that from two negative premises nothing follows, the same method can be used to show that if one premise in a valid syllogism is negative the conclusion is negative. For, suppose the conclusion is affirmative, then the negative premise in combination with the contradictory of the affirmative conclusion (which must be negative) constitutes a pair of negative premises—from which no conclusion follows. J. N. Keynes ascribes this also to de Morgan.

Again, concerning the two rules of distribution: given that if a term is distributed in the conclusion of a valid syllogism it must be distributed in the premise containing it, we can show that in a valid syllogism the middle term is distributed at least once. For, if it is undistributed twice, indirect reduction places the contradictory of one of these as the new conclusion, and contradiction alters the distribution of all terms, so that what was the undistributed middle term appears as a distributed term in the new conclusion—which is invalid. J. N. Keynes ascribes this to W. E. Johnson.

We are thus left with the following three rules requiring proof:

1. If in a valid syllogism a term is distributed in the conclusion it is distributed in the premise containing it.

2. From two negative premises no conclusion follows.

3. If in a valid syllogism the conclusion is negative, one premise is negative.

Popper's way of arriving at appropriate counter-examples (after defining distribution) is to consider up to three (a) non-empty and (b) non-overlapping classes: X, Y, Z.

Let us deal with rule 3. Suppose, if possible, both premises are affirmative. The conclusion is false, in fact self-contradictory, for the substitutions $S = X, P = X$. (X is chosen to be non-empty to ensure the falsity of a *particular* negative proposition.) The substitution $M = X$ renders the premises true, in fact tautologies (again X is chosen to be non-empty to ensure the truth of a particular affirmative proposition). Thus we have a counter-example to our supposition: hence, one premise is negative.

The methods of providing counter-examples for rules 1 and 2 are similar (and require more than the one class X). Rule 2 is established by considering two cases, one where the conclusion is affirmative, and one where it is negative. Rule 1 is somewhat less simple, and hinges on forming an appropriate sum of two classes.

It also occurred to me that one could construct and prove similar rules for the validity of syllogisms whose terms might be empty classes. They consist of one rule of quantity and two of distribution:

I. In a valid syllogism, if the conclusion is particular, then one of the premises is particular.

II. If the middle term is distributed twice, no conclusion follows.

III. In a valid syllogism, if a term is undistributed in the conclusion, then it is undistributed in the premise containing it.

Popper's way of proving Rule 1 is as follows. Suppose, if possible, that both premises are universal. Now the substitutions $P = XY, M = XY$, $S = XY$ render the universal premises true; for since X and Y do not overlap, XY is empty. So that universal propositions with XY as subject and as predicate, whether affirmative or negative, are tautologies. But they render the conclusion false, for a particular proposition with an empty class as subject is self-contradictory.

Rule II is easily derivable from this by the ordinary methods the textbooks use for proving the "corollaries" and Rule III by indirect reduction.

These rules exclude nine classically valid moods with weakened conclusions or strengthened premises.

No doubt all this is not of overwhelming importance. But it is of some interest to see that rules left unproved for many centuries should be susceptible of proof in the end, and to see an illustration of the unsuspected power of the counter-example method of proof.

Lucasiewicz gives two "axioms of rejection" which suffice to exclude the classically invalid moods. They are not of the same sort as the traditional ones, and I have not entered into the relationship between the two sets.

IV

In Popper's handling of the two fields, there is a specially interesting connection between logic and scientific method at one or two places.

It is universally recognized that no hypothesis in empirical science can be completely certain beyond the possibility of refutation. This is part and parcel of the outlook of a scientist. It is interesting to find, however, that this point can be established from logic. From theorems showing the equivalence of tautologies and demonstrable statements it follows that only tautologies are in the strictest sense provable. From this it follows that no statement in empirical science is demonstrable.

V

Returning to the idea of the counter-example, let us enquire into the bearing this has on scientific procedure. According to Popper, an empirical hypothesis is not verifiable but is falsifiable, and this is because falsity is transmitted from the conclusion to at least one of the premises but truth is not. Hence, to verify the conclusion of a scientific inference contributes in no way to our knowing whether the premises are true or not. In other words all we can seek, where a scientific theory is concerned, is a counter-example.

There is an interesting parallel between Popper's notion of a counter-example and his theory of refutability in science, because this theory of refutability hinges basically on the fact that an invalid inference can be characterized directly, whereas a valid one cannot. An invalid one can be characterized either by its having true premises and a false conclusion or by its susceptibility to this arrangement if appropriate substitutions are made. Thus, at the outset of investigating validity, what can be found is a theory of invalidity. Now in formal logic this has the immediate consequence that an inference is valid if it is not invalid, so it is possible to step straight away from the theory of invalidity to a theory of validity. Nonetheless, the point I am making is that the crucial step is the negative one, the critical one, of attempting to show an inference to be invalid; and it is this same general idea that has underlain Popper's approach to science, which, as is well known, consists in the attempt to be critical in general and in particular to falsify hypotheses.

VI

From the interpretation of logic either as a set of tautologies or as a set of rules of inference it follows that logic lacks content, or is in a certain sense trivial. But Popper has taken the trivialization of logic some distance, and this in two ways.

It will be remembered that in *Principia Mathematica* Russell was able to found a logic on five primitive ideas and five primitive propositions, and it will be remembered further that Scheffer reduced the five primitive ideas to one and Nicod reduced the five primitive propositions to one. This is an improvement in elegance and economy of assumption. Popper likewise retained one primitive idea, namely the idea of deducibility, which is not very different from Scheffer's primitive idea, which was incompatibility; though Popper's is, intuitively at any rate, more central. But Popper takes a further step of reducing the number of primitive propositions from one to nought, which is a most elegant achievement.

It is for this reason that one of his papers is entitled "Logic without Assumptions."[3] At first sight this must seem strange and unplausible because one would surely have to start from some assumption or axiom, but on the other hand if we take seriously the idea that logic is essentially empty or trivial, then intuitively it ought to contain no undemonstrable assumptions.

This has very interesting overtones in connection with one of the great metaphysical problems of the past.

It will be recalled that Descartes ushered in one great stream of philosophy in terms of innate ideas, and the procedure both of himself and of the whole intellectualist school was to attempt to derive conclusions of a metaphysical kind by rational deduction from clear and distinct ideas. To put this story extremely briefly, their view was that knowledge is essentially analytical. This reached the height of absurdity when Wolff, successor of Leibniz, attempted to show that the number of windows in your room could be proved analytically. It has generally been regarded as a defect, a logical defect, of the Cartesian procedure that it had to start, and inevitably had to start, from an unproved assumption. This might be some clear and distinct idea of the self, or it might be the *causa sui* of Spinoza, but whatever it was, if it could be proved it could only be proved by deducing it from prior premises. Hence, to demonstrate anything philosophical there would have to be an assumption at some point. This defect Popper has obviously shown not to be so serious as it seems, because it is possible to have a logical system without a demonstrable starting point. Such a conclusion does not, of course, save the Cartesian or intellectualist philosophies. It is one thing to show that a logical system of deduction may start without a demonstrated premise, but Popper's conclusion would show that no significant conclusion can be derived, only an empty or trivial one. So in this sense the ancient criticism would survive; and if a

---

[3] K. R. Popper, "Logic without Assumptions," *Proc. Arist. Soc.*, 1947, 47.

conclusion of any significance is to be attempted, then an assumption would have to be inserted in some premise.

VII

In connection with this history, a much greater conclusion may be brought out. We have just seen that the outcome of the intellectualist school was the impossibility of deriving conclusions about the world by purely logical means. This made a deep impression upon Kant. On the other hand Kant was equally well aware that the alternative method of procedure—the only alternative known at that time, that is, to build up knowledge upon observations of facts—was equally bankrupt, because, as Hume showed, skepticism was the only possible conclusion to be drawn upon this basis. This is one of the most dramatic positions in the whole history of philosophy. Two alternative methods, which appear to be mutually exclusive and exhaustive, pointed to the conclusion that knowledge was either vacuous or unattainable. Now Kant, faced with this, sought a way out because he believed that Euclidean geometry was absolutely true and also believed the same of Newtonian mechanics. His problem then was, how could these things be absolutely true in view of the historical impasse which really did seem to show that knowledge was impossible? Hence Kant's problem—how is knowledge possible?—and his tremendous if strange and untenable attempt to solve it.

What I wish to draw attention to is this. Popper's methodological theory about scientific procedure, his theory about scientific theories, is a theory about the nature of knowledge which in fact constitutes an answer to the Kantian dilemma.

Popper's theory of knowledge, as is widely known, is very simply that all we ever have at the very best is a set of hypotheses; however well authenticated, they are never more than that. The best-established piece of knowledge may have to be modified or even given up. We may retain it only so long as our attempts to test it have failed to refute it. Now what is salient about this in the present context is that it is a conception of knowledge that does not include the ingredient of certainty at all. So long as we are seeking knowledge that is certain, we shall try no doubt to attain it either by analytical or empirical means or by some desperate measure like the Kantian *a priori* synthetic proposition. But once we give up this conception of knowledge altogether and are content with a conception of knowledge that contains no ingredient of certainty, then we are no longer tempted by the idea of a tautological system with unproved assumptions, which has been one of the unattainables of philosophers; nor shall we attempt to build up certain knowledge brick by brick upon observations. Thus the Kantian problem is solved by being shown not to arise in the form in which it presented itself to Kant. And Kant's belief that geometry and Newtonian mechanics are certain comes to be replaced by something different. Geometry taken as a piece of pure mathematics becomes tautological and says nothing about the world; taken as a piece of physics it is open to correction and so is Newtonian mechanics. But this is harmless so long as we are satisfied with the Popperian conception of knowledge that always remains hypothetical. I would add that I do not think this achievement of Popper's has been widely grasped. I would maintain that the problem of knowledge and skepticism is one of the most central of all philosophical problems of all ages, including the present. Indeed it seems to me probable that it is the *central* problem. It comes up acutely in the form in which it presented itself to Kant and it comes up in a great deal of contemporary philosophical writing, where philosophers are concerned at being defeated in the search for knowledge by the challenge of a skeptic. Popper's solution disposes of this problem.

Personally I think that, imposing though it is to have solved one of the great philosophical problems, there is a more important consequence of this solution for scientific research. So long as the idea lurks in philosophers' minds that some form of certainty may be attained, it is likely that some scientists at least may become imbued with the same idea and this must be fatal to scientific progress, or at least to the contributions that those scientists might have made. It is therefore of living value, as opposed to disposing of a dead problem, that living scientists are in a position to realize that certainty of knowledge is not only unattainable but is not needed, and that their most imaginative and venturesome researches depend upon a different conception of knowledge.

VIII

I wish to take this opportunity of acknowledging my debt to Karl Popper as a philosopher and as a friend. I have learnt a great amount from him, which I rate most highly. Yet I do not allot this top place. I have found with him a freedom of mind, which allowed me to range over ideas unhampered by the prevailing philosophical outlook. This freedom was not new to me, as I had known it from my first teacher, Macran, but it was tending to become inhibited by the insidious power of logical analysis and positivism. There must be many who have grown up under these influences and have never known this sense of freedom—which some might think somewhat necessary to a philosopher. Popper restored it for me. As a friend over many years he has shown a humanity and a warmth that are not customarily associated with the profession.

II

THE CRITICAL APPROACH TO LOGIC AND MATHEMATICS

CHAPTER 9

The Elimination of Variables by Regular Combinators[1]

BY HASKELL B. CURRY

Introduction

IN A RECENT PAPER[2] Quine has commented on the nature of variables and the possibility of eliminating them from logic. Such an elimination of variables was achieved by Schönfinkel[3] and has formed an essential feature of what has since been called combinatory logic.[4] But Quine finds the Schönfinkel theory unacceptable. He makes, in effect, two criticisms. In the first place he says that the Schönfinkel operators "operate on themselves and one another, whereas our [i.e., Quine's] six operate only on the original predicates and the predicates thence derived by the operators." In the second place he says that Schönfinkel operators "presuppose an abstract universe equivalent to that of higher set theory, whereas ours make no ontological demands. . . ."

The purpose of this paper is to present some considerations concerning combinatory logic which this paper by Quine has suggested. These considerations are partly opposed to the criticism by Quine, partly not. Thus,

---

[1] This paper was written in connection with a research project supported, in part, by a grant from the National Science Foundation.

[2] Quine [VEA]. (For explanation of the bracketed citations, see the Bibliography following this paper. When no author's name is given the reference to a work by Curry, or Curry and Feys, is understood. Ed.).

[3] Schönfinkel [BML]. Quine cites this in a footnote on p. 346.

[4] For a full account and references see [CLg].

in regard to Quine's second criticism : interpreted in a certain way,[5] I disagree with it; and I think it is necessary to begin by explaining, in §II, the reasons for this disagreement. But Quine's first criticism is quite another matter. There is indeed some interest in developing that part of the theory of combinators which has the sort of interpretation which Quine envisages. We shall see here that this can be done within the present framework of combinatory logic. This leads to a study of what have been called regular combinators. A part of this paper may thus be regarded as developing Quine's suggestions further in the same direction.

II.   Ontology of Combinatory Logic

We shall first turn to the question of the ontology of combinatory logic, because it is necessary to be clear about it in view of the constant reference to ontological questions in the work of Quine and philosophers of similar persuasion. The discussion will refer to combinatory logic as it is presented, for example, in [CLg], Chapter 6. The argument will not involve anything which has not already been said elsewhere; but in the present context it is more important to be clear about it than to avoid repetition.

Combinatory logic deals with a certain domain of objects. In order to be completely noncommittal about the nature of these objects, I have called them *obs*. These obs form an inductive class;[6] they are generated from certain primitive obs, called atoms, by a binary operation called *application*. This means that every construction starting with the atoms and using the application operation produces an ob; that every ob has such a construction; and that obs constructed in different ways are not identified, but are considered distinct as obs.

In [CLg] a particular symbolism for obs is adopted. The conventions for the system are, in fact, stated essentially as follows : The atoms consist of three atomic combinators, **I**, **K**, **S**, together, possibly, with additional atoms used in forming extensions. Among these last are certain indeterminates[7] e_1, e_2, ... If X and Y are obs, the ob formed from X and Y

---

[5] Of course it is not clear in what way Quine meant his "presupposes an abstract universe," and the interpretation discussed in §II may not be that which he intended. However, it is one possible interpretation which needs to be disposed of. Other possibilities are mentioned incidentally in §VII.

[6] This term was introduced in Kleene [IMM], p. 258. For a discussion see [FML], §2A.

[7] For discussion of the nature of indeterminates see [CLg], §2C. The need for additional atoms which are not indeterminates arises in illative combinatory logic and other extensions of the underlying theory; the illative obs mentioned in §VI are examples.

in that order—called "the application of X to Y"—is (XY). The obs formed by this operation from **I**, **K**, **S** are called *combinators*; obs formed from **I**, **K**, **S** and additional atoms are called *combinatory* obs. More generally, if $\mathfrak{A}$ is any class of atoms, the combinatory obs formed by application from atoms of $\mathfrak{A}$ are called combinations of $\mathfrak{A}$. It is further specified that parentheses may be omitted according to the rule of association to the left; and that **B**, **C**, **W**, . . . etc. are certain specific combinators.

For the present purposes it is important to realize that the conventions just made do not say that '**I**,' '**K**,' '**S**' are atoms, or that the application of 'X' to 'Y' is '(XY).' To understand them this way would be a grave error. According to standard conventions in regard to single quotes an expression in single quotes is a name for that expression. Thus '**I**,' '**K**,' '**S**,' 'X,' 'Y' are capital letters, and '(XY)' is a sequence formed by writing the capital italic forms of the twenty-fourth and twenty-fifth letters of the alphabet inside a pair of parentheses. These various symbols are not the objects being talked about, but they are symbols which we use in a certain technical way. In fact the conventions must be understood as saying that we are going to use '**I**,' '**K**,' '**S**' as names of atomic combinators, 'e_1,' 'e_2,' . . . , as names of indeterminates, and perhaps other letters as names of additional atoms. Having given the names of two combinatory obs, the name of the application of the first of them to the second is to be formed by writing a left parenthesis, then the name of the first ob, then that of the second ob, then a right parenthesis. Letters such as 'X,' 'Y,' 'Z,' 'U,' 'V,' 'M' are used as pronouns for designating unspecified obs; whereas 'x,' 'y,' 'z,' etc., are to be for unspecified indeterminates. Such conventions are made in the description of almost any mathematical system; and although a certain symbolism is described, that symbolism is to be used, not talked about.

Thus the conventions specify how the obs are to be named. What, then, are the obs themselves? All that has been said about them is that they form an inductive class—that they are formed from certain initial ones by a binary operation, and that obs formed in different ways are to be regarded as distinct. Nothing more than this is said, because nothing more is relevant. One may, if one likes, go on and specify unique objects for the obs; and any way of doing this such that the above conditions for uniqueness are fulfilled is called a *representation*. But everything which is said about the system as such will be unaffected by the choice of representation, or by a change of representation once chosen.

In much of current logic it is fashionable to represent systems in the expressions (or words) of an "object language," here called the *O-language*. For combinatory logic this means that the obs are taken as finite sequences (*O-words*) of letters (the *O-letters*) chosen, in any order and multiplicity, from a fixed list (the *O-alphabet*). Simply for the sake of conforming to this fashion, I shall exhibit such a representation, choosing arbitrarily one among the multitudes possible.[8] Let the O-alphabet consist of the two letters '*' and 'c.' Let the atoms be the words

$$\text{'*}cc\text{,' '*}c\text{*}cc\text{,' '*}c\text{*}c\text{*}cc\text{,' '*}c\text{*}c\text{*}c\text{*}cc\text{,' } \ldots$$

If X and Y are obs (O-words), let XY be the O-word formed by writing first '*,' then X (not 'X'!), then Y. Such a representation satisfies all requirements.[9]

It is worth noting, in passing, some properties of this representation. Not all O-words are obs: for instance, '*,' 'c,' '*c*,' 'c**c' are not. The atoms form a sequence A_n such that A_1 is '*cc' and A_{n+1} is formed by writing '*c' before A_n. We can identify **I**, **K**, **S**, respectively with A_1, A_3, A_5, and e_k with A_{2k}; then the A_{2k+1}, for $k>2$, can be reserved for additional atoms. Composite obs begin with at least two stars; these indicate a unique construction in accordance with the principles of a Łukasiewicz notation.[10] Thus **KI** is '**c*c*cc*cc,' while **S(KS)K** is the O-word

$$\text{'***}c\text{*}c\text{*}c\text{*}c\text{*}cc\text{**}c\text{*}c\text{*}cc\text{*}c\text{*}c\text{*}c\text{*}c\text{*}cc\text{*}c\text{*}c\text{*}cc\text{'}$$

Now what sort of ontology is implied by all this? Clearly only that which is presupposed in any kind of system based on concatenation. We do suppose that there are such things as letters and words, and that these may be identified in their various occurrences; thus among the following three inscriptions

$$\text{**}c\text{*}c\text{*}cc\text{*}cc \qquad \text{**}cc\text{*}c\text{*}c\text{*}cc \qquad \text{**}c\text{*}c\text{*}cc\text{*}cc$$

we can recognize that the first and third are instances of the same word (viz., **KI**) whereas the middle one is a different word (viz., **IK**). For certain higher epitheoretical purposes we make the further assumption that we can ignore the limitations of time and space; there are obs whose names are so long that it is impossible to write them down. Thus we are admitting the first two of the three stages of abstraction described by Markov and his associates.[11] But the third stage, where we treat entities involving infinitely many steps as existents, does not enter into combinatory logic at all, at least in its foundations. This third stage of abstraction characterizes nonconstructive mathematics, in particular higher set theory; whereas the first two stages are admitted in all forms of constructive mathematics. Thus

---

[8] The representation is taken from [DTC]. The basic ideas are due to Chwistek and Schönfinkel.

[9] This representation uses an O-language with only two letters in its alphabet. Alternatively one could use an O-language with only one letter. In that case the words are essentially numbers, and we have a "Gödel representation." (Cf. [FML], §2C4.) The idea of taking the Gödel numbers as the objects of a formal system occurs in recent work by Feferman, in particular his [AMG]. Cf. also Mostowski [SUF], p. 26.

[10] See, for example, [CFS], §4.

[11] For this my authority is Shanin [LPA]. The first two abstractions are mentioned in Markov [TA1] §§7,11.

it can hardly be said that combinatory logic presupposes an abstract universe equivalent to that of higher set theory.

So far we have been considering only uninterpreted combinatory logic. But combinatory logic is motivated by an interpretation. In fact the application operation is so called because it formalizes the operation of applying a function f to an argument-value a to form the function value fa (in ordinary mathematics $f(a)$). If f is an n-place function we can interpret fa as the $(n\text{-}1)$-place function formed by fixing on a as the value of the first argument.[12] Clearly this gives an interpretation for XY only when X is a function and Y is in the range of its first argument. The fact that there are thus obs which have no interpretation is not an objection to combinatory logic, any more than it is in other mathematical and physical systems where an analogous situation occurs. The interpretable obs can be picked out by additional considerations, such as those of the theory of functionality. Since combinatory logic does not make any assumption as to the existence of interpretations for its obs, it implies no ontological commitment beyond that already mentioned.

Be that as it may, there is certainly interest in developing partial systems in which only combinations which are interpretable from some standpoint or other are admitted. The system proposed by Quine is such that the "operators"—which are partly combinators, and partly combinations of combinators and other logical notions—are applied directly to predicates. That is, in this system XY is only defined when X is such an operator and Y is interpreted as predicate with a fixed domain (as in first order predicate calculus) and with a finite number of arguments ranging over that domain, this number being supposed known when Y is given. In this paper, without accepting the ontological argument which Quine advances in favor of his system, I shall deal with some of the formal consequences of his approach. I aim to show its relation to previous work on combinatory logic, and to suggest ways in which the approach can be modified and extended.

III. Regular Combinators

Before we proceed to the discussion of this theme it will be convenient to review some facts and conventions concerning combinatory logic as formulated in Curry and Feys [CLg].

Let $\mathfrak{H}(\mathfrak{A})$ be a system of combinatory logic based on a certain set $\mathfrak{A}$ of atoms, including the combinatory atoms and perhaps others, and let $x_1, x_2, \ldots, x_n$ be indeterminates which are not in $\mathfrak{A}$. Then $\mathfrak{H}(\mathfrak{A}; x_1, \ldots, x_n)$, or more briefly $\mathfrak{H}(\mathfrak{A};x)$ shall be the system formed by ad-

---

[12] This idea is due to Schönfinkel [BML].

joining $x_1, \ldots, x_n$ to $\mathfrak{H}(\mathfrak{A})$ as additional atoms. Any ob $\mathfrak{M}$ of $\mathfrak{H}(\mathfrak{A};x)$ can then be thought of as a function over $\mathfrak{H}(\mathfrak{A})$, viz., that one whose value for arguments $X_1, \ldots, X_n$ is the ob $\mathfrak{M}^*$ obtained by substituting[13] $X_1, \ldots X_n$ respectively for $x_1, \ldots x_n$ in $\mathfrak{M}$. Now it is a thesis of combinatory logic that this function can be represented by an ob M of $\mathfrak{H}(\mathfrak{A})$ itself, in the sense that the equation

$$Mx_1 \ldots x_n = \mathfrak{M} \tag{1}$$

can be derived, using the usual properties of equality (viz., reflexivity, symmetry, transitivity, and the replacement property), from the equation schemes

$$\mathsf{I}X = X \tag{2}$$

$$\mathsf{K}XY = X \tag{3}$$

$$\mathsf{S}XYZ = XZ\,(YZ) \tag{4}$$

which are assumed to hold for all obs X, Y, Z. The ob M is a combination formed from $\mathsf{I}, \mathsf{K}, \mathsf{S}$ and the atoms, other than $x_1, \ldots, x_n$ which occur in $\mathfrak{M}$.[14] However M is not, in general, uniquely defined by these conditions; and one of the problems of the theory of combinators was to find additional axioms, called the combinatory axioms, such that if there were two obs M satisfying (1) they could be shown to be equal within $\mathfrak{H}(\mathfrak{A})$ itself. This property I shall call the *extensionality property*.

In case $\mathfrak{M}$ is constructed from $x_1, \ldots, x_n$ alone, without using any other atoms, it is called a pure combination (of $x_1, \ldots, x_n$). In that case M will be a combinator. Combinators M of that kind are known as *proper combinators*.

If the proper combinator is such that the first indeterminate appears isolated in the initial position of $\mathfrak{M}$, then M is called a *regular combinator*. In that case it is convenient to use 'f,' rather 'x_1,' for the first indeterminate, and to write (1) in the form:

$$Mfx_1x_2 \ldots x_m = f\mathfrak{m}_1\mathfrak{m}_2 \ldots \mathfrak{m}_n \tag{5}$$

where $\mathfrak{m}_1, \ldots \mathfrak{m}_n$ are pure combinations of $x_1 \ldots x_m$. If a regular combinator M satisfies an equation of the form (5) it will be said to have the *order m and degree n*.[15]

---

[13] This substitution is meant in the sense that whenever we use x_i in the construction of $\mathfrak{M}$, we use X_i in the homologous place in the construction of $\mathfrak{M}^*$. Cf. [CLg], §2C2.

[14] This is meant in the sense that we are only interested in such M's. Of course with K present it may be possible to have an M satisfying (1) for which extraneous atoms are cancelled by (3). In the extensionality property, presently to be considered, there is no restriction on M.

[15] This definition of order and degree is suitable for regular combinators. For combinators in general a slightly different definition is made in [CLg] §5C3; that definition would give the M of (5) the order $m + 1$.

Evidently a regular combinator can be thought of as an operator transforming a function f into another function; if f is an n-place function and M is of order m and degree n, Mf will be an m-place function whose value for m given arguments will be the same as the value of f for certain combinations of those arguments. A regular combinator thus satisfies part of the condition imposed by Quine. Furthermore the transformations on predicates f allowed by Quine are all special cases of transformations which can be accomplished by regular combinators. It is therefore expedient to investigate the formation of regular combinators as independent entities without having to have reference to the theory of combinators in general.

For this purpose there is need of (i) certain primitive regular combinators, and (ii) certain operations for combining regular combinators to form other regular combinators. Reserving until later the discussion of primitives, we turn here to the discussion of the operations. Evidently the application operation is not relevant here; for if X and Y are regular combinators, XY is not necessarily a regular combinator; for example, **K** and **I** are regular, but **KI** is not. But there are two operations which come to mind as suitable for forming regular combinators from regular combinators: a binary one, called composition or composite multiplication, and symbolized by an infixed centered dot; and a unary one, called deferment, and symbolized by a prefixed 'β.' The interpretation of these is to be such that the following :

$$(X \cdot Y)f = X(Yf) \tag{6}$$

$$\beta Xfa = X(fa) \tag{7}$$

hold where X and Y are regular combinators, and f, a are suitable operands. In the full theory of combinators $X \cdot Y$ is **B**XY and βX is **B**X.[16] From the meanings of these operations it is natural to assume that the composite product is associative and distributive with respect to β; thus, for all regular combinators X, Y, Z :

$$X \cdot (Y \cdot Z) = (X \cdot Y) \cdot Z \tag{8}$$

$$\beta(X \cdot Y) = \beta X \cdot \beta Y \tag{9}$$

We shall understand X^k to be the product of k factors each identical to X, and β^k as the operation of iterating β k times.

Among regular combinators one can distinguish several species. If such a combinator satisfies (5) we may regard it as transforming a left sequence $x_1 x_2 \ldots x_m$ into the right sequence $\mathfrak{m}_1 \mathfrak{m}_2 \ldots \mathfrak{m}_n$. If the right sequence differs from the left only in that certain parentheses have been inserted,

---

[16] Cf. [CLg] §5D.

then M is called a (regular)[17] *compositor*. On the other hand if each $\mathfrak{m}_i$ is the same as some x_j, so that there are no parentheses (assuming association to the left) on the right, then M is called a *variator*. The variators may be further subdivided as follows : If the right sequence is a permutation of the left, M is called a *permutator*; if the right sequence is the same as the left except that certain x_i are repeated without change of order, M is called a *duplicator*; if the right sequence differs from the left only in that certain x_i are omitted, M is called a *cancellator*; and if the right sequence is identical to the left, M is called an *identificator*.

In the theory of [GKL][18] it is shown that, given $\mathfrak{M}$ having the form of the right side of (5), there is a corresponding M of the form

$$\mathfrak{K} \cdot \mathfrak{W} \cdot \mathfrak{G} \cdot \mathfrak{B} \tag{10}$$

where $\mathfrak{K}, \mathfrak{W}, \mathfrak{G}, \mathfrak{B}$ are respectively a cancellator, a duplicator, a permutator, and a compositor; if one imposes certain conditions on each of these four components, they are uniquely determined. A combinator satisfying these conditions is called *normal*. The extensionality property is established by showing that every combinator satisfying (5) can be shown to be equal to the normal one. We seek to do the same thing for the independent theory of regular combinators.[19]

Evidently variators can be thought of as transformations of a single predicate or function which takes the place of f. In other words, if we have two categories of obs, J (individuals) and H (propositions),[20] and if the $x_1, \ldots , x_m$ are thought of as unspecified elements of J, whereas f is thought of as a function from J^m (the set of m-tuples of J) to H, then a variator M satisfying (5) will be an operator transforming f into a function from J^m to H. But, if compositors are present, the indeterminate at the beginning of each subordinate parenthesis must represent a function; and if $\mathfrak{m}_i$ is such a subordinate parenthesis headed by, say, g, then f itself must

---

[17] Each of the kinds of combinators here defined is a subclass of the regular combinators, so that the word 'regular' may be used in connection with any of them for the sake of emphasis. In some cases irregular combinators with analogous properties can be defined, but that is not relevant here.

[18] For a summary see [CLg], §6S2.

[19] The reader will no doubt have noticed that there are two senses of 'regular combinator.' In the first sense a regular combinator is any combinator for which an equation of the form (5) holds in the general theory; in the second sense it is one obtained from the primitive regular combinators by a construction using only the stated operations. The second sense is that intended in the independent theory. Since it may be shown that the second sense is a specialization of the first, and that the two senses are equivalent in the general theory, it is hardly necessary to fuss about the distinction.

[20] Note that a proposition is thus a certain kind of ob (i.e., O-word). Whether these O-words are or are not sentences in some interpretation of the O-language is not relevant to the system as formal system. For this reason the word 'proposition' is preferred here to 'sentence.'

be a function whose ith argument ranges over the value domain of g. This more complex situation is reserved for §VI.

IV. Variators

In the situation of Quine (loc. cit.) the operators operate on one or more "predicates" to form another predicate, not necessarily with the same number of arguments. Here we can understand an n-place predicate as a function from J^n to H as in §III. There are two kinds of such operations. The first kind are simply combinators; the second kind are combinations of combinators with operations from ordinary logic, viz., negation, conjunction and existential quantification. In combinatory logic such operators are called "illative," and the branch of combinatory logic which has to do with them is called "illative combinatory logic." I shall postpone all illative considerations till §VI and devote attention to the combinators.

These combinators are variators. Since they convert a predicate to another predicate, a second operation can be applied to it and so on. Thus we may get combinations such as:

$$X_1(X_2(X_3 f))$$

If we define the composite product so that $(X \cdot Y)$ is an operation such that (6) holds, this can be expressed as:

$$(X_1 \cdot X_2 \cdot X_3)f$$

Thus we have to deal with regular variators generated by composite multiplication from the primitive ones. It is clear that (8) holds.

This situation was studied in some detail in [ALS]. The method there is essentially as follows: Let us introduce as new atoms combinators $\mathbf{C}$ and $\mathbf{W}$ satisfying the conditions (analogous to (2), (3), (4)):

$$\mathbf{C}XYZ = XZY \tag{11}$$

$$\mathbf{W}XY = XYY \tag{12}$$

Then let $\mathbf{C}_k$, $\mathbf{W}_k$, $\mathbf{K}_k$ be defined for each k as follows:

$$\mathbf{C}_1 = \mathbf{C} \qquad \mathbf{C}_{k+1} = \beta\mathbf{C}_k$$
$$\mathbf{W}_1 = \mathbf{W} \qquad \mathbf{W}_{k+1} = \beta\mathbf{W}_k$$
$$\mathbf{K}_1 = \mathbf{K} \qquad \mathbf{K}_{k+1} = \beta\mathbf{K}_k$$

If we take all these as primitives, it is easily shown (as in [GKL], §IIC5) that every permutator can be expressed as a product of the $\mathbf{C}_k$, every duplicator as a product of the $\mathbf{W}_k$, every cancellator as a product of the $\mathbf{K}_k$. Consequently every variator can be defined on the basis of $\mathbf{C}$, $\mathbf{W}$, $\mathbf{K}$ by use of the composite product and the β operation; or by use of the composite product alone on the basis of all the $\mathbf{C}_k$, $\mathbf{W}_k$, and $\mathbf{K}_k$. It is possible to eliminate the β operation if we introduce the combinators $\mathbf{C}_{[k]}$[21] defined by :

$$\mathbf{C}_{[1]} = \mathbf{C} \qquad \mathbf{C}_{[k+1]} = \beta\mathbf{C}_{[k]} \cdot \mathbf{C} \tag{13}$$

Then $\mathbf{C}_{[k]}$ represents the cyclic permutation which moves x_{k+1} to the first position. Since $\mathbf{C}_{[k]}$ has period $k+1$, its inverse $\mathbf{C}_{[k]}^{-1}$ is $\mathbf{C}_{[k]}^{k}$. Let X be a regular variator such that (5) holds. Then we have

$$\beta Xfzx_1 \ldots x_m = X(fz)x_1 \ldots x_m$$
$$= fz\mathfrak{m}_1\mathfrak{m}_2 \ldots \mathfrak{m}_n$$
$$= \mathbf{C}_{[n]}f\mathfrak{m}_1 \ldots \mathfrak{m}_nz$$
$$= (X \cdot \mathbf{C}_{[n]})fx_1 \ldots x_mz$$
$$= (\mathbf{C}_{[m]}^{-1} \cdot X \cdot \mathbf{C}_{[n]})fzx_1 \ldots x_m$$

Thus it would be appropriate to define :

$$\beta X = \mathbf{C}_{[m]}^{-1} \cdot X \cdot \mathbf{C}_{[n]} \tag{14}$$

We could then define all variators in terms of composite product with $\mathbf{C}_1$, $\mathbf{W}_1$, $\mathbf{K}_1$, and all $\mathbf{C}_{[n]}$ as primitives.

The theory of Quine has several peculiarities. In the first place there is no $\mathbf{K}$, and thus cancellations are not allowed. The same thing is true of [ALS], and probably for the same reason.[22] In the second place Quine uses $\mathbf{C}_{n-1}$ and $\mathbf{W}_{n-1}$ instead of $\mathbf{C}_1$ and $\mathbf{W}_1$. This has the disadvantage that one needs a separate operation for each n. Quine speaks of the whole group of $\mathbf{C}_n$ as one operation, and indeed, in view of the absence of cancellation, the value of n is uniquely determined as soon as f is known; but there is no single equation of form (5) for all of these operations independent of f. The fact that the effect of one of Quine's operations on a function f requires information about f, so that f is not strictly an indeterminate, is a disadvantage from the purely combinatory point of view.

V.   The Algebra of Regular Combinators

The question of axioms to ensure the extensionality property will be our next subject. Quine completely ignores this question; and in this

---

[21] This is the D_{k+1} of [ALS]. Cf. [CLg], §5E2, p. 171.

[22] The possibility of cancellation does not occur to one naturally from the present approach. It came from Schönfinkel directly; and I discovered Schönfinkel's [BML] in a literature search while [ALS] was in process. I was able to cite Schönfinkel, but not to incorporate his idea in the text. For the reluctance of Church to accept $\mathbf{K}$ or anything connected with it, cf. [CLg], §3S3.

respect he is on a par with Schönfinkel. In [ALS] a set of axiom schemes was given in terms of C_k and W_k (for all $k = 1, 2, \ldots$) as primitive regular combinators; and extensionality follows from these axioms, although certain details of the proof were left till [GKL] and [ATC]. In [GKL] and all later publications, including [CLg], extensionality is proved for combinators in general, and thus for regular combinators in particular. From the method of [GKL] (but not from that of [CLg]) it is easy to extract an axiom set and proof of extensionality which applies to regular combinators generated by the composite product and the β operation. This axiom system will be given here.

In order to provide for compositors, it is necessary to add to the primitive regular combinators C, W, K of §IV the combinator B mentioned in the discussion following (6) and (7). This B satisfies the scheme

$$B XYZ = X(YZ) \tag{15}$$

By omitting everything related to B in the following set, one gets a set suitable for variators; by omitting everything related to K also, one gets a set suitable for the variators without cancellation, which were considered by Quine as well as in [ALS].

In the following it is supposed that the usual rules for equality hold (viz., symmetry, transitivity, and replacement; reflexivity is a consequence of Ic). The statements under I and II are axiom schemes in which 'X,' 'Y,' and 'Z' designate unspecified regular combinators. The equations (2), (3), (6), (7), (11), (12), and (15) are not part of the formal algebra, but are to be used in interpreting the regular combinators as operations on functions.

AXIOMS FOR REGULAR COMBINATORS

I. General

    a. $X \cdot (Y \cdot Z) = (X \cdot Y) \cdot Z$

    b. $\beta(X \cdot Y) = \beta X \cdot \beta Y$

    c. $I \cdot X = X$

II. Commutative axioms—special cases of the principle that if X is of order m and degree n, then:

$$\beta^m Y \cdot X = X \cdot \beta^n Y \tag{16}$$

    a. $X \cdot I = I \cdot X$

    b. $\beta^2 X \cdot B = B \cdot \beta X$

    c. $\beta^2 X \cdot C = C \cdot \beta^2 X$

    d. $\beta X \cdot W = W \cdot \beta^2 X$

    e. $\beta X \cdot K = K \cdot X$

III. Permutative axioms—special cases of (14), but written so that the right side is closer to being normal.

a.[23] $$\beta I = I$$

b. $$B \cdot C = \beta C \cdot C \cdot \beta B$$

c. $$\beta C \cdot C \cdot \beta C = C \cdot \beta C \cdot C$$

d. $$C \cdot W = \beta W \cdot \beta C \cdot C$$

e. $$C \cdot K = \beta K$$

IV. Transmutative[24] axioms

a. $$B \cdot W = \beta W \cdot W \cdot \beta C \cdot \beta^2 B \cdot B$$

b. $$B \cdot K = K^2$$

c.[25] $$C \cdot C = \beta^2 I$$

d. $$W \cdot C = W$$

e. $$W \cdot \beta W = W^2$$

f.[26] $$W \cdot K = \beta I$$

The proof of the extensionality property for regular combinators can be carried out by making minor modifications of the proof in [GKL], §IID.[27] The principle of the method is that if one has a composite product having two adjacent factors which are not in the order prescribed for the normal form, one can replace the pair of factors by a group of factors which is more nearly in normal form. If the two factors act on non-overlapping segments in the sequence of arguments, the interchange can be

---

[23] By virtue of this axiom all identificators are equal to I. If one were to omit this axiom and adjoin

$$\beta^2 I \cdot B = B$$
$$\beta^2 I \cdot C = C$$
$$\beta I \cdot W = W$$
$$\beta I \cdot K = K$$

one would have a weak form of extensionality principle corresponding to the "β-sense" of equality in [CLg]. One would need to add an identificator to the normal form (10), and the β's on the right in IVc and IVf would be essential.

[24] All the axioms of II, III, IV have a permutative character. The term 'transmutative' was introduced in [GKL] to cover all cases of quasi interchange which were not covered by II. It thus included the analogues of III as well as IV.

[25] If IIIa is adopted one can have I on the right; but otherwise the β^2 may be essential. Cf. the footnote to IIIa.

[26] Cf. the preceding footnote.

[27] Some details are taken care of in [ATC]. For the situation described in the footnote to IIIa see Rosser [MLV].

made by one of the axioms II; in all other cases one of the other axioms can be applied.

VI.   Generalizations

We now consider operations which do not come under the foregoing treatment. This includes not only regular combinators, such as compositors, which operate on two or more functions, but also operators with an illative component.

Let us begin with the three illative operators which Quine admits. These are negation (Neg), derelativization (Der), and cross product (denoted by an infixed '×'). Using the signs '=,' '¬,' '∧,' for equality, negation, and conjunction in the usual sense, and assuming that f is an m-place predicate and g is an n-place predicate, we can state the meaning of these as follows:

$$(\text{Neg } f)x_1 \ldots x_m = \neg\, (fx_1 \ldots x_n)$$

$$(f \times g)x_1 \ldots x_m y_1 \ldots y_n = (fx_1 \ldots x_m) \wedge (gy_1 \ldots y_n)$$

$$(\text{Der } f)x_1 \ldots x_{m-1} = (\exists x_m)fx_1 x_2 \ldots x_m$$

Now the trouble with these operations is the same as that with respect to the $\mathbf{C}_n$, $\mathbf{C}_{[n]}$, etc.; viz., that the arguments f, g have to be classified as to number of arguments in advance, and the effect of the operations depends on that information. From a certain formal point of view there are infinitely many distinct operations involved. We can indicate these by subscripts, so that Neg_m is the negation operation on an m-place function (or predicate). It is desirable to exhibit these as combinations of a single illative notion and combinatory operations.

For the operations on a single functional argument this is easy. Let Γ and Σ be, respectively, the negation of a proposition and the existential quantification of a one-place predicate—in Quine's notation, as above modified, Neg_0 and Der_1, respectively. Then we can identify Neg_m and Der_{m+1} with $\beta^m\Gamma$ and $\beta^m\Sigma$ respectively.

For operations on two functional arguments the situation is more complex. Let Λ be the ordinary propositional conjunction. Then the conjunction of $fx_1 \ldots x_m$ and $gy_1 \ldots y_n$ will be:

$$\Lambda(fx_1 \ldots x_m) (gy_1 \ldots y_n)$$

Now in terms of combinators we have:

$$\Lambda\, (fx_1 \ldots x_m) (gy_1 \ldots y_n) = \beta^m \Lambda fx_1 \ldots x_m (gy_1 \ldots y_n)$$

$$= \mathbf{B}^n\, (\beta^m \Lambda fx_1 \ldots x_m)\, gy_1 \ldots y_n$$

$$= \beta^m \mathbf{B}^n\, (\beta^m \Lambda f)\, x_1 \ldots x_m gy_1 \ldots y_n$$

$$= (\mathbf{C}_{[m]}^{-1} \cdot \beta^m \mathbf{B}^n \cdot \beta^m \Lambda)\, fg x_1 \ldots x_m y_1 \ldots y_n$$

Hence if we set

$$T_{mn} \equiv \mathbf{C}_{[m]}^{-1} \cdot \beta^m \mathbf{B}^n$$

we have

$$(f \times g)_{m,n} = (T_{mn} \cdot \beta^m \Lambda)\, fg$$

An alternative method of procedure would be to use a family of combinators associated with the combinator $\mathbf{S}$ (see (4)).[28] In [Clg], §5E3 there was defined a combinator $\mathbf{S}^{[m]}$ such that

$$\mathbf{S}^{[m]} fg x_1 \ldots x_m = f x_1 \ldots x_m \, (g x_1 \ldots x_m)$$

In fact we have

$$\mathbf{S}^{[1]} = \mathbf{S} \qquad \mathbf{S}^{[m+1]} = \mathbf{S} \cdot \beta \mathbf{S}^{[m]}$$

Let U and V be two variators such that for any f

$$Uf x_1 \ldots x_m = f u_1 \ldots u_p$$

$$Vg x_1 \ldots x_m = g v_1 \ldots v_q$$

Then we should have

$$\Lambda(f u_1 \ldots u_p)\,(g v_1 \ldots v_q) = \mathbf{S}^{[m]}\,(\mathbf{B}^m \Lambda\,(Uf))\,(Vg) x_1 \ldots x_m$$

Thus we should have, reverting to the above,

$$(f \times g)_{m,n} = \mathbf{S}^{[m+n]}\,(\mathbf{B}^{m+n} \Lambda\,(\beta^m \mathbf{K}^n f))\,(\mathbf{K}^m g)$$

The first of these methods is essentially that suggested in [GKL], §IIE4, p. 833. The second is of more recent origin; it comes from studies of "formalizing combinators" beginning with [UQC] and extending up to the publication of [CLg]. It seems much more natural than the first; in fact there is more occasion for combining functions with some or all of their variables identified than with them totally distinct.

The second method suggests further that it would be expedient to add $\mathbf{S}$ rather than $\mathbf{B}$ to the list of primitive regular combinators. Then $\mathbf{B}$ would become definable, since one would conclude from the extensionality property that

$$\mathbf{B} = \mathbf{S} \cdot \mathbf{K}$$

Conversely, of course, $\mathbf{S}$ is definable in terms of $\mathbf{B}, \mathbf{C}, \mathbf{W}$ since

$$\mathbf{S} = \beta \mathbf{W} \cdot \mathbf{C} \cdot \beta \mathbf{B}$$

But if $\mathbf{S}$ is taken as primitive it is to be expected that there would be some more suitable definition of normal combinator, and some more convenient axiomatic characterization. What these are I do not know.

---

[28] Such combinators might be called distributors.

Again it is suggested that the illative operations, such as Γ, Σ, and Λ, be regarded, not as operators, but as functional arguments for operations. This means that we must envisage a more complex functional structure than was the case in §IV. Suppose we have a category J of obs which we call individuals, and another category H which we call propositions; then in §IV all argument functions were propositional functions of individuals, i.e., functions from J^n to H. But in the present situation we have propositional functions of propositions, and even, in the case of Σ, a propositional function of a predicate argument. But there seems no good reason why this would not be done. If it is done then we should replace $\beta^m\Gamma$, $\beta^m\Sigma$, $\beta^m\Lambda$ in the foregoing by $\mathbf{B}^m\Gamma$, $\mathbf{B}^m\Sigma$, $\mathbf{B}^m\Lambda$, respectively.

It is clear that the foregoing considerations can be applied to much more general situations. We could have conditions where there are more than two fundamental categories, and where there are functions from any cartesian product of categories to any other category, and that such functions form a new category. In such a situation the combinators are to be thought of as operators which combine obs in various categories to form a result in some other category in very complicated ways. In fact the situation becomes too complex to handle by the methods used here. One needs then the theory of functionality. This is a branch of combinatory logic in which there is an ob $\mathbf{F}$ which is such that if ξ and η represent (can be interpreted as) categories, $\mathbf{F}\xi\eta$ represents the category of functions f which, when applied to an argument in ξ, give a value in η. Then $\mathbf{F}\xi(\mathbf{F}\eta\zeta)$ will represent the category of functions g which, when applied successively to a first argument in ξ and a second argument in η, give a value in ζ, and so on. Thus if J and H are categories as above, the category of one-place predicates is $\mathbf{F}JH$, of two-place predicates is $\mathbf{F}J(\mathbf{F}JH)$, of three-place predicates is $\mathbf{F}J(\mathbf{F}J(\mathbf{F}JH))$, of a binary descriptive function $\mathbf{F}J(\mathbf{F}JJ)$, etc.; whereas Γ, Σ, and Λ belong respectively to the categories $\mathbf{F}HH$, $\mathbf{F}(\mathbf{F}JH)H$, and $\mathbf{F}H(\mathbf{F}HH)$. General rules for the categories of combinators can also be given; thus for all ξ, η, ζ, $\mathbf{I}$ is in $\mathbf{F}\xi\xi$ and $\mathbf{B}$ is in $\mathbf{F}(\mathbf{F}\eta\zeta)$ $(\mathbf{F}(\mathbf{F}\xi\eta)\ (\mathbf{F}\xi\zeta))$; and an axiomatic system is developed for deriving such.

In the presence of such a theory of functionality such restrictions as that to regular combinators seem unnatural. It is much better to admit the full theory of combinators, and then to say that we are not claiming the existence of a direct interpretation for obs which do not belong to our scheme of significant categories. Situations similar to this exist in modern physics.

VII. Concluding Remarks

It is evident that the theories described at the end of §VI can be used in the formulation of higher set theoretical formalisms of the dizziest sort.

But the fact that they can be used for that purpose does not mean that they presuppose an extra-fancy ontology in any reasonable sense. They can probably also be used in formulating situations where no ontological questions enter, as in the linguistic situations described in [LAG]. What sort of ontology will be suited to the intended interpretation of a given theory will depend on the assumptions made in the illative part of the theory—the conditions under which one can infer that Σ f holds, etc. Combinatory logic can be applied to ontological situations of various sorts, and, except in a very broad sense, it does not impose restrictions on the type of ontology which one can assume.

Bibliography

Works are cited by abbreviated titles in brackets as listed below. When these citations are made without author's name (explicitly or in the context) it is understood the author is Curry, or Curry and Feys. Journals are abbreviated according to the practice of Mathematical Reviews (Providence, R. I.).

CURRY, H. B.: [ALS] "An Analysis of Logical Substitution." *Amer. J. Math.* 51:363–384 (1929).

[ATC] "Some Additions to the Theory of Combinators." *Amer. J. Math.* 54:551–558 (1932).

[CFS] "Calculuses and Formal Systems." *Dialectica* 12 in (Festschrift in honor of P. Bernays; also published separately as "Logica; studia Paul Bernay's dedicata" by Editions de Griffon, LaNeauveville, Switzerland), 249–273 (1958).

[DTC] "The Deduction Theorem in the Combinatory Theory of Restricted Generality." *Logique et Analyse*, 3ᵉ Annee (1960), pp. 15–39.

[FML] *Foundations of Mathematical Logic* New York, 1963.

[GKL] "Grundlagen der kombinatorischen Logik." *Amer. J. Math.* 52: 509–536, 789–834 (1930).

[LAG] "Some Logical Aspects of Grammatical Structure." *Amer. Math. Soc. Proc. of Symposia on Applied Math.* 12 (The structure of language and its mathematical aspects): 56–68 (1961).

[PKR] "The Paradox of Kleene and Rosser." *Trans. Amer. Math. Soc.* 50:454–516 (1941).

[UQC] "The Universal Quantifier in Combinatory Logic." *Ann. Math.* (2) 32:154–180 (1931).

CURRY, H. B. and ROBERT FEYS: [CLg] *Combinatory Logic.* Amsterdam, 1958.

FEFERMAN, S. C.: [AMG] "Arithmetization of Metamathematics in a General Setting." *Fund. Math.* 49:35–92 (1960).

KLEENE, S. C.: [IMM] *Introduction to Metamathematics.* Amsterdam and Groningen, 1952.

[LDR] "λ-definability and Recursiveness." *Duke Math. J.* 2:340–353 (1936).

MARKOV, A. A.: [TA1] "Teoriya algorifmov (Theory of algorithms)." *Trudy Math. Inst. Steklov* 38:176–189 (1951). English translation by Edwin Hewitt, *Amer. Math. Soc. Transl.* 15:1–14 (1960).

MOSTOWSKI, ANDRZEJ: [SUF] *Sentences Undecidable in Formalized Arithmetic; An Exposition of the Method of Kurt Gödel.* Amsterdam, 1952.

QUINE, W. V.: [VEA] "Variables explained away." *Proc. Amer. Philos. Soc.* 104: 343–347 (1960).

ROSSER, J. B.: [MLV] "A mathematical logic without variables." *Ann. of Math.* (2) 36:127–150 (1935) and *Duke Math. J.* 1:328–355 (1935).

SCHÖNFINKEL, M.: [BMLg] "Über die Bausteine der mathematischen," *Logik. Math. Ann.* 92:305–316 (1924).

SHANIN, N. A.: [LPA] "O nekotorikh logicheskikh problemakh aritmetiki." (On some logical problems of arithmetic). *Trudy Mat. Inst. Steklov*, No. 43. Moscow, 1955.

CHAPTER 10

On Popper's Use of the Notion of Absolute Logical Probability

BY WILLIAM KNEALE

IT IS ONE of Popper's main theses in his *Logic of Scientific Discovery* that the degree of confirmation or corroboration attained by a scientific hypothesis should always be distinguished sharply from the probability discussed in mathematical textbooks, i.e. from probability in that sense in which we may be said to give the probability of a proposition when we state what chance there is of its being true. This thesis of Popper's, with its corollary that induction cannot be justified in the calculus of probabilities, seems to me to be not only correct but very important for an understanding of scientific method, and I remember that I was influenced especially by the argument of Popper's original *Logik der Forschung* when I suggested in my *Probability and Induction* of 1949 that it would be convenient sometimes to speak of the acceptability of scientific hypotheses rather than of their probability. But there are two features in Popper's presentation of his case which seem to me less fortunate.

The first is a purely verbal point. When Popper contrasts acceptability, or degree of corroboration, with probability, he does not always make clear that he refers only to probability in a technical sense in which it is a measure of chance. On the contrary, he goes out of his way sometimes to dwell on the apparent paradox of his assertion that a scientific hypothesis which is highly acceptable because it has survived rigorous testing may nevertheless be one of low probability. He writes for example in *L.S.D.* (p. 270): "If we compare these views of mine with what is implicit in (inductive) probability logic, we get a truly remarkable result. According to my view, the corroborability of a theory—and also the degree of corroboration of a theory which has in fact passed severe tests—stand both as it were in inverse ratio to its logical probability." The result is indeed worthy of remark, but some of the surprise which the reader feels when he first comes upon it may be due to a misunderstanding of the word 'probability.' In ordinary life, when we say that a proposition is probable, we mean primarily that it is one which a rational man would accept at least provisionally. Unless, therefore, we read Popper's work with great care and notice that he is using the word 'probability' in a technical sense, we may be led to suppose that he is maintaining the absurdity that acceptable hypotheses are not acceptable. In fact he means only to say that the propositions acceptable in science as explanations are not such as to leave open many logical alternatives.

The second point is of more importance, and it is this I wish to discuss. When Popper connects the high acceptability of a successful scientific hypothesis with its low absolute logical probability, he thinks that what he says marks a fundamental difference between laws of nature and laws of logic. For in his view, laws of logic, unlike supposed laws of nature, can be taken as certain precisely because they have maximum logical probability and null content, i.e., are not open to refutation by experience. This is essentially the doctrine of Wittgenstein in his *Tractatus Logico-Philoso-phicus,* but it is interesting to see that Popper has tried to retain it even after he has come to realize in his *Logic of Scientific Discovery* that laws of nature are not, as Wittgenstein thought, propositions of accidental universality. In *L.S.D.* (p. 438), when talking of natural necessity, Popper says: "Unlike Kneale, I regard 'necessary' as a mere word—as a label useful for distinguishing *the universality of laws* from 'accidental' universality. Of course any other label would do just as well, for there is not much connexion here with logical necessity. I largely agree with the spirit of Wittgenstein's paraphrase of Hume: 'A necessity for one thing to happen because another has happened does not exist. There is only logical necessity.' " Naturally I do not, for my part, want to deny that laws of logic have the maximum value in a scale of logical probability; nor yet that they, unlike suggestions of natural law, are safe from refutation by experience. But I think that Wittgenstein's influence has prevented many philosophers of science from appreciating the real complexity of their situation, and I want to maintain in particular that what Popper calls absolute logical probability is only one of a number of conceivable probability ratings, each of which might with equal justice be called absolute. When this has been realized, it no longer seems plausible to say, "There is only logical necessity," or to suggest that there is no more than a punster's link between logic and natural necessity.

In his *Tractatus* (5.15), Wittgenstein introduces the notion of probability by saying that the probability which one proposition gives to another is the ratio of the number of their common truth-grounds to the number of the truth-grounds of the first. Here, by a 'truth-ground' of a proposition is meant a distribution of truth-values to the truth-arguments of the proposition such as would verify the proposition. It follows, of course, that a tautology has the probability 1, and a self-contradiction the probability 0, in relation to any other proposition whatsoever, and it can at least be made to seem plausible that there should be elementary propositions which give to each other the probability 1/2. Wittgenstein does not speak of absolute probability, but within his scheme it is easy to introduce the notion by saying that the absolute probability of a proposition is the ratio of the number of its truth-grounds to the number of all the distinguishable distributions of truth-values to its elementary truth-arguments. Alternatively, since a tautology excludes nothing but self-contradiction, we may say that the absolute probability of a proposition is its probability in relation to any tautology. Popper sometimes adopts this course, as when for example in *L.S.D.* (p. 318), he writes the equation $p(x) = p(x,xx)$. Given these definitions, it is reasonable to say that the absolute probability of any supposed law of nature is less than any assignable fraction. For, as Popper argues in *L.S.D.* (p. 365), any such law may be assumed to entail an infinity of distinct nontautological propositions about the various individuals of the actual universe, and from this assumption it follows within the theory of probability under consideration that its probability must be infinitesimal.

As might be expected, Popper argues for his thesis without relying on the most dubious part of Wittgenstein's doctrine, namely, the assumption of logical atomism according to which each elementary proposition gives to each other elementary proposition the probability 1/2. But it is essential to his argument that the absolute probability which he contrasts with acceptability or degree of corroboration should be a notion of pure logic. For he agrees with Wittgenstein that necessity in a strict sense belongs only to the tautologies of logic, and against any view such as mine he wishes to maintain that natural laws, so far from being necessary in the same sense as tautologies, actually approximate to self-contradictions in a scale of falsifiability. That is why he tells us that every hypothesis of natural law has an absolute logical probability of 0 (*L.S.D.*, pp. 366, 372). According to his way of thinking, capacity for corroboration is the same as falsifiability, and falsifiability is at its maximum when absolute probability is at its minimum. It is incorrect, however, to assume that absolute logical probability, as he understands it, is either complementary or inverse to falsifiability.

In order to make this clear let us consider first the phenomenological law that nothing can look both red and green all over at the same time to the same person. This can be supposed to entail an infinity of distinct nontautological propositions about the various individuals of the universe, and by an argument exactly like that which Popper uses in relation to natural laws, its absolute logical probability can be shown to be infinitesimal, though in fact its truth is known *a priori*. Wittgenstein was rightly worried about phenomenological necessities and tried on two different occasions to exhibit them as logical, but without success. In his *Tractatus* (6.3751), he said that it was logically impossible for two colors to be at one place in the visual field, and then undertook to reveal the formal contradiction in any such supposition, but could do no better than talk irrelevantly about physics. In a paper called "Logical Form," which he contributed to the *Aristotelian Society Proceedings* (Supplementary Volume IX [1929]), he made a new start, but showed by his argument that he had in effect abandoned the notion of logic with which he worked in his *Tractatus*. During the past thirty years many of his followers have used the word 'logic' in a very wide sense which allows them to say without more ado that it is logically impossible for two colors to be at the same place in the visual field, but Popper shows due respect for the old tradition according to which logic is concerned with the form of propositions as distinct from their subject matter, and he would therefore, I think, not wish to maintain that within his scheme the proposition I have called a phenomenological law is to be regarded as a tautology. If, however, he takes the opposite course, it seems that he is committed to the curious thesis that it is falsifiable in the highest degree.

Admittedly there is some disagreement among the professionals about what is to be included in logic, but Popper's work on the definition of formative signs suggests that he takes a strict line and does not wish to include the theory of sets. If this is so, we can repeat our argument with new examples. For it is evident that according to the narrow definition of 'logic' any universal thesis of the theory of sets which is not valid by the rules of quantification theory (i.e., of predicate logic or lower functional calculus) must have an absolute logical probability less than any assignable fraction. Are we, then, to say that all such themes are in principle falsifiable? There has been a lot of debate about some of them, in particular about the axiom of choice, but I do not think that any mathematician would be satisfied by the suggestion that even the most contentious of them should be classified with empirically testable hypotheses of natural law. For the axiom of choice is satisfactory enough in relation to finite sets, and if it is to be disproved at all, that result will be achieved only by consideration of infinite sets, which are certainly not given in experience. If, on the other hand, we hesitate to go so far, and say instead that such theses are falsifiable only in the very weak sense of not being guaranteed by purely logical considerations, we reduce the importance of the notion of falsifiability and make it useless for Popper's general purposes.

In each of the cases we have just considered the absolute logical probability of a universal proposition is found to be vanishingly small, though the proposition does not appear to be falsifiable by experience. But we discover a situation that is even more curious if instead of confining our attention to logical possibilities we take for our universe of discourse the field of all well-formed statements. When we present a tautology like $\sim (P \ \& \sim P)$ as a truth of logic, we say in effect that certain propositions of this field are necessarily false, but in order to reject them as logically impossible we must first recognize them as possible objects of thought. In his *Tractatus* (3.02–3), Wittgenstein says "The thought contains the possibility of the state of affairs which it thinks. What is thinkable is also possible. We cannot think anything unlogical, for otherwise we should have to think unlogically." If this means only that we cannot without absurdity think of the logically impossible as possible, it is obviously true. On the other hand, it is obviously false if it means that no one can ever say sensibly $\sim (P \ \& \sim P)$, where the significance of the whole depends upon the significance of what comes after the main negation sign. And it is at least bad psychology, as we all know from unhappy experience, if it means that no one can even mistakenly think of the logically impossible as possible. Candidates in elementary logic examinations frequently make mistakes, and when it is said that they do so through failure to think, the word 'think' is used in a special way to mean what would ordinarily be meant by 'think logically.' Now I do not want to suggest that every possible object of thought is possible without further qualification. But just as something which is phenomenologically or mathematically impossible may be logically possible in the sense of not being excluded by the laws of truth-function theory or quantification theory, so what is logically impossible may be categorically or syntactically possible, i.e., not excluded by the laws of type theory or whatever it is that makes the distinction between propositional sense and gibberish. For my own part, I can suggest no plausible system of measurement for the field which contains all well-formed statements, i.e., none which appears as plausible as Wittgenstein's metric for the field of the logically possible. But it seems evident that if there were a metric of the wider field, it would by analogy allow for the construction of an argument to show that the sub-field left open by the laws of logic, i.e., by universal propositions of such forms as $(P) \sim (P \ \& \sim P)$, was an infinitesimally small part of the whole. Perhaps it is not worth discussing whether in that case we could usefully say that the laws of logic had an absolute syntactic probability of 0, but I see no reason to hold that such talk would be much odder than declaring all hypotheses of natural laws to have an absolute logical probability of 0.

Returning now to our previous examples, by parity of reasoning we may say that, if it makes sense to talk of measure in such connections, true mathematical generalizations have an absolute mathematical probability of 1 and true phenomenological generalizations an absolute phenomenological probability of 1, though both alike have an absolute logical probability less than any assignable fraction and *a fortiori* an absolute syntactical probability which is also infinitesimal. For a graphical representation of the situation let us suppose first that the universe of all statements which are syntactically permissible is symbolized by a circle. Then within this we can indicate the field of logically possible propositions by a smaller concentric circle, shading the rest of the area of the first circle with vertical lines to show that it stands for the field of the logically impossible. Obviously we cannot draw an inner circle whose area is less than any assignable fraction of the outer circle : but if we agree to overlook this defect of the scheme, we can go on to draw two still smaller concentric circles to represent respectively the field of the mathematically possible and the field

of the phenomenologically possible. In the accompanying figure I have used horizontal shading to indicate mathematical impossibility and skew shading to indicate phenomenological impossibility. Since there are, of course, contexts in which we should not admit as thinkable everything which is syntactically permissible but should be inclined to treat some stricter criterion of possibility as all-important for our purposes, we may in practice take any one of the circles in our figure as the representation of our propositional universe and then, if we have a sufficient motive, try to elaborate in the field it indicates a measure of absolute probability. Having chosen the field we propose to regard as our propositional universe, we shall, of course, ignore anything which lies outside it, and so long as our prevailing interest continues we shall be inclined to ignore also any difference of status between propositions which all alike fall within our propositional universe. Thus if we are interested primarily in logical possibilities, we may think of the universal propositions which together delimit the field of the phenomenologically possible as just a group of propositions that happen to be true. If this is our attitude, we shall say that measurements of sub-fields within the field they determine are only measurements of relative probability. But at another time we may take the field of phenomenological possibility as our propositional universe, and then we shall give a different account of absolute probability. For the philosopher it is important to recognize what happens when men change their standards for application of the word 'thinkable,' and to realize that among propositional universes the narrowest has a special claim to attention because there can be no reason but contingent fact for preferring any sub-field that lies within it.

Some of this may be acceptable to Popper. It is true that he has not distinguished between wider and narrower notions of the thinkable; but he may be willing to do so. For he is well aware that mathematics cannot be reduced to logic, if logic does not extend beyond quantification theory; and this is enough to illustrate my thesis that what is logically possible may nevertheless be unthinkable by some stricter standard. There still remains, however, the question whether natural laws, as he is now prepared to conceive them, may be supposed to determine a propositional field which is related to the field of the phenomenologically possible as that is related to the field of the mathematically possible and also as the field of the mathematically possible is related in its turn to the field of the logically possible. If my argument is correct, what I have shown so far is that the infinitesimal size of the absolute logical probability of supposed laws of nature is not a conclusive reason for assuming, as Popper does, that they cannot be laws of necessity in any strict sense. What I wish to show now is that his new conception of natural laws requires him to go on to draw the parallel I have suggested.

In order to distinguish hypotheses of natural law from merely accidental generalizations Popper now says that the former are to be conceived as holding for all worlds which differ from the actual world, if at all, only in initial conditions (*L.S.D.*, p. 433). This seems to me a good way of explaining what we mean by essential universality, and I welcome it. But I do not think it is consistent with his view that there is none but a verbal link between natural necessity and logical necessity. When in ordinary life we consider the intrinsic credibility of a story—for example, that told by a witness in a court of law or by an angler in a fishing inn—we may be said to be interested in the absolute natural probability of the story in question. That is to say, we then take for our propositional universe the field of possibility delimited not only by logical, mathematical, and phenomenological laws but also by such natural laws as we suppose ouselves to have discovered. If someone tells us, for example, that he caught twenty large fish in an hour, we raise our eyebrows higher than we should raise them if he said only that he had caught ten. But in such a context we do not feel or show any surprise at all if he says that water puts out fire or that bodies attract each other with a force directly proportional to their masses but inversely proportional to the square of the distance between them, unless indeed we express some surprise that he should think it worth while to repeat propositions which we are all prepared to accept as truisms. And the reason for our attitude is simply that in contexts where we are concerned with the intrinsic credibility of historical statements we think of accepted scientific generalizations in exactly the way Popper would have us think of them, namely as principles of delimitation for the narrowest universe of thinkables, i.e., the field containing all those sub-fields and only those sub-fields that differ in respect of initial conditions.

Aristotle spoke of the necessary as that which admits of no alternatives, and it seems to me that according to his definition laws of logic, mathematics, phenomenology, and natural science may all alike be described as laws of necessity, because all alike can be conceived as principles for the delimitation of the thinkable by exclusion of pseudo-alternatives which our language allows us to formulate. I admit, of course, that in the enunciation of natural laws we cannot, as in the other cases, rely solely on our familiarity with customary rules of usage for words. When we engage in scientific research, we suggest in effect that the restrictions commonly regarded as essential to the meaningful use of language may be usefully supplemented by others of a sort we could scarcely have learned at the beginning. Because of the way in which they are introduced, these later restrictions must always remain provisional, unless indeed they are transformed into scientific rules by a modification of the language to which they were originally attached in an external fashion. But there is nothing in all this to show that the word 'necessity' has different meanings in the phrases 'natural necessity' and 'logical necessity.'

CHAPTER 11

Aristotle's Theory of Modal Syllogisms and its Interpretation[*]

BY NICHOLAS RESCHER

I. Preface

IT IS, I BELIEVE, as difficult to overrate as it is to attain sound historical perspective. But wherein does the value of historical understanding lie? Surely not—as historicists urge—in somehow furnishing knowledge about the future. History is not a crystal ball in which the shape of things to come may be discerned. The value of history in relation to the future lies in its providing not *knowledge,* but *wisdom.* Historical understanding does not tell us what will happen, but rather it prepares us better to react with intelligence, and balanced perspective, and good sense, to whatever may happen, no matter what this may be.

---

* I was stimulated to enter upon the line of thought leading to this study of the interpretation of modalities in Aristotle when Mr. R. S. McCall of McGill University was kind enough to send me his cognate study, McCall [AMS]. (For references of this type see the Bibliography at the end of this paper.) By providing an axiomatization of Aristotle's modal syllogistic in the manner of Lukasiewicz's well-known axiomatization of Aristotle's assertoric syllogistic, McCall demonstrates the internal coherence and consistency of Aristotle's theory. This induced me to rethink the question of the *interpretation* of Aristotle's concept of modality. Being engaged on a study of Arabic logic, under the generous sponsorship of the National Science Foundation, I was aware that the Arab logicians regarded the modal inferences of *Anal. Pr.* as part and parcel of the theory of scientific reasoning of *Anal. Post.* This led me to wonder if it might not be possible to find in the treatment of modality of *Anal. Post.* the clues needed for an interpretation of the modal syllogistic of *Anal. Pr.* The present study is the natural outgrowth of this line of thought.

The present essay in intellectual history will not, I hope, prove wholly valueless from this didactic standpoint. It affords a vivid illustration of the pitfalls and difficulties in achieving a correct grasp of the thought of another mind on the basis of a transposition of its ideas into our own patterns of thinking, no matter how effective and successful these may be for other purposes. Specifically I hope to show in detail how (and why) the attempts of the past generation to reconstrue Aristotle's modal syllogistic on the basis of a straightforward, direct application of the methods of modern symbolic logic have met with utter failure. This result alone, I cannot but feel as one who numbers himself among symbolic logicians, is not devoid of instructiveness and interest.

II.   Historical   Excursus

Aristotle's theory of modal syllogisms has been a source of trouble to interpreters almost from the day of its enunciation. We know that Theophrastus of Eresos (fl. 330 b.c.), Aristotle's student and his successor as head of the school, as well as Eudemos of Rhodes (fl. 320 b.c.), another star pupil, contradicted their master's views on this point.[1] We know too that many Aristotelians of late Greek times rejected Aristotle's theory, and that this rejection militated against its transmission to the Syriac and Arabic students of Aristotelian logic.[2] From a most interesting Arabic source we learn that Eastern Christianity took a dim view of Aristotle's treatment of modalities. Explaining why the Syriac and Arabic scholars prior to his own time had occupied themselves primarily with only the first few sections of Aristotle's two treatises on Analytics, the Arabic philosopher and logician al-Fârâbî (fl. 910) wrote :

And so the Teaching of Aristotelian logic . . . remained in Alexandria, until the Christian emperor concerned himself about it and the bishops met and deliberated as to what part of the Teaching should continue and what should be abolished. They decided that there should only be instruction in the Books of Logic[3] up to the end of the figures of the assertoric syllogisms,[4] and not in that which comes after this, for they were of the opinion, that

---

[1] Bochenski [FL]. pp. 116–18.

[2] The earlier Syriac and Arabic translations of *Anal. Pr.* stopped at I, 7 (Meyerhof [VANB], p. 394). Although the Syriac translation of Aristotle's Organon commenced around A.D. 600, *Anal. Post.* was not translated until around 850 (by Hunain ibn Ishaq) ; and although Arabic translation of the Organon began around 820, *Anal. Post.* was not translated until around 900 (by Abû Bishr Mattâ ibn Yûnus).

[3] In its Late Greek–Syriac–Arabic structuring, the Aristotelian Organon of logic consisted of nine parts, in the following order: *Isagoge* (Porphyry), *Categoriae, De Interp., Anal. Pr., Anal. Post., Topica, Soph. Elen., Rhetorica, Poetica.*

[4] In Arabic, *al-ashkâl al-wûjudiyyah*; i.e., the figures of the assertoric, as opposed to the modal syllogisms.

herein lay a danger to Christianity, but that aid towards the victory of their faith resided in that part in which they permitted instruction. The public (exoteric) part of the Teaching was therefore confined to this limit, while the study of the rest was carried on in private (esoterically), until the rise of Islam much later.[5]

Nor did Aristotle's theory of modal syllogisms fare much better in Western, Latin Christendom. Medieval schoolmen, despite some interest in propositional modalities, generally took little notice of modal syllogisms, and had little use for Aristotle's views on this point; by and large they seemed to care no more for *Analytica Posteriori* than had their Eastern co-religionists in earlier days.[6] As a result, modern manuals of "traditional" or "Aristotelian" logic, under the influence of the Scholastic tradition, pass over the theory of modal syllogisms in utter silence.[7]

Modern students of Aristotle's logic, both classical scholars and logicians, are also sharply critical of his theory of modal syllogisms. Representing the former group, W. D. Ross endorses Theophrastus's major criticism of Aristotle and adds some others ([APPA], pp. 46–47), taking the view that "Aristotle's doctrine is plainly wrong" (*ibid*, p. 43). As for the latter group, A. Becker ([ATM], pp. 37–43) charges Aristotle's theory of modal syllogisms with using two diverse and mutually incompatible constructions of modal categorical propositions. This charge was endorsed by I. M. Bochenski ([AFL], pp. 57–58), who also ([LT], p. 99) endorsed the critical view put forward by J. Lukasiewicz ([AS, II]).

The current status of scholarly opinion regarding Aristotle's modal syllogistic is neatly epitomized in the article on "Logic, History of" contributed by Benson Mates to the 1958 edition of the *Encyclopaedia Britannica*:

We have confined our attention to the assertoric syllogism, i.e., to such syllogisms as contain only assertoric components. Actually much the greater portion of *Prior Analytics I* is devoted to modal syllogisms, which contain problematic or apodictic premises or both. Aristotle develops the theory of such syllogisms in a manner similar to that used for the assertoric syllogism.

---

[5] Quoted in Meyerhof [VANB], p. 394. For an analysis of al-Fârâbî's entire report, and an evaluation of its wider bearing upon the history of logical studies among the Arabs, see the writer's paper "Al-Fârâbî on Logical Tradition," *Journal of the History of Ideas*, vol. 24 (1963).

[6] This is not to say that those scholars who (like Albert the Great) wrote commentaries on *Anal. Pr.* did not treat of this part of Aristotle's logic; but their purely expository work, following in the footsteps of the Arabs, did not make its way into those medieval discussions that represented the active development of logical doctrine.

[7] See for example Welton [ML], one of the best modern manuals. Propositional modality is accorded a brief discussion (vol. I, pp. 192–95); nothing whatever is said of modal syllogisms. Over 100 years ago a French logician said: "J'ai été frappé de voir que la théorie des propositions modales est bannie de toutes nos logiques, et que, même dans les commentateurs anciens ou les philosophes du moyen âge, elle n'a fait, pour ainsi dire, aucun progrès." (Rondelet [TLPM], p.v.).

However, his treatment is incomplete and contains many errors. . . . His contribution to modal logic is undoubtedly important, but a great deal of work remains to be done before it can be regarded as well understood. [Vol. 14, p. 318, col. 2]

In summary, I think it fair to say that no other part of Aristotle's logic has encountered as much criticism, opposition, and even hostility as has greeted his theory of modal syllogisms throughout the two-and-a-third millennia of its existence.

And yet there can be no doubt that Aristotle himself regarded this theory as an essential and central part of his logic. A look at the table of contents of Book I of *Analytica Priora*, where Aristotle's theory of the syllogism is developed, is sufficient to show that he himself viewed the theory of the pure assertoric syllogism (presented in sections 4–7), which alone survived in the mainstream of the "Aristotelian" tradition, as merely an heuristic preliminary that needed to be developed in preparation for his major concern, the theory of modal syllogisms (presented in sections 8–22). And when the whole Aristotelian theory of the syllogism is regarded *in context,* and viewed as preparatory for Aristotle's theory of scientific reasoning (i.e., inference from necessary premises) as presented in *Anal. Post.,* it becomes clear indeed *why* his primary concern is with modal syllogisms, and not with the pure assertoric syllogisms in the manner of the bulk of his successors.

The object of the present paper will be to offer the results of a systematic study of Aristotle's theory of modal syllogisms. The aim of this inquiry is to answer three questions: (1) What did Aristotle himself have in mind in his theory of modal syllogisms? (2) Is his development of this branch of logical theory consistent with his concept of modality, or is it replete with mistakes, as modern critics have alleged? (3) Is Aristotle's concept of modal syllogisms adequate as a theory of modal inference which serves the purposes of his own conception of modality, or does it fail in its own purposes, as was alleged by Aristotle's immediate peripatetic successors? In brief, our aim is a synoptic exposition and evaluation of Aristotle's theory of modal syllogistic.

III.   Outline of Aristotle's Theory of Modal Syllogistic

In this section is given a bare outline of the main features of Aristotle's theory of modal syllogistic, the exposition following, in the main, the magisterial treatment of W. D. Ross [APPA].

In his theory of modal syllogisms, Aristotle envisages the usual categorical propositions (A, E, I, and O) as being subject to qualification by three modalities besides their usual assertoric mode—the modalities of necessity, possibility, and contingency (a contingent proposition being one neither necessary or impossible). These two elements in combination give rise to the groups of modal categorical propositions shown in Table 1.

TABLE 1

Type	*Modality*	
	A (*assertoric*)	N (*apodictic*)
A	All S is P.	All S is necessarily P.
E	No S is P.	No S is necessarily P.
I	Some S is P.	Some S is necessarily P.
O	Some S is not P.	Some S is not necessarily P.

Type	*Modality*	
	P (*possible*)	C (*contingent*)
A	All S is possibly P.	All S is contingently P.
E	No S is possibly P.	No S is contingently P.
I	Some S is possibly P.	Some S is contingently P.
O	Some S is not possibly P.	Some S is not contingently P.

For convenient reference we shall identify the various types of modal categorical propositions by placing the appropriate modal indicator as a subscript to the A–E–I–O classifier (except in the assertoric case where the subscript becomes superfluous). Thus A_N indicates a proposition of type "All S is necessarily P," E_P of type "No S is possibly P," and so on.

Aristotle nowhere discusses directly the matter of the construction or interpretation of modal categorical propositions. He apparently regards this as something which ought to be more or less self-evident on the basis of various concrete examples. He does, however, present a detailed discussion of logical interrelationships among such propositions, the upshot of which will be given in brief outline.

The *contradictories* of modal categorical propositions are determined in accordance with the following rules:

1. The contradictory of an assertoric modal categorical proposition is determined in the well-known way (O is the contradictory of A, I of E, E of I, A of O).
2. The contradictory of X_N is X'_P where X may be A, E, I, or O, and X' is the (assertoric) contradictory of X.
3. The contradictory of X_P is X'_N.
4. The contradictory of X_C does not lie within the framework of modal categorical propositions.

The conversion of modal categorical propositions is carried out in a manner parallel to that used for ordinary assertoric categorical propositions (i.e., A_N converts by limitation to I_N, E_N converts to E_N, I_N converts to I_N, O_N does not convert, etc.), except that E_C is not convertible, and O_C is convertible.

Certain interrelationships of entailment obtain among modal categoricals (where X may be A, E, I, or O)

1. X_N entails X_A.
2. X_A entails X_P.
3. X_C amounts to (i.e., in the sense of mutual entailment) the conjunction of X_P and X'_P where X' is the [assertoric] contradictory of X). Consequently X_C amounts to X'_C.

These few indications effectively summarize the totality of the information provided by Aristotle with respect to the nonsyllogistic part of the logic of modal categorical propositions.

Let us turn, then, to the syllogistic logic of modal categorical propositions. This is a matter of the greatest interest to Aristotle, and Book I of *Anal. Pr.* informs us of his views in the greatest detail. The results of that discussion are summarized here in Table 2.

Table 2 provides a convenient and compact synoptic summary of Aristotle's theory of modal syllogistic. It represents the fundamental datum with which any interpretation of Aristotle's theory of modal syllogistic inference must come to grips.

Several aspects of this tabulation warrant brief comment.

(1) Aristotle often comments on the *invalidity* of syllogisms and a good deal of this information is available from the tabulation by indirection. For example, the fact that an 'A' occurs at IV-1, instead of an 'N,' is bound up with the fact that Aristotle dismisses A_A A_N A_N–1 as invalid.

(2) A blank is ambiguous. Sometimes it means that no valid conclusion results, but sometimes it simply represents Aristotle's silence on the case.

(3) It is strikingly obvious that Aristotle consistently tries to apply the second and third figure justification procedures of the purely assertoric case (number I) in the cases of other modalities. Only when this procedure fails does he resort to some other, "abnormal" mode of justification.

(4) While there seem to be no errors of commission of any obvious kind, there are some few errors of omission. In some cases when the "normal" (purely assertoric) justification procedure fails, a hunt for some "abnormal" justification procedure would bear fruit. For example, in case VIII-5 a *reductio ad absurdum* argument would justify E_C A_N E–2 in terms of case VI-4.

TABLE 2

Modalities of the Conclusions of Valid Modal Syllogisms	I Major—A minor—A	II Major—N minor—N	III Major—N minor—A	IV Major—A minor—N
First Figure				
1. Barbara (AAA)	A(PS)	N(PS)	N(PS)	A(PS)
2. Celarent (EAE)	A(PS)	N(PS)	N(PS)	A(PS)
3. Darii (AII)	A(PS)	N(PS)	N(PS)	A(PS)
4. Ferio (EIO)	A(PS)	N(PS)	N(PS)	A(PS)
Second Figure				
5. Cesare (EAE)	A(C to I-2)	N(C to II-2)	N(C to III-2)	A(C to IV-2)
6. Camestres (AEE)	A(C to I-2)	N(C to II-2)	A[C to IV-2]	N[C to III-2]
7. Festino (EIO)	A(C to I-4)	N(C to II-4)	N(C to III-4)	A(C to IV-4)
8. Baroco (AOO)	A(R to I-1)	N[Ecthesis]	A(R to III-1)	A[R to I-1]
Third Figure				
9. Darapti (AII)	A(C to I-3)	N(C to II-3)	N(C to III-3)	N [C to III-3]
10. Felapton (EAO)	A(C to I-4)	N(C to II-4)	N(C to III-4)	A(C to IV-4)
11. Disamis (IAI)	A(C to I-3)	N(C to II-3)	A[C to IV-3]	N[C to III-3]
12. Datisi (AII)	A(C to I-3)	N(C to II-3)	N(C to III-3)	A(C to IV-3)
13. Bocardo (OAO)	A(R to I-1)	N[Ecthesis]	A[R to I-1]	A(R to IV-1)
14. Ferison (EIO)	A(C to I-4)	N(C to II-4)	N(C to III-4)	A(C to IV-4)

KEY: *Categorical propositions:* A, E, I, O.

Modalities: N(apodictic), A(assertoric), C(contingent), P(possible).
Syllogistic justification procedures: PS (perfect, i.e., self-evident, syllogism); C(conversion); R(*reductio ad absurdum*); CC(complementary conversion).
The "normal" justifications for syllogistic conclusions (i.e., those which proceed on analogy with case I) are indicated in *parentheses*, the "abnormal" ones in *brackets*. Additional syllogisms justifiable by complementary conversion are not tabulated.

SOURCE: The information here tabulated is taken from Ross [APPA], opposite p. 285, where the appropriate references to Aristotle's text are given.

A cursory glance at the tabulation shows clearly why Aristotle assigns a preferred status to the first figure of the syllogism. It plays an axiomatic role for him, for in this case alone does he countenance an appeal to the intuitive justification of syllogisms as "self-evident." The other figures are regarded as having syllogistic validity only when this is demonstrable by an appropriate justifying argument. I cannot in this regard forbear from quoting the brilliantly illuminating analogy due to Paul Henle:

In general, Aristotle's attack on the syllogism is like the present day attack on the mathematical problem of integration. Certain expressions are readily integrable and the problem is to reduce as many forms as possible to them. Similarly, for Aristotle, certain forms of the syllogism are transparent—

TABLE 2

V Major—C minor—C	VI Major—C minor—A	VII Major—A minor—P	VIII Major—C minor—N	IX Major—N minor—C
C(PS)	C(PS)	P[R to III–13]	C(PS)	P[R to III–13]
C(PS)	C(PS)	P[R to III–11]*	C(PS)	{ A[R to III–4] P[*a fortiori*] }
C(PS)	C(PS)	P[R to III–14]	C(PS)	P[R to II–6]
C(PS)	C(PS)	P[R to III–12]*	C(PS)	A[R to III–2]
—	—	P(C to VII–2)	—	{ A[R to III–4] P[C to VII–2] }
—	P[C to VII–2]	—	{ P[C to IX–2] A[R to VI–3] }	—
—	—	P(C to VII–4)	—	A(C to IX–4)
—	—	—	—	—
C(C to V–3)	C(C to VI–3)	P(C to VII–3)	C(C to VIII–3)	P(C to IX–3)
C(C to V–4)	C(C to VI–4)	P(C to VII–4)	C(C to VIII–4)	A(C to IX–4)
C(C to V–3)	P[C to VII–3]	P[C to VI–3]	P[C to IX–3]	C[C to VIII–3]
C(C to V–3)	C(C to VI–3)	P(C to VII–3)	C(C to VIII–3)	P[C to IX–3]
C[CC to V–11]	P[R to II–1]	—	P[R to II–1]	A[R to VII–1]
C(C to V–4)	C(C to VI–4)	P(C to VII–4)	C(C to VIII–4)	A(C to IX–4)

NOTE: We omit here to indicate those syllogisms which can be validated by complementary conversion, i.e., by use of the fact that X_c amounts to X'_c. In this way, but *in this way only*, can an assertorically invalid syllogism (e.g., one with two negative premises) become validated.

* See Ross; but there is apparently some mistake in the indicated justification.

their conclusions are evident from the premises—they need no further steps to validate them and so they are perfect. His task is to reduce all other forms to the perfect. (Henle [FFS], p. 100.)

Only first-figure syllogisms will qualify here—no doubt in part because the first figure is the most "scientific" in that here alone can a conclusion be had that is both universal and affirmative (see Ross [APPA], pp. 67-68).

IV. Difficulties in Aristotle's Theory

As has already been intimated, students of Aristotelian logic have, from the very first, objected to the master's conception of modal syllogistics. The objections of *modern* students relate to the particular features of their proposals for interpreting modal categorical propositions, and will have to be dealt with in their proper place below. The objections of *ancient* students are of two types, those relating to theological epistemology on the one hand, and those based on considerations of scientific epistemology on the other.

The objections of the former category need detain us but a moment. Bound up as it is with the guiding concept that all genuine, i.e. "scientific," knowledge must be based upon demonstrative reasoning involving self-evident premises, Aristotle's theory of modal syllogistic was *doctrina non grata* to the theologians who feared that this concept of the nature of knowledge would demote religious truth to some inferior plane of knowledge. Their dislike of the epistemology of *Anal. Post.* was carried forward to those sections of *Anal. Pr.* that appear an essential preliminary to it.

More serious are the objections to Aristotle's modal syllogistic offered by the early peripatetics. Theophrastus and Eudemus criticized Aristotle's modal syllogistic, not because of a dislike for his conception of knowledge based on an ideal of geometric demonstration, but because they thought to find some technical flaws in their master's theory. They offered little objection to Aristotle's theory of opposition and immediate inference for modal categorical propositions, but concentrated heavy fire on his theory of modal syllogistic.

Specifically, Theophrastus objected to Aristotle's view that the modality of a conclusion could be *stronger* (in the N–A–C–P scale) than that of a premise on which it was based. It is wrong, argued Theophrastus, to hold that there are valid syllogisms of the type A_N A A_N-1, in which one of the premises on which a *necessary* conclusion is based is merely *assertoric*. Theophrastus "corrected" Aristotle's modal syllogistic by insisting that, just as with quantity (universal over particular) and quality (affirmative over negative), so with modality : the conclusion of a syllogism must follow the weaker premise (*peiorem partem*). This view prevailed in the peripatetic tradition over that of Aristotle, in so far as attention continued to be given to this "incorrect" part of the master's logic.

V.   The Becker Interpretation

In their studies of Aristotle's theory of modal syllogisms, modern logicians have brought the machinery of (modal) symbolic logic to bear in an effort to provide a viable interpretation of Aristotle's modal categorical propositions. This line of approach was pioneered in 1933 in A. Becker's dissertation [ATM]. (Becker's proposed interpretation was endorsed by I. M. Bochenski in [AFL], pp. 57-61, despite some earlier doubts in [LT], pp. 96–98).

Becker proposed ([ATM], p. 19) to construe Aristotle's modal categorical propositions in terms of the group of symbolic interpretations shown as Table 3.

TABLE 3

Group I

	Apodictic (X is N)	*Possible (X is P)*	*Contingent (X is C)*
A_X	$(x)(Sx \supset \square\, Px)$	$(x)(Sx \supset \Diamond\, Px)$	$(x)(Sx \supset C\, Px)$
E_X	$(x)(Sx \supset \square\, {\sim}Px)$	$(x)(Sx \supset \Diamond\, {\sim}Px)$	$(x)(Sx \supset C\, {\sim}Px)$
I_X	$(\exists x)(Sx\ \&\ \square\, Px)$	$(\exists x)(Sx\ \&\ \Diamond\, Px)$	$(\exists x)(Sx\ \&\ C\, Px)$
O_X	$(\exists x)(Sx\ \&\ \square\, {\sim}Px)$	$(\exists x)(Sx\ \&\ \Diamond\, {\sim}Px)$	$(\exists x)(Sx\ \&\ C\, {\sim}Px)$

The only nonstandard symbol here, namely 'C,' is a modal operator of contingency, such that 'Cp' amounts to '${\sim}\square p\ \&\ {\sim}\square\, {\sim}p$.'

This interpretation however, as Becker himself realized fully, fails to accord with Aristotle's statement about the validity of modal inferences in various fundamental respects. First of all, many modal syllogisms fail (on Becker's proposed interpretation) to have the validity status explicitly assigned to them by Aristotle. (To cite just one of many possible examples, Aristotle holds that E_N I O_N–2 is valid, a claim falsified by the proposed interpretation). Even more importantly, Aristotle insists—deliberately, explicitly, and repeatedly—that the "usual" rules of immediate inference hold for modal categorical propositions, with minor exceptions explicitly stressed by him :

  (i) An A_x proposition is convertible by limitation to I_x.
  (ii) An E_x proposition is convertible, *except* when X is C.
  (iii) An I_x proposition is convertible.
  (iv) O_c is convertible.

None of these rules is forthcoming on the proposed interpretations of Group I.

Becker proposes to remedy these discrepancies by recourse to an ingenious expedient which I cannot but regard as both desperate and farfetched. He puts forward the hypothesis that Aristotle is himself uncertain or confused over the meaning of modal categorical propositions, and oscillates between the interpretations shown in Group I, above, and a second construction.[8]

TABLE 4

Group II

	Apodictic (X is N)	*Possible (X is P)*	*Contingent (X is C)*
A_X	$\square\ (x)(Sx \supset Px)$	$\Diamond\ (x)(Sx \supset Px)$	$C\ (x)(Sx \supset Px)$
E_X	$\square\ (x)(Sx \supset {\sim}Px)$	$\Diamond\!\!\!/\ (x)(Sx \supset {\sim}Px)$	$C\ (x)(Sx \supset {\sim}Px)$
I_X	$\square\ (\exists x)(Sx\ \&\ Px)$	$\Diamond\ (\exists x)(Sx\ \&\ Px)$	$C\ (\exists x)(Sx\ \&\ Px)$
O_X	$\square\ (\exists x)(Sx\ \&\ {\sim}Px)$	$\Diamond\ (\exists x)(Sx\ \&\ {\sim}Px)$	$C\ (\exists x)(Sx\ \&\ {\sim}Px)$

[8] See [ATM], pp. 42–43 (for the apodictic case), and pp. 59–65 and 83–91 (for possible and contingent modal categorical propositions).

According to Becker's hypothesis, Aristotle bears in mind this second group of interpretations when thinking of the rules of immediate inference.

On Becker's hypothesis, the many "errors" of Aristotle's discussion of modal syllogisms are inherent in his procedure in analyzing the validity of syllogisms. As is well known, this procedure of Aristotle's consists of two steps : the first to justify the first figure syllogisms (using, according to Becker, the interpretations of Group I), and the second to justify the syllogisms of the other figures (i.e., 2 and 3 only) by means of the rules of immediate inference and opposition (which are based on the interpretations shown in Group II).[9] As a result, Aristotle's theory of modal syllogisms is full of mistakes, because his analysis of the validity of such syllogisms involves the conflation of the two groups of interpretations.

As this brief outline of Becker's approach to the interpretation of Aristotle's modal categorical propositions graphically indicates, the interpretation requires, and indeed is founded upon, the charge that Aristotle is confused as to the meaning of modal categorical propositions, and oscillates between two distinct and divergent constructions of such propositions.

VI.   Failure of This Approach

The discussion of the preceding section yields the negative result that neither of the two sets of symbolic interpretations of modal categorical propositions considered (Groups I and II) provides a construction of such propositions whose consequences answer to Aristotle's assertions about the validity of modal inferences. The possibility remains open that by adopting Becker's basic idea of a construction in terms of (modal) symbolic logic, changing only the details of his proposed interpretations, a suitable interpretation (i.e., one whose consequences agree with Aristotle's claims) might be devised. Unfortunately this ray of hope is soon extinguished, for a detailed analysis of the alternative possibilities leads to the discouraging conclusion that no combination of the even partially plausible symbolic interpretations of this sort can fulfill the required conditions. Let me narrow the scope of my discussion for the remainder of this section to apodictic categorical propositions, assuring the reader that the same essential result obtains, *mutatis mutandis,* when contingent and possible categorical propositions are admitted to view.

In a search for an interpretation of Aristotle's apodictic categorical propositions a host of alternative *prima facie* possibilities come to view. What might seem to be the most promising are shown in Table 5. From this standpoint, Becker's two interpretations (namely $A_N^4 - E_N^6 - I_N^4 - O_N^6$

---

[9] By themselves this second group of interpretations would not lead to Aristotle's theory of modal syllogisms. They would not, for example, validate A_N A A_N-1.

TABLE 5

A_N	E_N
1. $\Box\,(x)\,(Sx \supset Px)$	1. $\Box\,(x)\,(Sx \supset \sim Px)$
2. $(x)\,\Box\,(Sx \supset Px)$	2. $(x)\,\Box\,(Sx \supset \sim Px)$
3. $(x)\,(\Box\,Sx \supset \Box\,Px)$	3. $(x)\,(\Box\,Sx \supset \sim\Box\,Px)$
4. $(x)\,(Sx \supset \Box\,Px)$	4. $(x)\,(\Box\,Sx \supset \Box\,\sim Px)$
5. $(x)\,(\sim\Box\,\sim Sx \supset Px)$	5. $(x)\,(Sx \supset \sim\Box\,Px)$
6. $(x)\,(\Box\,Sx \supset Px)$	6. $(x)\,(Sx \supset \Box\,\sim Px)$
7. $(x)\,(\sim\Box\,\sim Sx \supset \Box\,Px)$	7. $(x)\,(\sim\Box\,\sim Sx \supset \sim Px)$
	8. $(x)\,(\Box\,Sx \supset \sim Px)$
	9. $(x)\,(\sim\Box\,\sim Sx \supset \Box\,\sim Px)$

I_N	O_N
1. $\Box\,(\exists x)\,(Sx\,\&\,Px)$	1. $\Box\,(\exists x)\,(Sx\,\&\,\sim Px)$
2. $(\exists x)\,\Box\,(Sx\,\&\,Px)$	2. $(\exists x)\,\Box\,(Sx\,\&\,\sim Px)$
3. $(\exists x)\,(\Box\,Sx\,\&\,\Box\,Px)$	3. $(\exists x)\,(\Box\,Sx\,\&\,\sim\Box\,Px)$
4. $(\exists x)\,(Sx\,\&\,\Box\,Px)$	4. $(\exists x)\,(\Box\,Sx\,\&\,\Box\,\sim Px)$
5. $(\exists x)\,(\Box\,Sx\,\&\,Px)$	5. $(\exists x)\,(Sx\,\&\,\sim\Box\,Px)$
	6. $(\exists x)\,(Sx\,\&\,\Box\,\sim Px)$
	7. $(\exists x)\,(\sim\Box\,\sim Sx\,\&\,\sim Px)$
	8. $(\exists x)\,(\Box\,Sx\,\&\,\sim Px)$

and $A_N^2 - E_N^2 - I_N^2 - O_N^2$) appear as merely two of over two thousand $(7 \times 9 \times 5 \times 8 = 2520)$ possibilities.

It is readily shown, however, that none of the combinations of possible interpretations of this sort accord with the requirements set by Aristotle's discussion. Consider the E_N possibilities. Only in the first three cases is conversion possible, as Aristotle requires. And yet none of these three cases validates Aristotle's claim that E_N A E_N–1 is valid. Similar lines of argument can be developed against the various foregoing interpretations for A_N, I_N and O_N, in each case without having even to resort to "mixed" syllogisms in which more than one type of apodictic categorical proposition is involved.[10]

---

[10] A detailed check of the validity of apodictic modal syllogisms subject to the various possible interpretations enumerated above was carried out on an LGP-30 computer at Lehigh University during the week of April 10, 1961, using a program devised by Mr. Gerhard Rayna of the Lehigh Department of Mathematics. This check substantiated in detail the conclusion that no such interpretation will accord with more than a fraction (at most 80 per cent) of Aristotle's assertions about the validity status of the applicable immediate and syllogistic inferences. (The machine-time for this work was made available through the generosity of the Institute of Research of Lehigh University.)

The program of this machine calculation was based on the availability of only one rule regarding the necessity-operator, viz., from p to infer p. Clearly the validity status of certain syllogisms would be affected by the addition of other auxiliary assumptions, such as, for example $(p \supset q) \supset (\Box\,p \supset \Box\,q)$. But there are arguments available

As I see it, the foregoing result is not to be construed as a decisive proof that Aristotle's theory of modal syllogisms is based upon errors or confusions. Rather, it constitutes *very* powerful presumptive evidence to the effect that Aristotle's theory of modal categorical propositions is not to be interpreted in terms of the particular type of symbolic representation deriving from Becker's proposed interpretations. It appears to be necessary to look in a wholly different direction if we are to obtain a tenable interpretation of Aristotle's theory of modal syllogisms.

VII. The Lukasiewicz Interpretation and Its Failings

The second (1957) edition of Jan Lukasiewicz's important book on *Aristotle's Syllogistic* devotes a substantial monograph (chaps. VI–VIII) to Aristotle's theory of modal logic, a topic not touched on in the first (1951) edition of this work. In line with the strategy of the approach discussed in the preceding section, Lukasiewicz treats modal categorical propositions as the result of prefixing a propositional modal operator to an ordinary, assertoric categorical proposition. Now, as was shown in the preceding section, this approach, when based on a system of modal logic embodying rules of modal propositional logic of the customary sort, cannot yield an adequate interpretation of Aristotle's treatment of modal syllogistic. Recognizing this, Lukasiewicz attempts the construction of an extraordinary, peculiarly Aristotelian set of modal rules:

. . . modern logicians have not as yet been able to construct a universally acceptable system of modal logic which would yield a solid basis for the interpretation and appreciation of Aristotle's work. I have tried to construct such a system, different from those hitherto known, and built up upon Aristotle's ideas. The present monograph on Aristotle's modal logic is written from the standpoint of this system. (p. 133.)[11]

That Lukasiewicz's system fails utterly to provide an appropriate setting for Aristotle's conception of modal logic is decisively illustrated by one of its theses—one on which Lukasiewicz himself puts the greatest emphasis—viz., that *no apodictic proposition is true* in this system (see pp. 169–72, 205–8). I am utterly at a loss to comprehend how Lukasiewicz

---

(too long to be detailed here) that the effect of the addition of such rules, which is to make valid certain otherwise invalid arguments, always brings with it undesired consequences by rendering valid certain syllogisms (in the various interpretations) which, on Aristotle's view, ought not to be so.

In checking this, and several other points, I was assisted by my student Mr. Richard K. Martin, whom I wish to thank for his help.

[11] This system was published separately by Lukasiewicz under the title "A System of Modal Logic," *The Journal of Computing Systems*, vol. I (1953), pp. 111–149.

could have convinced himself that Aristotle's modal syllogistic, whose *ratio essendi* is the construction of a theory of apodictic inference, is properly to be understood on the basis of a theory of modality whose centrally characteristic feature is the denial of truth to apodictic propositions.

Like the efforts of Becker and other interpreters, the proposals of Lukasiewicz do not lead to a consistent construction of Aristotle's assertions, and like his predecessors, Lukasiewicz puts the blame on Aristotle. Aristotle's modal syllogistic, charges Lukasiewicz, "looks like a logical exercise which in spite of its seeming subtlety is full of careless mistakes and does not have any useful application to scientific problems" (p. 181). Lukasiewicz's discussion is replete with references to supposed errors committed by Aristotle and of "rectifications" of his "serious mistakes." (See almost any page of chap. VIII, pp. 181–208.)

VIII.   A New Interpretation

Since it appears that the approaches made by Becker and Lukasiewicz to the interpretation of Aristotle's theory of modal syllogisms constitute blind alleys as regards the possibility of interpreting Aristotle's discussion as it stands, without introducing numerous "corrections," it would seem that a wholly new approach is in order. I want now to suggest such a new way of construing Aristotle's theory of modal propositions and modal syllogisms.

The new approach I wish to propose rests upon the following conception : Let us conceive of the modality of a statement not as an explicit integral facet of the statement itself—as is done in construing "All S is necessarily P" as "$(x)(Sx \supset \Box Px)$" or as "$(x)\Box(Sx \supset Px)$"—but rather as a way of according a certain *status* to the "ordinary" proposition "All S is P." In this manner, modality is to be construed an analogy with the assertion sign " $\vdash$ " in modern logic—i.e., as a means for indicating the status of a proposition, not as an operator for changing propositions about. Modality thus no longer enters explicitly into the formulation or into the asserted content of a proposition, but (like truth and falsity) represents a means for categorizing or classifying propositions *ab externo,* in such a way that the categorization of the proposition is by no means apparent by an inspection of the statement in which this proposition is formulated.

On this conception, then, the ordinary, unmodalized categorical propositions A, E, I, O can be qualified by four modalities : A(assertory, "is actual"), P(problematic, "is possible"), N(apodictic, "is necessary"), and C(contingent, "is neither necessary nor impossible," where the impossible is that whose negation, i.e., contradictory, is necessary). We may indicate the modal status of a categorical proposition by an appropriate qualifying prefix, as shown in Table 6.

TABLE 6

Modality

Type of Categorical Proposition	Assertoric	Probable	Apodictic	Contingent
A	A:A or simply A	P:A	N:A	C:A
E	A:E or simply E	P:E	N:E	C:E
I	A:I or simply I	P:I	N:I	C:I
O	A:O or simply O	P:O	N:O	C:O

From the standpoint of this conception of modal categorical propositions, the main problem of settling the logic of such propositions is to
specify the rules of immediate inference to govern (I) conversion, (II)
negation, and (III) mutual interrelations of entailment and equivalence.
This can be done in all instances in strict accordance with Aristotle's
explicit strictures, namely as follows :

TABLE 7

Rules of Immediate Inference for Modal Categorical Propositions

I. Conversion Rules	Modal Status (Left Unchanged)			
	A Assertoric	N Apodictic	P Possible	C Contingent
1. Conversion of an A-proposition	no	no	no	no
2. Conversion by limitation of A to I	yes	yes	yes	yes
3. Conversion of an E-proposition	yes	yes	yes	no
4. Conversion of an I-proposition	yes	yes	yes	yes
5. Conversion of an O-proposition	no	no	no	no

II. Rules for Contradictories (i.e., for Negation)	Modal Status			
	A	N	P	C
1. Contradictory of an A-proposition	A:O	P:O	N:O	Nothing
2. Contradictory of an E-proposition	A:I	P:I	N:I	in the
3. Contradictory of an I-proposition	A:E	P:E	N:E	framework
4. Contradictory of an O-proposition	A:A	P:A	N:A	

III. Mutual Interrelations

1. N:X entails A:X.
2. A:X entails P:X.
3. C:X amounts to (~N:X) & (~N:X') and so to (P:X) & (P:X'), where X' is the
   contradictory of X.

In the present approach, this handful of basic rules of inference adequately embodies Aristotle's intuitions about the immediate inference relationships among modal categorical propositions. But so far we have been traversing relatively well-known ground within the area of essential agreement among interpreters. We come now to the second and controversial part of our task, the theory of syllogistic inference.

The rules for syllogistic modal inferences fall into two groups: (1) Rules specifying the conditions of syllogistic validity for the first figure only; and (2) rules for determining the validity of syllogisms in the other (two) figures through a derivation process going back to first-figure syllogisms. The rules of the second kind are essentially the same for modal as for "ordinary" unmodalized categorical syllogisms (viz., conversion and *reductio ad absurdum,* with the addition of ecthesis). The sole point of difference lies in the possibility of complementary conversion (i.e., in the fact that C : X amounts to C : X'). Only the first-figure rules need therefore concern us.

The rules for syllogistic modal inference *in the first figure only* can be indicated by the stipulating that the modality of the premises has nothing to do with the *type* (A, E, I, or O) of conclusion drawn, but enters in only in determining the modal status of the conclusion. Here Aristotle's basic intuitions can be summarized, I submit, as shown in Table 8. The guiding principle underlying this tabulation is clearly the thesis that, in the basic first-figure cases, the modality of conclusion follows that of the major premise. I am myself inclined to place considerable weight on the fact that Averroes, the most faithful and best-informed Aristotelian of medieval times, grasped this principle with admirable clarity when he criticized the *peiorem* rule of Theophrastus and Eudemus, and endeavored to vindicate Aristotle's own principle: *modus conclusionis sequetur modum propositionis maioris* (see pp. 65–66 of the classic Junta edition [Venice, 1503] of Aristotle's works with Averroes' commentaries).

TABLE 8

Rules for Syllogistic Modal Inferences in the First Figure

	Major Premise	*Minor Premise*	*Conclusion*
1.	A	A	A
2.	N	A	N
3.	A	N	A
4.	C	C	C
5.	C	A	C

The few basic rules listed in Table 8 entail corresponding rules for other modal combinations (in the first figure), of which the important cases (i.e., the ones treated by Aristotle) are shown in Table 9.

TABLE 9

	Major Premise	Minor Premise	Conclusion	Mode of Derivation
6.	N	N	N	*a fortiori* from 2, above
7.	A	C	P	derivative from other cases
8.	C	N	C	*a fortiori* from 5, above
9.	N	C	A or P	derivative from other cases

The meaning of this tabulation can be summarized by the following rule:

In a modal syllogism of the first figure (i.e., an assertorically valid one) the modality of the conclusion is determined as follows: (i) if the modality of the major premise *is not* stronger than that of the minor premise (i.e., not prior in the $N-A-P-C$ scale), then the modality of the conclusion simply follows that of the major, and (ii) if the modality of the major premise *is* stronger than that of the minor premise, then *the modality of the conclusion is always stronger than that of the minor premise* (but need not be as strong as that of the major premise).

It is this rule assigning the determining role to the major premise which represents—in contrast to Theophrastus's *peiorem* rule—Aristotle's basic intuition into the logic of modal syllogistic inference.

Given the rules tabulated above, Aristotle's entire theory of modal syllogisms (as summarized in Table 2) can be derived. The modes for the derivative justification of syllogisms needed for this purpose are precisely those needed in the purely assertoric case for the reduction of second- and third-figure syllogisms to those of the first figure (with the sole exception of the additional—but rarely used—principle of "ecthesis"). And it is noteworthy that Aristotle's treatment of the nonassertoric cases is patterned deliberately as closely as possible upon the purely assertoric case.[12]

In this conception of Aristotle's theory of modal syllogistic, then, the modalities are means for the *ab externo* assignment of a certain *status* to ordinary (assertoric) categorical propositions, and the theory of modal syllogistic amounts in effect to the stipulation of a handful of rules governing the basic operations of immediate and syllogistic inference. In fact, of all the machinery of modern symbolic logic, it is the notion of a rule of inference that comes closest to representing an analogue of Aristotle's concept of modality. Such a rule as the ordinary *modus ponens*, "If ⊢ A and ⊢ A ⊃ B, then ⊢ B," may be construed: "If A is of the status *Asserted*, and A ⊃ B is of the status *Asserted*, then the conclusion B is of the status *Asserted*." Analogously we construe Aristotle's endorsement of, for example, A_N A $A_N - 1$ as a perfect syllogism as follows: "Given three propo-

---

[12] See the illuminating discussion of Henle [FFS], pp. 98–100.

sitions conforming to the AAA–1 pattern, then if the first (major) is of the status *Necessary* and the second (minor) is of the status *Asserted,* then the third (conclusion) is of the status *Necessary.*" Modality, for Aristotle behaves, we believe, in a way closely analogous with the concept of assertion, ⊢ , introduced by Frege into modern symbolic logic. Unlike the Lewis-type modalities which guided Becker in his proposals for interpreting Aristotle's modal syllogistic, the prefix ' ⊢ ' does not enter into the content of a proposition and effect a recasting of its meaning, but simply signalizes a certain *status* of an otherwise unaffected proposition.

This concept of the character of Aristotle's modal syllogisms is at variance with the views of some recent interpreters of Aristotle's syllogistic who have insisted that the Aristotelian syllogism is to be construed as a complex (hypothetical) statement—a thesis, as Lukasiewicz puts it—rather than as a rule of inference as the present conception would suggest. (See Lukasiewicz [AS, II], chap. II; unchanged from the first edition.) It must be borne in mind, however, that this view was worked out solely on the basis of an analysis of Aristotle's discussion of purely assertoric syllogisms, which leaves the modal syllogistic wholly out of account.

It is strongly suggested by the foregoing considerations that Aristotle had a coherent and internally self-consistent theory of modal categorical inferences, a theory which he developed in a manner free, in its essentials, not only from errors of commission (i.e., self-contradiction) but also, with only trivial exceptions, free from errors of omission (i.e., incompleteness). This theory represents the natural and inevitable development of certain basic insights into the nature of modal inference. Given the consistency of Aristotle's theory, it is only on these basic rules that valid critical disagreement with Aristotle *could possibly* be based. Let us therefore turn to the task of subjecting to critical scrutiny these basic premises of Aristotle's theory of modal inference.

IX.   Aristotle's Theory of Apodictic Inference as Basis for His Conception of Modal Syllogistic

From Alexander of Aphrodisias (as cited in Lukasiewicz [AS, II], p. 187) we learn that Theophrastus, Eudemus, and their followers disagreed with their teacher's theory of modal syllogisms, objecting to his claim that a necessary conclusion could be obtained from syllogisms some (i.e., one) of whose premises is merely assertoric. The conclusion must follow the weaker premise (*peiorem partem*), they insisted. Aristotle's insistence on the validity of syllogisms of the type of $A_N A A_N - 1$ struck his main followers as an error. We cannot claim to understand Aristotle's theory of modal inference unless we see his reasons for adopting the view he does, rather than Theophrastus's *peiorem* rule.

The key to Aristotle's theory lies, I am convinced, in viewing the theory of modal syllogisms of the *Analytica Priora* in the light of the theory of scientific reasoning of the *Analytica Posteriora*. This fundamental fact was seen clearly by those ancient Aristotelians who, in dropping the study of *Anal. Post.*, also broke off with *Anal. Pr.* in the middle of the first book. (In fact it was reflection on this facet of the Arabic treatment of the Organon that opened up to me the line of interpretation I am about to explain.)

I wish to submit that Aristotle's theory of modal syllogisms has its developmental and conceptual roots in Aristotle's concept of demonstration. To understand the role and nature of the Aristotelian theory of modal syllogisms generally, and of apodictic syllogisms in particular, we must, I believe look to his concept of scientific reasoning in *Anal Post.*, for it is here that we learn clearly and explicitly how Aristotle conceives of the job and function of apodictic reasoning. Let us then adopt Aristotle's own functional and teleological approach, and attempt to understand his theory of modal syllogisms in terms of the theory of scientific inference of which it forms an integral part. The following quotations should prove sufficient for our purposes.

By demonstration I mean a syllogism productive of scientific knowledge, a syllogism, that is, the grasp of which is *eo ipso* such knowledge. Assuming then that my thesis as to the nature of scientific knowing is correct, the premises of demonstrated knowledge must be true, primary, immediate, better known than and prior to the conclusion, which is further related to them as effect to cause. Unless these conditions are satisfied, the basic truth [of a demonstration] will not be "appropriate" to the conclusion. (71*b*17-24.)

Since the object of pure scientific knowledge cannot be other than it is, the truth obtained by demonstrative knowledge will be necessary. And since demonstrative knowledge is only present when we have a demonstration, it follows that demonstration is an inference from necessary premises. (73*a*20-24.)

Demonstrative knowledge must rest on necessary basic truths; for the object of scientific knowledge [i.e., that which is known by demonstration] cannot be other than it is. Nor attributes attaching essentially to their subjects attach necessarily to them . . . . It follows from this that the premises of the demonstrative syllogism must be connections essential in the sense explained . . . . We must either state the case thus, or else premise that the conclusion of demonstration is necessary and that a demonstrated conclusion cannot be other than it is, and then infer that the conclusion must be developed from necessary premises. For though you may reason from true premises without demonstrating, yet if your premises are necessary you will assuredly demonstrate—in such necessity you have at once a distinctive character of demonstration. That demonstration proceeds from necessary premises is also indicated by the fact that the objection we raise against a professed demonstration is that a premise of it is not a necessary truth . . . . A further proof that the conclusion must be developed from necessary premises is as follows. Where demonstration is possible, one who can give no account which includes the cause has no scientific knowledge. If, then, we suppose a syllogism in which, though A necessarily inheres in C, yet B, the middle term of the demonstration, is not necessarily connected with A and C, then the man who argues thus has no reasoned knowledge of the conclusion, since this conclusion does not owe its necessity to the middle term; for though the conclusion is necessary, the mediating link is a contingent fact . . . . To sum up, then; demonstrative knowledge must be knowledge of a necessary nexus, and therefore must clearly be obtained through a necessary middle term; otherwise its possessor will know neither the cause nor the fact that his conclusion is a necessary connection. (74b5–75a15)[13]

These excerpts from Book I of *Anal. Post.* make abundantly clear that Aristotle's interest in modal syllogisms derives from his concept of scientific reasoning, and are the natural result of his concept that truly scientific, i.e. demonstrative, reasoning is pre-eminently present only in arguments which establish a necessary conclusion.

In *Anal. Pr.*, then, Aristotle does not (not yet? no longer?) hold the view (later to be espoused by Theophrastus and Eudemus) that a necessary conclusion must rest exclusively on necessary premises.[14] The view of scientific reasoning underlying the discussion of apodictic syllogisms in *Anal. Pr.* seems to be that (1) the major premise lays down a necessary rule of some sort, and (2) the minor describes some special case which has been shown by observation or induction to fall under this rule, so that (3) the conclusion is justified that this special case necessarily conforms to the rule. The paradigm of such reasoning is:

Law (necessary rule):	All B's are A's. [All twinkling things are distant.]
Special case (observation):	All C's are B's. [All stars are twinkling things.]

Explained consequence (necessary result):	All C's are A's. [All stars are distant.]

---

[13] Note that in these discussions in *Anal. Post.* Aristotle adopts the view that a conclusion is necessary only when *both* premises are necessary. This, of course, is a view he does not espouse in *Anal. Pr.*, holding there that it suffices (with first-figure syllogisms) for the major premise alone to be necessary. No statement could be more sharply explicit than the following: All demonstrative reasoning proceeds from necessary or general premises, the conclusion being necessary if the premises are necessary and general if the premises are general. (*Anal. Post.*, 87623–25.)
I am not aware that any Aristotelian student, modern or ancient, has noted, let alone assessed, the significance of this change of Aristotle's conception of necessary inference as between the *Prior* and the *Posterior Analytics*.
[14] The question of the temporal relationship between *Anal. Pr.* and *Anal. Post.* remains very much *sub judice*. An excellent survey of many of the relevant considerations is given by Ross in [APPA] (see part II of the Introduction). If it were true, as in

A demonstrative argument of this sort is also, for Aristotle, a piece of scientific reasoning because that feature of the minor term (the stars) which is to be explained (namely that they are distant) is justified in terms of possession of a property (namely twinkling) which provides a necessary, law-governed "bridge" over which passage to the desired conclusion can be effected.[15]

In short, Aristotle espouses the validity of A$_N$ A A$_N$ − 1 not on grounds of abstract formal logic, but on grounds of *applied* logic, on *epistemological* grounds. What he has in mind is the application of modal syllogisms within the framework of a theory of scientific inference along the lines of his own conceptions. We must recognize that it is Aristotle's concept that in truly scientific reasoning the relationship of major to minor premise is governed by the proposition :

major premise : minor premise :: general rule : special case

When we take note of this line of thought we see why Aristotle taught that the major premise of a modal syllogism can strengthen the modality of the conclusion above that of the minor premise. For a rule that is necessarily (say) applicable to all of a group will be necessarily applicable to any sub-group, pretty much regardless of how this sub-group is constituted. On this view, the necessary properties of a genus must necessarily characterize even a contingently differentiated species. If all elms are necessarily deciduous, and all trees in my yard are elms, then all trees in my yard are necessarily deciduous (even though it is not necessary that the trees in my yard be elms).

X. The Peripatetic Critics of Aristotle's Theory of Modal Inference

If Aristotle's theory of apodictic inference is as we have outlined it in the preceding section, and if (as it would appear) this theory is a plausible one, why did his star pupils (Theophrastus and Eudemus) criticize his views, views which they must have understood with perfect clarity and correctness? The answer, as I see it, lies in the fact that these old peripatetics effected a shift from Aristotle's *construction* of modal categorical propositions (a shift in which the later Aristotle may himself have participated.)

Ross's view, that *Anal. Post.* is later than *Anal. Pr.*, it would mean that Aristotle himself abandoned his theory of modal syllogistic and adopted the views that have come down to us in the name of his pupils.

[15] In general, Aristotle clearly intends this "law-governed bridge" to be provided by a *definition*. See Ross [APPA], p. 61, the first paragraph beginning on the page.

As counter-examples to Aristotle's acceptance of A_N A A_N – 1, Theophrastus gave three concrete arguments (see Ross [APPA], p. 41; but I produce the arguments in my notation):

(a)   N: All men are animals
       A [at sometime t]:   All moving-things are men
       _____
       N [at t]:   All moving things are animals

(b)   N: All literate-beings are possessors of scientific
                                     knowledge
       A [at sometime t]:   All men are literate-beings
       _____
       N [at t]:   All men are possessors of scientific knowledge

(c)   N: All walking-beings are moving-beings
       A [at sometime t]:   All men are walking-beings
       _____
       N [at t]:   All men are moving-beings

I think it is transparently clear from the very statement of these counter-examples that Theophrastus has shifted away from the ground of Aristotle's discussion. Aristotle takes (true) categorical propositions to represent timeless connections among species (groups). Thus "No S is P" for Aristotle requires that "No S is *ever at any time* or *under any circumstances* P," and comparably with "All S is P." Theophrastus, taking a large step in the direction of the Stoic practice of using temporal reference in the analysis of logical connections, permits the construction of universal propositions subject to temporal reference (*kata chronon*). (See Bochenski [LT], pp. 73, 80–81.) It is only to be expected that with this shift in the terms of reference, Aristotle's rules of modal inference will also have to be modified.

It is an interesting and debatable point to raise the question of the extent to which Aristotle himself construes the modality of necessity along temporal lines. Let me quote Lukasiewicz:

There is a principle of necessity set forth by Aristotle which is highly controversial. He says in the *De Interpretatione* that "anything existent is necessary when it exists, and anything nonexistent is impossible when it does not exist." This does not mean, he adds, that whatever does not exist is impossible: for it is not the same to say that anything existent is necessary when it does exist, and to say that it is simply necessary (19a23). It should be noted that the temporal "when" (*hotan*) is used in this passage instead of the conditional "if." A similar thesis is set forth by Theophrastus. He says, when defining the kinds of things that are necessary, that the third kind (we do not know what the first two are) is "the existent, for when it exists, then it is impossible that it should not exist" (Alexander, 156:29). Here again we find the temporal particles "when" (*hote*) and "then" (*tote*). ([AS], p. 151.)

This sort of Aristotelian discussion readily lends itself to a construction of the concept of necessity along Stoic lines: that is (absolutely) necessary which is existent or actual at all times. But whatever may be the extent to which Aristotle himself held such views at some point of his career, it would seem that the construction of 'necessity' that underlies the modal syllogistic of *Anal. Pr.* is not this temporal conception.[16] When the present study was already on its way to the printers, I became acquainted with J. Hintikka's article [NUTA], which presents the case for imputing to Aristotle himself the temporal construction of modal propositions. There is thus substantial warrant for a most careful scrutiny of the following possibility : at first, earlier in his career, Aristotle espoused a timeless (Platonic) conception of necessity (and modality as well), and at this point held to the modal syllogistic of *Anal. Pr.*, based on the major-premise rule. Later, he switched to a temporal (Stoic-style) conception of necessity (and modality), and then shifted to the modal syllogistic implicit in *Anal. Post.*, based on the *peiorem* rule.

To particularize somewhat on the above discussion let me consider some other Theophrastus-style counter-examples to $A_N A A_N - 1$ — examples in which temporal or spatial limitations have been covertly imported into the minor premise :

(d)   N: All wise men are philosophers
A: All sons of Nicomachus are wise men
_____
N: All sons of Nicomachus are philosophers

(e)   N: All rectangular buildings are quadrilateral structures
A: All Temples of Diana are rectangular buildings
_____
N: All Temples of Diana are quadrilateral structures

Examples of this type constitute serious difficulties for maintaining the universal, formal validity of $A_N A A_N - 1$. But Aristotle's conception is not destroyed by these counter-examples. It is a thing of spatio-temporal haphazard that such generalizations as the minor premises of (d) and (e) obtain. Aristotle's concept, based specifically upon "organic" relationships between large natural groups ("All stars are twinkling things," "All men are capable of laughter," etc.) is not refuted by a failure to obtain in cases of a type to which it was not intended to apply.

None the less, it is entirely possible that this restricted character of validity of the theory of apodictic inference led Aristotle to modify this theory, and himself to adopt the view (of *Anal. Post.* as well as of the later

---

[16] Seeds for the procedure of importing time into the construction of categorical propositions (and indeed the technical phrase *kata chronon* itself) can be found in *Anal. Pr.*, for example in 34*b*6–18. This passage is often taken to mean that categorical propositions are atemporal, but what Aristotle in effect says is that "All A's are B's" must be construed as "All A's are *always*, i.e., at all times, B's," so that time, rather than being excluded from our purview, is deliberately imported into it.

peripatetics) that a necessary conclusion is forthcoming only from necessary premises.[17]

If, however, Aristotle did abandon (or modify) the theory of modal syllogisms of *Anal. Pr.*, this was not due to any shortcomings of that theory from the standpoint of *formal* logic (i.e., to some sort of self-contradiction), but was due solely to considerations of a *material* (rather than formal) defect—to a failure to yield "plausible" results in certain presumably legitimate applications. The Theophrastus-style objections to the modal syllogistic of *Anal. Pr.* do not go against the formal structure of this theory, but proceed on *material* considerations, and impugn its suitability on grounds relating to the methodology of "scientific demonstration."

XI.   Reassessment

The time has come to bring our study of Aristotle's modal syllogistic to a close with a brief retrospective glance over the ground that has been covered.

We have seen that those modern students who have treated Aristotle's theory of modal syllogisms from the vantage point of modern symbolic logic have, virtually to a man, found this theory to be full of "confusions" and "mistakes." But we have seen too that these charges are based upon erroneous interpretations of the conceptions upon which Aristotle's modal syllogistic is based.

We have argued that the Aristotelian theory of modal syllogism rests on the conception that the modality of a modal categorical proposition does not enter into the content of what such a proposition asserts, but rather has to do with the logical *status* of the proposition. Given this conception, the rest of Aristotle's theory flows inevitably out of a small handful of consistent and reasonably plausible interpretive insights into the rules governing the logical relationships among the so modalized categorical propositions. There is no basis whatsoever for denying the internal consistency and tenability of Aristotle's assertions about the logic modal categorical propositions.

However, an objection can be made with substantial propriety, not against Aristotle's logical *formalism* but against the underlying *material* considerations (drawn from his epistemological theory of scientifically demonstrative reasoning) which guided the construction of his theory of modal syllogistic. Specifically this objection attaches to his governing role of the major premise in determining the modality of modal syllogisms (i.e., in the pre-eminent case of the first figure). While these material criti-

---

[17] It is possible too that this shift was motivated by recognition of difficulties of the type noted in Hintikka [AD].

cisms are strongest when we shift away from Aristotle's view of the prime domain of applicability of syllogistic reasoning (viz., natural groupings), they merit consideration (and indeed they may have led Aristotle himself to abandon in *Anal. Post.* the theory of apodictic reasoning which he detailed in *Anal. Pr.*). In any event, his star pupils, as we know, did abandon this theory in favor of the *peiorem* rule. But this step was taken on the basis of material rather than of formal considerations relating to the logic of modal inference. The criticism of modern scholars (classicists and logicians alike) notwithstanding, from the standpoint of purely formal logic Aristotle's discussion of modal syllogistic stands in need of no other than trivial amendments.

Bibliography

Becker [ATM]. Albrecht Becker. *Die Aristotelische Theorie der Möglichkeitsschlüsse.* Berlin, 1933.

Bochenski [NHPM]. I. M. Bochenski. "Notes Historiques sur les Propositions Modales." *Revue des Sciences Philosophiques et Théologiques*, vol. 26 (1937), pp. 673–692.

Bochenski [LT]. I. M. Bochenski. *La Logique de Théophraste.* (Publications de l'Universite de Fribourg en Suisse, N.S., No. 32). Fribourg en Suisse, 1947.

Bochenski [AFL]. I. M. Bochenski. *Ancient Formal Logic.* Amsterdam, 1951.

Bochenski [FL]. I. M. Bochenski. *Formal Logic.* Notre Dame, 1961. [A revised translation by Ivo Thomas of the German edition, *Formale Logik*, Freiburg/München, 1956.]

Dominczak [JM]. Stanislas Dominczak. *Les jugements modaux chez Aristote et les Scolastiques.* Louvain, 1923.

Henle [FFS]. Paul Henle. "On the Fourth Figure of the Syllogism." *Philosophy of Science*, vol. 16 (1949), pp. 94–104.

Hintikka [NUTA]. Jaakko Hintikka. "Necessity, Universality, and Time in Aristotle." *Ajatus*, vol. 20 (1957), pp. 65–90.

Hintikka [AD]. Jaakko Hintikka. "An Aristotelian Dilemma." *Ajatus*, vol. 22 (1959), pp. 87–92.

Joseph [IL]. H. W. B. Joseph. *An Introduction to Logic.* Second edition. Oxford, 1916. Reprinted in 1950.

Kapp [GFTL]. Ernst Kapp. *Greek Foundations of Traditional Logic.* New York, 1945.

Lukasiewicz [AS, II]. Jan Lukasiewicz. *Aristotle's Syllogistic.* Second edition. Oxford, 1957.

Maier [SA]. H. Maier. *Die Syllogistik des Aristotles.* Three volumes. Tübingen, 1896–1900.

Mates [EB]. Benson Mates. Article on "History of Ancient Logic" in *Encyclopaedia Britannica* (1956 printing), vol. 14, pp. 315-20.

Meyerhof [VANB]. Max Meyerhof. "Von Alexandrien nach Baghdad." *Sitzungsberichte der Preussischen Akademie der Wissenschaften* (Philosophisch–historische Klasse), vol. 23 (1930), pp. 389–429.

McCall [AMS]. R. S. McCall. "Aristotle's Modal Syllogisms." Mimeographed article dated March 19, 1960. Privately circulated, and since printed in expanded form (*Aristotle's Modal Syllogisms*, Amsterdam, North Holland Press, 1963).

Prantl [GLA]. Carl Prantl. *Geschichte der Logik im Abendlande*; Bd. I. München 1855. [Unchanged photomechanical reprint, Graz, 1955.]

Rondelet [TLPM]. Antonin Rondelet. *Théorie Logique des Propositions Modales*, Paris, 1861.

Ross [APPA]. W. D. Ross. *Aristotle's Prior and Posterior Analytics*. Oxford, 1949. (Text edition with magisterial Introduction and Commentary.)

Welton [ML]. J. Welton. *A Manual of Logic*. Second edition. Two volumes. London, 1896.

von Wright [EML]. Georg H. von Wright. *An Essay in Modal Logic*. Amsterdam, 1951. (Note the brief references to Aristotle's modal syllogistic on pp. 82–83.)

CHAPTER 12

Logical Terminology
and Theory of Meaning

BY GEROLD STAHL

A WELL-CHOSEN TERMINOLOGY will simplify the work and help to avoid obscurities. The fact that there is no uniformity in logical terminology must be considered as indicating that none of the currently used terminologies is completely satisfactory and may be taken as an excuse for the following exposition in which a terminology, in certain points different from the usual ones, is introduced.

In spite of maintaining rigorously the formalistic position, we will consider the logical calculus not as a language but as a formal construction (of logicians and mathematicians). We make the following distinction: A calculus is exposed in a language, but we are calculating with something which is denoted by part of a language. This distinction is made in agreement with the theory of levels of language, which is used here in a relative sense ('relatively lower level,' etc.), as also later is the simplified theory of types ('relatively lower order,' etc.). We will base the following treatment *on the language* of the calculus and consider the denotations, starting from this language. Proceeding in this form, the formal character will be extended from the language to the denotations.

An expression is a finite sequence of typographical signs. A language is considered here as a set of expressions. Certain sub-sets of a language which only consist of theorems are called "systems." It is possible to list (explicitly one for one or by rules) certain expressions which will be called "elementary names" or "symbols";[1] certain others, which are constructed (by rules) from the former ones, will be called "complex names."

---

[1] We speak of symbols in the logical sense indistinctly if they consist of one or more typographical signs; e.g., "a_1" is a symbol. The standpoint of written language is usually prevailing in logic, and we will conserve it in this exposition.

We call the denotation of a name the "formal object," and these formal objects are, according to our statement above, what we are calculating with. We make the convention that every name denotes something which need not be real. (It is not the task of formal logic to determine this; thus logic has a higher degree of generality and may be applied to any imaginary field.)

Only the distinction between that which denotes and the denotations is made here. A third category of terms, such as "sense," etc., and the corresponding intensional distinctions will not be introduced.

There may be two or more names which denote the same formal object, as, for example, "Scott" and "the author of *Waverley*." In this case, Scott is identical with the author of *Waverley* (this corresponds to a rigorous adoption of the extensional point of view)[2] but has two different names : "Scott" and "the author of *Waverley*."

The symbols of the usual logical systems will be classified principally into symbols of variables and symbols of constants. The symbols of variables denote variables, whereas we understand by "variable" that which (in different connections) is identifiable with different fixed formal objects (the constants) that are the values of the variables.[3] We may also say : A symbol of a variable is that which (in different expressions) is substitutable by different symbols of constants.

The symbols of constants denote constants, i.e., fixed formal objects. It is possible to construct systems made up of expressions which do not contain certain types of symbols of constants. In this case special composed expressions (the descriptions) made it possible to denote the referred constants.

We classify now the symbols of variables and constants as propositional symbols, connectives, individual symbols, p-functional (or predicative) symbols, and operators.

What we call "proposition" (in the extensional sense), is the formal object denoted by a propositional symbol (of a constant).[4] In historical languages, we have sentences in the place of propositional symbols. We will say, for example, that the propositional symbol "p_a" denotes the same proposition as the (English) sentence "The snow is white." The symbol

---

[2] There may be different (rational) associations and correspondingly different intensions by using different names for the same object, but *in formal logic* we will not need the intension and the corresponding distinctions at all, if without them the results (the truth values of the sentences) do not change. There is no known case in the current logic literature where we could not disregard this distinction. Some especially interesting cases will be shown in the following text.

[3] Supposing that the variable is 'free.' In the case that it is 'bound' one is not interested in its particular values but calculates with the corresponding range.

[4] The formal object denoted by a propositional symbol of a variable will be called "propositional variable."

"p_a" has in the logical system the same function as the quoted sentence in English; we consider it as a translation from English into the language of the logical system. It is necessary to distinguish among :

1. The symbol "p" *denotes, names, symbolizes,* or *represents* (all terms treated here as synonyms in logic) a proposition (the proposition is in the lower level, the symbol in the higher one).
2. The symbol "p" *is a synonym*[5] *of,* or *abbreviates* (is a shorter synonym of), a sentence (of the same language and the same level).
3. The symbol "p" *is a translation* of, or *expresses,* a sentence (of another language).[6]

The range (k) of the respective variables (propositional variables) may have any finite or infinite number of propositions (practically $1 < k \leq \aleph_0$). The number of propositions with which one calculates is one of the essential characteristics of the logical system; it determines its valency. The preferential logical system is the two-valued one which refers to two propositions. Both propositions have a great number of names. We will give some examples in mathematical language, English (about history and biology), and logical language. One of the two has the names "$2 + 2 = 4$," "Socrates was a Greek philosopher," "All men are mortal," "T," "0," etc.;[7] the other has the names "$2 + 2 = 5$," "Socrates was a Chinese philosopher," "Some men are immortal," "F," "1," etc. There are also—and here begins the work of science—a great number of names of propositions of which we do not know which proposition they denote (i.e., whether they are true or false).

We classify the propositions (this is of interest especially for the systems which are not two-valued) as "designated" (the sentences of which are to be asserted) and "not designated."

It would not be possible to apply exclusively extensional considerations if there were cases in which a replacement of one name by another denoting the same formal object changed the truth-value of the corresponding sentence (i.e., changed the proposition). Contrary to the affirmations of Carnap,[8] an extensional treatment is possible if there appear modalities (without having recourse to a higher level, as for example, ". . ." is pos-

---

[5] *Logical* synonyms and translations (based on the sameness of extension) must not be confounded with the *linguistic* ones, because for the latter the sameness of intension is needed.

[6] We distinguish, according to the level of language, between (1) the proposition denoted by "p," or simply the proposition p (without quotation marks); and (2) the sentence translated into or abbreviated by "p," or simply the sentence "p" (with quotation marks).

[7] In other words, this is the proposition whose sentences (names) are considered as true (are considered synonyms of the truth-value "truth").

[8] *Einführung in die Symbolische Logik,* Vienna, 1954, p. 40; *Meaning and Necessity,* Chicago, 1947, § 11; (*The Logical Syntax of Language,* London, 1937, § 69, 70).

sible). Following Carnap, the two sentences "It is raining"—supposing it does not rain—and "It is raining and not raining" are both false.[9] Now, "It is impossible that it is raining and not raining" is true; whereas by replacing the other name for the same proposition we get the sentence "It is impossible that it is raining," which is false. To resolve this problem, there are two possibilities:

1. If n-valued ($2 < n \leq \aleph_0$) modal systems are used. "It is raining" denotes another proposition (has another truth-value) than "It is raining and not raining." For this reason the replacement cannot be effected, and the contradiction disappears.

2. If we use a two-valued logic with quantifiers of the type "in all circumstances," etc., "It is raining" ("Fa"; "a is a circumstance in which it is raining") and "It is raining and not raining ("$Fa \cdot \sim Fa$") are only sentences and replaceable as such. If they are preceded by expressions such as "It is impossible that . . ." ("There does not exist a circumstance in which . . .") there appear new and completely different sentences with quantifiers "$\sim(Ex)Fx$" and "$\sim(Ex) . Fx . \sim Fx$" which do not contain "Fa" and "$Fa . \sim Fa$."

The connectives appear usually only as symbols of constants and denote so-called "truth-functions" or "connections" for one or more propositions. The values of the respective arguments (propositional variables) are propositions (denoted by "T," "F," etc.), the values of the truth-functions are also propositions.

The most frequently used connectives denote the disjunctive, conjunctive, conditional, and bi-conditional (or identity) connections between two propositions (binary truth-functions), and the complement connection (a unary truth-function).[10] In general, one speaks of "n-ary" truth-functions and accordingly of "n-ary" connectives.

So far we have spoken only about truth-functional constants, but it is also possible to calculate with truth-functional variables. The range of a truth-functional variable depends on the range of the propositional variables. If the latter is k (if the system is k-valued) the former will be k^{k^n}, where n is the number of the arguments of the truth-function.

We call "individual" the formal object (constant or variable) denoted by an individual symbol such as "a," "b," or "x," "y," and we agree to consider the individuals as of the relatively lowest order (according to the simplified theory of types). The individuals may be—corresponding to

---

[9] The corresponding propositions are (apparently) identical (the term frequently used for the corresponding sentences is "equivalent"), and their names are therefore replaceable.

[10] For the corresponding composite sentences in the higher level, one uses in the last three cases the names "implication," "equivalence," and "negation."

the completely relative character of this theory of meaning—things, persons, ideas, etc., but also classes, etc.[11] (if there is nothing of a lower order in relation to them) and names, sentences, etc. (if we *mention* them, using a language of higher level).[12] It must be understood, that there is no reference to "reality."

The range of the individual variables is sometimes called "universe of discourse (of the first order)." It may include any number of individuals. According to the axiomatization of the usual functional system it must contain at least 'one' individual; for a system with an infinity-axiom it must contain at least $\aleph_0$ individuals.[13]

We call "propositional function" or simply "p-function" the formal object (constant or variable) denoted by a p-functional symbol such as "F," "G," etc., or (if we want to denote it with its arguments) by a symbolization such as "$\lambda xy \ldots (Fx,y \ldots)$," etc.

One distinguishes among unary p-functions, binary p-functions, etc., where instead of "unary p-functions" we may also say "classes," "sets,"[14] "attributes," or "properties" (all used here as synonyms in logic); instead of "binary p-functions" also "binary relations"; instead of "ternary p-functions" also "ternary relations," etc.

The values of the corresponding arguments are individuals or p-functions; the values of the p-functions are propositions. The range of a p-functional variable depends on the range of the propositional variables (k) and on the range of the individual variables (r). If the p-functional variable is unary and of first order, then its range contains k^r p-functions; if the variable is of order m, the range contains $k^{k \cdots k^r}$ (with m "k") p-functions. An n-ary p-function, with $n > 1$, may be reduced to a unary p-function, according to the Wiener-Kuratowski method.

Suppose we have a binary p-function which has, for some ordered pairs of its argument-values, a designated proposition as its value. We then say that there exists between the first members and the second members of *these* pairs a unary "specific function," the argument-values of which are the values of the second argument of the p-function (they are the second members of the pairs), and the values of which are the values of the first argument of the p-function (they are the first members of the pairs). Practically, one speaks only of a unary specific function if the correspond-

---

[11] The classes are, naturally, not classes *for this system* but only individuals; the corresponding holds for names, sentences, etc.

[12] A detailed treatment will be found in G. Stahl, "Le problème de l'existence dans la logique symbolique," *Revue Philosophique de la France et de l'Etranger*, 1960, N. 1, pp. 97–104.

[13] Possible modifications are treated in G. Stahl "Les univers du discours et les calculs correspondants," *Annales de l'Université de Paris*, 27, No. 4, Oct.–Dec. 1957.

[14] If one works with the theory of types, the distinctions between classes and sets are not necessary.

ing binary p-function is one–many or one–one. The indicated correspondence holds in general between n-ary p-functions and $(n-1)$-ary
specific functions.[15]

All operators in use are symbols of constants. The most important of
them are the quantifiers which denote "quantifications." These may be
considered as relations between the ranges of the corresponding variables
and p-functions.

We will now show in some examples (taken from Church)[16] how to
avoid triple distinctions (among "name," "sense," and "denotation") of
the so-called "oblique use of names," distinguishing only between name
and denotation. The problem here (as well as in the example of Carnap)
is that replacing one name by another which denotes the same formal
object changes (apparently) the truth-value of the sentence in which the
replacement is made, in certain cases. If this were so effectively, special
artifices (such as triple distinctions, etc.) would be justified. To show that
there is no need for them, we will restate the sentences in a more convenient
manner, the way one does in many cases of formal treatment of sentences
in common language which have nothing to do with theory of meaning. It must be adjoined—as Church shows—that, using the name-relation
(or mentioning names in general), we must also indicate the language part
of which is mentioned. This last indication, however, will be given only if
it is not evident which is the language in reference.

1. "George IV wished to know whether Scott was the author of
   Waverley" (where "Scott" and "the author of *Waverley*" have
   oblique occcurrences—i.e., replacing them by other names of the
   same object, for example, "the author of *Waverley*" by "Scott,"
   changes the truth value of the sentence) will be restated in the
   known manner into "George IV wished to know whether 'Scott was
   the author of *Waverley*' is a true sentence" (where "Scott was the
   author of *Waverley*" is different from "Scott was Scott"; they are
   two different *formal objects*—the principal sentence is in meta-
   language—because they consist of other sequences of letters).[17]

---

[15] We have, for instance, the following equivalences where "F" denotes a binary
p-function and "f" the corresponding unary specific function:

$$\lambda xy\ (Fx,y)\ ab \equiv p$$
$$(\lambda y\ (fy)\ b = a) \equiv p$$

In order that F be satisfied and the identity-expression in the second equivalence
exact, it is necessary that p (the value of F) is a designated proposition.

[16] *Introduction to Mathematical Logic*, Princeton, 1956, pp. 8, 27.

[17] "Seneca wrote that man is a rational animal" will not be restated into "Seneca
wrote 'Man is a rational animal'," but into "Seneca wrote, expressed in English
(according to linguistic translation), 'Man is a rational animal'," or into "Seneca
wrote in Latin 'Rationale enim animal est homo'."

2. "Schliemann sought the site of Troy" (where "the site of Troy" has an oblique occurrence and is in this form not replaceable by another name of the same object) will be restated into "Schliemann sought the site of Troy under the name 'the site of Troy'."

The considerations applied to the foregoing are the following : Every language may have different names which denote the same object (*synonyms*) and which are mutually replaceable; (but not the names of the names if the names are different). Different languages may have, at least in certain sections, corresponding names for the same object (*translations*), as, for example, "homo" and "man," or "Rationale enim animal est homo" and "Man is a rational animal."

The synonymity represents a problem in modal contexts as Quine[18] especially has pointed out. The truth-values of "Necessarily if there is life on the evening star, then there is life on the evening star," and of "9 is necessarily greater than 7," do not (apparently) remain unchanged if we replace "evening star" by "morning star" or replace "9" by "the number of planets." To resolve this problem, there is, as in the case of Carnap, no need to use a higher level (" '9 is greater than 7' is necessary" or " '9 is greater than 7' is necessarily true") nor a special and very complicated "intensional ontology."

There is only one request : Synonymity (as any other affirmation in formal science) must be deduced (from axioms or premises). Until it has not been deduced (or stipulated by definition) that two names denote the same formal object, we will leave the possibility open that there may be two different objects. The evening star satisfies the p-function to be seen in the evening, the morning star to be seen in the morning. "Evening star" will not be replaceable by "morning star" if it has not been deduced that the corresponding objects are identical, i.e., that they satisfy the same p-functions.[19] If this has been deduced, the truth-value of the first sentence remains unchanged when the replacement is effected.

The same holds for "9" and "the number of planets," which is a very characteristic example because for a long time in the history of astronomy the number of planets was different from 9. The number of planets must be considered as undetermined so long as its identity with any number has not been deduced.

So there is no need for Quine's alternative of an "idealistic universe" of "concepts" or "sense of names" without concrete objects, numbers, and

---

[18] *From a Logical Point of View*, Cambridge, Mass., 1953, VIII.

[19] Naturally this refers directly only to what may be called "modality *de re*" and only indirectly (supposing certain implications) to the modality of propositions, while in the treatment in the higher level one might speak of "modality *de dicto*." See G. Stahl "General Considerations about Modal Sentences," *Zeitschr, f. math. Logik u. Grundl. d. Math.*, Berlin, Bd. 5, H 3/4, 1959, pp. 280–290.

classes. (But there is no place either for a universe of primitive realism.) The universe one works with is the formally fixed universe of the corresponding science.

Identity will coincide with necessary identity, because only in this form is it, for modal logic, in agreement with the usual definition of identity. To use the name "identity" for a relation which does not fulfill this condition would not be justified, and any replacement would be excluded.

With all this has been sketched a logical terminology which has the disadvantage of being in some cases different from the usual ones (but in the case of the term "variable," for example, more in agreement with the mathematical practice) and has the advantage of reducing the entities considerably. The possibility was shown of formulating, in agreement with these considerations, a simple theory of meaning (for the field of logic) which is exclusively extensional.[20]

---

[20] Naturally the terminology here exposed permits also an intensional treatment.

III
THE CRITICAL APPROACH
TO SCIENCE

CHAPTER 13

The Nature of Scientific Problems and Their Roots in Metaphysics*

BY JOSEPH AGASSI

ACCORDING TO POPPER'S PHILOSOPHY the perfect division of labor in research would soon stop scientific progress. His view explains why in the history of science many investigators have concentrated on a handful of problems. The problem arises: How did investigators coordinate their choice of scientific problems? By what criteria did the bulk of investigators of a given period decide which problem was fundamental or important?

There exist a variety of such criteria, but one criterion stands out as the most important. Those scientific problems were chosen which were related to metaphysical problems of the period; those scientific results were sought which could throw light on topical metaphysical issues.

My aim is to present this as a historical thesis. I do not contend that scientific interest devoid of metaphysical interest is in any sense illegitimate or inferior. Investigators may wish to study a small part of the universe without bothering to study the universe as a whole, without even bothering to ask how their partial picture integrates with man's picture of the universe as a whole. Yet I contend, firstly, that very frequently problems, theories, and experiments which are traditionally regarded as important are highly relevant to the metaphysics of their time; and secondly, that my first contention provides a solution to the question of how the choice of scientific problems is coordinated.

---

* I wish to acknowledge with gratitude the great assistance I received from I. C. Jarvie, Kenneth Topley, and J. W. N. Watkins.

[189]

This is all I wish to assert in the present essay. I shall discuss problems of demarcation of science, of pseudo-science, and of metaphysics, mainly to dispel some vulgar errors concerning metaphysics (namely the identification of it with pseudo-science) and its role in the scientific tradition. I shall argue that metaphysics can progress—not so much in order to defend metaphysics as to expound my view of metaphysics as a coordinating agent in the field of scientific research.

I.   Scientific Research Centers Around a Few Problems

Since there are more scientific problems to be studied than researchers to study them, a complete avoidance of overlap between projects is quite possible. The more the number of existing problems exceeds the number of researchers, the more one would expect the actual case to tend naturally toward the ideal of complete absence of duplication. But the facts are quite otherwise. Here are two historical examples where numerous obvious problems have been ignored. Diffusion is a phenomenon with instances widespread in physical nature : river water rapidly mixes with the oceans' waters, smoke with the atmosphere, salt with soup. Until the late eighteenth century no one paid any attention to this phenomenon and the scores of problems it raises. Priestley seems to be the first who studied it; Dalton concentrated on it for a while. Yet though Dalton's study received great publicity, only a handful of thinkers worked on diffusion before the celebrated studies of Maxwell rendered it an integral part of physics. My second example is elasticity, which was left almost entirely unstudied between the days of Hooke and of Young but was studied more and more seriously in the nineteenth century, only to be relegated in the twentieth century to the borders of applied mathematics and technology.

Whether concentration of intellectual power on a few problems is advantageous or a waste has hardly been studied because of misconceptions about science. Popper's theory of science answers this question unambiguously : perfect division of scientific research work will quickly bring scientific progress to an end. This theory makes the "friendly-hostile cooperation" between individuals crucial for progress. Some offer new ideas, some offer criticisms of these ideas, some offer alternatives to these ideas; if they all worked on different problems there could be no cooperation. Robinson Crusoe would be unable to sustain the development of science, because of his limited capacity to criticize himself and thus to get out of the routine of his way of thinking.

The existence of a variety of problems to be solved, and the fact that newcomers to science have a great variety of reasons which draw them to science, would by itself render science almost Crusonian. But by some process which has not yet been studied or even noticed, the more a person's interest develops, the nearer it approaches the interest of other students of the same field. Somehow interests coordinate themselves. And my problem now is what is this means of coordination (though I shall not discuss here the way by which individuals learn to apply it).

Undoubtedly, there exists a variety of coordinating factors. New economic and political needs, new mathematical or experimental techniques, offer new avenues which are sometimes explored. Yet, by and large, there are minor and often secondary factors—secondary, because developments of techniques and of their fields of application often follow interests. By and large, widespread scientific interests may be shown to be connected with some metaphysical problem of the day. It is my contention that whatever the starting point of a person's interest in a science, the more that person's interest develops the closer it approaches the general interest, the interest which dominates the tradition in that science, and that this general interest springs from, and flows back to, metaphysics.

Most philosophers and historians of science would vehemently oppose this view. Descartes, as is well known, developed a philosophical theory in which metaphysics provides the framework for science. His ideas were greatly improved by Kant, but this was the last significant effort in this direction; for good reasons or bad Kant's idea has been universally rejected. In this essay I wish to rehabilitate metaphysics as a framework for science, but within the framework of Popper's critical philosophy.[1]

My view is this. Metaphysical theories are views about the nature of things (such as Faraday's theory of the universe as a field of forces). Scientific theories and facts can be interpreted from different metaphysical viewpoints. For example, Newton's theory of gravitation as action at a distance was interpreted by Faraday as an approximation to a (future) gravitational field theory. An interpretation may develop into a scientific theory (such as Einstein's gravitational field theory) and the new scientific theory may be difficult to interpret from a competing metaphysical viewpoint. Metaphysical doctrines are not normally as criticizeable as are scientific theories; there is usually no refutation, and hence no crucial experiment, in metaphysics. But something like a crucial experiment may occur in the following process. Two different metaphysical views offer two different interpretations of a body of known fact. Each of these inter-

---

[1] This essay contains deviations from Popper's own views—as expressed in his classical *Logik der Forschung*. These deviations permit Popper's philosophy to accommodate the view that metaphysics is a framework for science. I do not think, however, that Popper himself will widely disagree with the content of this essay; indeed, I am happy to acknowledge much of it to his guidance, in lectures and in frequent and lengthy private discussions over a period of seven years, including the period of my graduate studies under his supervision.

pretations is developed into a scientific theory, and one of the two scientific theories is defeated in a crucial experiment. The metaphysics behind the defeated scientific theory loses its interpretative power and is then abandoned. This is how some scientific problems are relevant to metaphysics; and as a rule it is the class of scientific problems that exhibit this relevance which is chosen to be studied.

II. The Anti-metaphysical Tradition Is Outdated

My own interest in physics originates from a very early interest in metaphysics; the present essay may be no more than a projection of my own case history into the history of science at large. In my undergraduate days I used to resent the hostility toward metaphysics displayed by my physics teachers; my present view is in a sense an inversion of theirs. They derided all metaphysics as the physics of the past; I extol some metaphysics as the physics of the future. But I wish to be fair to their view, and perhaps the best means to arrive at a fair attitude to a doctrine is to try to see it in its historical perspective.

Francis Bacon's anti-Aristotelian-metaphysics, which was the first fanfare of the modern positivists, was very valuable. In launching an attack on Aristotelian metaphysics, he overenthusiastically took it to be an attack on all metaphysics. This was an exaggeration, and a very understandable and effective one at a time when Aristotelian metaphysics reigned supreme. Then came the victory of Copernicanism and of the Galilean-Cartesian metaphysics. This development admittedly altered the situation. From then onward Bacon's exaggerated idea might have been profitably cut down to size by studying the difference between Aristotle's bad metaphysics and Descartes's good metaphysics. Yet this is debatable, since at that time there was still a need to encourage experimentation rather than speculation. Moreover, throughout the seventeenth and eighteenth centuries metaphysics was closely linked with religion; and religion had to be banned from scientific discussions for very obvious social and political reasons. Since the early nineteenth century both of these factors have become negligible, but other factors have taken their place; fortunately for the positivist knight-errants, there was the task of slaying such awful metaphysical dragons as the Hegelians and the existentialists. Unlike Aristotelianism, positivism has not been useless during its period of obsolescence. It is still fighting bad metaphysics, under the somewhat absurd guise of fighting metaphysics as such.

In addition to being an overzealous criticism of irrationalist metaphysics, positivism has also served the rationalist metaphysician. Metaphysics can easily degenerate into pseudo-science by providing a framework for *ad hoc* explanations instead of scientific ones. The Baconian-positivist attack on metaphysics as *ad hoc* or pseudo-scientific helped the good metaphysician by putting him on guard against irrational practices.

It is unfortunate that the merits of positivism are so often exaggerated, since positivism is conducive to ignorance. I have met physicists who know about only one metaphysician—Hegel—and only one detail concerning him—that he said when a doctrine of his turned out not to accord with facts "so much the worse for the facts." Rarely has anyone paid more dearly for a silly joke.

It is not my purpose here to disprove positivism but I feel I have to stress that in this essay I am speaking of good metaphysics while intentionally ignoring bad metaphysics, after having acknowledged the partial justice of the positivist attack on it. Every field of human activity ought to be judged by its very best, and it is time to notice that examples of bad metaphysics do not show that all metaphysics is bad. One can show that all metaphysics is bad, but only after abandoning the ordinary or traditional meaning of the word 'metaphysics.' This word is used by Hegelians and by positivists to signify the theory of the cosmos as a whole, of the very mystery or essence of the universe. In his *Tractatus* Wittgenstein accepted Newton's metaphysics as a framework for physics, but he did not call it 'metaphysics'; he considered 'the mystical' alone to be the subject matter of metaphysics. The positivists, the Hegelians, and the mystics, rightly claim that the mystical is unexpressible. This is a point which Russell rightly considered (in his *Mysticism and Logic*) trivially true. Metaphysics in the sense of a theory of the mystical is hence impossible. My own use of the word 'metaphysics' in the present essay is in its traditional and much narrower sense. Metaphysical doctrines are to be found, first and foremost, in Aristotle's *Metaphysics,* especially in Book Alpha : all is water; atoms and the void; matter and form; etc. There are a variety of sets of first principles of physics. Do these belong to scientific physics? Are they entailed by scientific theories? Are they useful for scientific research? I think they do not belong to scientific physics (though in principle they might). Metaphysical ideas belong to scientific research as crucially important regulative ideas; and scientific physics belongs to the rational debate concerning metaphysical ideas. Some of the greatest single experiments in the history of modern physics are experiments related to metaphysics. I suggest that their relevance to metaphysics contributes to their uncontested high status. And yet, I contend, the metaphysical theories related to these experiments were not parts of science. This raises the problem of what kind of relation between a given theory and observable facts renders that theory scientific.

III.   A Historical Note on Science and Metaphysics

The term 'speculative metaphysics' and the term 'speculations,' when used as synonyms for 'metaphysics' (by Boscovitch, Faraday, and others), indicate the view that metaphysical doctrines are products of the imagination, in contrast with scientific theories which are—allegedly—products of inductive inference from facts. It was indeed this view which led to the tradition of divorcing science from metaphysics. The first modern positivist, Francis Bacon, presented the two methods, of induction and of speculation, as irreconcilably opposed to one another. The proper inductive investigation, he proclaimed, can be conducted only in the absence of all preconceived notions. Those whose minds are full of speculations are entirely unfit for proper scientific experiment and observation, much less for theorizing inductively : they are biased in favor of their speculations, and this bias makes them ready to observe only those facts which verify their speculations and unwilling to observe those facts which refute them. Consequently, they achieve not the truth but the reinforcement of their own preconceived opinions, and their biases thus become prejudices and superstitions.

Bacon's violent opposition to metaphysics was less violent than the ultra-modern one. His opposition to metaphysics was merely an opposition to its method; it was not an opposition to the abstract character of metaphysics but to the leaping to metaphysical conclusions. By developing science properly, by starting with observation and then slowly developing theories by gradually increasing the abstractness of knowledge, by ascending the inductive ladder properly without skipping any step, Bacon held, we shall end up with the most fundamental theory, namely, with scientific metaphysics. This metaphysics will be scientific because it will have been achieved, not by the speculative method, but by the inductive method.

Scientific metaphysics was later defended by Descartes and by Kant, each of whom considered his own metaphysics to be a body of certain, and hence scientific, knowledge. Their idea of certitude differed from Bacon's; it was based on *a priori* reasoning rather than on inductive inference. Consequently they viewed metaphysics as the beginning, not the end, of scientific inquiry. But both in viewing science as certain, and in taking it for granted that metaphysics must be scientific or perish, they barely differed from Bacon. It was William Whewell, the disciple of both Bacon and Kant, who first defended unscientific metaphysics from a scientific point of view.

In Whewell's view scientific doctrines do not emerge inductively from facts; they are first imagined and then verified empirically. And he considered his own (Newtonian-Kantian) metaphysics *a priori* valid, namely, demonstrable independently of empirical evidence. In accepting Kant's *apriorism* he rejected Bacon's view that all preconceived ideas are verifiable by virtue of their being prejudices, contending that much as people had sought to verify Newton's optics, much as they were prejudiced in its favor, they ultimately rejected it. His problem was how to explain why assent to Newton's mechanics was justifiable and assent to Newton's optics unjustifiable. He wished to find out the proper canon of verification and show that Newton's theory of gravity, but not Newton's optics, had conformed to it.

In brief (and in a slightly improved version), Whewell's canon can be put thus : proper verification is the result of severe tests. The procedure of severe testing is this : First try to explain known facts and state your explanatory theory as explicitly as possible. Then try to deduce in a rigorous manner from the theory a new prediction of observable facts. Then, and only then, decide by observation whether this prediction is true or false. If the prediction is false then the theory is obviously false too; if the prediction is true then the theory obviously explains the new facts without adjustment ("adjustment" being a suitable alteration or addition). In the latter case, Whewell declares, the theory is verified. Newton's theory of gravitation had been severely tested, and consequently the result of the tests could either refute it or be explained by it without any adjustment. In contradistinction, Newton's optics never stood the risk of a test and hence never explained a single new fact. Many new facts were alleged to be explicable by Newton's optics. Even Laplace had endorsed this allegation. Yet upon a simple and clear examination, which Whewell executed in a most masterly fashion, each of these new facts turned out to be explicable not by the original theory but by the adjusted theory.

Both Bacon and Whewell were interested in the problem of the demarcation of science. But their interests stemmed from different roots. Bacon considered Aristotelianism, which was then the academic metaphysics, to be the chief impediment to the advancement of learning. Whewell viewed Newton's metaphysics, which was by then the academic metaphysics, as demonstrable. His problem was not metaphysics but the overthrow of the allegedly verified Newtonian optics. Thus, while Bacon demarcated science mainly from metaphyhics, Whewell demarcated science mainly from pseudo-science.

Since, according to Whewell, science begins by the invention of explanatory hypotheses, he was all for every possible source of inspiration. And he viewed all (reasonable) metaphysics as such a possible source. He gave a striking example for this. Kepler had developed his scientific hypotheses, Whewell maintained, in an attempt to carry out Plato's metaphysical program as outlined in his *Timeus*. This idea of Whewell's was so revolutionary that this great philosopher is now almost entirely forgotten because Mill and his followers condemned him as an intuitionist. (This charge is, of course, quite untrue. Whewell relied not only on intuition but also on Kantian transcendental arguments and on empirical tests.)

Initially, Popper's interest in the problem of demarcation was similar to Whewell's, though his examples were different; it was Marxism and Freudianism which he viewed as pseudo-scientific. His demarcation of science may be contrasted with Whewell's thus : Whewell demands that a scientific theory be testable and emerge triumphant from the tests, while Popper merely demands testability. Neither of them is hostile to metaphysics, and both contend that metaphysics is sometimes important as a source of scientific inspiration. A remnant of positivist prejudice may perhaps be detected in Popper's lumping together (like Bacon and unlike Whewell) of a few kinds of nonscientific theories, including metaphysics, pseudo-science, and superstition, under the one label 'metaphysical.' Though I dislike this label, I do not think it matters beyond leaving some ambiguity concerning the difference between metaphysics and pseudo-science.

IV.   Pseudo-science Is Not the Same as Non-science

Popper's idea (pseudo-science is untestable) is a marvel of simplicity. It explains why no matter how bad a pseudo-scientific doctrine is, its proponent may regularly win debates. It resolves the conflict involved when we feel obliged, against our own better judgment, to take a theory seriously because its proponents seem to be entirely undefeatable. It amounts to a proposal not to embark on the game before fixing its rules, before deciding in advance what kind of argument, if any, would be capable of defeating the proponent of a theory, and determining not to try to defeat him if he turns out too evasive to be vincible. As Whewell has pointed out, no kind of argument will defeat the proponent of any theory if he is allowed to adjust even minor details of his theory in an *ad hoc* fashion. On this Whewell and Popper are agreed. Yet wonderful as Whewell's ideas about pseudo-science are, by demanding too much from science he threw out the baby with the bath-water.

According to Whewell, scientific theories must also have withstood test. Consequently, he viewed as pseudo-scientific those theories which falsely claim to have withstood test. This leaves unclassified those theories which are testable but have been obviously refuted. As Whewell considered these to be neither scientific nor metaphysical, he confusedly implied that they are pseudo-scientific, especially when they are submitted to recurrent readjustment and retest. According to Popper such theories are scientific, for he only demands testability; according to Whewell they could not be considered scientific, and so he held them in contempt. He knew that Newton's optics had been falsely held to have been verified. Yet he did not see that as long as verification was considered a hallmark of respectability, the immense respect for Newton gave these false claims an immense appeal. But if the requirement of Whewell and his predecessors of a respectable scientific theory is too stringent, is not Popper's requirement of a respectable scientific theory, namely, a high degree of refutability, a trifle too lax?

Traditionally, a variety of characteristics have been attributed to science. Popper accepts some of these attributes, such as high explanatory power, high informative content, abstractness, generality, precision, and simplicity; he rejects others, such as obviousness and verifiability. He seems to have claimed in his *Logic of Scientific Discovery* that the characteristics in the first group are all reducible to one, to testability. This is his justification for requiring only this one characteristic of a theory before labelling it 'scientific.' I have little doubt that Popper will fully agree that the spurious simplicity of some monistic doctrines (such as Marxism or mechanism) rather than their spurious explanatory power has deluded some people into regarding them as scientific. Simplicity, however, is traditionally viewed (since Leibniz) as the paucity of assumptions relative to the amount of factual information they explain, so that there is no need to differentiate between simplicity and high explanatory power for the purpose of demarcation. And Popper would say the same concerning explanatory power, which, in his opinion, increases with refutability. For my part, I consider that the various characteristics of science are less often dependent on each other than Popper suggests. But I still side with Popper in viewing spurious refutability, rather than, say, spurious simplicity, as the chief characteristic of pseudo-science, and for two reasons. First, whatever else may characterize a scientific theory, the very acceptance of the proposal that scientific theories are agenda to be tested renders Popper's proposal to check whether a doctrine in question is testable or only spuriously testable a matter of supreme practical importance. Second, the claim of pseudo-science is the claim for empirical character. And empirical character is nothing else but empirical refutability, as I shall soon explain. Thus, Popper's demarcation between science and pseudo-science does not require any amendment even on the assumption that he has erred in correlating the various characteristics of science. As to his characterization of science as such, it requires a reformulation if, as I think, his way of correlating the various characteristics of science is in error. I think we have to characterize scientific theories not only by their refutability, but also by their simplicity, high explanatory power, etc. This has an immediate bearing on the problem of selection of scientific problems and of scientific theories which is the topic of the present essay. According to Popper we always look for the most easily refutable theory. In my opinion this is not the case.

V. Popper's Theory of Science

Popper's arguments for his claim that empirical character is empirical refutability are very compelling. Logically, observation reports can contradict theories but not entail them in any way. Philosophically, Popper's view is the doctrine of learning from experience as a special case of learning from mistakes, of the critical method. Socially, it presents students of nature as human rather than as unerring supermen. Historically, it opens wide vistas of new studies of the history of science uncharted by the modern science textbook. Popper's greatest contribution to the philosophy of science seems to me to be rooted in the simple idea that since empirical character is empirical refutability, scientific research is a special case of Socratic dialogue. But I deny that the empirical character of science is all that makes science what it is.

It is not difficult to find empirical developments, i.e., empirical refutations, outside the field of science. Thales's metaphysical doctrine ("all is water") was refuted empirically when water was first decomposed; Moebius (as I. Lakatos would say) may have refuted empirically the mathematical theory "all surfaces have two sides"; Faraday refuted empirically some spiritualistic superstitions; Marx's prophecy about the geographical location of the socialist revolution has been refuted by his Russian followers; and this amounts to the refutation of his materialism since it entails the valuelessness of imaginative ideas; the very important philosophical doctrine about the universality of common sense (which even Duhem still advocated) is empirically refutable by comparative studies. Necessarily, either such cases should be viewed as scientific or Popper's proposal should be considered inadequate. My choice is the latter: I propose to use Popper's convention as a convention concerning the empirical character of science, not concerning empirical science as such. There is no difficulty in admitting that daily experience, as well as some developments of mathematics (or metaphysics, or any other field of intellectual or practical development), manifest a certain empirical character, even though they do not belong to empirical science. Empirical science manifests its empirical character more systematically than mathematics, and it manifests other characteristics as well, which are lacking in mathematics.

But what about the claim that theories manifesting empirical character, i.e., refutable theories, also necessarily manifest the other characteristics of science, i.e., they have informative content, explanatory power, simplicity, abstractness, generality, and precision? I simply reject this claim.

As I have said earlier, I interpret a great deal of Popper's discussion in his classical work to be an attempt to support this claim. I consider the value of that part of his discussion as a valid criticism of his opponents and as stimulating heuristic material, but as very far from being a finished product.

To maintain my thesis I must contradict Popper here. He would say that research is conducted toward the finding and the testing of highly testable hypotheses, whereas I say that it is very often conducted toward the finding and the testing of metaphysically relevant hypotheses. And as a rule, I shall later show, research tends to begin with which hypotheses have a low degree of testability or are not testable at all. Consequently investigators often have to use great ingenuity to test a barely testable hypothesis, and even first improve a hypothesis to the point of rendering it testable to some degree. If the aim of science were merely producing testable hypotheses and then testing them such procedures would be irrational. But the aim of science, or rather the aims of science, are different.

The aim of science is to attempt to comprehend the world rationally, as we all agree (including the positivists who should disagree). But this is too vague. What is the rational method and what is comprehension? Rationality, said Popper, is manifest in empirical tests. He later generalized this : the rational method is the critical method. Is metaphysics rationally debatable? Yes. I shall argue that the study of a hypothesis of a low degree of testability is often conducted with a view to criticizing some metaphysical theory upon which it may have some bearing. So much for rationality. As to comprehension, Popper views it as deductive explanation, and he has suggested that explanatory power goes with refutability. I deny that explanation is the only method of comprehension. As I shall show later, the attempt to coordinate our various explanations within one metaphysical framework is not explanation, yet it is, in some weaker sense, an attempt at comprehension. Moreover, I deny that explanatory power is always dependent on refutability. Already in the last section of his great book Popper has noted that some theoretical systems may have some explanatory power and yet be untestable. I have already mentioned examples of refuted theories of little or no explanatory power.

Degrees of testability are, I think, of little practical importance. All that matters is that we may test in at least one way an interesting theory. According to Popper, there are two factors contributing to the degree of testability of a theory, the number of possible events which may refute that theory, and the probability of each potential refutation. To my mind the possibility of observing the next refuting event is all that matters, not the number of possible refutations. As it is the number of all excluded possibilities which is the content of the hypothesis, content is not the same as practical testability. *Ad hoc* explanations have some empirical content yet are untestable. Explanatory power is not content, and not even truth-content (i.e., that part of a theory's content which is true), but I should say (in agreement with Leibniz's idea as I understand it), known-truth-content (i.e., the overlap of a theory's content with the class of true observation-reports). And high explanatory power is not the sole characteristic of a satisfactory explanation. As I have learned from Popper himself, a satisfactory explanation must be independently testable. Thus, Weyl's theory which unifies Maxwell's and Einstein's has a high explanatory power and a high degree of testability, but no known independent testability, and thus it is not considered scientific. Simplicity depends not only on explanatory power and the paucity of parameters, as Popper mentions in his early work, but also on depth, as he now says. Nor does abstractness go together with universality : Boyle's law is more general but less abstract than the theory of consumers' demand, and the Heitler-London theory is more abstract but less general than Schroedinger's theory.

The result is pluralism : we may admire one theory for its boldness, another for its explanatory power, another for its elegance; and yet another, I suggest, for the light it throws on some topical metaphysical issues.

There seem to be very good reasons for Popper's correlation of a higher degree of testability with a higher degree of explanatory power, etc., and these reasons are of heuristic value. One reason of Popper's is this : If one theory explains another theory, it is obviously not less refutable than the other. If one theory explains another theory as a first approximation, then it is more precise, and a higher degree of precision goes together with a higher degree of testability. This is so because a more precise theory excludes more (logically) possible states of affairs, thereby possessing both a higher informative content and a better (a priori) chance of being refuted, or a lesser a priori probability. These arguments are valuable but insufficient and partly incorrect.

In his classical paper "The Nature of Philosophical Problems and Their Roots in Science" Popper has given an admirable account of Pythagoras's metaphysics and the history of its refutation. When I read this excellent essay I decided to study under Popper; so the title of the present essay adverts to his, partly for sentimental reasons. Yet, perhaps because my prejudice in favor of metaphysics came first, I was unhappy about his taking Pythagorean metaphysics to be scientific. Since his reason was that this metaphysics was refuted, I was bound to examine his refutability criterion for the demarcation of science. I now propose his empirical refutability criterion to be the criterion of empirical character, not of empirical science as such. Empirical science is the set of highly informative and simple explanations which exhibit independent empirical character—satisfactory explanations, for short. I owe this idea to Popper himself : in his lecture courses Popper presents science rather in this way than in the way he does in his classical *Logic of Scientific Discovery*.

VI. Superstition, Pseudo-science, and Metaphysics Use Instances in Different Ways

Bacon justified his lumping together metaphysics with superstition and pseudo-science by saying that the method of them all is that of marshalling verifying or confirming examples or instances and persistently ignoring counter-examples or refuting instances. This is much too coarse a characterization; to refine it we must first notice a few of the different roles that instances may play in intellectual activity.

The role of an instance may be solely presentational: we understand an abstract idea better when we are told how to apply it to concrete cases. So long as the purpose of an instance is elucidatory, an author is at liberty to choose his instances so as to avoid a discussion concerning their truth or falsity, and the more obvious the instance the better. One should either take a presentational instance for granted or use another in its stead. The moment an instance is sufficiently significant to be not easily dispensable, it has additional roles.

The most important role of instances is their role as refuting instances. This is the crux of Popper's solution of the problem of induction: learning from experience is learning from a refuting instance. The refuting instance then becomes a problematic instance, namely, an instance which ought to be explained by a new theory. The last important role of instances is that of showing how high is the explanatory power of a proposed theory. Perhaps one may consider the instances explicable by a theory as problematic for those who wish to propose an equally good alternative to it. This would explain why usually previously refuting and/or problematic instances are presented as explained instances of a theory though that theory explains many other instances as well. So much for instances in science.

A common, though by now highly suspect, role is played by instances which Bacon has called 'clandestine.' A clandestine instance hints at a possible truth. For instance, a miraculous recovery may be due to unknown causes or due to the excellence of the doctor in whose charge the patient was at the time. If we accept an instance as clandestine, we need not at once accept the theory it points to (in our example, that Dr. X is excellent), but we are well advised to investigate the matter seriously. And the more clandestine instances there are that suggest a particular theory, the more seriously we should take the theory.

The most obvious characteristic of the superstitious is their serious approach to clandestine instances; the root of this lies in their want of a critical attitude. Not all errors are superstitions, only those concerning which we cannot conceive that we may be critical towards them.

In this sense of 'superstitious,' medieval empirical research was largely superstitious. The taking up of clandestine instances, hints which Mother Nature has mercifully thrown in our way, was quite routine procedure then. In modern times, mainly under Galileo's and Boyle's impact, this has been outlawed.

This immediately raises the question of the difference between a problematic instance, which requires explanation, and a clandestine instance, which should be ignored. The chief difference between them, I think, is that of attitude both toward theory and toward fact. When we have a problematic instance we first try to explain it and leave the question of the truth or falsity of our explanation to be discussed in a critical fashion afterward; whereas following a clandestine instance we hope to find the truth even though we may not fully understand it or fully formulate it to begin with—even though, that is to say, we are not capable of subjecting it to rational discussion straight away. And the same applies to facts. The fact constituting a clandestine instance, being a wondrous hint, should be taken seriously at once; whereas a problematic one should be capable of critical examination, and hence it must be repeatable.

Yet a critical attitude is but a necessary condition. While it is true that unrepeatable facts are useless, too many repeatable ones are left unstudied. Footprints in the sand are as repeatable as one could wish, yet science says precious little about them. In my view the ignored phenomena are those which our metaphysical frameworks are too poor to interpret (in the sense discussed below). They are too problematic. The same applies to theories, like elasticity theory, which are too difficult to incorporate within the existing metaphysical frameworks, and hence are not scientifically interesting.

Next comes the confirming or verifying instance. Whenever someone marshals instance after instance, challenging you to examine their truth, it is on the tacit assumption that if his instances are true his theory is also true. If you admit his instances and yet reject his theory he will marshal more instances. If you prove impervious to all his instances he will proclaim you unreasonable.

Confirming instances play the same role today as clandestine instances played in the Middle Ages. They play the role of clandestine instances for the uncritical audience and explained instances for the less uncritical audience. They are usually unsatisfactorily explained instances, yet the poor explanations are overlooked by audiences who are impressed because they are striking clandestine instances. For my own part I prefer to view all confirming instances as explained ones. For presenting an unsatisfactory explanation is still an attempt to explain, an attempt at a rational procedure; marshalling clandestine instances is plainly irrational.

To take an example. If someone throws a child into the river, Adler would interpret this act as one of self-assertion. And he would say the same if someone else rescues the child from the river. Thus, says Popper, opposite modes of behavior toward others are both somehow covered by Adler's doctrine. Hence it is no explanation. Adler's doctrine plus one of a given set of additional hypotheses, selected to suit each of the different cases, will indeed explain each action. But then all these explanations are *ad hoc*. The feeling is conveyed that many cases have been strikingly explained by one single hypothesis because Adler has claimed, in effect, that these instances are indications of self-assertion, clandestine instances for his theory.

An example to this effect which has greatly impressed me is Freud's story of a married woman who unthinkingly signed her maiden name. Freud interpreted this as an unconscious expression of suppressed discontent with her husband and, indeed, he triumphantly added, a few months after her pen slipped in that ominous fashion the poor lady was divorced. This is pseudo-science at its worst; it is a glaring case of a clandestine instance thinly masked as explained instance. Since some married women divorce their husbands without having accidentally used their maiden names, and since other married women use their maiden names by mistake without ever asking for a divorce, clearly in this special case Freud erroneously claimed that the error and the divorce were explained by his theory of slips of the pen. Yet it does appear as if this theory spectacularly explains the unexpected relation between a slip of pen and a divorce.

The mark of pseudo-science is the use of confirming instances. The practitioner of pseudo-science, unlike the superstitious, is not surprised by criticism. On the contrary, he is often painfully aware of the existence of critics; he is only too ready to meet his critics and argue with them. He will claim in the argument that every relevant case is an instance of his theory, that his critics' challenge can easily be met, that the critics do not see the immense explanatory power of his view simply through being so hostile toward it. When his explanations are scrutinized, however, it will be seen that the critic's facts are explicable not by the theory itself, but by the theory plus some additional hypothesis. Usually the additional hypothesis is so trite and plausible that one hardly notices its having been added, and those who make a fuss about it are prone to be successfully dismissed as mere pedants. Yet the great ease with which the pseudo-scientist so impressively explains all phenomena rests on these trivial (and usually acceptable) additions, not on the original theory.

Popper has accepted the claim of the pseudo-scientist that he can interpret all phenomena. He has stressed (in his "Personal Report") that since pseudo-science can interpret any conceivable (relevant) phenomenon it is not refutable by any conceivable phenomena, and hence it is untestable or unscientific. This is very neat, and quite important, yet perhaps it ought to be more explicitly stressed that though pseudo-scientific doctrines have high interpretative power, they have low explanatory power. This characteristic pseudo-science shares with metaphysics.

When Thales said that all was water, he provides a few instances for his doctrine, instances which led Aristotle to hint, and Bacon to assert, that he had based his metaphysical doctrine on facts by using the inductive method (to wit, that his metaphysical doctrine was to some degree scientific). Thales used the freezing and the evaporation of water as examples of his doctrines. He also claimed that solid deposits left in kettles by boiled water, and solid deposits in river-mouths, were instances of water turning into solids.

It is difficult for me to say what would be Thales's answer to such questions as why can we not turn a whole bulk of water into a piece of chalk. Quite possibly Thales, being the first metaphysician, was partly superstitious and partly (in some sense) a pseudo-scientist, and also (as Aristotle states) partly a mythologist; I do not know. Yet I suspect he was really none of these. I imagine he was asked such questions and in reply simply confessed his ignorance. Descartes's answer, and Newton's and Faraday's (whose doctrines I shall soon discuss), however, are clear and straightforward : we are not unaware of the lacunae in our doctrines, they would say, and we shall try and find some scientific theories to deal with your question in due course.

There is a similarity and a difference between pseudo-science and metaphysics. Freud's theory of the slips of the pen, like Descartes's and Newton's and Faraday's (if not also Thales's) metaphysics, sketch possible explanations. Metaphysics may be viewed as a research program, and the false claims of pseudo-science as the result of confusing a program with the finished product.

One corollary of this is that metaphysics can degenerate into pseudo-science. This corollary seems to me to be true, and exemplified by Aristotle's metaphysics, which becomes appallingly *ad hoc* when applied to phenomena, as in his *De Caelo*. I find the following corollary more interesting : it may be possible to elevate a pseudo-scientific theory to the rank of metaphysics. The first step in this direction is to strip it of its pretentiousness by making its logic clear. Expurgated, Freud's theory may be viewed as an interesting metaphysics of psychology.

As instances of a metaphysical doctrine are not clandestine or even confirming, what kind of instances are they? Thales's instances, I think, served two purposes : one presentational, and one to show that his doctrine, be it true or false, is not as fantastic as it sounds. Newton's metaphysics, which asserts that the universe consists of atoms with their associated conservative central forces, was instantiated by his theory of gravity. This instance served a more significant role than a merely presentational one. It illustrated the potentiality of his metaphysics and thus constitutes a challenge to construct instances of that metaphysics which are satisfactory explanations of all known physical phenomena. I shall call such instances 'conforming instances.'

Since Newton's metaphysics does not specify what central force causes gravitation, Newton's theory of gravity does not follow from Newton's metaphysics; it is not an explained instance. Otherwise Newton's metaphysics would be refuted by the refutation of his theory of gravity, which it was not. Newton's metaphysics does not follow from his theory of gravity : the one asserts that all phenomena are governed by central forces, whereas the other is confined to fewer phenomena. Generally, a metaphysical doctrine neither entails nor follows from any of its conforming instances. Nor does it follow from the set of all its conforming instances unless it may be assumed that the set is exhaustive. Since such assumptions are testable, the metaphysical doctrines in question would follow from scientific theories, and thus they could legitimately claim scientific status.

This is the ideal case. To my knowledge it has never been achieved. The doctrine that arrived closest to this ideal was Newton's metaphysics as it appeared around 1800. Yet the ideal had an immense driving force. The debate about metaphysical doctrines often concerns their status, and this often leads to the development of scientific instances conforming to them, or to the discussion of whether such developments are possible. Thus the desire to render a metaphysics sceintific leads to viewing it as a scientific research program whose satisfactoriness is open to critical discussion. To illustrate this I shall discuss in the next section the possible unsatisfactoriness of such research programs, and in the following section their possible satisfactoriness.

VII.   Metaphysical Doctrines Are Often Insufficient Frameworks for Science

The methodology of this and the next section is a generalized Cartesian methodology, and the generalization I am offering is possible only within Popper's framework. Descartes's metaphysics (which was an improvement on Galileo's), was a clockwork view of the universe. It explained almost nothing; it was not intended to explain anything. Descartes claimed that any scientific hypothesis which he could endorse must be one which conformed to his metaphysics. He added that explanatory hypotheses conforming to his metaphysics could always be found. Boyle made the same claim concerning his own semi-Cartesian metaphysics, and so did Newton concerning his own metaphysics (in his preface and the *Scholium Generale* to *Principia*). But this repeated claim of the metaphysician is often false. It may be argued that his doctrine allows insufficient room for explanation, that it provides too narrow a framework. When this is felt to be the case, the demand for a new metaphysical framework arises. Metaphysics which stagnates in scienceless (or uncritical) cultures, is progressive in scientific ones. It progresses then because existing metaphysical doctrines are felt to be constricting frameworks, and thus unsatisfactory.

Thales's doctrine aimed at explaining (physical and chemical) diversity and change by assuming an underlying and unchanging unity. Any such approach runs the danger of being too successful and thus self-defeating. For the assumption of an unchanging unity leads to regarding observable facts as illusory. This was the magnificent discovery of Parmenides, and it was this discovery that made him deny the existence of diversity. We have no right to despise him for having preferred his own logic to common sense, for having proclaimed appearances to be illusory; rather we should admire his dazzling logical acumen. But for him we might not have had Leucippus and Democritus. The greatest novelty of their atomic doctrine is that it expressly allowed for both unity—of the atomic character of matter—and diversity—of the atoms' shapes, sizes, and spatial order. To put it in quasi-ancient idiom, atomist metaphysics is a program to explain the many not by the one but by the few; it is thus more accommodating than the metaphysics of Thales.

(Of Parmenides's other great logical discovery, of the nonexistence of the void, I cannot speak here beyond saying that it was the cornerstone of the theory of space developed by Leibniz, Faraday, and Einstein. Further details of the story of Parmenides and Democritus, as well as the story of the downfall of the Pythagorean program, the demonstration of its narrowness, and its rectification by Plato, as well as the relation between Plato's program and Euclid's geometry, have been admirably presented by Popper in the paper already mentioned. The great role played by both Democritus's and Plato's programs in the seventeenth century have been beautifully told by Koyré. The relation between Leibniz's program and Einstein's scientific theory of space is discussed in Einstein's exciting preface to Jammer's *Concepts of Space*.)

My next example of an unsatisfactory metaphysics is Cartesian metaphysics, which contained the thesis that all (non-inertial) motion was due to push. The example for this was the suction pump whose (pull) action had been scientifically explained as due to atmospheric pressure (push). Lifting a jar seems to be pulling it upwards, but in fact it is pushing the jar upwards by the handles. Now this last example was implicitly criticized by Newton. If the jar is strong, or contains light material, it can be lifted by its handles; otherwise, pushing its handles upward hard enough will only constitute lifting the handles while leaving the jar itself on the floor. We must admit, then, that lifting it by its handles not only involves pushing the handles upward but also pulling the jar itself upward with the aid of the attractive forces which keep the jar and its handles connected—by the forces of cohesion. This example justifies Newton's claim (preface to *Principia*) that his program was in the first place more accommodating than Descartes's, and so could be more fruitfully adopted even if ultimately we should return to Descartes's program.

But Newton's metaphysical program, too, was so naive, that one may wonder how it was accepted for so long. Assuming, with Coulomb, that electric forces act solely between electric charges and that gross or ordinary matter is subject to the forces of gravity and cohesion only, why then does the charge remain on the charged body and pull it along when moving toward, or away from, another charge? This question (which was raised in 1800 with the discovery of electrochemistry) clearly indicates that gross matter is in some sense electrical. Yet so strongly impressed were people with Newtonianism that twenty years after Faraday had produced wonderful scientific theories which incorporate the supposition that gross matter has some electric characteristics, these theories were almost unanimously ignored (the exceptions were Kelvin and two other, rather minor, physicists). Those statements in the *Encyclopaedia Britannica* of the eighteen forties and fifties which appear to allude to Faraday's theories are certainly contemptuous and derisive.

It is not accidental that Boltzmann explained in 1885 (in a letter to *Nature*, p. 413) the general opposition to Maxwell by the general adherence to Boscovitch's, not to Newton's, program. Boscovitch had modified Newton's program to permit one material particle to dispose a variety of forces. This he did because he had discovered that otherwise the program would not accommodate any explanation of the phenomena of elastic collisions. But his program became popular only after Faraday imposed his view that gross matter had electric properties.

Incidentally, the indifference to Maxwell which worried Boltzmann shows that even Boscovitchian metaphysics may be highly dangerous; but it is a truism that any idea may become dogma.

So far I have only spoken of the requirement that metaphysical doctrines be sufficiently wide frameworks to accommodate possible future scientific theories. In the next section I shall speak of the requirement that metaphysics be inspiring and lead to the development of scientific theories.

VIII. The Role of Interpretations in Physics

A statement of fact or a scientific hypothesis restated in terms of a new metaphysical doctrine is a new interpretation. New interpretations are only too often unsatisfactory explanations of the original statements—for example, interpreting a motive as a sex motive, or survival as due to a high degree of fitness, or change of a person's pattern of behavior is due to physical change in the brain. But the logic of interpretations is made clearer, and so is their possible usefulness, when we take examples from physics.

The handles of a jar stick to the jar. To repeat this in Newtonian terms, the particles of the jar are attracted (by some central forces, that is) to those of the handles. Is this restatement a circular or a satisfactory explanation? We do not know. How small are these particles? What is the magnitude of these forces? One may try an estimate on the basis of known facts —perhaps the force needed to tear the handles off. Or perhaps it is easier to measure the force of cohesion by observing a drop of water hanging on to a solid surface, where cohesion counteracts gravity. Or perhaps it is still easier to observe a drop of water in a tube, where the balance of cohesion and gravity is perfect, and where the weight of the drop and the area of the contact surface are more easily calculable. It is easy to develop the first step of Laplace's theory of capillarity by thinking along this line: restate the connection between the inner diameter of the glass tube and the height of the water-column (or mercury surface) in it in Newtonian terms. As Newtonian forces are central, it would follow at once that the narrower the glass tube the higher the water column (and the lower the mercury level) in it. And the relative curvature of the fluid surfaces will be equally easily explicable. This is a particularly fortunate interpretation.

Another example: Newtonianism forces us to view lightly as either particles or waves in an elastic medium. Each of these interpretations leads to obvious questions which may be given testable answers. Faraday's metaphysics, to take another example, which views the universe as one field of forces, invites the view that light consists of vibrations of the lines of force in empty space. Faraday himself considered light to be waves of the magnetic field of force. For decades he tried to test this hypothesis and failed. And Faraday's interpretation of the electric current as the collapse of an electric field is another example of his failure. Tyndall rightly declared that Faraday's theory of the current was unsatisfactory. But he was too eager to reject it offhand (being a dogmatic Boscovitchian); by further specification Poynting soon rendered it highly satisfactory.

Interpretations apply not only to facts but also to theories. Faraday accepted Coulomb's Newtonian theory of electrostatic forces, but reinterpreted it in his field conception. His interpretation seemed unsatisfactory, and he was painfully aware of this. He succeeded in rendering it satisfactory by looking for curved lines of electric forces, which his interpretation of Coulomb's theory, though not Coulomb's theory itself, allowed for. He thus found that electric lines of force curve in the presence of dielectrices, i.e., materials like glass or sulphur. It is no accident that Coulomb denied the possibility of dielectricity: he was a Newtonian. Nor is it accidental, I think, that Cavendish failed to publish his own discovery of dielectricity: he wished to work on it further and reincorporate it within Newton's metaphysics, and he died before accomplishing this formidable task. That this task could be performed with but a slight deviation from Newton's program Faraday knew, and he outlined ways of doing it, without however being able to do so himself for want of mathematical technique. The technique had been provided by Poisson, and Faraday said as much, but he was too neurotic about mathematical symbols to write them down on paper. Shortly afterwards Liouville was in a quandary because Poisson had, on his death-bed, asked him to make Poisson's own work the topic of a prize essay, and Liouville felt understandably apprehensive in view of Faraday's discovery which seemed to him not to fit Poisson's Newtonianism too comfortably. Kelvin, who was a young lad then, related all this in a letter which he wrote to his father from Paris, and he added a description of how relieved Liouville was to hear that Kelvin could interpret Faraday's discovery in an almost Newtonian fashion by using Poisson's own method. This was Kelvin's first published paper.

But there was no escape from Faraday's inspirations. Kelvin's theory of the dielectric assumed gross matter to possess electrical properties; his theory was not Newtonian but Boscovitchian. It soon transpired that Boscovitch's program needed modification. Gauss and Weber tried, and the attempt continued until 1905. By then it was clear that the program had to be given up; it looked as if Faraday's program had won out at last. Yet this program too was abandoned very soon after. It was deterministic and determinism had to be abandoned.

IX. The History of Science as the History of Its Metaphysical Frameworks

The world is full of well known yet unstudied phenomena, of often heard but seldom debated theories. Historians of science all agree that some theories—Copernicus's, Maxwell's—and some experiments—Oersted's, Michelson's—are of supreme scientific importance.

That Oersted's experiment was of metaphysical significance is obvious in view of the supreme prestige Newton's metaphysics enjoyed at the time. The greatest problem in physics between 1820 and 1905 was, could there be a (satisfactory) Newtonian (or semi-Newtonian) explanation of Oersted's experiment? Study of this problem led to Newtonianism losing its interpretative power. It soon transpired that the only unrefuted satisfactory explanation of Oersted's experiment was Maxwell's, and it became an urgent task for Newtonians to interpret fields in accord with Newton's metaphysics, which means—since for Newton forces are attached to matter and since Maxwell's equations are not invariant to Galileo's transformations—with the aid of the assumption that space is full. A scientific version of this assumption was refuted by Michelson and Morley. In 1904 Kelvin still hoped that another Newtonian or Boscovitchian interpretation of electrodynamics could be found; but though a few shared his hope no one did anything about it, especially since his misgivings about Maxwell's theory were not shared by others. Undoubtedly, Maxwell's theory was so significant because it was a satisfactory explanation which conformed to Faraday's metaphysics. Undoubtedly Planck's theory became so important in 1905 when Einstein showed its conflict with Maxwell's theory because it seemed a major breakaway from Faraday's program.

I do not know why the significant events in the history of science should be metaphysically significant, but I have so far found it almost always to be the case. I suggest the theory that significance with respect to (pure) science is usually significance with respect to science's metaphysical frameworks. It is understandable that if metaphysical frameworks are research projects they should be taken very seriously, but why should all (pure) research projects be geared to a few metaphysical doctrines? Indeed, I think most research projects are not intended, at least not consciously intended, to be relevant to the dispute between the few competing metaphysical doctrines of the day. Yet those projects viewed later as significant show a capacity to throw light on current metaphysical issue. I can see no other explanation of the situation but that it is essentially metaphysical interest which gives (purely scientific) significance to this part of science rather than to that; hence, most (pure) scientists are more interested in metaphysics than they seem to be.

There are many studies which are not directly related to metaphysics. Take the continuum theory; it is the study of properties of matter, especially elasticity, on the assumption that matter is continuous. This study belongs to applied mathematics or technology rather than to pure science because it is based on a metaphysically unacceptable assumption. Its value for pure physics becomes apparent only when it is shown to throw light on an important scientific problem related to metaphysics. Indeed, since the Newtonian interpretation of the wave theory of light is the theory of the elastic ether, the rise of the wave theory caused immense efforts to be made to create any theory of elasticity whatever which might be used as a tool to render the ether theory scientific. Prior to that, the effort to develop a theory of elasticity were strictly in the Newtonian mode. We see how a significant plan of scientific research was first directly and then indirectly metaphysically relevant, and later it lost all relevance and with it all significance. Present day aerodynamics interests only few non-aeronauts, but it will interest more of them if it will reveal some bearing on existing metaphysical issues.

But what about scientific work unrelated to metaphysics? Let us take two examples. Jenner's study, his attempt to refute some village superstition, was highly idiosyncratic. Possibly it was connected with Bacon's idea that superstitions are dangerous to science, and yet as hardly anyone except Jenner undertook such researches, his work may well be viewed as idiosyncratic. The device of vaccination, which resulted from his study, was for long chiefly of practical value. The mechanistic interpretation of vaccination is identical with the theory of antibodies. It is thus a metaphysical theory. In the popular literature it is often presented pseudo-scientifically. Biochemists have used it as a program and found scientific instances which conform to it; they are still searching for others. This story shows how one idea entered into the mainstream of science because it fitted a metaphysical framework.

My second example is the discovery of the asteroids. It is insignificant. It refuted Hegel's doctoral dissertation, but this was of no value in any case. It refuted Kepler's metaphysics, but nobody had ever taken notice of this metaphysics. It agrees with Bode's law, but this law is related to no metaphysics. The discovery is insignificant because it has no direct or indirect relevance to topical metaphysics. It may, however, become significant, if asteroids are going to play some role in a future cosmogony.

CHAPTER 14

On the Problem of Truth and Understanding in Science

BY DAVID BOHM

IF ONE WERE TO ASK for a statement of the essence of Professor Popper's contribution to the theory of scientific research, it could probably best be expressed in the sentence : "Science is a part of the search for truth." To one who has not gone very deeply into the question this may seem self evident. However, on thinking more carefully one finds that the problem of truth is a slippery one, full of pitfalls. It is therefore perhaps not so surprising that scientists and philosophers of science have been attracted to a wide range of suggestions with the aim of avoiding this problem. To achieve this, they have tried to refer the essential content of science to something less vague and difficult to grasp, less "metaphysical," more solid, definable, and "positive" than the elusive notion of truth.

As a typical example of such efforts one can take the rather common idea that scientific theories are in essence just convenient and useful ways of summarizing, ordering, and organizing what we know of nature through past experiences. This point of view, in which the notion of truth is in effect replaced by that of utility and convenience, has most systematically been embodied in the positivist philosophies that were current during the nineteenth century (notably those of Mach and Comte). Later exponents of the positivist philosophy were, however, no longer satisfied with these ideas, but still felt that the notion of truth had to be replaced by something else that could be stated definitely and explicitly. Thus, the Vienna School proposed that the essential content of scientific theories was in their *verifiability*, i.e., in the *procedures* by which it could be seen that they are true. A closely related point of view had meanwhile arisen in America, pragmatism, in which the notion of truth was replaced by that of "what works." Scientific theory was thus conceived as nothing more than an instrument in man's efforts to control nature. Later there arose the allied view of operationalism, which identifies the content of a theory with statements about operations that a scientist can carry out with the aid of suitable equipment.

The tendency to try to replace the notion of truth by something else that can be given an explicit definition has continued into many more modern forms, which, however, need not concern us here. This tendency has penetrated very deeply into all thinking about science, both by the layman and by the scientist himself. For example, even the very form of theories in physics frequently reflects the commonly accepted presupposition that the essential purpose of physical laws is to enable us to predict the results of experiments. Once again, the question of truth is avoided, and it is replaced by something else on which one can put one's finger more easily, i.e., *prediction*.

Professor Popper has consistently criticized the point of view described above. Indeed, it was just in his efforts to show the inadequacy of the suggestions of the Vienna School to substitute verifiability for truth that he was led to formulate his by now very well known thesis that no theory can ever be verified conclusively by any number of observations, however large, and that in fact, the question of *falsifiability* is in many ways more relevant than that of verifiability. Professor Popper also pointed out, of course, that it is just as impossible to falsify a theory conclusively as it is to verify it. But his statements concerning the impossibility of conclusively falsifying a theory are frequently overlooked, with the result that his views are distorted into what is, in effect, a continuation of the Vienna School in a modified form. Thus, it is now often asserted of his position that he maintains that the essential content of a scientific theory is not in its verifiability but, rather, in its falsifiability. In fact, however, Professor Popper consistently stresses that the essential question to be studied is that of truth. While admitting that this is a very difficult question, he has shown throughout his whole life's work that if we do not tackle it, we will not even be able to make a start on the question of what is the essential point of scientific research.

It is with the above general situation in mind that I should like to go a bit more deeply into the problem of truth and its place in science. This problem is so closely allied with that of *understanding,* however, that it seems to be necessary to consider them as inseparable. Since there has already been a great deal of discussion of truth and comparatively little concerning understanding, it will be helpful to begin with understanding and then to return to the problem of truth later.

Let us start by asking ourselves a simple question : What is understanding ? In response to this question, there immediately arises another question : Can understanding ever really be defined ? For example, suppose that one were to give an exhaustive definition of understanding. At the end, one could ask : "Have you understood ?" In this way one shows that no definition of understanding is actually possible, which does not presuppose that the hearer *already* understands what is meant by understanding (even though he cannot put it into words). Indeed, a person who did not understand what is meant by understanding could not enter into a conversation, and he would hardly be called human in any respect whatsoever.

We see then that understanding is implicit, in the sense that it is logically prior to all words and thoughts. The recognition that the most fundamental things are implicit and incapable of being directly and positively asserted is, however, not entirely new. Thus, in geometry, it is well known that an explicit definition of points and lines and other such basic concepts is impossible. Rather, one begins by simply stating their existence, while their properties are only implicit in the axioms that are assumed to relate them. The choice of axioms is guided to some extent by vaguely and intuitively grasped common experiences with space, but finally, the properties of the points, lines, and other entities must be *explicated* (or made explicit) by a working out of the consequences of the axioms. And when this is done, the results are in general not a mere recovery in a more precise form of what was already known in common experience, but rather, they are often in many ways new and even surprising.

We may say then that in *fundamental* questions, it is always necessary to approach the object of our studies obliquely, by implication rather than by positive assertions and definite conclusions, so that the positivist goal of trying to state precisely what is happening is never appropriate in a really deep problem. And no problem can be deeper than that of what is meant by understanding. For every form of human activity, including science, mathematics, and philosophy, presupposes understanding. We are therefore led to study this problem in an oblique and indirect fashion. Guided by our over-all experience, let us try to find some general characteristics which throw light on a few aspects of understanding, at least by implication.

It is helpful to approach the problem by considering the example of the circle. Thus, at first people studied circles empirically, amassing many items of information about them (e.g., that the circumference is approximately $22/7$ times the diameter, that certain chords are related to the diameter in certain ways, etc.). This may be called the stage of associative thinking; for the circle is treated simply by associating together a large number of remembered properties that were discovered empirically. Of course, such a mode of thinking is good enough for a wide range of practical purposes. Indeed, it is by a similar type of thinking that we get through a great deal of everyday life.

At a certain stage, however, a new geometrical theory was developed, and a circle was regarded as a curve traced by a point moving equidistant from a fixed point, while a straight line was regarded as the result of moving a small straight segment in its own direction, etc. Later, when this geometrical theory is explained to a particular individual, at first he tries to bring together the various parts of the argument. Then suddenly he says "I see," meaning by this that he *understands*. But what is it that is seen? In what is generally described as a very sudden process, a "click" or a "flash," one grasps the basic principle of the circle, which is to say, one sees it as *a totality*. Of course, this does not mean that one immediately knows the whole set of properties of a circle exhaustively and in full detail. Rather, it means that one sees the essential process by which the circle is generated, along with its various parts and aspects, which are now treated as sides of this totality, so formed that they are automatically in their proper relationships. In this way, one not only grasps the essential character of *each* circle, but also of *all* circles.

With the coming of understanding, then, there is (at least in the field under discussion) what amounts to a revolutionary change in the mode of thinking, in which the earlier associative thought about this field is set aside, and in which the field is comprehended as a totality. However, the fact that understanding involves such a fundamental change in thinking is perhaps generally not fully realized consciously, or at least, its importance is frequently underestimated. We can bring out this importance a bit more sharply by considering a hypothetical device, which we shall call an "understanding machine."

Of course, it is common knowledge that calculating machines can now be made which can be programmed to do many tasks previously requiring human thought and attention. If one considers the possibility of an indefinite increase in the number of elements in such a machine (along with new and improved principles of construction and operation), one is quite naturally led to ask whether it would not be possible eventually to reproduce *all* of the properties of human thinking. It seems quite plausible that machines that could in effect "learn" by association could eventually be made to reproduce or perhaps even to excel human abilities in this field. But could a machine also come to "understand" in the sense that it could obtain a new "vision of totality"? For example, could it "learn" the principles of a science such as physics along with the basic experimental facts that are known at a given time, then "criticize" these principles, discover their weak points, and in a sudden "click" or "flash" (or in a series of them) "see" how newer and better theories could be developed? If it could do this, then it could "grasp" or "comprehend" its own system of programming and "suggest" new systems and perhaps even new principles for its own construction and operation.

Let us now suppose that we wish to develop such an "understanding machine." Our first task would be to define what is "understanding"; for if we do not know what it is that we are trying to reproduce, we can hardly design a machine that will function as planned. But as we have seen, every definition of understanding is necessarily only partial, and at best implicit. For it leaves out of the account the very action of understanding that is needed on the part of the proposer and the hearer before this definition has any meaning. Moreover, it also fails to include a potential infinity of new ways of understanding that can develop in the future. So, at least in our present state of understanding, it would seem to be impossible to design an understanding machine, whereas we can in all probability eventually design a machine that reproduces the essential characteristics of associative thinking quite well.

It seems clear, then, that understanding is something radically and fundamentally different from associative thinking. However, it is only in the context of the fact that understanding actually takes place that the question of truth can arise. For if there is no understanding of what is being thought and said, what meaning can there be to asking whether it is true or false? But when we understand deeply enough, then, as we can see by actually experiencing it as a fact, there is the possibility of perceiving the truth or falsity of what is being thought and said.

Truth, like understanding, is basically implicit. If one were to give a definition of truth, at the end one could say "And this is the *true* definition of truth," thus showing that the hearer had to understand the meaning of truth before communication was even conceivable. This means that all efforts to reduce science to some definite, explicit, "positive" characteristics are bound to fail, because none of these makes sense except in a context in which the undefinables, 'truth' and 'understanding,' are present from the very beginning.

The need for the falsifiability of theories, as emphasized by Professor Popper, already shows that the question of trying to determine the meaning of truth cannot properly be treated by the direct approach. Thus, if every really acceptable theory must be falsifiable, then it seems almost certain that, in time, such a theory *will actually be falsified* as more accurate experiments are done in broader domains and in new contexts. (As has, in fact, already happened to an extremely large number of theories in every branch of science.) In other words, it is hardly to be expected that any theory with a real predictive content will not in some way turn out to be false. But if *all* our theories are thus very likely to be false, then where is truth, and what is it that we are really searching for?

Here, once again, we must recall that there can be no explicit definition of truth, and that light on questions of this kind can best be obtained by proceeding in an oblique and indirect fashion. The first thing to do is to recognize that while almost anything that we say in our theories may eventually be shown to be in some way false, these theories will nevertheless continue to express a great many *true relationships*. For example, in the early phases of the development of physics, gases were often treated as perfectly continuous distributions of matter. Yet, even though the falsity of that conception is implied by the atomic theory developed since, the relationships of temperature and pressure obtained in terms of the former treatment are true (at least within a certain degree of approximation). True relationships can never be obtained from self-contradictory theories. For example, though we write the self-contradictory statements, $x = 2x$, $y = 2y$, this contradiction will not matter in the quantities which depend only on the ratio x/y.

It is clear that the appearance of some experiments confirming a theory still leaves plenty of room for false implications in the theory in question, as well as for self-contradictions which are subtle enough to have escaped notice thus far. The falsity of such implications and the existence of such contradictions may be shown up by future experiments. As a rule, however, experiments are *by themselves* sufficient for doing this only in very specific and narrowly delimited problems. Thus, it is well known that, with regard to theories having a broad domain of applicability, it is almost impossible to find a set of experiments that clearly and unambiguously show the theory to be false, because it is always possible that a modification of some inessential feature could bring the theory back into agreement with experiment. For this reason it frequently happens that a set of experiments which can be used for falsifying the theory may be available for a long time, while their true significance is not realized. In such cases, the crucial step in falsifying the older theory comes with an *understanding* of the implications of these experiments.

The development of the theory of relativity furnishes a very good example of the above point. Thus, the Michelson-Morley experiment and many others had been available for some time, but there had been a widespread and not unnatural feeling that some change in the theory of the ether or some modification in our ideas concerning the mode of propagation of light could eventually account for these results. Einstein's basic contribution was his new way of understanding the problem. For he saw that the space-time coordinates are not something self-existent and wholly independent of the laws of physics, but rather that these coordinates must express *relationships* between physical entities (e.g., as manifested in the readings of clocks and measurements made with rulers). As a result, he was able to see that the essential point at issue in all these experiments was that the space-time frame of reference and the laws of physics must be interconnected very fundamentally. Therefore, it was through understanding the whole situation in a new way that it was possible for Einstein to perceive the falsity, not of some specific and detailed aspect of the old theory, but rather of the *Newtonian conception of absolute space and time in its totality*.

In a similar way, most of the experimental facts underlying the quantum theory had been available for a number of years when Niels Bohr saw a new way of understanding these facts on the basis of the notion of discrete energy levels in atoms and discontinuous processes of transition between these levels. Once again, it is clear that Bohr's new vision of the laws of physics falsified not a particular aspect of classical theory but *the conception of continuous motion as a whole*. Other examples of less comprehensive developments in our theories will occur to the reader. However, he will readily see that in most cases it is not the experiment itself that falsifies earlier theories and conceptions; rather, it is some new understanding which arises in response to reflection on the *total* situation, experimental and theoretical.

The question of seeing just what is false is particularly important to clarify. For when one realizes that the older point of view is false, one can also see in what way it is false, and in what way it continues to have truth in it. Thus, in the case of the theory of relativity, it was evident that while the Newtonian conception of absolute space and time is false in a basic sense, nevertheless at velocities small compared with that of light, it leads, at least within a certain level of approximation, to true relationships in a great many physical processes. (Similarly with regard to the connection between classical and quantum mechanics.) It may therefore be said that in addition to perceiving what is true and what is false, we can see the truth *in* the false (true relationships) and the truth *about* the false (the reasons why it is not true). When we go into the question of what is the real significance of falsification, it then becomes clear that in a fundamental problem much more is involved than merely to show that a given theory does not agree with experiment. For what actually happens is that the perception of the truth in the new conceptions and of the falsity of the old conceptions takes place in the same step, at the very moment of understanding. Thus, truth and understanding are really inseparable. Perhaps it could best be put by saying that they are themselves two sides of a totality.

In view of this essential connection between truth and understanding, what can we say about the question of an absolute truth? Is there an absolute truth which is, as it were, something accomplished, definite, and finished? If there is, then it would not be unreasonable to suppose that our theories can approach this truth. Such a light might even be infinite, so that man could never actually reach it or even get nearer to it, in any absolute sense. Nevertheless, at least he could still always be moving in its direction, or constantly accumulating more and more of it.

If there is not such an absolute truth, the idea arises quite naturally that perhaps truth is only relative—for example, to the state of mind of a particular person, to the tastes, habits of thought, and general culture of a particular group during a particular epoch, or to the particular procedures that the main body of scientists may have chosen to adopt in their work in order to get results of a kind that they desire. It seems clear that all the various philosophical points of view such as positivism, pragmatism, operationalism, etc., do in fact try in this way to refer truth to something else and thus end up in effect by assuming that it is *only* relative. But a purely relative truth is in reality no truth at all, since it finally depends neither on itself alone nor on some more fundamental truth, but rather on something completely extraneous to the question of truth that happens to have been regarded as important by a particular person or by a particular group of people.

It seems that we are faced with a very difficult problem. The notion that truth is *only* relative simply makes no sense, for then there would be no reason to choose one criterion of what is important over another. Indeed, even the most thorough-going positivists, pragmatists, and others of similar schools are trying to maintain that *their* notion of the relativity of truth is the true one, and in this way they bring in absolute truth by the back door. (If they did not wish to suggest that at least their ideas on the subject were really true, there would evidently be no point in their saying or doing anything whatsoever concerning this problem.) On the other hand, any attempt to avoid the arbitrariness of the notion that truth is only relative by asserting that there is an absolute truth is at bottom self-contradictory. For since there is no way to prove that this assertion is itself really true, it finally reduces to a belief or a hope, the foundation of which is only in the mental or emotional state of a particular individual or group of individuals. Therefore, the person who criticizes those who hold to the purely relative character of truth for basing everything on subjective ideas, preferences, and beliefs is doing just the same thing if he asserts the absolute character of truth.

We see that there is something paradoxical in the effort to regard truth within the framework of the system of categories, absolute versus relative. Perhaps, then, the difficulty can be resolved only if we *begin* by seeing the falsity of this whole way of considering the problem. For as we saw earlier, truth (like understanding) does not seem to be approachable through positive assertions with regard to its character. Let us therefore try once again to approach it more indirectly.

The problem described above seems to arise, at least in part, in the effort to refer truth to something fixed, definite, and final, either a subjective criterion which we are to choose once and for all, to be applied in research, or else an objective truth that is supposed to exist somewhere "out there" in a finished form, and which we are supposed to approach step by step, or to accumulate bit by bit. But it may be that truth is none of these. Perhaps it is something that has no fixed and final forms or limits within it, so that it cannot be known in its totality nor approached nor accumulated nor even referred to some definable criterion by which it can be recognized. Instead, what may happen is that both truth itself and the methods and criteria for establishing it must be understood afresh from moment to moment, because everything is always changing, so that the problem is, in some respects, fundamentally new on each occasion on which the question of truth is to be considered.

Of course, a great many aspects of the total situation that confronts us from moment to moment (and in which we are) do actually repeat their essential characteristics, at least on a large number of occasions and for a wide variety of conditions and contexts. To the extent that this happens knowledge established on the basis of past understanding will continue to be valid. However, the repetition of past situations is never complete. Indeed, any particular characteristic, no matter how firmly and securely established it seems to be, can change fundamentally, often in a totally unexpected way. Such changes are in fact always occurring naturally, as things move and develop of their own accord, as well as because of human activities. In this regard, it must be remembered that in the particular activity of scientific research, the *indefinite* repetition of the same experiment under identical (or essentially identical) conditions is of very little interest, even in those cases in which such a repetition is actually possible. Indeed, since theories are expressed in the form of universal laws, scientific research is always directed towards testing these laws by the study of their operation under new conditions, in new contexts, and within new degrees of approximation. Moreover, these studies are carried out with the aid of experiments involving new techniques, new kinds of apparatus, and new modes of investigation. As a result, not only because of the general nature of the world and of human activities in this world, but also because of the peculiar nature of scientific research, the conditions in the total situation confronting us from moment to moment are certain *not* to undergo indefinite and effectively identical repetition.

It is basically because of the externally changing character of the world and of our experiences with it that the question of falsification of existing theories is so important. For it then follows that the validity of past knowledge can in general be only *partial* (in the sense that it contains some true relationships). When a theory is falsified, the fact that its validity is only partial is what is actually established. But now, we can see more thoroughly into the full implications of this feature of scientific research. For because nothing remains completely identical (even in its essential characteristics) with what it was previously, an understanding that is adequate to the situation confronting us at a given moment will, in certain ways, cease to be adequate for the next. Truth will therefore slip out of our grasp, unless we are continually alert and attentive to the ever-changing total situation, being ready always to perceive the falsity of our older ideas in newer conditions and contexts and continually to develop new ideas that are appropriate to new situations. Thus, we are led to consider truth, not as fixed and finished, but, rather, as coming into being anew from moment to moment.

It is clear that truth as conceived above is not subjective, in the sense that it would have meaning only in relation to the arbitrary ideas, preferences, goals, and feelings of various individuals or groups of individuals. Nevertheless, to the actual situation prevailing in a given moment there will be special standpoints and perspectives, as well as a general "world view," which are peculiarly appropriate for the understanding of what is in that moment. If one holds too strongly onto these, then one may misunderstand something that arises in the next moment, and thus lose part of the truth. In this regard, a new situation or problem may even require basically new general methods of investigation and general criteria for acceptable theories. For example, consider what is usually called "the scientific method." Is it possible, once and for all, to define exhaustively what this is? Evidently not, because this method has itself evolved, and is still evolving, in response to our being confronted with ever new kinds of problems. The scientific method of today contains aspects that were not present several centuries ago, and it seems very likely that in a few more centuries it will be very different again in many respects from what it is now. But this means that it must be *continually* changing. Perhaps the change is imperceptible in the short run but, nevertheless, it is clearly a real one.

The notion that we can completely separate the *mode* of understanding from the *object* of understanding is then evidently false. If our understanding is to follow its object and thus to remain true, it is necessary that even our methods of research, criteria for truth, and general perspectives and "world view" shall be free to change from one occasion to another. For it must be seen that in each concrete situation and problem, the proper (i.e., the true) mode of understanding is an essential and indispensable part of what is to be meant by truth in that situation and problem. Both the notion of a fixed and definite absolute truth and that of a fixed and definite method for establishing and understanding truth are therefore not valid. Rather, each truth must contain within it the true mode for its being established and understood, in such a way that, without this mode, that truth has no meaning.

If truth has the character described here, it may be that our whole mode and level of understanding things in general, which has evolved in response to a certain range of specific kinds of problems, practical and theoretical, is not adequate to the problem of understanding truth itself; i.e., of grasping *the basic principle of truth*. To do this would require of us not the mere repetition of some set of words, but rather (as in the case of the circle, discussed earlier) a real act of comprehension, in which truth would be seen as a totality—coming into being, as it actually does, from moment to moment, but always with radical differences in its essential characteristics. It seems evident that to understand truth in this way would be an extraordinarily difficult task. But if this is the way that truth really is, then such an understanding may well be just what is needed for seeing its basic principle. Our customary excessively narrow and limited approach to the problem may therefore be what is responsible for the confused and self-contradictory nature of most of our ideas on the subject.

We may illustrate what is meant here by a kind of analogy. Generally speaking, science develops by a series of small steps of understanding that may be compared to the lighting of candles, each of which illuminates some small domain in the surrounding darkness for a short period of time (until it burns out). Every now and then there is a flash of understanding, which, like lightning, illuminates a whole field of study, and so brings about a fundamental change in our basic conceptions in that field. It is with the memory of what was seen during these occasional flashes that we usually try to direct our steps beyond the regions rendered visible by our candles. Meanwhile, however, the situation may have in some ways changed fundamentally, so that this earlier vision is no longer completely appropriate. To deal with this problem, perhaps it will be necessary eventually to emerge into a kind of understanding analogous to the steady sunshine, in the light of which there can be a full and adequate response in each moment to the *whole* of the ever-changing situation because no feature is, as it were, allowed to fall into a shadow dark enough to obscure its significance.

Such an unrestricted mode of understanding would probably enable us at each moment to see truth as an integral whole. On the other hand, our present modes of understanding are not only too rigid and fixed to follow the actual situation confronting us in its rapid and often unexpected changes, but also, being generally fragmented, specialized, and otherwise only partially valid, they must introduce a further kind of basic falsity by splitting the whole truth into pieces. Naturally, these kinds of falsity are not important in a wide variety of problems of relatively limited scope, because (as we have seen quite generally in science, for example), conceptions which are demonstrated to be false under new conditions or in a broader domain can still express true relationships under an approximate repetition of older conditions, or when the domain under consideration is restricted to what it was formerly. Nevertheless, it does not seem reasonable to suppose that a mode of understanding which inherently and unavoidably breaks truth into pieces and fixes the form as that of the pieces is likely to be adequate for the task of grasping truth as a totality in the kind of essential movement and process by which it is always coming into being anew, without fixed limits or characteristics.

Of course, it must be recognized that there exists a widespread feeling that truth really does not have the ever-changing and elusive character that we have suggested for it here. Or if it is admitted that truth is in fact an elusive and difficult notion, it is often felt that as far as scientific research is concerned, we can avoid such questions in the manner discussed in the beginning of this article; i.e., by basing everything on something else that is easier to grasp in a definite way (such as verifiability, falsifiability, intrumentalism, etc.). To a certain extent, such feelings arise out of a very natural desire to put things on some solid foundation, which can be asserted positively once and for all, so that we can then get on with our other tasks, knowing that at least we are secure in our foundations. But as we have seen, there is good reason to suppose that this cannot be done. If this is so, we may be confusing ourselves by the effort to do the impossible; and as a result, we may be making our actual position more insecure than it really needs to be. In any case, what task could be more important than the development of a mode of understanding that will be adequate to the problem of perceiving what is the meaning of truth and how all our activities must be related to it?

CHAPTER 15

The Mach Principle

BY P. W. BRIDGMAN

THE EXPLICIT WORDS "Mach Principle" seem to have been first used by Einstein in connection with the field equations of general relativity, in accordance with which the inertia of local masses is determined by the distribution of mass throughout the entire universe. There is nowhere, either in Mach or Einstein or elsewhere, any sharp formulation of precisely what the principle is, but the usual implication is that there must be *some* sort of connection between local inertial phenomena and distant mass distribution, leaving latitude for different sorts of precise connection according to the particular theory involved. It is the necessity for some sort of connection that is generally understood by the principle.

Perhaps the most striking physical phenomenon which seems to demand the principle is the gyrocompass, which persistently points to the same star as the earth rotates beneath it. Other manifestations are the Foucault pendulum and the deviation of a falling body from the plumb line. The average physicist is not able to contemplate the gyrocompass without the almost irresistible impulse to say "The stars must have *something* to do with it." On the other hand, the idea that a distant star can have any effect on local phenomena has gravely offended the physical intuition of men like Eddington, Whittaker, Whitehead, Bertrand Russell, and Hermann Weyl. Whitehead says, "I cannot persuade myself to believe that a little star in its twinkling turned around Foucault's pendulum in the Paris Exhibition of 1851." Eddington says (*Space, Time and Gravitation,* 1920 ed., p. 153) "I doubt whether anyone will persuade himself that the stars have anything to do with the phenomenon. We do not believe that if the heavenly bodies were all annihilated it would upset the gyrocompass." In saying this Eddington was guided by more than mere intuition, for he adds in the next sentence that precise calculation shows that the centrifugal force could not be produced by the motion of the stars, so far as they are known. He does not, however, indicate the nature of the "calculation."

Mach enunciated the ideas which are at the basis of his "principle" in the course of a critique of Newton's treatment of his laws of motion. Mach strenuously objected to Newton's recognition of two kinds of space and time, relative and absolute. Mach made an essentially operational analysis of the situaton. If absolute space and time by definition have no identifiable parts, then there is no method by which change with respect to them can be established and it is meaningless to talk about absolute position or duration. This point of view was applied in detail to Newton's celebrated examples of the pail of water rotating under a twisted rope and two spheres rotating about a mean center. Newton was trying to show by these examples that there was something "absolute" about rotation. Mach showed that there was nothing in the situation which could not be described in terms of rotation with respect to the fixed stars, and that therefore Occam's razor and Mach's own principle of scientific economy demand that we talk about the situation only in terms of reference to the fixed stars. And if we do talk about the phenomena of rotation in terms of the fixed stars, then we may say that in a certain sense the phenomena of rotation are "determined" by the stars.

It seems to me that Mach did Newton less than justice. Newton did not say everything he had in the back of his head, and he was groping for something which he either could not or was not willing to express. It must be remembered that Newton's activities were by no means exclusively scientific or mathematical, but there was a side of his thinking, his religious thinking, which occupied more time and effort than his scientific thinking, and which he was pathologically reluctant to share with his fellows. This is to a certain extent understandable when we remember that at that time anti-trinitarian beliefs were punishable as a criminal heresy. In one place Newton speaks of absolute space as the "sensorium of God." Against this background Newton's insistence on an absolute time and space becomes more comprehensible—he was not intellectually stultifying himself in thinking that these might "exist." It was for Newton a most pressing question whether human beings are in any way capable of acquiring any understanding of these attributes of Deity. Newton recognized the great difficulty of acquiring such understanding through *physical* means. He says "It is indeed a matter of great difficulty to discover and effectively distinguish the *true* from the apparent motions of particular bodies; for the parts of that immovable space in which the bodies actually move do not come under the observation of our senses. *Yet the case is not altogether desperate, for there exist to guide us* certain marks, abstracted partly from the apparent motions, which are the differences of the true motions, and partly from the forces which are the causes and effects of the true motions." Newton then goes on to consider what these "certain marks" are in one specific system, that of two spheres rotating about a common center tied together by an elastic cord. He shows that there are measurements which can be made entirely within the system, such as the tension of the connecting cord, from which both the magnitude and the direction of the "absolute" rotation can be determined. Mach would describe this rotation as being with respect to the fixed stars, but it is not necessary so to do; it can be described completely in terms of measurements made within the system. Newton says in concluding: "In this way we might find both the quantity and the direction of the circular motion, considered even in an immense vacuum, where there was nothing external or sensible with which the globes could be compared."

It seems pretty evident that Newton was hoping that at some time in the future other physical methods might be discovered which would carry further the clue offered by our present knowledge of rotational motion and perhaps give us measurable knowledge of absolute space and time. It would be hard to imagine more pregnant words than Newton's "Yet the case is not altogether desperate, for there exist to guide us . . . ." All this is to be taken by way of understanding Newton, not of justifying him in the light of present day ideas.

Let us now turn from Newton to consider what it was that Mach had in mind. He showed that any phenomena which we might be inclined to call phenomena of absolute rotation can be described in terms of rotation with respect to the fixed stars. What significance did he ascribe to this? Did he regard it as pretty fundamental, or did he regard it as more or less of an artifact? The indications are that he regarded it as pretty fundamental. If so, then Einstein and other cosmologists are justified in crediting to him a new idea and thus speaking of the "Mach Principle," in spite of the fact that Mach's formulation was never precise.

It would almost appear that Mach, as well as Newton, was groping for something that he was never able to satisfactorily express. He says that until fundamentally new sorts of experiment are available it will be expedient provisionally to regard all motions as determined by these bodies (that is, the distant masses). He never attempts to give any mechanism by which the distant bodies could act on local bodies, or to make more precise what he means by "determined." At times he expresses himself with a vagueness which he would probably characterize as "metaphysical" if he had found it in another. Thus: "But we must not forget that all things in the world are connected with one another and depend on one another, and that we and all our thoughts are also a part of nature."

It was Mach's idea not only that the phenomena of mechanics *could* be described in the frame of reference of the fixed stars, but that it was *necessary* so to describe them. He says (*Mechanics*, 1919 English ed. p. 546, appendix): "I abandoned these attempts [i.e., to use only locally defined coordinate systems] because I was convinced that we only *apparently* evade by such expressions references to the fixed stars and the angular rotation of the earth . . . . In point of fact, it was precisely by consideration of the fixed stars and the rotation of the earth that we arrived at a knowledge of the law of inertia as it now stands, and *without these foundations* we should never have thought of the explanations here discussed." In another place (*Mechanics,* p. 235) he says: ". . . we see that even in the simplest case, in which we apparently deal with the mutual action of only *two* masses, the neglecting of the rest of the world is *impossible.* Nature does not begin with elements, as we are obliged to begin with them."

Mach's mathematical analysis took a form which led to his attitude and conclusions very naturally. From the very beginning he used Newton's law of motion in the conventional form : force equals mass times acceleration. "Acceleration" always appears without qualification, and is in a certain sense treated as an absolute, yet of course acceleration has meaning only in a frame of reference. This frame of reference has to be a full three-dimensional coordinate system, in which the acceleration appears as a *vector.* Mach never speaks explicitly of a "frame of reference," but nevertheless he had to provide one for his accelerations. Such a frame is at hand as a sort of average over all the masses in the universe, and Mach gives a mathematical expression for this average. So long as such a frame is used, the laws of mechanics obviously cannot even be formulated without reference to all the masses in the universe. Apparently it was this unique frame that Mach had in mind in all his thinking. He made no use of the fact, which must have been well known to him, that given any frame in which the Newtonian laws hold, then there is a triply infinite set of other frames in which they also hold. Mach's tactics here are radically different from those of Einstein in his special theory. Einstein was content merely to say that Galilean frames of reference (that is, frames in which Newton's laws of motion are valid) exist, without attempting to give a specific example of such a frame, which it is to be presumed could be found by the method of trial and error if in no other way. Mach, on the other hand, felt the necessity of exhibiting such a frame, and in following this impulse, which under other circumstances can be regarded only as admirable, was led into a subtle logical error. For, given the possibility of other equally valid frames, the cogency of Mach's position rests on the assumption that none of these other frames can be defined in purely local terms, but that *all* valid frames must be defined in terms of all the masses in the universe.

The point of my present argument is that there *are* valid frames which can be specified in purely local terms. My refutation of Mach's position begins with the denial of his contention ". . . we see that even in the simplest case, in which we apparently deal with the mutual action of only *two* masses, the neglecting of the rest of the world is *impossible."*

Our starting point is the concept of force; this we regard as an aspect of the "interaction" between bodies. "Interaction" is a loose word, which acquires sufficient precision only in the context of our whole experience.

Roughly, two bodies are in interaction when the observable behavior of either in the absence of the other is different from its behavior in its presence, all other things being equal. This means that in this present argument we do not recognize the so-called "motional" forces, such as centrifugal or Coriolis forces. We recognize two sorts of force between interacting bodies. There are forces when bodies are in mutual contact, which may be described and measured in terms of the distortions of the regions in contact, and there are forces between separated bodies. A special case of the first kind of force is the force between two bodies connected by a stretched thread. We may define the unit force as the force exerted by a standard thread when it is stretched by a standard amount. We can build up from this, by coupling threads in parallel and by stretching them different amounts, a method of giving a numerical value to the force exerted by any known thread stretched by a known amount. By means of threads, forces may be exerted between the members of any pair of bodies, separated by any distance. Having command of the stretched thread technique, we can then study the interaction between separated bodies with no tangible coupling between them. We establish in the first place that in general any such interaction between separated bodies drops off rapidly with the distance. At small distances of separation we can measure these forces of interaction in terms of stretched threads tied to scaffoldings in such a way as to suppress any motion due to the other sort of force. We presently arrive at the notion of gravitational (and other, such as electrical) forces between separated bodies, and we are able to specify when these forces become to small because of the magnitude of the separating distance that they may be neglected for our purposes.

We assume now a swarm of massive particles, moving in any arbitrary manner whatever with respect to each other, but all of them "force-free," that is, with no material connections with each other and all sufficiently remote from each other so that gravitational or other distant interactions are negligible. We study the distances between all possible pairs of particles as a function of time. We can imagine this done perhaps by taking successive instantaneous photographs from different vantage points. The distance of separation of two particles can be measured, given only the pair—it does not require a frame of reference to do this. Neither does it require a frame of reference to measure time : we can suppose this done in a prerelativity context by a single observer with a single clock, and we can neglect effects due to the velocity of light, or we can make naive first order corrections for them if we insist. In general, the observer will find that the members of any arbitrary pair of bodies in the swarm at first approach and then recede. That is, the time derivative of the distance of separation will change sign, which means that the distance of separation (a scalar) will be accelerated in time (non-vanishing second time derivative). Our observer now directs his attention to the scalar *acceleration* of all pairs of bodies. He finds that these scalar accelerations for certain pairs of bodies sometimes vanish. He makes the surprising discovery that if this scalar acceleration is at any time zero for any particular pair it continues to be zero for all time, past and future. Among the many bodies of the swarm the observer may discover one or more such that to each one of them there correspond at least three other bodies that the scalar accelerations vanish. (If the observer does not find a body exactly satisfying this condition, we may assume that he finds one satisfying it to a sufficient degree of approximation so that he can set up a hypothetical body which would satisfy it.) Having now one body coupled to three others with no scalar accelerations, the observer makes the further spectacular discovery that the three angles between the lines joining the body with the three others remain invariant. (It does not need a frame of reference to measure the angle between two intersecting lines.)

The three directions thus found, with the particle at their point of intersection, together determine a Galilean frame of reference. In this Galilean frame the motion of any particle of the swarm may be described as a vector, and since its motion is force-free the vector acceleration of all particles will vanish. The Galilean frame which our observer has thus found is by no means unique; he might be fortunate in finding other particles coupled with other particles with no scalar acceleration, or if he did not find such particles he could construct them. All such Galilean frames preserve a fixed orientation with respect to each other and move with uniform relative velocity.

If the statements in the last two paragraphs are not intuitively obvious, they may be simply proved by working the conventional Newtonian equations backward.

Having now a Galilean frame of reference we can discard the condition that the particles be force-free and study the behavior of a swarm with arbitrary forces acting between the particles. We shall presently be led to attach a characteristic parameter to each of the particles, its mass. These masses may be determined in terms of entirely self-contained measurements made on the particles grouped separately in pairs. For, contrary to Mach, each pair of particles has a self-contained mechanics, expressible completely in terms of measurements made within the pair, with no reference to other bodies. Let us now isolate for further study some particular pair of bodies. For this study we start the pair in a certain initial condition. In this initial condition they are to be in the first place force-free (we have the criterion for this in sufficient distance from other bodies), and in the second place, they are to be so manipulated, by trial and error if necessary, that their distance of separation is constant in time. We now allow a mutual force to act between the members of the pair by stretching a thread between them. We study the behavior when this force varies, something which we can do because we know how to measure the force. We

discover the exact analogue of Newton's laws of motion in terms of parameters which have operational meaning given the pair only. There is a
single mass parameter, M, characterizing the two members of the pair
together, and there is a single geometrical parameter, s, the distance of
separation of the members of the pair. The equation of motion is:

$$F = M \frac{d^2 s}{dt^2}$$

where s is the distance of separation, M the joint mass parameter, and F
the force, taken positive if it tends to separate the members of the pair.
Since s, t, and F are independently knowable, M is determinable through
the equation by appropriate measurements. As in ordinary mechanics, the
pair has a "momentum," $M \, ds/dt$, such that momentum is equal to Ft, a
kinetic energy $\frac{1}{2} M \, (ds/dt)^2$, and a conservation law such that kinetic
energy plus the negative of the work done by the force ($F \, \Delta \, s$) is conserved.

By studying the behavior of the three possible pairings of three bodies,
the mutual mass parameters may be made to give the conventional Newtonian masses, m, of the individual bodies. In particular, given M_{12}, M_{23}
and M_{31}, we find

$$m_1 = \frac{2M_{12}M_{23}M_{31}}{M_{23}(M_{12}+M_{31}) - M_{12}M_{31}}$$

with similar expressions for m_2 and m_3 by advancing the subscripts.

Having the individual masses we are now in a position to develop the
whole of conventional Newtonian mechanics for extended bodies, and all
this without any reference to distant masses. In particular, we are in command of all the phenomena of rotating bodies. We can construct gyrocompasses, and shall find that two gyrocompasses originally parallel retain their
parallelism, or, if not originally parallel, they retain unaltered the original
angle between their axes. Furthermore, there are unique connections between the gyroscopic effects and other independently measurable physical
effects. In particular, such bodies are in a state of internal stress, the
presence of which can be demonstrated by making suitable cuts in the
material of the bodies, and which can be measured by such means
as elastic distortions or optical double refraction. All these effects have
meaning and are measurable in terms of operations localized entirely
within the system. It is never necessary to go outside the system, which
constitutes a logically closed universe of its own. The discovery that a
gyroscope persists in pointing toward the same distant star is an interesting
discovery, but use of this fact is not necessary to develop a self-consistent
and exhaustive treatment of all the phenomena within the system. Among
the internal parameters there will be one equal to the angular velocity
with respect to the stars. This aspect of the situation was described in the
quotation from Newton.

Internal to the system itself there are certain types of behavior reminiscent of the behavior of the local gyrocompass with respect to the stars. Consider the original swarm of force-free particles, and within this swarm consider one of those particles which served as the origin of a possible Galilean triplet. Paired with each such particle is another particle travelling away from it with uniform scalar velocity. As time goes on the distance between this third particle and the paired particles to this paired particles will increase indefinitely. A gyroscope at the origin of the particular Galilean triplet will continue to point toward the receding member of the pair. When this receding particle gets sufficiently remote it will function as a "star," and a gyroscope situated anywhere within a pre-assigned distance from the origin (not only at the origin itself) will continue to point toward it within a pre-assignable margin of error. The same is true for the receding particles originally determining the other two coordinate directions. It is true also of all the other possible coordinate triplets. Eventually we end with a system of very distant bodies, in all possible directions, toward any one of which the axis of a local gyroscope will continue to point.

Forgetting now the genesis of the coordinate system, we see that any sufficiently distant system of force-free bodies functions as "stars." The actual stellar universe functions as such a system to the extent that the stellar objects are force-free. "Force-free" means that they exert no detectable forces on local objects and exert no mutual forces on each other. Experience shows that they are sufficiently far away to exert no detectable force on local objects. The only criterion that we can apply to prove that they are mutually force-free is that they exhibit no mutual accelerations. In practice this means that they exhibit no relative rotations, since accelerations in the line of sight are in general beyond experimental detection. These requirements correspond to what we do in practice. Since our galaxy does exhibit rotational phenomena, our galaxy cannot be included in the universe with respect to which the gyroscope maintains its orientation. It is the universe of distant nebulae with respect to which the gyroscope maintains its direction.

The situation at which we have arrived is riddled with paradox. We have in effect been saying that the stars appear to control the behavior of a local gyrocompass for the reason that they are so far away that they can exert no influence on it. On the other hand, no matter how far away the star, it has to have *some* connection with the local gyrocompass, for the axis of the compass has to point to *some* star, and to that extent the star is connected with the gyrocompass, willy-nilly. One can even rationalize the situation, for what could be more natural or simple than for the compass to continue pointing toward the same star? The principle of sufficient reason seems to demand it.

We would appear to be in the presence of two closely related but logically distinct points of view. On the one hand we have the naive physicist who, when he sees the gyrocompass pointing persistently toward the same star feels under compulsion to say "the stars must have *something* to do with it," and on the other hand we have the sophisticated Eddington who cannot persuade himself that the stars can have anything to do with it. However, Eddington himself would have to admit that the star and gyrocompass are "connected" in so far as one points at the other, but he would probably maintain that this cannot be an "effective" connection. He would probably be hard put to formulate precisely what constitutes an "effective" connection, but he would doubtless insist that there must be something with physical significance in the background even if it is difficult to formulate it. For otherwise any two entities whatever are tautologically "connected" merely by virtue of being in the universe of discourse. We may perhaps call this sort of connection a "nominal" connection.

The thesis of this essay is that the persistent orientation of the gyrocompass with respect to the universe of fixed stars need be evidence of no more than a "nominal" connection between local inertial phenomena and the distant universe, whereas the Mach principle, as commonly understood, takes this as evidence of an "effective" connection.

It is especially to be noted that the argument cannot be inverted. We have not shown that an "effective" connection between local inertial phenomena and the distant universe is impossible. There may be such an "effective" connection, but such a connection can be established only by detailed and specific argument—general considerations such as implied in the Mach principle are irrelevant. As a matter of fact there are at present in the literature a few suggestions as to the possibility of an "effective" connection between local and distant inertia which are more or less detailed and specific. Thus there has been an attempt to exploit the dimensionless combinations of the various universal constants, including Hubble's constant for the expansion of the universe, which has only recently come within our control. Or there have been attempts to exploit what at first sight is a purely verbal principle of the relativity of acceleration, but which curiously turns out to have a physical basis in that a radiated accelerational wave attenuates as the first power of the distance rather than the second power. Up to the present all such suggestions are somewhat vague and are not yet capable of experimental verification.

In addition to our specific result that there is no logical justification for connecting the inertial behavior of local bodies with the total mass in the universe, I think our considerations have somewhat simplified the general problem of understanding the inertial behavior of matter. Instead of considering the behavior of any body in the context of the entire cosmos, we need consider it only in the context of a local swarm of force-free particles which establish a local Galilean frame. It has been shown in detail how to set up this frame. In this frame every individual particle continues to move for all time with uniform velocity with respect to the frame. This suggests some sort of "effective" connection with the frame, not merely a "nominal" connection. But what it can be that makes this connection "effective" we have no inkling. We have not exorcized the miraculous from the inertial behavior of matter by restricting ourselves to the local arena, and Newton's first law of motion, when we roll it under our tongues, still remains incredible.

CHAPTER 16

Phenomenological Theories

BY MARIO BUNGE

WHENEVER THE NEED for a new theory is felt in some field of factual science, both the theory builder and the metascientist are confronted with the problem of choosing the *kind* of theory that should be tried next. Shall the next endeavor be in the direction of increasing detail and depth (growth of the population of theoretical entities)? Or shall it eschew speculation on what goes on in the innermost recesses of reality and focus, on the other hand, on data fitting, with the sole help of fairly directly observable variables? In other words, shall *the* future theory be representational or phenomenological, shall it be conceived as a more faithful picture of reality or only as a more effective tool for summarizing and predicting observations? Both tendencies—the representational and realistic on the one hand, and the phenomenological and instrumentalistic on the other—have had their defenders ever since Democritus and Plato.

Phenomenological theories—like thermodynamics and S-R psychology—are often commended on account of their acknowledged generality, and at other times because of their alleged philosophical virtue of not trespassing on the description of phenomena, by abstaining from introducing dubious occult entities such as atoms or will. Unfortunately, the recommendation is not necessarily a wise one, and even if it were it would be difficult to practice owing to the ambiguity of the term 'phenomenological.' In the following the distinctive features of phenomenological (or black box, or behaviorist) theories will be investigated, and their merits and demerits will be pointed out. The net result will be that black boxes are necessary, but not sufficient, and that black-boxism is apt to hinder the progress of knowledge.

I. Scientific Theories as Boxes

Both scientific theories and their referents have often been likened to devices in the form of boxes with external dials that can be manipulated.[1] The dials correspond to the "external" variables representing observable properties, such as the size and direction of motion of visible bodies; the pieces inside the box correspond to the "internal" or hypothetical variables, such as elastic strain and atomic weight. If, in order to run the box, only the dials have to be manipulated, we have a *black box* theory—a handy name coined by electrical engineers to describe the dealing with certain systems, such as transformers or resonant cavities, *as if* they were units devoid of structure. If, in addition to the handling of the dials representing the external variables, we are required to meddle with a hypothetical inner mechanism described with the help of "internal" variables (hypothetical constructs), we are confronted with what may be called a *translucid box* theory. Black box theories are also called *phenomenological*; and translucid box theories may be called *representational*.

Distinguished representatives of the class of black box theories are:

(*i*) *Kinematics,* or the study of motion without regard to the forces involved—which study is left to dynamics, a typically translucid box theory.

(*ii*) *Geometrical optics,* or the theory of light rays, which makes no assumption about the nature and structure of light—a problem approached by physical optics, a representational theory.

(*iii*) *Thermodynamics,* which makes no assumption about the nature and motion of the constituents of the system, a problem which is dealt with by statistical mechanics, a translucid box theory.

(*iv*) *Electric circuit theory,* in which each element in a circuit is treated as a unit devoid of internal structure; this structure is the subject of field theory and electron theory.

(*v*) *Scattering-matrix theory* in atomic and nuclear physics, which focuses on the measurable characteristics of the incoming and outgoing particle fluxes; the corresponding translucid box theory is the usual hamiltonian quantum theory, which postulates definite interactions among the particles.

(*vi*) *Classical chemical kinetics,* which dealt with reaction velocities and evaded the question of reaction mechanisms.

---

[1] See, e.g., J. L. Synge, *Science: Sense and Nonsense* (London: Cape, 1951), Chap. 2, and Warren Weaver, "The Imperfections of Science", *Proc. Amer. Philosophical Soc.,* Vol. 104 (1960), p. 419.

(*vii*) *Information theory,* which ignores the kind and structure of the elements involved (transmitter, channel, etc.), as well as the meaning of the messages transmitted.

(*viii*) *Learning theory* in behaviorist psychology, which eschews every reference to physiological mechanisms and mental states.

Black box theories are, then, those whose variables are all external and global, whether directly observable (like the form and color of perceptible bodies) or indirectly measurable (like temperature and potential difference). Translucid box theories, on the other hand, contain in addition references to internal processes described by means of indirectly controllable variables that do not occur in the description of ordinary experience; examples of such hypothetical constructs are electron position, wave phase, gene, and subjective utility. None of these concepts can be handled in the same way as the external variables, although they are often objectifiable in a more or less tortuous fashion usually involving some sophisticated theory. In short, black box theories focus on the *behavior* of systems and, particularly, on their observable inputs and outputs. Translucid box theories do not regard behavior as an ultimate but attempt to explain it in terms of the *constitution and structure* of the concrete systems concerned; to this end they introduce hypothetical constructs that establish detailed links among the observable inputs and outputs.

II.   Some Misunderstandings

The terms 'black box,' 'external,' and 'nonrepresentational,' which are all equivalent, seem preferable to 'phenomenological,' a highly equivocal qualifier. In fact, 'phenomenological' suggests description of phenomena (experiential facts) rather than of *objective facts*; it even suggests a theory cast in phenomenalistic language—the nonexistent language of *sensa* dreamt of by some philosophers. But no scientific theory is merely a summary of phenomena, or even of objective facts; and no scientific theory dispenses altogether with diaphenomenal or transcendent terms, i.e., terms such as 'mass' and 'nation,' which represent entities and properties not given in ordinary experience. Thus thermodynamics—the paradigm of phenomenological theory—is not concerned with describing heat phenomena, but rather very general properties and laws, with the help of high-level constructs such as energy and entropy. *A fortiori,* no scientific theory has ever been cast in purely phenomenal terms such as secondary (sensible) qualities : hardly anybody is interested in my private sensations. The task is rather to account for the world—including that part of the world we call our private experiences—by means of objective theories.

Further possible misunderstandings related to black box theories should be cleared up before proceeding to a more detailed analysis. First, 'black box' refers to a kind of *approach* rather than to a subject; it suggests that over-all behavior rather than internal structure is being dealt with—without implications concerning the nonexistence of a structure. Hence, 'black box' or 'phenomenological' should not be equated with, say, 'macroscopic.'[2] One and the same macroscopic entity can alternatively be approached as a unit or as a system of independent or interdependent parts; and microscopic systems, like nuclear particles, can be treated either as black boxes or as complex systems.

Second, black box theories do not all contain "external" or observable variables alone. Current and voltage, the chief variables of electric circuit theory, are not directly observable; their values are inferred from pointer readings with the help of theory. Nor are particle momentum and state function—the main variables of the S-matrix theory—directly observable either. What is essential in the black box approach is not so much the restriction to observables—a restriction that would render theorizing impossible—as the *interpretation* of all nonobservable variables either as merely computational auxiliaries devoid of concrete reference,[3] or as characteristic of the system as a whole. Thus entropy, which in statistical mechanics is in most cases a measure of microscopic disorder, is treated by thermodynamics as a convenient abbreviation for certain relation between the heat content and the temperature of the system. Using a terminology familiar to psychologists,[4] we might say that black box theories cannot fail to contain *intervening variables*, that is, variables mediating between the input and the output; translucid box theories, on the other hand, contain in addition *hypothetical constructs,* i.e., variables referring to unobserved entities, events, and properties.

A third common misleading assertion is that all phenomenological theories are nonfundamental or derivative. It is true that macroscopic black box theories do not appeal to the properties of the "fundamental" constituents. Thus, classical elasticity theory treats solids as continuous media, without regard to their atomic structure. But phenomenological theories of "fundamental" particles, such as the one in which the "strange-

---

[2] See, on the other hand, the classical account of A. d'Abro, in *The Decline of Mechanism* (New York, Van Nostrand, 1939), p. 91, "In the phenomenological theories, our attention is restricted to the macroscopic properties which appear on the commonplace level of experience."

[3] Thus, e.g., Ernest W. Adams, "Survey of Bernoullian Utility Theory," in Herbert Solomon (ed.), *Mathematical Thinking in the Measurement of Behavior* (Glencoe, Ill: The Free Press, 1960), p. 158: "From the behavioristic point of view, the analysis of the mental processes involved in decision simply functions as a heuristic guide to the construction of theories whose scientific meanings lie entirely in their observable consequences."

[4] Kenneth McCorquodale and Paul E. Meehl, "Hypothetical Constructs and Intervening Variables," *Psychological Review,* Vol. 55 (1948), p. 95.

ness" parameter plays a key role—are counter-examples to the equation Phenomenological = Nonfundamental, which must therefore be rejected.

Fourth, black box theories are not purely descriptive devices. No scientific construction is legitimately called a theory if it does not provide explanations in the logical sense of the word, i.e., subsumptions of singular statements under general statements. What is true is that black box theories afford only *superficial* explanations, in the sense that they do not supply interpretations in terms of events and processes within the system concerned. (We shall return to this question in § IX.)

Fifth and last, black box theories are not incompatible with causality. Thus the theory which regards organisms as units pushed hither and thither by external stimuli is both causal and phenomenological.[5] Moreover, that behaviorist theories must have a causal ingredient follows from the definition of efficient cause and the definition of behavior as the set of responses to changes in the environment. That a knowledge of causes does not supply the mechanism is dramatically illustrated by present pathology : a fairly advanced etiology of cancer is consistent with a stubborn ignorance of the mechanisms triggered by the acting causes. Consequently, Black box ≠ Noncausal.

A closer look at the role of "internal" variables should bear out the above contentions.

III.   Structure of Black Box Theories

Any quantitative scientific theory concerning the transactions of a system with its environment may be summed up in the following symbolic relation :

$$(1) \qquad\qquad O = M\,I$$

where 'I' designates either the initial state of the system concerned or the set of stimuli (input), 'O' stands either for the final state or the set of responses (output), and 'M' summarizes the properties of the box. In black box theories the "mechanism" connecting I and O will be left unspecified; that is, 'M' will be just a symbol (e.g., an operator) performing the syntactical linkage among the input data I and the output data O. In translucid box theories, on the other hand, 'M' will concern the constitution and structure of the box—in short, 'M' will represent the mechanism responsible for the overt behavior of the box.

---

[5] See Mario Bunge, "Chance, Cause, and Law," *American Scientist*, Vol. 49 (1961), p. 432, and *The Myth of Simplicity* (Englewood Cliffs, N. J.: Prentice-Hall, 1963), Chap. 11.

Three classes of question can be raised in relation with Equation (1):

(*i*) *The problem of prediction:* given the input I and the kind of box (i.e., M), find the output O.

(*ii*) *The inverse problem of prediction:* given the output O and the kind of box (i.e., M), find the input I.

(*iii*) *The problem of explanation:* given the input I and the output O, find out the kind of box—i.e., determine M.

The contrast between representational and nonrepresentational theories does not occur in connection with the first two problems as sharply as with the third one. If only a black box theory is available or wanted, the problem of explanation (*iii*) will be solved by computing the inverse, I^{-1}, of the input since, in accordance with (1), we have

$$(2) \qquad\qquad M = O\,I^{-1}$$

The completion of this task will coincide with the building of the black box theory; or, if the latter is at hand, the particular question will then be answered. But this is only a *first stage* in theory construction, and in the applications of the theory if the translucid box approach is adopted, in which the *interpretation* of M in descriptive terms is wanted. This interpretation involves the hypothesizing of entities that make up M, and the assignment of specific (physical, biological, etc.) *meaning* to all the parameters, otherwise uninterpreted, that usually plague the phenomenological theories.

In other words, a "mechanism" linking I to O is wanted in the translucid box approach. Now, no double column of input data and output data ever points unambiguously to the mechanism symbolized by 'M.' If it is beyond our senses, which is not the case of our watch but is certainly the case of our genetic equipment, such mechanism must be invented. And such an invention does not call for more or even better observation but an effort of the imagination[6]—and this has often been a source of distrust of representational theories. Once the "mechanism" has been invented, tested, and found satisfactory (i.e., not so far refuted), the translucid box theory is regarded as "established"—until new notice. Needless to say, the "mechanism" M need not be mechanical or visualizable; it may be a field, or a chain of chemical reactions, or a system of social relations. What characterizes representational theories is not visualizable models but the assumption that the theory itself is a model of the whole system referred to by the theory, including the system's insides.

---

[6] See Karl R. Popper, *The Logic of Scientific Discovery* (1935; London: Hutchinson, 1959), pp. 31–32, and Bunge, *Intuition and Science* (Englewood Cliffs, N. J.: Prentice-Hall, 1962), Chap. 3.

IV.   Some Limitations of Black Box Theories

The task of analyzing and interpreting the symbol M mediating between inputs and outputs is not always completed : the form of the relation (1) can often be ascertained, but the nature of the mechanism may remain unknown; we say then that we can account for the behavior but not for the structure of our box. When we stop half-way, leaving M unspecified in descriptive (e.g., physical) terms, we have a black box theory.

According to the above account, then, phenomenological theories occur primarily, though not exclusively, in the *early* stages of scientific theory construction, namely, in the accomplishment of the *fact-fitting* task. A fuller understanding of the relation between I and O will only be obtained by filling the skeleton '$O = M\ I$' with a definite "mechanism." This is not merely a psychological requirement, a need to satisfy the urge to understand what has been accurately described. It is a scientific requirement : black box theories are *incomplete,* since they leave the insides of boxes in the dark. A desideratum of the representational (realistic, non-conventionalistic) approach is to *derive* M from assumptions concerning the constitution and structure of the box. And such a derivation usually leads to the realization of inadequacies, or at least limitations, in the phenomenological approach—as in the case of the discovery of statistical fluctuations of thermodynamic and electric circuit variables.

The derivation of M from fundamental laws, and particularly the expression of the coefficients occurring in M in terms of fundamental constants, involves the introduction of "internal," and sometimes even "hidden," variables. In fact, from a logical point of view the "mechanism" represented by M consists of a net of relations among intervening and ostensible variables. Now, according to positivism, operationalism, phenomenalism, and conventionalism, "internal" variables are parasitic and must consequently be eliminated. Yet, to the extent to which the above analysis is adequate, internal variables are not only psychologically and heuristically instrumental, but their introduction increases the depth and coverage of theories, while at the same time increasing their risk of refutation—which, according to Popper,[7] amounts to enhancing their content and testability.

The history of factual science can be construed as a sequence of transitions from black box to translucid box theories, notwithstanding a few occasional reversals of that main trend. The Copernican revolution amounted to introducing "internal" variables describing, not the apparent motion, but the real path of celestial bodies. The field physics that replaced action-at-a-distance theories involves unobservable field strengths, and,

---

[7] See Popper, *op. cit.,* Chap. IV.

what is worse from a phenomenalistic viewpoint, it involves field potentials. Statistical mechanics, which explains phenomenological laws such as Boyle's, employs some of the transcendent predicates characteristic of atomic physics. The quantum theory deals with unobservables such as particle position and momentum (originally regarded in matrix mechanics as mere computational auxiliaries), in addition to essentially unmeasurable properties like wave phases and virtual states. And recent attempts concerning a subquantum-mechanical level involve the introduction of deeper level variables, for the time being hidden, accounting for the random behavior of microscopic systems.[8] Finally, genetics enables us to deduce Mendel's phenomenological laws, much as neurology attempts to provide the mechanism linking stimuli and responses. Both epistemology and history refute, then, the claim that phenomenological theories are the highest type of scientific systematization.

So far we have dealt with black boxes, translucent boxes, and the transition between the two. Are there not intermediate species—semitranslucent box theories? The following three sections will show that there is room for a comparative concept of the blackness of theories, i.e., that theories can be ordered, as it were, according to the degree of light they shed on the structure of their referents.

V.   Semiphenomenological Theories in Electromagnetism

Maxwell's theory of the electromagnetic field has often been called phenomenological,[9] presumably on the ground that, properly interpreted, it rendered *mechanical* models of the field dispensable. In this case, 'phenomenological' was taken to mean "nonmechanical." And since in Kelvin's time it was widely held that only mechanical models could produce satisfactory explanations, it was concluded that Maxwell's theory was descriptive rather than explanatory—whence it constituted a triumph of the descriptive, antiexplanatory policy of positivism.[10]

Now, it is true that Maxwell's theory is nonmechanical. It does not

---

[8] See David Bohm, *Causality and Chance in Modern Physics* (London: Routledge and Kegan Paul, 1957).

[9] E.g., Gian Antonio Maggi, *Teoria fenomenologica del campo elettromagnetico* (Milan: Hoepli, 1931).

[10] Incidentally, Kelvin's work on resonant circuits, within the frame of the phenomenological circuit theory, was a historical link between Maxwell's prediction of the existence of electromagnetic waves, and Hertz's empirical confirmation of it. This is the more interesting in view of Kelvin's opposition to field theory.

deal primarily with the motion of particles but with the structure and motion of a nonobservable and imponderable kind of matter, viz., the electromagnetic field. But this theory is far from accounting for the field *ab extrinseco,* regarding it as a black box of which only the terminals are scrutable. Far from this, Maxwell's equations are the classical laws of the electromagnetic field *structure* (as determined by the curl and the divergence of the field intensities at each point). Every event concerning macroscopic fields in void—save typically quantal effects like the "creation" of pairs of particles out of the field—can be explained on the basis of the field structure. The field's *behavior,* as manifested through the motion of charged and magnetized bodies, is determined by the field's *structure,* and this is not precisely what characterizes behaviorist or black box theories.

Furthermore, although Maxwell's equations do not require any of the complicated ether machineries that had been fancied by Maxwell himself and by other British physicists toward the end of the last century, it does provide us a *nonmechanical model,* namely, the patterns of force lines that surround and interconnect charged and magnetized bodies. Thus we can picture or visualize the electrostatic field between the plates of a condenser, or the radiation field around an antenna, in terms of oriented lines. In short, Maxwell's theory deals with the structure of its object and it affords an interpretation of electromagnetic processes. Why then should it be called phenomenological?

Compare now Maxwell's field theory with the action-at-a-distance theories. Both pre-Maxwellian field-less electrodynamics (Ampère, Gauss, and Weber) and post-Maxwellian theories of direct interparticle action (Tetrode, Fokker, and Wheeler and Feynman) are black box theories, to the extent to which they do not enquire into the "mechanism" of electrodynamic interaction. They postulate no intervening variables (field intensities and potentials) linking, say, the observable motion of two charged bodies. What they aim at is the computation of the observable net effects of one body upon another. Compared with this set of theories in the Newtonian (or rather Amperian) spirit, Maxwell's is a paradigm of translucid box theory.

What about the picture of the electromagnetic field proposed by the quantum theory of fields? In a sense it is a translucid box theory, too, and even more translucid than the classical theory, since it accounts for the fine-grained structure of the field. A consideration of the electrostatic field will suffice to show the differences among the three approaches under consideration. The classical action-at-a-distance theory describes this field with the sole help of Coulomb's law. Classical field theory accounts for it with the help of Poisson's equation, which subsumes Coulomb's law and enables us to draw the picture of force lines and equipotential surfaces. Finally, quantum electrostatics suggests the following picture: the charged particles are surrounded by unobservable virtual photons, and electrostatic interaction is the outcome of the emission and reabsorption of these virtual field quanta.[11]

Whatever be our belief in the reality of virtual photons the point is that, as compared with this dynamic picture of electrostatic interaction, Maxwell's static account looks semiphenomenological even though, as we saw, it does deal with the field structure and does provide a picture of sorts. In return, quantum electromagnetic theory does not focus on the production and propagation of wave fields. Its central question is not "How does such and such a field originate and propagate?" but rather "How many field quanta with a given momentum and polarization direction are there in a given spacetime volume?" In this respect, current quantum field theory comes close to the accountant's view of nature triumphant in thermodynamics. At any rate, the moral seems to be this : some boxes are more translucid than others when looked at from certain angles. Or, rather, the degree of "phenomenologicalness," or blackness, of theories varies according to the respect that is being considered. Consequently, one should not say "x is more phenomenological than y," but, rather, "x is more phenomenological than y in the respect z."

VI.   Semiphenomenological Theories in Quantum Mechanics

It is often maintained that quantum mechanics is a phenomenological theory, although it is not clear why this claim is made. The contention seems false : quantum mechanics not only contains intervening variables but, just as in the case of Maxwell's theory, *all* its basic variables are intervening in the sense that only derivative quantities (such as eigenvalues and averages) can be contrasted with the results of present-day experiments; at the same time they are hypothetical constructs in the sense that their referents are hypothetized entities and properties. Thus, if one wishes to obtain the frequencies of the light emitted by certain atoms, or the collision probability of two particles of a certain kind, one does not get them by processing information about instrument readings, but one starts by setting up a microscopic model described by basic formulas, such as hamiltonians, wave equations, commutation relations, "causality conditions," and so forth. Such equations relate variables that are far removed from the empirical data concerning the phenomena to be ultimately accounted for. In fact, they contain, say, the number, charge, mass, wave length, and phase

---

[11] The virtual quanta are not free; they are bound to the "bare" charged bodies. When an electron is stopped, it "shakes off" some of the virtual quanta surrounding it— a much more pictorial explanation of the Bremsstrahlung than the classical explanation. See, e.g., W. Heitler, *The Quantum Theory of Radiation*, 3rd ed. (Oxford: Clarendon Press, 1954), p. 146.

of the colliding particles—and such data are never collected by mere observation but must be either inferred or hypothesized. Yet there are cases where a phenomenological approach can be *grafted* onto this essentially translucid box theory; from this union, *semi*phenomenological microphysical theories are born. Let us recall a few important recent cases.

The problem of nuclear forces can be approached either field-theoretically (meson theory of nuclear forces) or by directly hypothesizing evidence-fitting nucleon-nucleon potentials—which is the gist of the so-called phenomenological or empirical theory of nuclear forces. This name is justified to the extent that the theory keeps silent concerning the nature of the fields that are supposed to be described by the respective force functions; in other words, no field equations are postulated in this approach, but what would normally be regarded as the solutions of field equations are tried haphazardly. In short, the whole aim of this approach is to find force functions fitting the empirical data—a target that can always be attained by manipulating a large enough number of parameters.

But in order to compare the assumptions with the empirical evidence one has to introduce the hypothesized potentials into a wave equation; i.e., one has to graft the phenomenological assumptions onto a basically nonphenomenological theory, such as ordinary quantum mechanics. This is why this theory of nuclear forces deserves to be called *semi*phenomenological rather than phenomenological. It is only the single hypotheses, concerning the various possible law forces, that deserve to be called phenomenological. Incidentally, the testability of the semiphenomenological theory of nuclear binding is pitifully low, since its testable consequences are practically insensitive to large variations in those hypotheses, both in the form of the force functions and in the numerical values of the parameters. This low assailability is characteristic of the black box approach.

A second example of a phenomenological approach within the representational (but only partially picturable) scheme of quantum mechanics is the scattering-matrix theory. This approach was originally proposed in 1943 on the philosophical ground that quantum mechanics is overdescriptive, since it says far more than what can be checked by experiment (e.g., it speaks about the average velocity of a single electron within an atom).[12] In the frame of the *S*-matrix approach, the problem of the interaction among particles, such as a nuclear reaction, is not solved by hypothesizing interactions and by postulating detailed particle motions but by regarding the scattering region as a black box into which certain particles impinge along certain channels and out of which the same or other particles emerge in a lawful but indeterminate way.[13] The force (or, rather, inter-

---

[12] Werner Heisenberg, *Zeits. f. Physik*, Vol. 120 (1943), pp. 513, 673; *Zeits. f. Naturforschung*, Vol. 1 (1946), p. 608.

[13] See John M. Blatt and Victor F. Weisskopf, *Theoretical Nuclear Physics* (New York: Wiley, 1952), pp. 313, 517 ff.

action) problem is bypassed in this approach because the original problem is reformulated in the following way : A set of widely separated (hence practically free) particles is given at the beginning of the process (schematically, in the infinitely remote past), and another set of equally separated (hence practically free) particles is given at the end of the process (in the infinitely distant future). Nothing is asked about the particle motion and the interaction mechanism. Only the incident and the outgoing fluxes are involved in this treatment, along with very general requirements such as flux conservation and the time priority of the input over the output[14]—a postulate usually, and wrongly, called causality condition.

The S-matrix approach is phenomenological and at first blush it complies with the positivist requirement of abiding by what can be measured. But this last is an illusion. The S-matrix theory (and the accompanying dispersion relations technique) is not an *independent* theory; it does not replace quantum mechanics but, rather, supplements it. Even though after the theory was formulated independent techniques for *computing* the scattering matrix were devised,[15] the basic concepts of linear and angular momentum, and state function, which occur in the theory are meaningful in the wider context of the quantum theory alone. The S-matrix theory, then, is a *semi*phenomenological system, even though the original intention, i.e., the approach, was altogether phenomenological.

Similarly, one can apply the scattering-matrix method to waveguide circuits,[16] a procedure which for the electrical engineer (not so for the physicist) has the advantage of saving him the effective computation of the field within the cavities. Yet this approach employs the conceptual framework of Maxwell's theory, if only because the general (unspecified) expressions for the field at the waveguide junctions are required. Incidentally, the S-matrix approach can be generalized so as to apply to systems of any kind,[17] which shows once again that phenomenological theories are not characterized by a subject but by an approach.

---

[14] The central equation of the S-matrix theory is $\psi(\infty) = S\psi(-\infty)$, where '$\psi(-\infty)$' designates the amplitude of the initial state of the system and '$\psi(\infty)$' the amplitude of its final state, while 'S' designates the scattering operator. The corresponding S-matrix gives the probability amplitudes of the possible transitions.

[15] Originally, the calculation of the S-matrix was based on the usual quantum-mechanical (Hamiltonian) formalism, whence it consisted in just refusing information about the intermediary processes. Only six years afterward a method for computing the S-matrix without employing that formalism was evolved. See, e.g. N. N. Bogoliubov and D. V. Shirkov, *Introduction to the Theory of Quantized Fields* (New York: Interscience, 1959), pp. 198 ff., and J. Hilgevoord, *Dispersion Relations and Causal Description* (Amsterdam: North-Holland, 1960).

[16] A. E. Pannenborg, *Philips Research Reports*, Vol. 7 (1952), pp. 131, 169, 270.

[17] Mario Bunge, "A General Black Box Theory," *Philosophy of Science*, Vol. 30 (1963), p. 343.

VII.   A Semiphenomenological Theory in the Elementary Particle Domain

Our third and last example of semiphenomenological theory will be the Gell-Mann and Nishijima scheme for "elementary" particles (1953–54). This theory is characterized by the *ad hoc* but surprisingly fertile hypothesis that certain particles, such as K-mesons and hyperons, have a new, nonvisualizable property called "strangeness." The hypothesis is *ad hoc* because, so far, the strangeness parameter has no physical interpretation and occurs in a single law statement, namely, the further *ad hoc* hypothesis of the strangeness conservation in the interactions that are supposed to give rise to certain metamorphoses, like the transformation of a lambda-zero hyperon into a proton and a pi-meson. On the other hand, the familiar properties of "elementary" particles are understood, if only partially. Mass is related to inertia and quantity of substance, charge to the coupling with the electromagnetic field, spin to some kind of internal rotation, and isobaric (or isotopic) spin the kind of nucleon. None of these variables occurs casually or in isolation. They all occur in more than one law statement concerning the structure of matter, so that they are quite firmly entrenched in physical theory. The strangeness hypothesis, on the other hand, is for the time being just a fact-fitting assumption that accounts for the existence and the behavior—not for the structure—of "elementary" particles.

Yet it would be false to say that the Gell-Mann and Nishijima theory is *altogether* phenomenological. After all, it does relate the new parameter to the old ones, namely, in the expression of the strangeness conservation law; and to this extent it makes sense, if at all, in relation with the general theory of quantum mechanics. One should rather say that the strangeness hypothesis is a phenomenological assumption, and that the theory as a whole is *semi*phenomenological. And one can hope (or dread, according to one's philosophy) that the theory will eventually be subsumed under a deeper, translucid box theory, which will derive all the features of the present Gell-Mann and Nishijima scheme from basic assumptions concerning the complex structure of the particles and fields involved.[18] Such a theory will presumably *interpret* the strangeness parameter, or its successor, in descriptive terms, just as the spin hypothesis provides an inter-

---

[18] Such an attempt is the extremely abstract theory of P. Hillion and J.-P. Vigier, "New Isotopic Spin Space and Classification of Elementary Particles," *Nuovo Cimento*, Vol. 18 (1960), p. 209. Internal state vectors corresponding to the internal structure of elementary particles of different kinds are introduced; strangeness, isotopic spin, and ordinary spin are all explained by certain motions in ordinary spacetime.

pretation of the new quantum number that had originally been introduced in an *ad hoc* or phenomenological way to account for the "anomalous" Zeeman splitting of spectral lines.

Our contention that semiphenomenological or semitranslucent box theories exist between the ideal extremes of black box and translucid box theories seems to be borne out. Let us now examine the scope of the black box approach.

VIII.   Scope of the Black Box Approach

The black box approach has the following peculiar traits.

(*i*) *High degree of generality*. Every black box theory is consistent with an unlimited number of conceivable mechanisms. The generality of black box theories can be carried to the extreme of almost *abstract* theories, i.e., systems containing almost no specific (interpreted) variables, whence they can be applied to a large class of systems.

(*ii*) *Global* or over-all character. Black box theories are at least partially nonlocal in the sense that they regard systems as units rather than as complexes of interacting localized parts.

(*iii*) *Simplicity*. By dispensing with structure details that would require the introduction of hypothetical constructs, and by leaving most parameters uninterpreted, black box theories are formally, semantically, and epistemologically the simplest possible.[19] This simplicity makes it rather easy to construct and apply phenomenological theories.

(*iv*) *Accuracy*. Phenomenological theories can be made to fit any set of data by adequately increasing or manipulating a number of parameters.

(*v*) *Safety*. Black box theories are the most firmly anchored to experience, hence the best protected against refutation. Thus circuit theory remains true, within its domain, despite the various changes in electron theory.

The above are, at the same time, both advantages and weaknesses of the black box approach. In fact, a high degree of generality, or lack of

---

[19] These kinds of simplicity (economy of forms, economy of presuppositions, and economy of transcendent concepts), as well as pragmatic simplicity, are examined in the author's *The Myth of Simplicity* (Englewood Cliffs, N. J.: Prentice-Hall, 1963), Chaps. 4 and 5.

specificity, reveals that the kind of system is not taken into account; non-locality shows that the probable complex structure of the system has been overlooked, or at least integrated; simplicity is a mark of shallowness[20]; finally, excellence of fit unaccompanied by depth, and safety, or certainty, take phenomenological theories dangerously close to irrefutability—which, according to Popper, is the seal of nonscience.[21] All such sinful virtues of black box theories, of course, stem from the avoidance of postulating "internal" variables. As a consequence of such a restriction certain questions cannot be answered because they are purposefully not asked. Questions like "How are the field and the particles inside an iron sphere?" "What is the nerve-and-muscle circuit responsible for an arc-reflex?" and "What are the social forces behind the last Latin American revolution?" simply do not occur within the framework of black box theories. Such questions call for deeper, representational theories, capable of affording suitable *interpretations* of facts.

IX.   Explanation and Interpretation

The fact that certain problems cannot be stated in the framework of phenomenological theories does not mean that black box theories afford no explanation, as is so often heard. Whenever a singular statement is deduced from statements about laws and circumstances, there is scientific explanation. Phenomenological theories do afford, then, scientific explanations (see § II). But explanations can be more or less deep. If the laws invoked in the explanation are just laws of coexistence and succession, the explanation will be superficial. This is the case of the explanation of a deed of an individual on the ground that he always does such things, or the account of the compression of a gas upon a pressure increase in terms of Boyle's law. We often need such superficial explanations, but we also need deeper explanations, such as those contrived in terms of the inner constitution and structure of the system concerned—the dynamical structure of a gas, the personality traits of an individual, and so forth.

A black box theory may provide a logically satisfactory explanation and prediction of a set of data, in the sense of their *derivation* from the theory and specific information. But it will fail to provide what the scientist usually calls an *interpretation* of the same data. Such an interpretation in descriptive terms is obtained when a "mechanism" is postulated and all the parameters are assigned to properties of the "mechanism" (see § III

---

[20] See Mario Bunge, "The Weight of Simplicity in the Construction and Assaying of Scientific Theories," *Philosophy of Science*, Vol. 28 (1961), p. 120, and *The Myth of Simplicity* (Englewood Cliffs, N.J.: Prentice-Hall, 1963), Chap. 7.

[21] See note 6, *supra*.

and IV, above). The "mechanism" will be the incessantly moving and colliding molecules in the case of thermodynamics, the interfering waves in the case of optics, the crystal lattice and the electron gas in the case of solid state physics, the nerve circuits and the associations in the case of behavior theory, the interplay of social groups and interests in the case of sociology and history, and so forth.

Note, however, that although black box theories do not intend to *provide* interpretations they do not *preclude* them; i.e., black box theories can be supplemented by representational hypotheses. This was realized as early as in the beginning of the nineteenth century, with relation to the disputes between the holders of the caloric theory and those of the atomic theory. Thus Fourier derived his equations of heat transmission by employing the molecular hypothesis as a heuristic tool, but he saw that "The truth of these equations [which were phenomenological] is not based on a physical explanation [interpretation] of the effects of heat. Whatever be the way in which one wishes to conceive of the nature of this element, whether one regards it as a definite material thing that passes from one part of space to another [i.e., as the caloric fluid], or as only a transmission of motion, one will always arrive at the same equations, because the hypothesis that one forms must represent the general and simple facts out of which the mathematical laws are derived."[22] Generally speaking, every phenomenological theory is consistent with a number of alternative hypotheses concerning the "mechanism" involved. In other words, a number of deeper, representational theories are consistent with any given phenomenological theory.

The contrast between kinematical and dynamical theories clearly illustrates the concepts of theory depth and explanation level. Any given set of changes can be studied either from a kinematic or from a dynamic point of view, according as a superficial or a deep explanation is possible or required. Thus, for example, chemical reactions may be studied as regards the reaction velocities (chemical kinetics) or as regards the reaction mechanisms (i.e., the processes whereby a given system gets transformed into another system). Just as in the case of mechanics, the second approach subsumes the former; if the reaction mechanism is known, or assumed, then the kinetics of the process can be deduced. But this greater depth of dynamical as compared with kinematical theories does not render the kinematical approach superfluous. It is only on the basis of some previous knowledge of certain kinds of motion that laws of force can be hypothesized, and only a detailed knowledge of chemical kinetics has made it

---

[22] Charles Fourier, *Théorie analytique de la chaleur* (1822), in *Oeuvres*, ed. by G. Darboux (Paris: Gauthier-Villars, 1888), I, p. 538. See also Bunge, *Causality: The Place of the Causal Principle in Modern Science* 2nd ed. (N. York: Meridian Books, 1963), pp. 77 ff.

possible to suggest mechanisms compatible with it. That is, even though the dynamical, nonphenomenological theory usually embraces the corresponding kinematical theory, the latter is always useful as an access to the former. Moreover, black box theories are often sufficient for certain purposes; thus most engineering work, for example, can get along without statistical mechanics.

In short, black box theories, though superficial, are necessary. They satisfy a genuine *scientific* desideratum, namely, that of having general and global pictures of real systems. Furthermore, phenomenological theories are useful because they constitute a bridge between the deeper, representational theories and empirical data. Unfortunately there is a further, illegitimate motivation of phenomenological theory construction, namely, the *philosophical* demand of renouncing transobservational or diaphenomenal concepts, i.e., not just the *methodological* discarding of the system's entrails but the *ontological* refusal to acknowledge the existence of such entrails. To this obscurantist philosophy I now turn.

X.   Black-Boxism

Let us call *black-boxism* the view that the conversion of black boxes into translucent boxes, by filling the former with definite "mechanisms," is neither necessary nor desirable. It would not be necessary, in this view, because black box theories give us all we could legitimately ask for, namely, tools for transforming sets of actual observational statements (evidence)· into sets of potential observational statements (predictions); and it would not be desirable to go beyond phenomenological theories because the introduction of occult entities and properties is unwarranted by sense experience, which is the highest court of truth. As is well known, this deliberate refusal to speculate on the "mechanism" hidden in the box, or even to grant its existence, has been voiced by eminent philosophers of the positivist tradition, as well as by eminent scientists—some of them famous for their contribution to translucid box theories.[23] That the view is inadequate, and even dangerous for the advancement of knowledge, should be clear from the following considerations.

(*i*) Historically, black box theories have usually emerged as *first steps*

---

[23] The history of the war of black-boxism (particularly, energetism) against translucid-boxism (particularly, mechanism and atomism) has been told several times. See Pierre Duhem, *La théorie physique*, 2nd ed. (Paris, Rivière, 1914), Abel Rey, *La théorie de la physique chez les physiciens contemporains*, 2nd ed. (Paris, Alcan, 1923), and Ernst Cassirer, *Substance and Function* (1910; New York: Dover, 1953). For more recent utterances of the point of view of black-boxism, see P. A. M. Dirac, *Proc. Royal Society* (A), Vol. 180 (1942), p. 1, and Heisenberg, *op. cit.*

in theory construction; as a rule they have been superseded or supplemented by representational theories subsuming the former and enabling us to account for more facts than the former.

(*ii*) Epistemologically, black box theories are *less complete* than the corresponding translucid box theories, if only because they emphasize behavior at the expense of structure. (Very often, phenomenological theories focus on the time course of processes, i.e., they are kinematical, and disregard the spatial characteristics of the system, which are necessary, though not sufficient, to account for the mechanism.) Representational theories have a richer content and lend themselves, as a consequence, to a greater variety and fineness of empirical test.

(*iii*) Logically, black box theories *stand somewhat apart from the rest of science*; as a consequence they do not enjoy the support of contiguous areas but have an almost exclusively "inductive" support (i.e., the support of empirical evidence). Now, a theory gains in corroboration if it is shown to be not only compatible with other theories but is in a way logically necessitated by them.

(*iv*) Pragmatically speaking, although black box theories have a wide coverage they have a *low fertility,* in the sense that they do not help us explore the still hidden aspects of reality. It is only by first suspecting and then assuming that there *may* be something beyond phenomena that we eventually succeed in actually discovering that invisible something. The hypothesizing of occult entities and properties is not bad in itself, since most of reality *is* hidden to direct sense perception. As Hertz saw, it is precisely the limited success of the attempt to establish direct relationships among observable phenomena that leads us to realize "that the manifold of the actual universe must be greater than the manifold of the universe which is directly revealed to us by our senses."[24] What is noxious to science is both the postulation of inherently inscrutable entities and the ban of controlled speculation.

From a logical point of view, black-boxism is akin to wholism and Gestaltism, insofar as these schools, too, wish to stop analysis, i.e., to limit reason. Epistemologically, black-boxism may be regarded as a mild form of phenomenalism, the philosophy that tries to reduce everything to experiential elements, such as sensations. But the kinship should not be exaggerated. Black-boxism just asks us, for example, not to postulate particle motions inside atomic nuclei to account for nuclear reactions, and enjoins us to do exclusively with information regarding the input and the output particle fluxes. Radical phenomenalism, on the other hand, would have us dispense with atomic nuclei altogether, on the ground that they are essentially nonsensuous fictions. To phenomenalism, the whole of microphysics is a fairy tale.

---

[24] Heinrich Hertz, *The Principles of Mechanics* (1894; New York: Dover, 1956), p. 25.

Contrarily to black-boxism, radical phenomenalism has never attracted theoretical scientists, because not even the simplest black box theory is couched in terms of sense data. All scientific theories, whether phenomenological or representational, are systems of physical object statements (physicalist propositions). All of them deal with what the scientists mean by 'phenomena,' not with the phenomena of philosophers (i.e., what immediately appears to the subject). Moreover, black box theories do not necessarily involve the denial of the independent existence of the boxes they describe, and they contain no reference whatsoever to the cognitive subject. Quite the contrary, one of the motivations of behaviorism is the avoidance of internal variables such as private sensations or feelings.

In short, neither phenomenological theories nor black-boxism support phenomenalist epistemology. The facts that phenomenalism, as a philosophy, has historically motivated some developments in black box theories (notably in thermodynamics), and that phenomenalists have often commended phenomenological theories, can perhaps be accounted for by the failure to realize that not even time, the least committal of all variables, is directly observable, and that the most epidermic of theories is made up of constructs, not of percepts.

Conclusion

Black boxes will be built, presumably, as long as general and global theories are prized and as long as the available representational theories are known to be inadequate. And translucid box theories will be built as long as the need is felt to explain black boxes, and as long as it is realized that boxes do have insides worth looking into. To ban translucid boxes, as demanded by traditional positivism, is to regard behavior as an unexplainable ultimate, to renounce its explanation in terms of constitution and structure, and to replace the metaphysics of unalterable substance by a metaphysics of function without anything functioning. In particular, black-boxism prohibits the construction of visualizable models—save the black box itself, the poorest of models. Now if models, whether handled as representations (iconic signs) or as analogues (allegoric signs), are renounced, we are prevented from exploiting their heuristic power. In addition, we are prevented from getting a glimpse—literal or symbolic—of the inner workings of the world. A sound metascience will require that not every kind of model but only altogether unfounded and untestable models should be avoided.

The ultimate aim of scientific theorizing is to build representational theories encompassing and explaining the corresponding phenomenological theories. Black-boxism, on the other hand, makes the false assumption that the sole aim of scientific theorizing is to systematize directly observable phenomena. The history of science suggests that this is only an immediate desideratum. The long-run aim of scientific theorizing is not to summarize scientific experience but *to interpret reality* and, in particular, to explain that part of reality coupled to the knower—i.e., the field of phenomena (in the philosopher's sense). Scientific theorizing may *initially* be motivated by the urge to understand what is observed, and is certainly *tested* by facts of this kind; but it does not succeed in its task unless it takes unobservable (but inferrable) facts into account. Not a direct linking among observable aspects but an account of the unobservable, and an interpretation of the observable in terms of the unobservable, has been the most ambitious and rewarding program of factual science, ever since it was formulated by Democritus. The realization of this historical fact helps abandon the quest for final certainty—one of the motivations of black-boxism—and espouse a critical variety of realism.

Yet to abominate black-boxism need not involve suppressing black boxes. On the contrary, a critical realist will have to grant that the world still is and will always be full of black boxes, and that research will never succeed in turning them all altogether translucid. To ban black boxes would be as obscurantist as to condemn translucid boxes. For, first, black boxes are unavoidable in the early stages of theorizing and are useful whenever details can be neglected or when over-all effects alone are studied; thus, whatever the kind of radiation (photons, electrons, neutrons, etc.) and the absorption mechanism, the law of radiation absorption will be exponential because in all cases what matters is the quantity of radiation remaining at a given depth. Second, black box theories should always be tried when the available assortment of translucid box theories fails— as was the case of behaviorist psychology confronting the sterility of introspectionism, and of the dispersion relations and allied theories confronting the failures of hamiltonian theories. Third, black box theories provide general and global accounts, and as such they are useful even long after they become subsumed under representational theories. Fourth, black box theories provide a test for the corresponding translucid box theories; thus a "depth" psychology dealing with subliminal psychical processes and inner motivations cannot be established as a science unless it satisfies the boundary condition of accounting for everything the behaviorist approach does.

What hinders the progress of knowledge is not the multiplication of black box theories but the philosophy that praises phenomenological theory as the highest type of scientific systematization and abuses representational theory. The harm done by this philosophy to physics and psychology is too great to permit remaining indifferent toward it. What must be tolerated or, even better, encouraged, is the proliferation of testable theories of all kinds, phenomenological or representational, kinematical or dynamical, cautious or bold—though keeping in mind that nonphenomenological theories and the realistic epistemology that stimulates their construction are ultimately to be preferred because they have the greater content, they take the bigger risk, and they are the most fertile : in short, representational theories satisfy best Popper's canons of good science.[25]

---

[25] I gratefully acknowledge a discussion with Professor Juan José Giambiagi (Departamento de Física, Universidad de Buenos Aires) on the scope of the dispersion relations theory (a phenomenological theory).

CHAPTER 17

The Simple Laws of Science and History

BY J. T. DAVIES

THEORIES OR "LAWS" have *no* logical status : by this approach Popper[1] eliminates drastically but effectively all the difficulties of the general deriving from the particular. According to this view, a theory may be inspiration, a leap of the imagination, or perhaps merely an analogy or a dream. But wherever the idea originates, whether it seems obvious or trivial, brilliant or stupid, it is a scientific theory if, and only if, it is an attempt to correlate experimental or observational quantities. These must be such that suitable experiments could conceivably be used to *disprove* the theory. The latter thus provokes scientific experiments or observations, and Popper stresses this leading role of theory, pointing out that we always make our experiments or observations with some preconceived idea (i.e., theory) of how the system might or might not behave. We may have at least the theory that it might behave unusually.

I remember vividly how Popper, while lecturing on this theme, strode up to the water tap on the demonstration bench, and, holding the tap, proposed the simple law that "water always flows out when I turn on the tap." This seemed obvious to the class, until Popper pointed out that some plumbers might have just started work, and that the water supply might therefore be interrupted. But during his lectures it never was! Indeed, such tendencies or propensities (e.g., for water to come out of the tap) have formed the subject of some of Popper's work in recent years.

Not only should a scientific theory be experimentally testable, however; it should preferably be *easily* tested. With the possibility that many

---

[1] K. R. Popper, *The Open Society and Its Enemies* (London: Routledge & Kegan Paul, 1945), especially pp. 249–50 and 253; and *The Logic of Scientific Discovery* (London: Hutchinson, 1959).

theories are silly and trivial, experiments that are carried out as *easily* as possible are desirable; and even more important, as *few* experiments as possible should be necessary in each check of the theory. If it were possible to disprove a theory only after millions of measurements, the scientific literature would be full of wild ideas remaining to be tested with a view to disproof. That this is not so is due to our desire for a simple theory (often called a 'law,' though the latter term is no more rigorous than the word 'idea'). Esthetically, most people prefer simplicity and symmetry, and apply the rule of never proposing a complicated theory if they can propose a simpler one. There is a more important argument for simple theories, however, and Popper has developed this definition of a "simple" theory as one involving few arbitrary parameters, and which is therefore relatively easily tested experimentally with a view to possible disproof. Even though certain functions (e.g., $y = A \sin x$, $y = Bx^{1.211}$ or $y = C \log x$) intuitively appear complicated, they each contain only one arbitrary parameter and are therefore relatively easily checked against experimental results.

Given that there are many simple theories or laws which have stood up to repeated testing, let us look a little further at some simple laws and see, first, how accurately certain of these simple laws are valid experimentally; second, how far the simple laws follow from our choice of systems to study; and third, whether the number of simple laws is surprisingly great.

Examples of Simple Laws

Among the "simple" laws, involving few parameters and small, integral powers, may be cited:

Avogadro's Law: For any substance the molecular weight in grams behaves as N separate kinetic particles in diverse phenomena. Here N is a constant, independent of the value of the molecular weight.

Inverse Square Law: The gravitational or electrical force between two bodies varies inversely as the square of their distance apart.

Constancy of Mass of Electron: All electrons (or protons) have the same rest mass: electricity is not found in random sizes.

Ohm's Law: The electrical current flowing through a body is proportional to the applied voltage.

Hooke's Law: The extension of a spring varies as the first power of the load applied.

Gas Law: The pressure of an "ideal" gas varies inversely as the volume in which it is confined.

Poiseuille's Law: The rate of flow of a given liquid through a tube is proportional to the pressure drop across the ends of the tube.

Kekulé's Hypothesis: Certain organic reactions occur as if part of certain organic molecules consisted of a ring of 6 carbon atoms, which comes intact through the reaction.

These simple laws correlate well the experimental data over wide ranges of variables, including different materials. The value of N, for example, is found to be 6.02×10^{23} for materials as different as pollen grains, hydrogen gas molecules, and crystalline salts. The value of N for the latter is found from X-ray analysis. Concordant figures are also obtained from such diverse studies as comparing the viscosity and diffusion of sugar in water, and measuring the amount of helium produced from radioactive elements. Again, the exponent in the "inverse square law" has often been found to be exactly -2, showing that certain chosen systems approximate closely to the mathematical model of completely isolated spherically symmetrical charges or masses. It is again to be remembered that such systems can relatively easily be found. Kekulé's hypothesis is as famous for its origin as for its success. Riding on the top of a London bus, he pondered over the difficulties of reconciling the properties of certain (aromatic) organic compounds with the simple relationships among many others. Tired of fruitless puzzling, he began to doze, and pictured in his mind a mass of snakes turning and twisting. Suddenly in his dream he saw one seize its own tail in its mouth and form a closed loop—and he awoke to postulate the closed loop of a 6 carbon-atom chain as the basic nucleus of the aromatic compounds.

This dream, illogical and nonquantitative as it was, provided a simple key to aromatic organic chemistry. It played a leading role in stimulating experiment. Modern investigations involving electron diffraction and X-ray studies have not revealed anything to disprove Kekulé's simple analogy, which thus seems valid over a wide range of phenomena.

Quantitatively, it is clear that any correlation can be "simple" provided that the range of variables is chosen to be sufficiently small. Hooke's law is a good example of this. In general,[2] one may demonstrate the effect of the range of variables on the simplicity (measured by the paucity of the adjustable parameters) by considering some function y which is altered by a change in the variable x. If the latter is increased from x_1 to $(x_1 + \Delta x)$, then Taylor's series gives:

$$y = f(x_1 + \Delta x)$$

$$= f(x_1) + \Delta x f'(x_1) + \frac{(\Delta x)^2}{2!} \cdot f''(x_1) + \cdots \qquad (1)$$

or

$$y = A + B\Delta x + C(\Delta x)^2 + D(\Delta x)^3 + \cdots \qquad (2)$$

where A, B, C, etc., are constants.

---

[2] J. T. Davies, *British J. Philos. Sci.*, Vol. 5 (1954), p. 191.

This equation shows that a correlation may always be made simple, though at the expense of its generality or accuracy, by neglecting the terms involving C, etc. Further, for *any* given values of C, D, etc., we may always make the correlating law appear simple by working over such small ranges Δx that the terms in $(\Delta x)^2$ and higher may be neglected. Hence a simple (two parameter) law of the form:

$$y = A + B \cdot \Delta x$$

can always be obtained provided that the range of variables is sufficiently small. Such linear "laws," often with $A = 0$, are found for the relation between the pressure of gas in a vessel and the mass of gas in that vessel; in electrochemistry; and for rheological systems.

The latter may be illustrated by typical plots (Fig. 1) of shear stress against the rate of strain. For "Newtonian" liquids the plot is rectilinear through the origin, the slope being called the viscosity, i.e.:

$$S = \eta D \tag{3}$$

where η is called the viscosity.

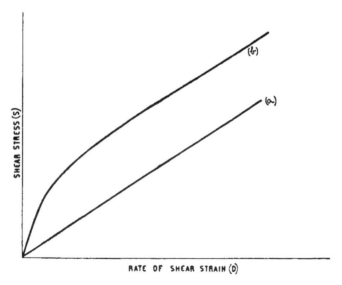

Fig. 1 Plot of shear stress against rate of shear strain for two liquid systems. Plot (*a*) is a straight line, for a liquid such as water obeying the simple "Newtonian" equation (3) with η constant. Plot (*b*) is for a solution of polymeric material.

For "simple" liquids, such as water and alcohol, η is constant, denoting that the work to make the necessary hole in the liquid for viscous flow to occur is unaffected by the flow process. The intermolecular forces vary as r^{-7}, but the intramolecular forces for such liquids vary as r^{-10}. If, however, the liquid consists of a solution of long, threadlike molecules, these may be uncoiled relatively easily, obeying a modified Hooke's law : these additional intramolecular forces are of longer range and much weaker, and consequently the flow process may affect the molecular configuration. These intramolecular effects correspond to larger values of B, C, etc., in equation (2), and the viscosity, identified with y of equation (2), will then be variable unless the flow rate (Δx) is made extremely small.

Simple Laws and Convention

We necessarily test the laws of physics by working over a limited range of variables; the simple gas laws, for example, have not usually been tested over enormous ranges of pressure, temperature, or time. Further, we usually choose relatively simple systems to investigate in detail, particularly those in which it is possible to keep all but one of the possible variables constant during the test.

A third aspect of convention is *analogy*. That analogy plays a large part in our theorizing is generally agreed. Although, as Popper has pointed out, arguing by analogy has no logical basis, simple analogies do often correlate physical data. Spengler emphasized that we often transpose concepts into physics; thus we speak of migrant ions, electrical and magnetic fields, flying and colliding gas molecules, electrical currents, and electromagnetic waves. We also speak of the "age" of the universe.[3]

To me it is always surprising that so many of these simple concepts stand up so well to repeated physical tests. We "scale down" many macroscopic concepts, and often find that they still work tolerably well. The concept of the molecule behaving as a minute billiard ball is another example of the same process.

The Abundance of Simple Laws

That this "scaling-down" is so often in accord with physical experience, and that there are so many simple laws of considerable accuracy, is due, I believe, to the wide separation of the various groups of forces occurring in an objective world. These forces occur in four bands, ranging from very strong forces of very short range up to much weaker, long-range forces. Here the range of a force between two bodies is measured by the dependence of the force on the distance r between them; a dependence on $1/r^{12}$ would denote a very short-range force, while a dependence on $1/r^2$

---

[3] J. T. Davies, *British J. Philos. Sci.*, Vol. 7 (1956), p. 129.

would denote a long-range force. The bands of forces are summarized in Table 1.

TABLE 1

Force	Type of Force	Variation with distance
Gravitational or electrostatic	Weak and long range	$\dfrac{1}{r^2}$
Inter-molecular attractive forces	Weak, rather short range	$\dfrac{1}{r^7}$
Intra-molecular attractive forces	Strong, very short range	$\sim \dfrac{1}{r^{10}}$
Forces within the nucleus	Extremely strong but of extremely short range	$\sim \dfrac{1}{r^{15}}$

This non-random distribution of forces must result in a relatively simple behavior of matter; and certainly no such behavior could be expected if there were a continuous distribution of forces. The band structure of the force spectrum is reflected in the consistency—in many diverse phenomena—of the "billiard ball" atomic model. The atoms remain sharply distinguished in the molecule, and the molecules (e.g. of alcohol) remain sharply distinguished from each other and from the solvent in which they may be dissolved. This simple behavior, a result of the intra- and inter-molecular forces being of different strength and ranges, is also responsible for the constancy of the Avogadro number N referred to above. The "billiard ball" model is thus very close to representing a molecule, and recent work with the electron-microscope and X-ray analysis has made it possible to photograph single molecules. Another simple result* of extensive measurements has been the constancy of the rest mass of the electron and proton. Again, very powerful forces, very short-range in their action, must be responsible.

The success of the "scaling down" process may now be seen to lie in the rather wide separations of the bands. Thus, compared with gravitation, interatomic forces are different in range by a factor of r^5; and a billiard ball remains roughly spherical in practice. If we now scale this model down, we see that compared with interatomic forces, intra-atomic forces are different in range by a factor of r^3, so that the success of the simple billiard-ball model of an atom is explicable. Scaling down through

---

* I am indebted to Prof. Popper for this example.

another band, we see that since the nuclear forces are shorter-range than intra-atomic forces by a factor of about r^5, the consistency with experiment of the billiard-ball model of the nucleus is explicable.

A strong tendency to symmetry in nature is another result of the "band" distribution of the attractive forces. There is a general tendency to contraction within each band; consequently gravitation keeps the stars roughly spherical, liquids are pulled by intermolecular forces into spherical drops, and matter is concentrated into symmetrical atomic nuclei.

Prediction and Extrapolation

We often use a theory or law for predicting what will happen in the future. This process has again no logical basis, but is equivalent to testing the theory.[4] Extrapolation of a simple law to extremely low or extremely great values of the variables cannot be expected to give even a good approximation, however, unless certain conditions are satisfied.

As an example of extreme extrapolation, we may cite the "winding up" of the Universe.[5] According to simple probability theory applied to molecules, occasional reversals of the Second Law of Thermodynamics should occur, with a decrease in entropy (a decrease in randomness) instead of the usual increase (which implies more randomness, or running down toward a final random state). Now such reversals should, on the simple theory of probability, be very infrequent, and though they have never been observed in macroscopic systems, these reversals have been suggested as a source of a "winding up" of the universe. For example, if on earth all the molecules of water in a lake happened at some moment to be in the ice configuration the lake would freeze solid, and in that moment heat could flow into this sink from neighboring bodies and energy could be obtained. Of course, the extrapolation of molecular probability to such large numbers of random molecules and to the time intervals calculated to be necessary for such a happening to be likely is a very long one; I shall examine the basis of the extrapolation, and suggest that the lake will never freeze solid, any more than the monkey at the typewriter will ever type Shakespeare's sonnets. In both examples new factors may operate in the very long time intervals considered, making invalid the simple probability law.

As a concrete example, consider the normal curve of error:

$$y = A \cdot e^{-b(x - x_1)^2} \tag{4}$$

---

[4] Popper, *The Open Society, op. cit.*
[5] Davies, *op. cit.*

To this simple mathematical function many observations can be fitted, at least in the region around the peak of the curve (Fig. 2) : y could be the

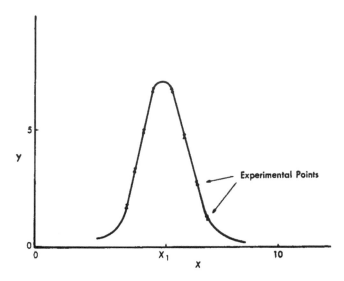

FIG. 2. The normal curve of error, represented by equation (4). Some typical experimental results are suggested.

number of craters on the moon of depth x in a given area. However, though the plot of equation (4) might fit the data around the peak of the curve, we know that there are, in any area we select, certainly no craters as deep as (say) 10,000 miles. Thus, while probability functions of the type of equation (4) are useful in covering the data near $x = x_1$, the function cuts off in practice when $(x - x_1)$ is very large and y becomes zero, as it never does in the simple equation (4); extrapolation of equation (4) to large values of $(x - x_1)$ may thus lead to absurd results.

Another example arises in the toxicity of reagents to living cells, or in hemolysis. In the latter, for example, y represents the number of cells of a resistance x to hemolysis, and here x is measured by the amount of hemolytic agent added. Such a distribution of resistances in a population is quite general, but members never have resistances greater or less than certain values, though the fit around the peak of the curve may be excellent.

These two examples show that extrapolation to very large values can involve absurdities if the extrapolated points are interpreted physically. I shall now suggest that the same sometimes applies in the interpretation of molecular probability theory. Let y in equation (4) represent the probability of x molecules in a lake of water being simultaneously in the ice structure at some moment in the time interval Δt. If $y \cdot \Delta t$ is constant, when y is very low we should, by equation (4), expect to wait a long time for a large number of molecules simultaneously to be in the ice structure. I believe that, just as with the craters and the hemolysis of the blood cells, this is absurd : the lake will *never* become momentarily solid in the present conditions. The extrapolation of equation (4) again has no physical significance, and macroscopic "winding up" never occurs.

If, however, equation (4) is rewritten :

$$Z = (1/(x - x_1)^2) \, (ln(A/y)) = b \qquad (5)$$

it is clear that when Z (which we may call the "improbability factor" because if Z is large y becomes exceedingly small at large values of x) is plotted against x (Fig. 3), a straight line is obtained parallel to the x axis at

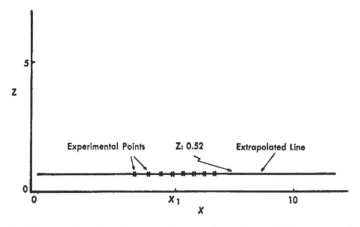

FIG. 3. The simplest functions, as in equations (4) and (5), suggest that Z should be constant. This straight line is extrapolated in the figure.

a distance b above it. Now many experimental results can be fitted onto the curve of equation (5), usually around $x = x_1$; but when we apply equation (5) to the craters or to the blood cells the improbability factor Z rises steeply to infinity at a certain value of x (x_2) which represents the largest x value ever found, at which y becomes zero. This is shown in Fig. 4. Thus, the extrapolation of the simplest curve (the straight line of Fig. 3) to very large values of x has no significance, and here it does not even furnish a useful approximation for testing theories. Additional factors have so modified the simple theory [equation (5)] that Z must now be a function of x and not a constant, and the straight line is valid only over a limited range. To make y cut off to zero at some value x_2, one must modify equation (4). For example, this may become :[6]

$$y = A \cdot e^{-b^1(x - x_1)^2} - \delta \qquad (6)$$

----

[6] *Ibid.*

where δ is very small and can be related to x_2 by substituting $x = x_2$, $y = 0$ in equation (6). From equation (6), Z, defined as in the first part of equation (5), is given by :

$$Z = b^1 + \frac{1}{(x - x_1)^2} \ln \left(1 + \frac{\delta}{y} \right) \qquad (7)$$

This shows mathematically the behavior in Fig. 4. The improbability factor Z is approximately constant only over a limited range; linear extrapolation beyond this is not even a good approximation.

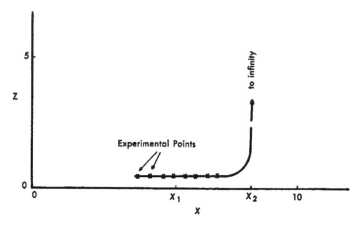

FIG. 4. The modified equations (6) and (7) (here $\delta = 0.1$) allow y to become zero at a certain value x. Thus the "improbability factor" Z may rise to infinity at this point, as is observed from hæmolysis and certain other experiments. This shows that extrapolation, as in Fig. 3 can be completely wrong.

Thus equation (6) could well cover the behavior of blood cells in hemolysis, the depth of craters, and possibly also the molecules of water in a lake; for 10^{28} is a very large value of x with which to extrapolate the simplest rule (Fig. 3).

Historical and Economic Laws

Are these laws "scientific"? To answer this question, we must note that historical or economic laws are usually put forward in the simple form

$$y = f(x) \text{ only,}$$

where y is the course of history (or of the so-called economic "cycles") as manifested by the behavior of a group of people, and x is one single quantity such as the investment rate, the economic system, the availability of

food, the climate, the popular enthusiasms of the people, or the strength of
character of the leader. Even in physics such simple functions, determined
by only one variable, are rare, so it is perhaps not surprising that the data
advanced to "support" any one of these particular interpretations are
often claimed by the adherents of any other theory to have been carefully
selected for the purpose. The difficulty arises in part from the belief that
it is possible to "support" or to "prove" a theory. Popper has pointed out
that a scientific theory can never be proved. Rather should one continue
to test it, "to try to disprove it." By this standard, any one of these simple
theories can be demonstrated to be inadequate. But, one may argue, supposing that one produced a theory of the form

$$y = f(w, x, z)$$

where w, x, and z refer to (say) three of the variables listed above. Could
this not be satisfactory? The difficulty here is that the relationship is now
more complex, and hence is very difficult to test; nevertheless, there may
be slow progress in correlating the variables. Even though "experiments"
are difficult, observations may suffice to test the theory (as is also true in
astronomy).

Possibly future generations will be in a better position than we are to
evaluate the complex relationships of human affairs. I believe they may
well make extensive use of popular opinion polls. For example, one could
propound some relatively simple law stating that wars (or concentration
camps) become feasible for some nation if (say) 5 per cent of its population
are cruel or ruthless enough to answer "Yes" to the question: "Do you
believe in exterminating the X's?" Again, one could assess from public
opinion polls, compiled before the event occurs, people's motives from the
answers to such questions as: "Do you want this policy because (a) it will
bring you prosperity, or (b) because it will bring you prestige, or (c) because
the Leader says it is good?"

Only by adopting Popper's principle of testing all theories, including
those of history, with data chosen to try to *disprove* them are we applying
scientific method to the subject. Even when our theory connects a wide
range of historical data it is never inevitable. It has no logical status, and
extrapolation to the future of what appears to be a highly complicated
function must be fraught with difficulties. Always are we conscious of our
environment, always are we as human beings interfering with it. The only
role of theoretical and economic theories is to lead us on to test them, and
by so doing to obtain a better understanding of what are the important
variables. We shall then be more able (like Popper's plumber) to "tamper
with the water supply," to control the future.

CHAPTER 18

The Neurophysiological Basis of Experience

BY J. C. ECCLES

IN HIS PREFACE to the 1959 edition of *The Logic of Scientific Discovery* Karl Popper states "I, however, believe that there is at least one philosophical problem in which all thinking men are interested. It is the problem of cosmology: *the problem of understanding the world—including ourselves, and our knowledge, as part of the world.*" In this *Festschrift* in honor of Karl Popper it is appropriate that I should here express my own thoughts on this philosophical problem, particularly since in doing so I will be able to show how greatly I am indebted to him. No attempt will be made here to correlate this simple account with the elaborate systems of epistemology developed by so many philosophers.

I am attempting to base my account on my own experience; hence I must of necessity adopt at the outset a purely personal or egocentric method of presentation, which may be called methodological solipsism. My conscious experience is all that is given to me in my task of trying to understand myself; and it is only because of and through my experience that I come to know of a natural world and so to embark on the attempt to understand it. Polanyi[1] has expressed this personal situation well: "Our tacit powers achieve these results by reorganizing our experience so as to gain intellectual control over it. There is one word which covers all these operations. They all consist in comprehending experience, i.e., in making sense of it; the word which covers them all is simply '*understanding*'."

Furthermore, I have to consider the totality of my conscious experiences, not only here and now, but in all my past. Because of the experiences that can be recalled in memory, and so re-experienced, I recognize my

---

[1] M. Polanyi, *The Study of Man* (London: Routledge & Kegan Paul, 1959), pp. 1–102.

unity and identity through all the past vicissitudes. It is memory that gives me that continuity of inner experience which belongs to me as a self; and this inner experience comprises not only my memories, but all the sequences of imagery, ideas, desires, volitions, and emotional feelings that characterize my waking life, and in addition includes my dreams and hallucinations. Sherrington[2] has written most movingly on the self, as may be illustrated by the following quotation :

This 'I,' this self, which can so vividly propose to 'do,' what attributes as regards 'doing' does it appear to itself to have? It counts itself as a 'cause.' Do we not each think of our 'I' as a 'cause' within our body? 'Within' inasmuch as it is at the core of the spatial world, which our perception seems to look out at from our body. The body seems a zone immediately about that central core. This 'I' belongs more immediately to our awareness than does even the spatial world about us, for it is directly experienced. It *is* the 'self.'

In contrast to this inner experience I have experiences that are derived from activation of my sensory receptors and that may be called perceptions. Some of these perceptions are readily recognized as resulting from activation of specific sense organs : sight from eyes; hearing from ears; smell from nose; taste from tongue; and touch, heat, cold and pain from the skin. These associations are so recognized by primitive man and by very young children. Scientific investigation has added to this primitive knowledge, showing that other sense organs are responsible for specific sensory experiences : joint and muscle receptors contribute to the sensation of the relative positions of the various bodily parts, e.g., of a limb; muscle and tendon receptors give information on the force stretching a muscle; vestibular receptors contribute to the sense of stability and rightness of the environment; there are deep pain receptors in most parts of the body. It is now generally assumed that osmotic receptors in the brain (the hypothalamus) give thirst, and there are probably also central receptors for the sensations of air hunger and food hunger.

It is solely from such perceptual experiences that I derive the concept of an external world of matter and energy, which is a world other than the world of my inner experience and which even includes my body, the "body image" of neurology. How, it may be asked, can my perceptual experiences give me such an effective knowledge of an external world that I can find my way around in it and even manipulate it with such success? So effective is this practical operation that I am not conscious of the problem in my whole experience of practical living; my body and its environment appear to be directly known to me.

However, physiological investigation reveals that all perception de-

---

[2] C. S. Sherrington, *Man on His Nature* (London: Cambridge University Press, 1940), p. 413.

pends on very complex processes of detection by sense organs and of transmission of signals (nerve impulses) from them to the brain. There is much neurophysiological evidence[3] that a conscious experience arises only when there is some specific cerebral activity. For every experience it is believed that there is a specific spatio-temporal pattern of neuronal activity in the brain. Thus, in the case of perception, the sequence of events is that of a stimulus to a sense organ causing the discharge of impulses along afferent nerve fibres which, after various synaptic relays, eventually evoke specific spatio-temporal patterns of impulses in the neuronal network of the cerebral cortex. The transmission from sense organ to cerebral cortex is by a coded pattern of nerve impulses that is quite unlike the original stimulus to that organ, and the spatio-temporal pattern of neuronal activity that is evoked in the cerebral cortex is again different. Yet, as a consequence of these cerebral patterns of activity, I experience sensations (more properly the complex constructs called percepts) which in my private perceptual world are "projected" to somewhere outside the cortex; it may be to the surface of the body or even within it, or, as with sight, hearing, or smell, to the outside world. However, as succinctly expressed by Russell Brain[4] : ". . . the only necessary condition of the observer's seeing colors, hearing sounds, and experiencing his own body is that the appropriate physiological events shall occur in the appropriate areas of his brain."

This direct relationship of brain activity to perception was first clearly seen by Descartes. It is immaterial whether this activity is caused by local stimulation of the cerebral cortex or of some part of the afferent nervous pathway, or whether it is, as is usual, generated by impulses discharged by sense organs. However, electrical stimuli applied to the sensory zones of the cerebral cortex evoke only chaotic sensations :[5] tingling or numbness in the skin zones; lights and colors in the visual zone; noises in the auditory zone. Such chaotic responses are to be expected, since electrical stimulation of the cortex must directly excite tens of thousands of neurons regardless of their functional relationships, thereby initiating a widely spreading amorphous field of neuronal activation quite unlike the fine and specific patterns that must be set up by input to the cortex from the sense organs. A familiar chaotic sensation, involving elements of touch, heat, cold, and pain, arises for similar reasons when a sensory nerve is directly stimulated, as when the ulnar nerve in the elbow (the "funny bone") is mechanically stimulated.

[3] E. D. Adrian, *The Physical Background of Perception* (Oxford: The Clarendon Press, 1947), and J. C. Eccles, *The Neurophysiological Basis of Mind: The Principles of Neurophysiology* (Oxford: Clarendon Press, 1953).

[4] W. Russell Brain, *Mind, Perception and Science* (Oxford: Blackwell Scientific Publications, 1951).

[5] W. Penfield and H. Jasper, *Epilepsy and the Functional Anatomy of the Human Brain* (Boston: Little, Brown & Co., 1954).

In response to sensory stimulation, I experience a private perceptual world which is an interpretation of specific events in my brain. Hence I am confronted by the problem of how this cerebral pattern of activity can give me a valid picture of the external world. Usually this problem is discussed in relation to visual perception. There seems to be an extraordinary difficulty in explaining how impulses from the retina, when relayed to the cerebral cortex, give rise to a picture of the external world with all its various objects in three-dimensional array and endowed with brightness and color. However, this epistemological problem has led to much philosophical confusion because it has been discussed on the assumption that visual perception is an inborn property of the nervous system. On the contrary; it is an interpretation of retinal data that has been learned through association with sensory information from muscles, joints, and skin. The three-dimensional world pictures which result from my visual experience are primarily based on perceptual data derived from movements and touch, and are the end product of a long effort of progressive learning by trial and error. As a well-trained adult it is difficult for me to realize that my earliest learning occurred in a cot with movement of limbs under visual observation; and that thereafter the field of visual education was extended by crawling, walking, and still other modes of locomotion which progressively extended my sphere of observation still further. I judge distance and space as distance and direction that I could travel, if I so wished; and thus I orient the world around myself. Thus my three-dimensional perceptual world is essentially a "kinesthetic world." It was initially bounded by the cot, but has since been enormously extended in range and subtlety.

The learning processes of early childhood are largely unremembered, but I can remember my many early efforts to evaluate distance and size, as well as my errors of judgment when confronted by strange environments where familiar clues were lacking. Fortunately, I do not have to rely on memories from infancy, for there are well-documented accounts by Senden[6] of adults who were given patterned vision for the first time by the removal of congenital cataracts from their eyes. They reported that their initial visual experiences were meaningless and quite unrelated to the spatial world that had been built up from touch and movement. It took many weeks and even months of continual effort to derive from visual experiences a perceptual world that was congruous with their "kinesthetic world" and in which, as a consequence, they could move with assurance. Similar evidence is provided by Riesen's[7] experiments on chimpanzees that

---

[6] M. von Senden, *Raum- und Gestaltauffassung bei operierten Blindgeborenen vor und nach der Operation* (Leipzig: Barth, 1932).

[7] A. H. Riesen, "The Development of Visual Perception in Man and Chimpanzee," *Science*, Vol. 106 (1947), pp. 107–108.

were reared in darkness and then transferred to an illuminated world. It took many months of training before they could skillfully use visual experience in guiding movement. A further illustration of the way in which learning can transform the interpretation of visual information is provided by Stratton's[8] experiences when a system of lenses was placed in front of one of his eyes (the other being covered) so that the image on the retina was inverted with respect to its usual orientation. For several days the visual world was hopelessly disordered. Since it was inverted, it gave an impression of unreality and was useless for the purpose of apprehending or manipulating objects. After eight days of continual effort on his part, however, the visual world could be sensed correctly and again became a reliable guide for manipulation and movement.

These experiments establish that, as a consequence of trial-and-error learning, the brain events evoked by sensory information from the retina are interpreted so that they give a valid picture of the external world that is sensed by touch and movement; i.e., the world of visual perception becomes a world in which we can move safely and effectively. Actually, this perceptual world is much more synthetic than we imagine; for example, it normally remains fixed and stable when the images on the retina are moved in the most diverse ways by my body, head, or eye movements. The kinesthetic information from all these movements as well as the sensory information from the vestibular apparatus are synthesized with the retinal information. The action of this automatic correcting device for visual perception is best appreciated when there are disturbances of vestibular function. Under such conditions there is gross movement of the visually perceived world, which gives rise to the sensation of vertigo.

Further problems are involved in attempting to understand how the brain events derived from the various sensory inputs can give me not only my own private perceptual world, but also experiences that are shared by other observers and which, therefore, I have come to regard as manifestations of an external world. Communication between observers serves to establish the existence of a world that is virtually identical to many observers. Its manner of operation is best illustrated by giving instances in which there are differences between observers. For example, it is easily established that many observers differ in their perception of colors. We have resolved this discrepancy not by rejecting color as an attribute of an external world common to all observers, but by classifying some observers as defective in color perception, i.e., as color blind to a varying degree. Rushton[9] has demonstrated the physiological basis of at least one form of color blindness by showing that it is attributable to deficiency of a retinal

---

[8] G. M. Stratton, "Vision without Inversion of Retinal Image," *Psychol. Rev.*, Vol. IV (1897), pp. 463–481.

[9] W. A. H. Rushton, "Visual pigments in the colour blind," *Nature*, Vol. 182, (1958), pp. 692–694.

pigment. Similarly, there are those with defects in taste or hearing to varying degrees. Again, a subject under the influence of an hallucinogenic drug, such as mescaline, experiences a wealth of imagery that is not shared by other observers close by. It is readily appreciated that such a discrepancy does not cast doubt on the validity of the external world as it is derived from the perceptual world. Instead, the exceptional experiences that occur under the influence of mescaline or in disordered cerebral function are classified as hallucinations. It will be realized that, when observers report one or another of these exceptional features of their perceptual worlds, the situation is customarily handled in a "common-sense" way so that the concept of a real external world independent of observers is preserved. It suffices merely to mention our attitude to such perceptual experiences as dreams and daydreams.

Finally, reference should be made to illusions. These differ from hallucinations in that they are experienced by all observers similarly placed; yet we recognize that, though belonging to our perceptual world, they give distorted or erroneous information about the external world. There are the illusions produced by such optical devices as mirrors, lenses, and microscopes; and the spatial illusions produced by objects observed in media of different refractive index, a common example being the bent appearance of a stick projecting across an air-water interface. Many additional examples of kinds of illusions are provided in the literatures of the special senses, psychophysics, and psychology; yet the concept of a real external world is not thereby impugned. Instead, most of the illusions have been explained as arising from some physical mechanism or from the properties of the receptor organs or of the central nervous system.

It would seem that the status of the external world is assured, for it has a reality that apparently transcends all the imperfections in the perceptual equipments of the observers. Thus, we have the concept of a contrast between the reality of the external or objective world on the one hand and the subjectivity of our perceptual experiences on the other, the former alone providing a sound basis for scientific investigation. However, this objective-subjective distinction is illusory, being derived from a misinterpretation and a misunderstanding, as has been convincingly argued by Schrödinger.[10] For example, he says:

> Without being aware of it and without being rigorously systematic about it, we exclude the Subject of Cognizance from the domain of nature that we endeavour to understand. We step with our own person back into the part of an onlooker who does not belong to the world, which by this very procedure becomes an objective world. This situation is the same for every mind and its world, in spite of the unfathomable abundance of 'cross-references' between them. The world is given to me only once, not one existing

---

[10] E. Schrödinger, *Mind and Matter* (London: Cambridge University Press, 1958).

and one perceived. Subject and object are only one. The barrier between them cannot be said to have broken down as a result of recent experience in the physical sciences, for this barrier does not exist.

The illusory nature of the objective-subjective dichotomy of experience is further illustrated by what might be called a spectrum of perceptual experiences.

1. The appearance of an object can be confirmed by touching it, and in this same manner can be sensed by other observers, the perception of the object thus achieving public status.

2. The pinprick of a finger can be witnessed by an observer as well as by the subject, but the pain is private to the subject. However, each observer can perform a similar experiment on himself and report his observation of pain, which in this way is shared and so achieves a public status.

3. The dull pain or ache of visceral origin cannot be readily duplicated in another observer, yet clinical investigators have provided a wealth of evidence on the pains characteristic of the visceral diseases, and even on referred pains, so that reports of visceral pain achieve indirectly a kind of public status. Similar considerations apply to such sensations as thirst or hunger.

4. Unlike the preceding three examples, mental pain or anguish is not a consequence of stimulation of receptor organs; yet again a kind of public status can be given to such purely private experiences, for there is a measure of agreement in the reports of subjects so afflicted. Similar considerations apply to other emotional experiences, such as anger, joy, delight in beauty, awe, and fear.

5. The experiences of dreams and of memories are even more uniquely private, belonging as they do even more exclusively to the realm of inner experience; yet again a kind of public status is established by the wealth of communication that there is between observers.

It can be claimed that all transitions exist between any two successive examples of this spectrum of experiences, which conforms very well with the postulate that every one of these various experiences is associated with a specific pattern of neuronal activity in the brain. Apparently such specific patterns can sometimes be evoked by electrical stimulation of the brains of epileptic patients. Penfield and Jasper[11] have given fascinating accounts of the way in which electrical stimulation of the temporal lobe of the cerebrum will evoke vivid and detailed memories of long-forgotten incidents.

The conclusion is that every observation of the so-called objective world depends in the first instance on an experience which is just as

---

[11] Penfield and Jasper, *op. cit.*

private as the so-called subjective experiences. The public status of an observation is given by symbolic communication between observers, in particular through the medium of language. By this same method of communication, our inner experiences, as illustrated in categories 4 and 5, can also achieve public status. We report such experiences to others and discover that they have like experiences to report to us. A large portion of our literature is concerned with such verbal communications of inner experiences, either of the author himself or of the characters that he so creates.

When I re-examine the nature of my sensory perceptions, it is evident that these give me the so-called facts of immediate experience and that the so-called "objective-world" is a derivative of certain types of this private and direct experience. This must not be interpreted as a purely idealistic attitude, for the implication is that the perceptual world is my symbolic picture of the "objective world," and thus resembles a map. This map or symbolic picture is essential so that I may act appropriately within this "objective world"; and, as we have seen, it is synthesized from sensory data so as to be effective for this very purpose. It is built upon spatial relations, but also gives symbolic information in terms of secondary qualities. For example, colors, sounds, smells, heat, and cold, as such, belong only to the perceptual world. Furthermore, it is part of my interpretation of my perceptual experience that my "self" is associated with a body that is in the "objective world," and I find innumerable other bodies that appear to be of like nature. I can exchange communications with them by bodily movements that give rise to perceptual changes in the observers—for example, by gestures, or, at a more sophisticated level, by speech that is heard and by writing that is read—and thus discover by reciprocal communication that they too have conscious experiences resembling mine. Solipsism becomes for me no longer a tenable belief. There is a world of selves, each with the experience of inhabiting a body that is in an "objective world" comprising innumerable bodies of like nature and a tremendous variety of other living forms and of apparently nonliving matter. This discovery leads to an enquiry into the way in which it has been possible to advance in knowledge of the "objective world" so that it can be claimed that much is intelligible and controllable. The enquiry concerns the nature of science. I do not, however, distinguish sharply between the methods of science and the empirical method of trial and error that has given mankind such a wealth of skill in practical living.

In discussing the nature of the scientific method, I shall continue to adopt a purely personal attitude whenever that is appropriate. When I consider the origins of my scientific life, I find that it gradually took shape within my ordinary life, being just a special example of my efforts to know and understand the world in which I find myself. It is in particular a special example of my effort to interpret my experience. My earliest memories already gave me a world in which I was immersed and which I was trying to understand and trying to control. Though it was a very limited world, as I now re-experience my earliest memories, the limitations did not then appear oppressive or restrictive. The boundaries were not barriers. It was possible to probe beyond them at will, but always within the limits of knowledge and understanding that had been won by others. I can remember in adolescence pondering on the problem of the frontiers of knowledge. What lay beyond these frontiers then appeared to be inaccessible to all but those endowed with scientific genius. I was prepared to accept the accounts that they gave and to try to understand, to appreciate, and to memorize them. At the most I could see discrepancies in the accounts of different scientists and attempt to see how they arose and how they might be resolved. My scientific life can properly be said to have begun when I made the conventional attempt to try out in a new situation a method of investigation that had been developed by others in investigations of somewhat similar situations. I developed as a scientist by working and living in a scientific laboratory, and in particular by the good fortune of collaborating with a great scientist—Sherrington. Then, much later, Karl Popper brought illumination into my groping scientific efforts.

Until 1944 I had succeeded moderately well in the conventional scientific manner with beliefs that may be categorized as follows: that hypotheses grow out of the careful and methodical collection of experimental data; that the excellence of a scientist is judged by the reliability of these developed hypotheses, which no doubt would need elaboration as more data accumulated, but which it was hoped would stand as firm and secure foundations for further conceptual development; that it is in the highest degree regrettable and a sign of failure if a scientist espoused an hypothesis that was falsified by new data, so that it had to be scrapped altogether. In the years preceding 1944 I had been occupied in a controversy concerning the role of specific chemical substances in transmission across the functional connections (synapses) between nerve cells or between nerve and muscle fibers. It was becoming apparent to me in the mid-1940s that the accumulating experimental evidence was telling heavily against my hypothesis that at synapses there was a quick initial electrical action and a slow later chemical transmitter action.

At this opportune time I learnt from Popper that it was not scientifically disgraceful to have one's hypothesis falsified. In fact, I was persuaded to formulate the electrical hypotheses of excitatory and inhibitory synaptic action as rigorously as possible so that they invited falsifications. They were in fact falsified some years later[12] in one of the first results of

---

[11] Eccles, *op. cit.*

the intracellular method of investigating synaptic action. Thanks to my tutelage by Popper I was able to accept joyfully this "death" of the "brain child" that I had nurtured for almost two decades, so that I was able to contribute immediately, both theoretically and experimentally, to the successful rival (Loewi-Dale) hypotheses of chemical synaptic transmission. My personal experience has shown that action in accord with Popper's view of scientific method has two great practical advantages. First, the development and precise formulation of an hypothesis so that it encourages attempts at falsification greatly economizes the experimental effort, and gives it significant direction. The experiments are designed so as to subject the hypothesis to the most rigorous testing. Diverse experimental procedures are not being tried simply in the hope that something interesting will turn up; usually such random probing is a wasteful procedure which has the great disadvantage of cluttering up the literature with reports of meaningless investigations. Second, when one's hypothesis has been falsified one should even rejoice, because in this dénouement science has been well served. We advance in scientific understanding by the experimental rejection of erroneous hypotheses, clearing the way for new conceptual developments.

In brief, I regard it as of prime importance that recovery from erroneous scientific beliefs is so easy and natural when one adopts the "Popperian" view of scientific method. It would be invidious to cite examples of eminent scientists who were unable to recover from scientific error, or were able to do so only after great travail.

The one scientific problem in which I have been deeply interested concerns the manner in which my "self" is related to the objective world. Scientific investigation (physiology) has shown how receptor organs receive information from the "matter–energy" world (outside our bodies or within them); from there it is transmitted along nerve fibers to the central nervous system, which in turn discharges impulses along nerve fibers to muscle, ending in a movement that can be regarded as the result of the initial input of information. However, as pointed out by Adrian[13] and Sherrington,[14] this account fails to bridge the gulf between my conscious experiences on one hand and the events in the matter–energy world on the other. How does the information that my sense organs relay to my brain give me perceptual experiences which of their very nature are more directly known than the matter–energy world, which in fact is merely a derivative from such experiences? Is any reconciliation possible between the direct experience that an act of will can call forth a muscular movement and the scientific account, whereby such a muscular movement results from an activity of nerve cells in the brain,

---

[13] Adrian, *op. cit.*
[14] Sherrington, *op. cit.*

relayed by nerve impulses to motoneurons and from them to muscles? It is my contention that these questions concerning the problem of brain–mind liaison have been wrongly posed.

By direct experience I know that my thought can lead to action, an idea which has been vividly expressed by Sherrington in the quotation given on page above. I can decide on a particular action, perhaps of the most trivial nature, and my muscular movements can be directed toward accomplishing that act. I have no experience of the manner in which my willing leads to action. Naturally, scientific investigation can be applied to study the sequence of events leading to movement, though such investigations are necessarily restricted to the material events : the responses of nerve cells, the synaptic transmission, the propagation of nerve impulses, the muscular activation and eventual contraction.

There would be no evidence supporting my belief that my body does carry out my willed movements. Curiously enough, the most compelling evidence for this belief comes when there is some failure in the control of movement. If I find that I cannot direct my muscular movements in some willed manner, I immediately recognize this as due to some disorder in my neuromuscular apparatus. I consult a neurologist; this is the reaction of all normal human beings in a civilized society. The belief that it is possible to exert a conscious control of movement is universally demonstrated by the response to any untoward limitation of this so-called "freedom of the will." "Freedom of the will" is a primary fact of experience and the formulation of the problem arising from this experience should be the inverse of its usual statement. The problem is to discover in the brain the functional properties that give it the requisite responsiveness so that, when I consciously will an action, I call forth responses that lead to the desired muscular movements. Speculations relating to the manner in which the active cerebral cortex may exhibit an extreme sensitivity and subtlety have been given in the last chapter of my book;[15] but it should be realized that these are merely initial tentative probings in relation to this hitherto intractable problem. It seems that both physics and physiology are too primitive to allow even the proper formulation of the problem, let alone its solution. One can surmise from the extreme complexity and refinement of its organization that there must be an unimagined richness of properties in the active cerebral cortex. In the meantime, belief in the freedom of the will is not impugned, though its mode of operation cannot at present be explained scientifically.

The key question in perception is : How can some specific spatiotemporal pattern of neuronal activity in the cerebral cortex evoke a particular sensory experience? A partial approach to an answer may be essayed by considering the neuronal activity of the cortex in states of

---

[15] Eccles, *op. cit.*

unconsciousness, i.e., when stimulation of sense organs fails to evoke a sensory experience. The electroencephalogram reveals that in such states there may be either a very low level of neuronal activity, as in coma, concussion, anesthesia, and deep sleep, or a very high level of stereotyped and driven activity, as in convulsions. On the other hand, the electrical activity of the waking brain indicates that a large proportion of the neurons is occupied in an intense dynamic activity of great variety. Under such conditions, I have postulated[16] that at any instant a considerable proportion of the neurons are passing through levels of excitation at which the discharge of an impulse would be problematical, such neurons being "critically poised" with respect to the generation of impulses. I have postulated further that consciousness is dependent on the existence of a sufficient number of such critically poised neurons, and consequently only under such conditions are willing and perceiving possible. However, it is not necessary for the whole cortex to be in this special dynamic state. There is clinical evidence that excision of a large part of the cerebral cortex does not interrupt consciousness; and in convulsions unconsciousness does not supervene until the convulsive activity has invaded a large part of the cortex.[17]

About a hundred years ago there was a period of intensive psychophysical investigation of perception by Helmholtz, Fechner, Hering, Brücke, and others. It was principally concerned with the manner in which light signals to the eyes are related to visual perceptions such as brightness, flicker fusion, after-images, and contrast phenomena. In recent years it has been shown that in these various conditions there are remarkable parallels between the responses of neurons in the occipital cortex and the visual perceptions reported in psychophysical experiments.[18] Evidently these procedures offer important methods for studying the relationship of neuronal activity to sensory perception. They have been unfashionable in an age that is dominated by behavioristic psychology, depending as they do on the reports of conscious experiences by the observers. However, as stated above, these reports achieve public status because several observers can be subjected to the same experimental procedures and their reports cross-checked, just as in any other scientific investigation.

If I reflect on the happenings during a scientific investigation, I find that there is incessant "traffic" between my conscious experiences and the objects and events in the external world. For example, from the framework of scientific knowledge I derive some ideas about what I should

---

[16] *Ibid.*

[17] Adrian, *op. cit.*, and Penfield and Jasper, *op. cit.*

[18] R. Jung, "Neuronal Integration in the Visual Cortex and Its Significance for Visual Information," in *Sensory Communication*, W. Rosenblith (ed.) (Boston: M.I.T. Press, 1961).

observe under certain experimental conditions. I plan these conditions and then by means of controlled movements proceed to actualize them. My observations or conscious experiences of the ensuing results are correlated and evaluated in rational and critical thought against my original ideas, further experiments are planned and executed, and so on. The consequence is that my scientific ideas or hypotheses are enriched, changed, or falsified. My scientific activity is thus seen to be essentially an affair of my rational and conceptual thought together with my exercise of willed movement and my sensory perception. Generalizing from this, I would define science as the thoughts of scientists in their efforts to understand the world of their experience; whereas in technology, "control" would have to be substituted for "understand." This idea of the personal involvement of the scientist has been very convincingly developed and illustrated by Polanyi.[19]

It is imperative, when speculating on the whole range of the problem of mind and matter, to avoid making apparently profound and compelling statements that in fact are merely wishful thinking. On the basis of evolutionary theory it is frequently stated by scientists that the association of mind and matter in the human brain must imply that there is a mental attribute latent in all matter; and that, as the organization of matter gradually became perfected in the evolutionary process, there was a parallel development of the mental attribute from its extremely primordial state in inorganic matter or in the simplest living forms through successive states until it reached full fruition in the human brain. This statement is often expressed as if it were scientifically established, which is certainly not true. It is a purely gratuitous assumption that inorganic matter or the simplest organism has some mental attribute that is refined and developed in the evolutionary process. Evidences of reactions to stimuli and of apparently purposive movements in animals are naively regarded as establishing that they have conscious experiences of the same nature as those that I, for example, experience directly, and that can only be known directly to each one of us.

We must remain agnostic with regard to the consciousness of animals. Statements of a progressive emergence of conscious mind during evolution are not supported by any scientific evidence. They are simply made within the framework of a faith that evolutionary theory, as it now exists, will, in principle at least, explain fully the origin and development of all living forms. There is, in fact, much evidence against the belief that there is a mental attribute in all matter, even in the organized matter of the central nervous system. As stated clearly by Adrian,[20] conscious experience stems from only one part of the body—the highest levels of the

---

[19] M. Polanyi, *Personal Knowledge* (London: Routledge & Kegan Paul, 1958).
[20] Adrian, *op. cit.*

brain—and even then only when the brain is in the right state of dynamic activity. Sentience in any part of my body is dependent on its functional nervous connection with my brain.

Conclusions

This brief survey of the relationship between neuronal activity on the one hand and conscious experience on the other has impinged on many philosophical problems. The principal problems, those of brain–mind liaison in perceiving and willing, remain unsolved, though their reformulation will serve to challenge all varieties of materialists, mechanists, and behaviorists. It is important for philosophers to realize that scientific knowledge is still in a very primitive state, and unfitted for the development of precise hypotheses on brain–mind liaison. But this present failure must not be interpreted as indicating that the problems are either meaningless or forever insoluble. We have to await developments in physics and physiology. At the same time it is important that there should be speculations[5] within the existing framework of knowledge in the hope that at least some insight will be achieved.

It can be claimed that the philosophical position outlined here has the merit of encompassing, in principle, all experience. Admittedly these philosophical speculations are at a very elementary level, but I believe that they are consistent within themselves and that the metaphysical suppositions are adequate for the conceptual developments. Such features have been conspicuously lacking in all the materialist and behaviorist philosophies, which arbitrarily reject much of experience and which are based on initial metaphysical assumptions, though metaphysics is later repudiated. Their basic suppositions are planned so that they lead to some caricature of man, to robot man or computer man or cybernetic man, not to the spiritual being or self that I apprehend myself to be. To many, such philosophies provide satisfactory explanations of man as viewed from the outside, but they fail abysmally when applied to man as seen from the inside, which is the privileged position each of us has in respect to his own self.

CHAPTER 19

Realism and Instrumentalism: Comments on the Logic of Factual Support

BY PAUL K. FEYERABEND

I

REALISM AND INSTRUMENTALISM provide two alternative interpretations of science and of factual knowledge in general. According to realism such knowledge is descriptive of (general or particular) features of the universe. According to instrumentalism even a theory that is wholly correct does not describe anything but serves as an instrument for the prediction of the facts that constitute its empirical content. Thus, considering Newton's theory of gravitation, a realist would remark that it teaches us of the existence, in addition to physical objects and their spatio-temporal behavior, of entities of an altogether different kind which cannot be directly seen, heard, or felt, but whose influence is still noticeable enough : viz., forces. An instrumentalist, on the other hand, will take the position that there are no such entities and that the function of words like 'gravitation,' 'force,' and 'gravitational field' is exhausted by their giving an abbreviated description of the spatio-temporal behavior of physical objects. He may even deny the existence of these objects and regard object words, too, as instruments, usable for the ordering and predicting of sense data. In the present paper I shall argue that realism is preferable to instrumentalism.

II

Such an argument is of interest only if the issue between realism and instrumentalism is more than just a quarrel about words. Some philosophers deny that it is. Nagel, for example, holds that "the opposition between these issues is a conflict over preferred modes of speech"[1] which cannot be resolved in an objective manner. I do not doubt for a second that there are versions of the problem which do possess this degenerate character. At the same time it seems to me that the instrumentalistic position of Proclus, of some astronomers of the early seventeenth century, and of Niels Bohr is prompted by much more substantial motives than the predilection for certain modes of speech. These thinkers offer *physical* arguments for their point of view. They attempt to show that a realistic interpretation of certain theories is bound to lead to results which are incompatible with observation and highly confirmed physical laws. Now if they are correct in this—and it will soon emerge that they are—then a realist cannot rest content with the general remark that theories just *are* descriptions and not merely instruments. He must then also revise the accepted *physics* in such a manner that the inconsistency is removed; i.e., he must actively contribute to the *development* of factual knowledge rather than make comments, in a "preferred mode of speech," about the *results* of this development. In addition he must offer methodological considerations as to why one should change successful theories in order to be able to accommodate new and strange points of view. An excellent example of this situation is provided by the arguments against the realistic interpretation of the Copernican hypothesis and by the attempts that were made in order to overcome these arguments.

III

According to the Aristotelian philosophy, which was the accepted basis for physical reasoning throughout the later Middle Ages, motion was to be understood as the actualization of a potentiality inherent in an object.[2] This theory resolved the difficulty of *monism*, first exhibited by Parmenides, which consisted in the fact that change was impossible in a monistic universe. For now we are dealing with at least two different kinds of being, potential and actual. This theory also implied some very plausible assumptions concerning the circumstances under which change might occur. Actualization of a potentiality is possible only with the help of a form that corresponds to the properties exhibited by the object after the change is over. Forms do not exist by themselves; they can be separated from matter in thought, but not in reality. Hence, the occurrence of

---

[1] *The Structure of Science* (New York: 1961), p. 152.

[2] We shall develop this point of view in the form which it received in the later Middle Ages and which is in some respects different from what is found in the Aristotelian opus itself. For such a later account concerning the *specific* theory of motion (see note 3, *infra*), see document 7.1 in M. Clagett, *The Science of Mechanics in the Middle Ages* (Madison, Wis., 1957); cf. also Clagett's own summary, pp. 421 ff.

change in an object requires the presence of another object which possesses the appropriate form. Everything that is moved is moved by something else. Every motion needs a mover, and this mover must be present in the close neighborhood of the changing thing, as action at a distance is impossible. Conversely, the state of an object that is not under the influence of forces is a state of rest. This is the Aristotelian "law of inertia."[3]

It is worth pointing out that this result is confirmed by our everyday experience: physical objects do not move unless they are hit, or pushed by other objects. Their natural state is indeed a state of rest. Note also the quasi-empirical character of some of the assumptions made in the course of the argument. This applies to the assumption of contact action and to the denial of the existence of separate forms. Empirical success (actual motions), theoretical success (solution of the "Parmenidean problem"), comprehensiveness (applicability to any kind of change), consideration of details (in this respect the Aristotelian theory was superior to the considerations of the atomists)—these are weighty arguments in favor of the Aristotelian point of view. There were undesirable aspects, too, such as the linguistic character of many Aristotelian arguments which make them irrelevant to the solution of problems of fact. However this need not deter us from properly appreciating the *result* of these arguments. And the result was a very interesting and successful empirical theory.[4]

---

[3] This *general theory* of motion and its law of inertia must be distinguished from the *specific theory* that deals with the motions, natural or forced, which actually occur in the universe. In the specific theory motions are called "natural" when there is no *outer* agent that can be used for explaining their occurrence. Adherence to the law of inertia stated in the text, above, makes it necessary in this case to introduce an "intrinsic form" such as the impetus, or the gravity of the moving object. From the point of view of the general theory, therefore, the "natural motion" of the specific theory is still a motion that occurs under the influence of forces; these forces, however, are frequently left unspecified (although not unnamed).

[4] Another objection might arise from the fact that the theory failed to give a satisfactory account of the motion of projectiles and of falling objects. Two points ought to be remembered in connection with this objection. First, that it *was* possible to account for both kinds of motion within the framework of the general Aristotelian theory of motion. The impetus theory gave such an account for the motion of projectiles; the theory of the inherent gravity of the heavy elements combined with the impetus theory explained the motion of falling objects. Second, the initial difficulties of the Aristotelian theory must not be taken as an indication of its "unscientific" or "metaphysical" character. *There is no single physical theory that is not beset by similar difficulties* (unless, of course, its defenders refrain from comparing it with the facts). Take Newton's theory of gravitation. It was about a century before the great inequality of Jupiter and Saturn and the secular acceleration of the mean motion of the moon were shown to be in accordance with Newton's law. And there exist still phenomena which resist explanation by the theory although they do not belong to the domain where relativistic effects become relevant. The same is true to an even greater extent

of general relativity. General relativity is a nonlinear theory which allows for straight-forward solutions only in the very simplest cases. And so it is with all theories. They are successful in a number of cases and will be regarded as revolutionary if these cases have been troublesome for a considerable time. However, there always exist other cases which are *prima facia* refuting instances of the theory but which are put aside, for the time being, in the hope that a favorable solution (favorable, that is, to the theory under consideration) will be forthcoming. Now if we postulate that a theory which is problematic because of the existence of *prima facia* refuting instances must not be used in cosmological arguments regarding the existence or non-existence of certain situa-tions, then we shall thereby eliminate not only the Aristotelian point of view *but every succeeding physical theory as well.* For example, we shall be unable to use the theory of relativity in arguments concerning the nature of space and time, we shall be unable to use the quantum theory in arguments concerning determinism, and so on. Conversely, if we agree to base our arguments on the best theory available at a given time, then we cannot escape admitting the validity of the considerations to be outlined in § IV.

Two additional comments are in order. First, that this problematic character of *every* scientific theory is very often kept from the eyes of the public, and even of students of the subject. Both popular presentations and textbooks dwell at length on the successes of a theory and hardly ever mention the much more interesting difficulties it faces. Some thinkers assert that this is a necessary evil, as only people who are firmly com-mitted to a theory will be able to work strenuously at overcoming whatever difficulties it may possess. This is plain nonsense. It amounts to saying that only those who have first been given an incorrect account of the theory will be able to show that the theory is true after all. And what happens if the theory should break down? Who will then be able to overcome his conditioning to this one theory and suggest something very different?

The second comment concerns the implications, for the theory of confirmation, of the situation just mentioned. The following method suggests itself: We divide the empirical content of the theory into three classes, viz., the events which support the theory, the events which disconfirm it, and the events which *prima facia* disconfirm the theory but whose "real nature" is left undecided pending further investigation. We regard a theory as refuted if a situation is found which belongs to the second class. We regard it as well supported if no such situation has been found (although there are *potential* falsifiers, i.e., class two is not empty for logical reasons), and if there are many facts belonging to the first class. A well supported theory may be regarded as *meta-physical* to a very high degree if there are also many facts belonging to the third class. This agrees well with part of the "nature of metaphysics," viz., that it tries to reveal a hidden nature that is different from what things seem to be at first sight. A meta-physical *theory* of this kind may degenerate into a *purely transcendent metaphysics* if matters are arranged in advance so that no potential falsifiers exist, i.e., if facts are assigned either to the first or to the third class, but never to the second class. The ideal scientific theory would be a theory that possesses no *prima facie* falsifiers. No such theory exists. Now it seems to me that in comparing theories for the purpose of explanation one should take into account the extent to which they are metaphysical. Considering two equally general theories with equally many nonfalsifying instances, the theory that possesses fewer *prima facie* falsifiers will have to be preferred, for it is more likely to already have achieved correct expression of the actual situation of the world. Or, if C, D, P are the three classes introduced above, and c, d, p their elements which have been actually assigned on the basis of observation, then of two theories T and T' for which $d = 0$ and $c + p = c' + p'$, the one with the greater c and the smaller p will have to be preferred. Considerations like this one would lead to a theory of con-firmation which would be closer to scientific practice than are the theories available at the present time.

IV

This theory has immediate implications regarding the motion of the earth. If it is correct, and if we also take into account some very simple facts, then we must conclude that the earth is at rest, i.e., that it neither rotates nor displaces itself in space. The reason is that only those things which are in direct contact with it, such as houses and human beings resting on its surface, would be carried along by the motion; whereas anything disconnected, such as birds, clouds, human beings jumping, would immediately assume their natural state of motion, namely, rest, and would therefore be left behind.[5] Considering that the birds are still with us, and that cannon balls unfortunately are not lost,[6] but hit their target most accurately, we must conclude that the earth cannot possess any motion whatever.

---

[5] It would be unhistorical at this place to refer to the relativity of location, velocity, and, perhaps, of all motion. Place, or position, in Aristotle has physical properties: ". . . the typical locomotions of the elementary bodies . . . show not only that place is something, but also that it exerts a certain influence. Each is carried to its own place, if it is not hindered, the one up, the other down . . ." (*Physics* 208b, quoted after the Ross edition, Oxford: 1930). These different properties of different locations enable us to distinguish them absolutely, and not only in relation to objects occupying them. The idea that the observed motions are prompted by *objects* in space (such as the earth) rather than by *positions* in space (such as the center of the closed universe) is an alternative theory whose physical advantages were realized only after the Copernican point of view had been generally accepted. To a certain extent general relativity implies a return to the Aristotelian notions.

[6] The cannon argument was frequently used. For a discussion from the point of view of a new, and not yet existing dynamics, cf. Galileo, *Dialogues Concerning the Two Chief World Systems*, tr. by Stillman Drake (Berkeley and Los Angeles, 1953), pp. 126 f. For a very clear statement of the Aristotelian position, cf. Buridan's "Questions on the Four Books on the Heavens and the World of Aristotle," Book II, Question 22, section 9, quoted from Clagett *op. cit.*, Document 101: "But the last appearance [that must adduced against a rotation of the earth] is more demonstrative in the question at hand. This is that an arrow projected from a bow directly upward falls again in the same spot of the earth from which it was projected. This would not be so if the earth were moved with such velocity. Rather, before the arrow falls the part of the earth from which the arrow was projected would be a league's distance away. But still the supporters would respond that it happens so because the air, moved with the earth, carries the arrow, although the arrow appears to us to be moved simply in a straight line motion because it is being carried along with us. Therefore we do not perceive that motion by which it is carried with the air. But this evasion is not sufficient because the violent impetus of the arrow in ascending would resist the lateral motion of the air so that it would not be moved as much as the air. This is similar to the occasion when the air is moved by a high wind. For then an arrow projected upward is not moved as much laterally as the wind is moved, although it would be moved somewhat." Can there be any doubt of the empirical character of this argument?

It is with reference to this argument that Ptolemy[7] criticizes "certain thinkers" who "have concocted a scheme which they consider more acceptable and they think that no evidence can be brought against them if they suggest for the sake of argument that the heaven is motionless, but that the earth rotates about one and the same axis from West to East, completing one rotation approximately every day, or alternatively that both the heaven and the earth have a rotation of a certain amount, whatever it is, about the same axis, as we said, but such as to maintain their *relative* situations. These persons forget however that, while, as far as appearances in the stellar world are concerned, there might perhaps be no objection to this theory in the simple form, yet to judge by the conditions affecting ourselves and those in the air above us such a hypothesis must seem to be quite ridiculous."[8]

Ptolemy here distinguishes, as is also done in classical physics, be-

---

[7] Quoted from Cohen-Drabkin, *Source Book in Greek Science* (New York: 1948), pp. 126 ff.

[8] Another proof derives from the doctrine of natural places (which is part of the specific theory of motion explained in note 3, *supra*). According to this doctrine, which is again very closely related to experience, the elements of the universe are distinguished by the places to which they tend to move: the earth moves toward the center, fire moves towards the circumference, water and air move to intermediate places. Sometimes these purely dynamical properties seem to be regarded as the sole *defining properties* of the elements. Earth is distinguished from fire not by its appearance, nor by the fact that the latter burns and the former cools, *but solely by the fact that it moves down whereas fire moves toward the circumference.* [For more details see F. Solmsen, *Aristotle's System of the Physical World* (New York, 1960), Chaps. 11 ff.] It has already been pointed out (note 5, *supra*), how this doctrine gives physical content to the notion of position and thereby to (a finite) absolute space. The proof against the motion of the earth derived from it (Ptolemy, quoted from Cohen-Drabkin, *op. cit.*, p. 126), runs as follows: "So far as the composite objects in the universe, and their motion on their own account and in their own nature are concerned, those objects which are light, being composed of fine particles, fly towards the outside, that is, towards the circumference, though their impulse seems to be towards what is for individuals 'up,' because with all of us what is over our heads, and is also called 'up,' points towards the bounding surface; but all things which are heavy, being composed of denser particles, are carried towards the middle, that is to the center, though they seem to fall 'down,' because, again, with all of us the place at our feet, called 'down,' itself points towards the center of the earth, and they naturally settle in a position about the center, under the action of mutual resistance and pressure which is equal and similar from all directions. Thus it is easy to conceive that the whole mass of earth is of huge size in comparison with the things that are carried down to it, and that the earth remains unaffected by the impact of the quite small weights (falling on it), seeing that these fall from all sides alike. . . . But, of course, if as a whole it had a common motion, one and the same with that of the weights, it would, as it was carried down, have got ahead of every other falling body, in virtue of its enormous excess of size, and the animals and all separate weights would have been left behind floating in the air, while the earth, for its part, at its great speed, would have fallen completely out of the universe itself. But indeed this sort of suggestion has only to be thought of in order to be seen to be utterly ridiculous."

tween the purely *kinematic* aspects of motion on the one side and the phenomena of *inertia* which are brought to light by motion on the other. *Kinematically* the motion of the earth is indistinguishable from a situation where the earth is at rest and the stellar sphere rotates in the opposite direction (we are now considering rotation only). However the motion of the earth, if it occurs, will also lead to *dynamical* phenomena of a kind that can be well described in advance. These phenomena do not occur. Hence, the earth does not move. One should note that this argument against the dynamical, or, as one could also call it, the absolute motion of the earth has exactly the same structure as the argument *for* the absolute rotation of the earth as derived from Foucault's pendulum and the variation of pendulum clocks from the equator to the pole.[9] The only difference lies in the law of inertia used. According to Aristotle a thing left to itself will remain at rest.[10] According to Newton it will move on a straight line with constant speed. At the time of Ptolemy the limitations of the Aristotelian physics had not yet been exhibited in an unambiguous manner. We have to conclude, then, that apart from the hypothetical character of *any* argument from physical principles the argument for the unmoved earth referred to by Ptolemy was impeccable.

V

It is this situation which we must keep before our eyes when embarking upon the evaluation of the controversy over the Copernican hypothesis. Seen in the light of the argument given above, the attempt to regard this hypothesis as a correct account of the actual situation in the universe amounts to upholding an unsupported conjecture in the face of

---

[9] As is well known, Newton distinguished relative, or apparent motion from absolute, or true motion, and he also pointed out that the latter can be recognized through its dynamical effect (bucket experiment). The argument between the defenders of a relational account of space (Leibniz was one of them) and the absolutists is the exact parallel to Ptolemy's argument in the text. Ptolemy points out that although there may be kinematic equivalence between two situations involving motion, there is yet no dynamical equivalence. *This belief is shared by the Aristotelians and by the supporters of absolute space.* The only difference is that, because of the different laws of motion, the Newtonians and the Aristotelians work with different equivalence classes.

[10] In note 3, *supra,* we have noted some difficulties encountered by this law and have mentioned the theory of impetus as one of the possible ways out. Would it not be natural, therefore, to replace the Aristotelian law of inertia by the corresponding law of the impetus theory and thereby remove one of the most decisive obstacles to the motion of the earth? This would indeed be a possible procedure. But this emphasizes rather than weakens a point we are going to make, viz., that in the case of the Copernican hypothesis the realistic position was not a matter of pure philosophy and still less a matter concerning "preferred modes of speech." A realist had to change the contemporary physics.

fact and well supported theory.[11] True, the heliocentric point of view gave a simpler explanation of the second inequality of planetary motion (the loops) than did the geocentric scheme. However it did not on that account alone lead to better predictions. Epicycles were still needed for the first inequality. The specific way in which Copernicus introduced these epicycles might improve the empirical adequacy of the theory. But *these* details were not bound to the heliocentric scheme. They were a mathematical technique which, like the technique of Fourier decomposition,[12] could be applied under the most varied circumstances. For example, they could also be added to the geocentric hypothesis, where they had originated in the first place. Hence, "if the tables newly to be calculated should prove to be superior to the Alphonsine tables, which were based on the Ptolemaic system, this would not . . . be due to the heliocentric hypothesis as such, but only to the superior quality of the details of the new system."[13] There was no independent evidence in favor of the heliocentric theory; this theory was, at least initially, a conjecture that had no foundation in empirical fact.[14] The only favorable remark that

---

[11] For "well supported" see note 4, *supra*. Copernicus, of course, was well aware of the dynamical difficulties connected with the motion of the earth and he therefore tried to introduce a dynamics of his own. The same is true of Galileo, whose main work may be described as the attempt to show that motion of the earth was not only dynamically possible, but even required. See his arguments in the Second Day of the *Dialogues*.

[12] If I remember correctly it was Norbert Wiener who has pointed to the similarity, from a mathematical point of view, of the technique of epicycles and the technique of the Fourier decomposition.

[13] Dijksterhuis, *The Mechanization of the World Picture* (Oxford, 1961), p. 249. Cf. also T. S. Kuhn, *The Copernican Revolution*, Modern Library Paperbacks (New York: 1957), p. 169: "When Copernicus had finished adding circles, his cumbersome sun-centered system gave results as accurate as Ptolemy's but it did not give more accurate results. Copernicus did not solve the problem of the planets." Professor Kuhn's book contains an excellent semitechnical account of the comparative efficiency of the Ptolemaic and Copernican systems.

[14] This, more than anything else, should exhibit the error in the assertion that science started when people stopped being impressed by theories and turned to observations instead. Galileo especially is often represented as a thinker who started from observations and strictly followed the Baconian method. According to Herschel, *The Cabinet of Natural Philosophy* (Philadelphia, 1831), p. 85, Galileo "refuted the Aristotelian dogmas respecting motion, by direct appeal to the evidence of sense, and by experiments of the most convincing kind." As regards Copernicus Herschel has this to say: "By the discoveries of Copernicus, Kepler, and Galileo the errors of the Aristotelian philosophy were effectually overturned on a plain appeal to the facts of nature." Now we have seen that there were no "facts of nature," there was no "evidence of sense" to which Copernicus *could* have appealed. Even worse, the "evidence of the senses" was against him and his theory. It was the Aristotelians who could quote "nature" in their favor. Also "one outstanding fact about the scientific revolution is that its initial and in a sense most important stages were carried through before the invention of the new measuring instruments, the telescope, and the microscope, thermometer and accurate clock, which were later to become indispensable

could be made was that it somewhat simplified calculations by a suitable coordinate transformation. This is a phenomenon well known from mathematical physics : there are many problems which admit of immediate solution once a proper choice of coordinates has been made and whose solution is very cumbersome in different coordinates. Such a choice and the resulting mathematical success does not imply that the coordinate system chosen has any *dynamical* preference over other coordinate systems; for example, it does not imply that it is an inertial system. After all, the solubility of a problem, the circumstances under which it can be easily calculated, depend as much upon the mathematical formalism used as upon nature. There may exist assymetries in the formalism which do not exist in nature. The fact that the problems of positional astronomy can be dealt with in a more simple manner in a coordinate system in which the sun is at rest therefore does not imply that the sun is actually at rest and that the earth moves.

Combining this with the argument in section IV above, we arrive at the result that if the Copernican hypothesis has any merit at all (judged from the point of view of the contemporaries of Copernicus) it lies in the fact that it allows for a more effective *calculation* of the position of the planets. It does not lie in its giving a new *and true* account of what goes on in reality. This is one of the ways in which the instrumentalistic interpretation of Copernicus was introduced. Considering the dynamical

---

for getting the accurate and satisfactory answers to the questions that were to come to the forefront of the sciences." [A. C. Crombie, *Mediaeval and Early Modern Science*, vol. II (Doubleday-Anchor Books), p. 122]. The "Galileo myth" too, according to which Galileo busily rushed around making experiments and "climbed the leaning tower of Pisa with one one-hundred-pound cannon ball under one arm, and a one-hundred-pound cannon ball under the other" [ironical remark in Butterfield, *The Origins of Modern Science* (London, 1957), p. 81] has been refuted by historical research. "In general" writes Dijksterhuis (*op. cit.*, p. 338), "one has always to take stories about experiments by Galileo as well as by his opponents, with some reserve. As a rule they were performed only mentally, or they are merely described as possibilities." More especially there is evidence which proves "the complete baselessness of the belief tenaciously maintained by the supporters of the Galileo myth, namely that he discovered the law of squares by performing with falling bodies a number of measurements of distance and time, and noting in these values the constant ratio between the distance and the square of time." (*ibid.*, p. 340). All this of course cannot prevent an inductivist like Professor Dingle from repeating (in the 1961 edition of the *Encyclopaedia Britannica*, vol. XIX, p. 95) that "Galileo discovered the law of falling bodies by measuring how the space covered varied with the time of fall." Which shows how difficult it is for an inductivist accurately to represent the facts of history.

It is also surprising that this feature of the origin of modern science (viz., its original *incompatibility* with facts and well supported theory) has not yet been taken into account by scientific methodology. None of the methodologies in existence today would have permitted Copernicus to interpret his theory in a realistic fashion. But of this later.

arguments against the motion of the earth, this interpretation would seem to have been unassailable.[15]

VI

At this point it is very important to emphasize that the argument does not at all depend on a general philosophical position considering the nature of our knowledge. No such sweeping assumption is implied or presupposed. The argument is concerned with a *specific theory*, viz., the heliocentric hypothesis. It is based upon facts and physical laws. It shows that considering these facts and these laws the hypothesis cannot be *true*; that it can at most be an *instrument* of prediction. It is arguments of this specific kind with which we shall be concerned in the present paper (which means, of course, that we shall have to find a method of justifying the invention of unsupported conjectures in the face of fact and well supported theory).

Now it is well known that arguments of this kind are not the only ones which occur in connection with this issue between realism and instrumentalism. The above refutation of the heliocentric hypothesis was not even the most popular one. A great deal of the opposition to Copernicus derived from the difficulty in making his ideas agree with the scriptures as interpreted by the church fathers (which interpretation had become binding after the council of Trent). There were also more philosophical considerations which were closely connected with the belief that only those theories whose validity had been established by a *proof* implying necessity could be said to be descriptive of reality. No such proof was available in the case of the Copernican hypothesis, nor did it seem likely that it would ever be forthcoming. Hence the hypothesis could at most be regarded as an instrument of prediction. This is the train of reasoning on which Bellarmine seemed to base *his* evaluation of the case. "If there were one real proof," he writes to Father Foscarini,[16] "that the sun is in the center of the universe, that the earth is in the third heaven, and that the sun does not go round the earth but the earth round the sun, then we should have to proceed with great circumspection in explaining passages of scripture which appear to teach the contrary . . . . But as for myself I shall not believe that there are such proofs until they are shown to

---

[15] To put it in different words: the dynamical arguments amount to a straightforward *refutation* of the Copernican hypothesis. If we take these arguments at their face value, then we must regard this hypothesis as false. This, of course, does not prevent the hypothesis from giving a correct account of some facts of astronomy. It is therefore still a good instrument of prediction.

[16] Letter of April 12, 1615, quoted in Santillana, *The Crime of Galileo* (University of Chicago Press, 1955), pp. 99 ff.

me. Nor is it a proof that, if the sun be supposed at the center of the universe and the earth in the third heaven, everything works out the same as if it were the other way round." Later historians have repeated the argument. "Logic was on the side of Osiander and Bellarmine," writes P. Duhem,[17] "and not on that of Kepler and Galileo; the former had grasped the exact significance of the experimental method, while the latter had been mistaken . . . . Suppose that the hypotheses of Copernicus were able to explain all known appearances. What can be concluded is that they may be true, not that they are necessarily true, for in order to make legitimate this last conclusion it would have to be proved that no other system of hypotheses could possibly be imagined which could explain the appearances just as well." There is a very definite epistemology connected with this attitude. Different kinds of knowledge are distinguished and different claims to reality allotted to each kind. Physics deals with causes, with substance, and the real constitution of things, and it is capable of *proving* the truth of its assertions; astronomy is concerned with prediction only, and may for this purpose introduce hypotheses which are actually false.[18] Predictive success in astronomy is therefore no indication of truth and of factual relevance. Proof alone is. It is impossible to explain here why this position is both untenable and undesirable. However it will perhaps be admitted that such an explanation will be of a general kind, that it will be a matter of pure philosophy, and that it will concern not only one theory but any possible theory : the reasons which prompted Bellarmine and Duhem to regard the instrumentalism of Osiander as "logically superior" to the realism of Galileo can be refuted by philosophical considerations.[19] *However it would be a great mistake to assume that thereby all the objections to the heliocentric point of view have been done away with.* The physical arguments of sections IV and V still stand unrefuted and require an answer.

It is very important to realize this complex character of the situation, for otherwise one will be satisfied too early and too easily. Thus a thinker who is acquainted with the epistemological arguments only will regard a refutation of these arguments, and the construction of an alternative epistemology which allows hypothetical statements to have realistic implications, as the completion of his task, and he may even be seduced into thinking that his epistemology once and forever settles the issue between realism and instrumentalism in favor of the former. Quite obviously such

---

[17] Quoted in Santillana, *op. cit.*, p. 107.

[18] For a more detailed account see Simplicius, *Commentary on Aristotle's Physics*, quoted from T. L. Heath, *Aristarchus of Samos* (Oxford, 1913), pp. 275–276. Cf. also Duhem, *La Théorie Physique, Son Objet, Sa Structure*, Chap. III.

[19] For an analysis and definitive refutation of this position see K. R. Popper, "Three Views Concerning Human Knowledge," *Contemporary British Philosophy*, Vol. III (1956), pp. 2 ff.

an attitude will not impress the "physicists" whose arguments have not even been touched, and it will thereby either create, or further contribute to, a very undesirable split between physics and philosophy. And this is precisely what happens today in microphysics.[20] There exist some very impressive physical arguments against a realistic interpretation of quantum theory. There are also more philosophical arguments trying to establish the same result, viz., that the quantum theory is an instrument of prediction from which no realistic consequences can be drawn. These philosophical arguments proceed from the assumption that only observational terms are candidates for a realistic interpretation, an assumption which can be refuted once and for all by philosophical reasoning.[21] Thinkers who are conversant with philosophy only will assume that this settles the matter—which is far from being true. "This situation," I wrote[22] when discussing the interpretation of the quantum theory, "accounts for the strangely unreal character of many discussions on the foundations of the present quantum theory. The members of the Copenhagen school are confident that their point of view with whose fruitfulness they are well acquainted is satisfactory and superior to a good many alternatives. But when writing about it, they do not draw sufficient

---

[20] Other examples are the kinetic theory of matter and the theory of relativity. The kinetic theory of matter of the late nineteenth century was attacked *both* by philosophical arguments which tried to show the undesirability, from an empirical point of view, of the abstract and unobservable notions introduced by that theory, *and* by physical arguments which pointed out that any kinetic theory will be inconsistent with the laws of the phenomenological theory, especially the second law (reversibility objection of Loschmidt; recurrence objection of Poincaré–Zermelo). A purely philosophical defence of atomism is therefore insufficient. In addition it must be shown how the inconsistency can be circumvented and to what extent it can really be regarded as an objection. When the theory of relativity first became known to a wider circle of people, including philosophers, it appeared that its main conclusions were the result of the positivistic attitude that things that cannot be measured do not exist. It was therefore immediately attacked by realists who believed that their defense of realism was at the same time a defense, and a complete rehabilitation, of the notions of absolute space and time. It was again overlooked that relativity rests on considerations of a much more substantial nature than the positivistic principle just quoted. But it must be admitted that in this case it was *the physicists themselves* who created the confusion. Thus Bridgman's widely read *Logic of Modern Physics* represents the theory of relativity as the transition to a new era where considerations of objective existence are replaced by considerations of measurability, and Niels Bohr has interpreted the theory of relativity in the same manner. The most outstanding example of the confusion referred to in the text, however, is presented by the discussions of the quantum theory.

[21] Such a refutation is contained in Professor Popper's paper referred to in note 19, *supra*. For a different account, cf. my paper, "Das Problem der Existenz theoretischer Entitaeten," *Probleme der Wissens chaftstheorie* (Vienna, 1960), pp. 35–72.

[22] For this quotation and a more detailed account of what follows, see my "Problems of Microphysics," *Pittsburgh Publications in the Philosophy of Science*, vol. I. (Pittsburgh, 1963).

attention to its physical merits but wander off into philosophy and especially into positivism. Here they become an easy prey to all sorts of philosophical realists who quickly . . . exhibit the mistakes in their arguments without thereby convincing them of the invalidity of their point of view—and quite justly so, for this point of view can rest on its own feet and does not need any support from philosophy. So the discussion between physicists and philosophers goes back and forth without ever getting anywhere."

It is imperative to avoid a vicious circle of this kind and to *attack the instrumentalistic position where it seems to be strongest; and that is where it is based upon specific factual argument rather than upon general philosophy.* However, before doing so I shall introduce the example of the quantum theory in addition to the example already discussed.

VII

Not long after Planck had introduced the quantum of action[23] it was realized that this innovation was bound to lead to a complete recasting of the principles of motion of material systems. It was Poincaré[24] who first pointed out that the idea of a continuous motion along a well defined path could not longer be upheld and that what was needed was not only a new *dynamics,* i.e., a new set of assumptions about the acting forces, but also a new *kinematics,* i.e., a new set of assumptions about the kind of motion initiated by these forces. Both Bohr's older theory and the dual nature of light and matter further accentuated this need. One of the problems arising in the older quantum theory was the treatment of the interaction between two mechanical systems.[25] Assume (Fig. 1) that two

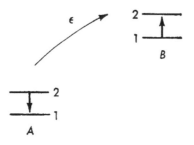

Figure 1

---

[23] I am here referring to what is known as Planck's *First Theory* in which both absorption and emission were regarded as discontinuous processes [*Verh. phys. Ges.* Vol. II (1900), pp. 237 ff.] and which also implies discontinuities in space [cf. Whittaker, *History of the Theories of Aether and Electricity,* vol. II (Edinburgh, 1953), p. 103].

[24] *Journal de Physique,* vol. II (1912), p. 1.

[25] Niels Bohr, *Atomic Theory and the Description of Nature* (Cambridge, 1932), p. 65.

systems, A and B, interact in such a manner that a certain amount of energy, ε, is transferred from A to B. During the interaction the system $A + B$ possesses a well defined energy. Experience teaches that the transfer of ε does not occur immediately, but takes a finite amount of time. This seems to suggest that both A and B change their state gradually, i.e., A gradually falls from 2 to 1, while B gradually rises from 1 to 2. However such a mode of description would be incompatible with the *quantum postulate* according to which a mechanical system can only be in either state 1 or state 2 (we shall assume there are no admissible states between 1 and 2), and is incapable of being in an intermediate state. How shall we reconcile the fact that the transfer takes a finite amount of time with the nonexistence of intermediate states between 1 and 2?

This difficulty was resolved by Bohr[26] on the basis of the assumption that during the interaction of A and B the dynamical states of both A and B cease to be well defined so that it becomes *meaningless* (rather than *false*) to ascribe a definite energy to either of them.[27]

This simple and ingenious hypothesis has so often been misrepresented that a few words of explanation are needed. First of all it must be pointed out that in the above formulation the term 'meaning' has not entered, as has been asserted by various critics, because of some connection with the now customary attitude of preferring semantical analysis to an investigation of physical conditions.[28] After all, there are well known classical examples of terms which are meaningfully applicable only if

---

[26] I shall not contend that this is the only way of getting around the difficulty, but it is a very reasonable physical hypothesis which has not yet been refuted by any of the arguments aimed against it.

[27] By the expression 'dynamical state' we refer to "quantities which are characteristic of the motion" of the system concerned (such as the positions and the momenta of its components), rather than those quantities which, like mass and charge, serve as a characteristic of the kind of system it is. For this explanation see Landau–Lifshitz, *Quantum Mechanics* (London, 1958), p. 2, as well as N. Bohr, *Atomic Physics and Human Knowledge*, p. 90; cf. also H. A. Kramer, *Quantum Mechanics* (New York, 1957), p. 62.

[28] It is to be admitted, however, that most derivations of the uncertainties, and especially those based upon Heisenberg's famous thought experiments, *do* make use of philosophical theories of meaning. Usually these arguments (and other arguments which proceed from the commutation relations of the elementary theory) only establish that inside a certain interval *measurements cannot be carried out*, or that the products of the mean deviations of certain magnitudes *cannot be ascertained* below Planck's constant h. The transition from this stage of the argument to the assertion that it would be *meaningless* to ascribe definite values to the magnitudes in this interval is then achieved on the basis of the principle that what cannot be measured cannot be meaningfully asserted to exist. This argument is, of course, unacceptable because the principle on which it is based is unacceptable. Moreover, it is liable to lead to a dogmatic belief in the result. For whereas a physical hypothesis such as the one discussed in the text will be accepted with caution, it is general habit to assume that philosophical considerations, and especially considerations flowing from a meaning criterion, possess a much greater binding force.

certain physical conditions are first satisfied and which become inapplicable, and therefore meaningless, as soon as these conditions cease to hold. A good example is the term 'scratchibility' (Mohs scale), which is applicable to rigid bodies only and which loses its significance as soon as the bodies start melting. Secondly, it should be noted that the proposed solution does not contain any reference to *knowledge,* or *observability.* It is not asserted that during the time of transfer *A* and *B* may be in some state which is unknown to us, or which cannot be observed. For the quantum postulate does not merely exclude the knowledge of, or the observability of, the intermediate states; it excludes these intermediate states themselves. Nor must the argument be read as asserting, as is implied in many presentations by positivist minded physicists, that the intermediate states do not exist *because* they cannot be observed. For it refers to a postulate (the quantum postulate) which deals with existence, and not with observability. It is here that the most misleading presentations occur. Physicists who have adopted the positivistic principle that things which cannot be observed do not exist turn the whole argument around and try to justify the indefiniteness of state descriptions by a combined reference to the fact that they cannot be observed and this principle. Philosophers immediately expose the fallacy of the argument (if they are anti-positivists, that is) and think that they have thereby shown the existence, or at least the physical possibility, of sharp states. This is, of course, not correct, for from the fact that a certain argument has been found to be fallacious it does not follow that a better argument does not exist. But this better argument is hardly ever used by the physicists, which creates the impression that positivism is indeed the only source of the peculiar features of the present quantum theory.[29]

The emphasis upon the absence of *predictability* is not satisfactory either. For this way of speaking would again suggest that we could perhaps predict better if we only knew more about the things that exist in the universe, whereas Bohr's suggestion denies that there *are* things whose detection would make our knowledge more definite. The third point concerns a suggestion for getting around the kinematics of ill-defined states which has often been made in connection with wave mechanics and which will be discussed in detail later in the present paper. According to this suggestion the difficulties which arise when we try to give a rational account of processes of interaction are due to the fact that the classical

---

[29] A philosopher who holds this belief is Professor M. Bunge, who writes in his *Causality* (Cambridge, Mass.: Harvard University Press, 1959), p. 328, that "the empirical indeterminacy characterizing the usual interpretation of the quantum theory is a consequence of its idealistic presuppositions. . . ." Similar sentiments have been expressed occasionally by Bohm, Kaila, Landé, and Popper. For a criticism, see again my "Problems of Microphysics," as well the final section of my review of Bunge's book in *Philosophical Review*, vol. 60 (1961), pp. 396–405.

point mechanics is not the correct theory for dealing with atomic systems, and the state descriptions of the classical point mechanics are not adequate means for describing the state of systems upon the atomic level. According to this suggestion we ought not to retain the classical notions, such as position and momentum, and make them less specific. What we ought to do is rather to introduce completely new notions which are such that when *they* are used states and motions will again be well defined. Now if any such new system is to be adequate for the description of the quantum phenomena, then it must contain means for expressing the quantum postulate, which is one of the most fundamental micro-laws; and it must therefore also contain adequate means for expressing the concept of energy. However, once this concept has been introduced, in the very same moment all our above considerations apply again with full force : while being part of $A + B$, neither A nor B can be said to possess a well defined theory; whence it follows at once that also the new and ingenious set of concepts will not lead to a well defined and unambiguous kinematics. Now if the new formalism should happen to work with functions, operators, and other mathematical tools which are unambiguous and precise from a mathematical point of view, then we shall have to conclude that this definiteness and absence of ambiguity has no correlate in the real world. In other words, *we shall have to interpret these mathematical tools in a purely instrumentalistic manner.* "It would," therefore, "be a misconception to believe," writes Niels Bohr,[30] "that the difficulties of the atomic theory [i.e., the indefiniteness of state descriptions demanded by the features of processes of interaction as well as by duality] may be evaded by eventually replacing the concepts of classical physics by new conceptual forms." This last remark will be of great importance in connection with the interpretation of Schrödinger's wave mechanics in section IX.

The empirical adequacy of the proposed solution is shown by such phenomena as the natural line breadth, which in some cases (such as in the absorption leading to states preceding Auger effect) may be quite considerable.

Its consequence is, of course, the *renunciation of the kinematics of classical physics* and an instrumentalistic interpretation for any future quantum theory that works with state descriptions which are well defined from the mathematical point of view. For if during the interaction of A and B neither A nor B can be said to be in a well defined state, then the change of these states, i.e., the *motion* of both A and B, will not be well defined either. More particularly, it will no longer be possible to ascribe a definite trajectory to any one of the elements of either A or B. If, on the other hand, the state function of some quantum theory should happen

---

[30] *Atomic Theory . . . ,* p. 16.

to develop in a well-defined fashion, then this development cannot have any real significance, it cannot correspond to any process in nature; that is, it can at most be regarded as an instrument for the prediction, different at different times, of observational results. This is a very forceful argument indeed in favor of instrumentalism, and it can be further supported by a detailed investigation of the properties of wave mechanics. Popper's assertion to the effect that "the view of physical science founded by Cardinal Bellarmino and Bishop Berkeley has won the battle without a further shot being fired"[31] therefore seems to me to be somewhat unjust.[32] Not only was the instrumentalistic position at the time of Copernicus supportable by arguments which were much stronger, at least for a contemporary thinker, than were the arguments flowing from Bellarmine's Platonistic epistemology. But modern physics has found *new* physical reasons why its own most important theory, that is the quantum theory, cannot be anything but an instrument of prediction. These reasons are of precisely the same character as were Ptolemy's: a realistic interpretation of the quantum theory is bound to lead to incorrect predictions. Admittedly, in the arguments usually presented these physical difficulties are almost buried beneath an unacceptable positivism. However, this does not mean that they do not exist and that no "further shot had been fired" since the time of Bellarmine and Berkeley.

One may now attempt to retain the idea of a well defined motion and merely make indefinite the relation between the energy and the parameters characterizing this motion. The considerations in the next section show that this attempt encounters considerable difficulties.

VIII

The reason for the difficulty is that the *duality of light and matter* provides an even more decisive argument for the need to replace the classical kinematics by a new set of assumptions. It ought to be pointed out, in this connection, that dealing with both light and matter on the basis of a single general principle, such as the principle of duality, may

---

[31] "Three Views Concerning Human Knowledge," *Contemporary British Philosophy*, vol. III (1956), pp. 2 ff.

[32] I completely agree with Professor Popper in his attack against instrumentalism. I also agree that one should not rest content with a theory which at most admits of an instrumentalistic interpretation but becomes false when interpreted realistically. Thirdly, I agree with Professor Popper that in the case of the quantum theory such an attitude is usually supported by the view that in any case theories are nothing but instruments of prediction. However, I also believe that within the quantum theory the instrumentalistic position has been forced upon the physicist by the realization that the current theory interpreted realistically must lead to wrong results; it is not merely a repetition of the philosophical idea that all theoretical thinking is of instrumental value only.

be somewhat misleading. For example, whereas the idea of the position of a light quantum has no definite meaning,[33] meaning can be given to the position of an electron. Also, no account is given in this picture of the coherence length of light. Omitting these details, however, we can now argue as follows :

It has been asserted[34] that the interference properties of light and matter and the duality resulting from them are but an instance of statistical behavior in general, which latter is then thought to be best explainable by reference to such classical devices as pin boards and roulette games. According to this assertion, the elementary particles move along well defined trajectories and possess a well defined momentum at any instant of their motion. It is sometimes admitted that their energy may occasionally undergo sudden and perhaps individually unexplainable changes. But it is still maintained that this will not lead to any indefiniteness on the *state* that experiences these sudden changes.

I shall now try to show that this assumption cannot give a coherent account of the wave properties of matter and of the conservation laws. It is sufficient, for this purpose, to consider the following two facts of interference : (1) Interference patterns are independent of the number of particles which at a given moment are dwelling in the apparatus; for example, we obtain the same pattern on a photographic plate whether we use strong light and short time of exposure or very weak light and a long time of exposure.[35] (2) The two-slit interference pattern is not simply the arithmetical sum of the patterns created by each single slit. It is quite possible for the two-slit pattern to possess a minimum in a place, say *P*, in which the one-slit pattern shows a finite intensity (see Fig. 2). The first fact allows us to neglect the mutual interaction (if any) between the particles. Considering the second, we may now reason as follows : If it is correct that each particle always possesses a well defined trajectory, then the finite intensity in *P* is due to the fact that some particle *E* wandered along *b* and ended up in *P*. As long as slit 2 remains closed there will always be some particles which, having passed slit 1, will travel along *b*. Now consider such a particle *E* and assume that it is about to enter slit 1. Now if we open slit 2 at this very moment we have thereby created conditions which are such that *E* must not arrive at *P*. Hence, the process of opening slit 2 must lead to a change in the former path of *E*. How can this change be accounted for?

---

[33] See, for example, E. Heitler, *Quantum Theory of Radiation* (Oxford, 1957), p. 65; D. Bohm, *Quantum Theory* (Princeton, 1951), pp. 97 ff.

[34] An example is A. Landé, "From Duality to Unity in Quantum Mechanics," *Current Issues in the Philosophy of Science* (New York, 1961), pp. 350 ff.

[35] With respect to light this was shown by Janossy. See the booklet edited by the *Hungarian Academy of Sciences* (1957), where previous experiments are reported as well as Janossy, *Acta Physica Hungarica*, vol. IV (1955), and *Nuovo Cimento*, vol. VI (1957).

It cannot be accounted for by assuming action at a distance. There is no room in the conservation laws (which remain valid also in the quantum theory) for energies deriving from such action. Furthermore,[36] the alleged action works not everywhere in space but only along those surfaces which in the wave picture are surfaces of equal phase, and reference to it is therefore nothing but a misleading way of bringing in the wave picture.

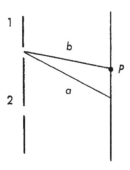

Figure 2

According to Popper[37] and Landé[38] the change of path of the *individual* particle is not in need of explanation. What *can* be explained, by reference to the change of physical conditions (opening of slit 2) is the emergence of a new stochastic process that leads to a new interference pattern. This position is indeterministic as it admits the existence of spontaneous individual changes and its indeterminism is about as radical as that of the Copenhagen point of view. It also shares with that point of view an emphasis on the importance of the experimental situation : Predictions are valid only under certain experimental conditions; they are not unconditionally valid. It differs from the Copenhagen interpretation in that it works with well defined trajectories. This being the case it must, and does,[39] admit that the conservation laws are valid only for large ensembles of particles in a certain situation and that they may be violated

---

[36] The idea of action at a distance has been discussed, and regarded as a possible explanation by Hans Reichenbach. See his *Philosophic Foundations of Quantum Mechanics* (Berkeley and Los Angeles, 1945), see 7. Action at a distance is not the solution Reichenbach himself adopts. For an evaluation of Reichenbach's analysis and of his own solution (three-valued logic), see my paper "Reichenbach's Interpretation of the Quantum Theory," *Phil. Stud.*, vol. IX (1958), pp. 47 ff.

[37] *Observation and Interpretation*, Körner (ed.) (London, 1957), pp. 65 ff.

[38] Landé's suggestions are in many respects similar to those of Popper. Originally [*Quantum Theory, A Study of Continuity and Symmetry* (New Haven, 1955), esp. pp. 24 ff.] Landé adopted the assumption of the indefiniteness of state descriptions. In his later publications he dropped this assumption.

[39] Professor K. R. Popper, private communication.

in the individual case. It is here that the difficulties arise. For energy and momentum are conserved in each individual case of interaction also for elementary particles.[40] This position must therefore be rejected *unless* it is developed in such detail that an account can be given of all those experiments which have convinced the physicists of the individual validity of the conservation laws. *Until* such a detailed account can be given (and nobody can say in advance that it is impossible!) we must again regard the assumption of the indefiniteness of state descriptions as the only satisfactory account. Let it be noted, by the way, that this restriction applies to all the arguments we are going to develop in favor of the assumption of the indeterminateness of state descriptions and the instrumental character of any quantum theory which describes states with the help of functions which are well defined from the mathematical point of view. All these arguments presuppose the validity of certain experimental findings such as the quantum postulate, the laws of interference, and the individual validity of the conservation laws—and the arguments shows that *given* these experimental findings we are forced to adopt instrumentalism.

This demonstrates what has been emphasized before, that the instrumentalism in the quantum theory is not a purely philosophical affair that can be disputed away by general arguments in favor of realism. The argument "quantum theoretical instrumentalism is a result of positivism; positivism is false; hence, we must interpret the quantum theory in a realistic fashion"[41]—this argument is completely irrelevant and also very misleading. It is very misleading because it suggests that a realist can at once interpret the ψ-function realistically, and that the reason why it was not so interpreted was only philosophical prejudice. And it is irrelevant because it does not proceed a single step on the way to resolving the *physical* difficulties which are connected with the realistic position in microphysics. A realistic alternative to the idea of complementarity is likely to be successful only if it implies that certain experimental results are not strictly valid. It therefore demands the construction of a *new theory*, as well as demonstration that this new theory is experimentally at least as valuable as the theory that is being used at the present time. This is a formidable task indeed, and a task that is not even realized as such by the purely philosophical champions of realism in microphysics.

But the situation is even more complex. We have already pointed out that physicists, too, use the very same philosophical arguments which philosophers think to be the only reasons for their having accepted instrumentalism. It is as if they did not really trust their very forceful physical considerations and as if they needed support from something more

---

[40] This was established by the experimental follow-up of the Compton effect, and especially by the experiments of Bothe and Geiger, and Compton and Simon.

[41] See note 29, *supra*.

"fundamental." Moreover it is the belief of many followers of the "ortho-dox" point of view that a theory which provides a realistic alternative to the present quantum theory is impossible, either for logical reasons or for empirical reasons. They therefore not only suggest an interpretation of the known experimental results in terms of indefinite state descriptions and instrumentalism. They also suggest that this interpretation *be re-tained forever,* and that it be the foundation of any future theory of the microlevel. This is another mistake.[42] Small wonder that philosophers find it difficult to unearth the valuable core of their argument from be-neath all the philosophical rubbish that conceals it.

IX

What we have said so far has immediate application to the interpreta-tion of wave mechanics. The older quantum theory, although experi-mentally very successful and also extremely useful in its power to unite a host of otherwise disconnected facts, was yet always regarded as un-satisfactory by many physicists. Its main fault was seen to lie in the manner in which it combined classical and nonclassical assumptions, making a coherent interpretation impossible. For many physicists it was therefore nothing but a stepping stone on the way to a really satisfactory theory, i.e., to a theory which could give us not only correct predictions but also some insight into the nature and dynamics of microscopic entities. It is quite true that Bohr, Kramers, Heisenberg, and others worked along very different lines. Their main objective was not the construction of a new physical theory about a world that existed independently of measure-ment and observation. Rather, they sought to construct a logical machinery for the utilization of those parts of classical physics which could still be said to lead to correct predictions. The inspiration for this lay no doubt in the surprising fact that many classical laws remained *strictly valid* even on the quantum level (for example, the laws of interference discussed in section VIII). This suggested that what was needed was not the elimina-tion and complete replacement of classical physics, but rather a modifica-tion of it. However that may be, the philosophical spirit behind the *Kor-respondenzdenken* was not shared by everybody. Thus de Broglie and Schrödinger tried to develop an entirely new theory for the description of the nature and the behavior of atoms, molecules, and their constituents. When this theory was finished it was hailed by many as the long-expected coherent account of the microlevel. The hypothesis of the indefiniteness of state descriptions, so it was thought, had only reflected the indefiniteness

---

[42] For a more detailed account of this mistake, see section 7 of my "Problems of Microphysics" (note 22, *supra*).

and incompleteness of the early theory, and it was now no longer necessary. More especially, it was assumed either that the states were now new, but well defined entities (the ψ-waves), or it was assumed that whatever incompleteness occurred was due to the statistical character of the theory, i.e., to the fact that wave mechanics was "primarily a variety of statistical mechanics, similar to the classical statistical mechanics of Gibbs."[43] These two interpretations still survive. I hope that our arguments in the preceding section have made it clear that such interpretation of wave mechanics is bound to lead to inconsistencies. The only presupposition of the hypothesis of indefinite state descriptions is the quantum postulate and the dual nature of light and matter (taken together with the individual conservation of energy and momentum). Both these facts are contained in the wave mechanics, which will therefore be equally in need of the said hypothesis and which will at most admit of an instrumentalistic interpretation. A closer analysis of the two main alternatives shows that this is indeed correct.[44]

In summary : Any attempt to give a realistic account of the behavior of the elementary particles is bound to be inconsistent with some very highly confirmed laws. Any such attempt therefore amounts to introducing unsupported conjectures in the face of fact and well supported physical laws. This is the main objection which is used today against the theories of Bohm, Vigier, de Broglie, and others. It is similar to the objections which were raised, at the time of Galileo, against the idea that Copernicus should be understood realistically.

X

To repeat : The objections to a realistic interpretation of the Copernican hypothesis and the objections to a realistic interpretation of quantum theory have this in common : they point out that such an interpretation, quite apart from not having any factual support of its own, is actually inconsistent with observation and well-confirmed physical laws. Now, in the case of the Copernican theory a new dynamics was found which was not only better than the Aristotelian dynamics—more detailed, and allowed for the motion of the earth—but even led to dynamical arguments *in favor* of at least part of earth's motion (rotation). The whole development was accompanied by the discovery of further difficulties for the Aristotelian scheme (sun spots, new stars, path of comets, etc.). Thus the persistence of the Copernicans was finally rewarded and the belief in the basic correctness of their point of view justified. The realistic position triumphed

---

[43] E. C. Kemble, *The Fundamental Principles of Quantum Mechanics* (New York, 1937), p. 55.

[44] For this, see §3 of my "Problems of Microphysics" (note 22, *supra*).

as the result of laborious research and seemed thereby to be proved as essentially correct. Is this not a splendid argument for realism? Does it not show that the realistic position encourages research and stimulates progress, whereas instrumentalism is more conservative and therefore liable to lead to dogmatic petrifaction? This consequence has been drawn by many thinkers. This is the philosophy that allowed Boltzmann and the other proponents of the kinetic theory to remain persistent in the face of the sometimes quite formidable objections of their opponents; and this is also the philosophy that inspires the contemporary critics of the Copenhagen interpretation. It is a very positive philosophy and a very optimistic philosophy. Personally I am very much inclined to accept it. At the same time there are grave difficulties, and even absurdities arise if one tries to think it through consistently. It is now time to face these difficulties, and perhaps to remove them.

XI

It is clear that the final success of the Copernican theory could not have been foreseen in the beginning. No set of methodological principles can ever guarantee the essential correctness of a theory that has just been introduced; this follows from Hume's investigations. And if the theory contradicts the accepted laws and facts then its future success would seem to be even less certain. The belief that such success is yet bound to occur, which inspired the Copernicans, can therefore be based neither on methodological considerations nor on factual argument. It is a metaphysical belief. Now if it was legitimate for the Copernicans to act on such a belief, then more recent metaphysical assumptions cannot be excluded either. For example, there is no reason why one should not now reintroduce Aristotelianism and hope for the best. The retort that this theory has already been given its chance does not count : the same was true of the hypothesis of the motion of the earth which was well known in antiquity and which was given up in view of the arguments of Aristotle and his followers. However, does not such an admission open the door to all sorts of wild speculations such as the hollow earth theory, Wilhelm Reich's Orgonomy, Dianetics,[45] astrology, and other crazy ideas? Is this not making illegitimate use of the success of Copernicus (and, so one might add, of Boltzmann, and Schrödinger[46])? Is there not a decisive difference between

---

[45] For an amusing description of these and other odd theories, see Martin Gardner, *Fads and Fallacies* (New York, 1952).

[46] For the very interesting history of Schrödinger's early attempts to solve the problem of atomic spectra see Dirac's letter in *The Scientific Monthly*, vol. 79 (1954), No. 4.

the Copernican conjecture and astrology? I agree : Copernicus has been successful, astrology has not been successful. But what I am talking about now is the attitude to be adopted *before* a theory has proved its fruitfulness. The objection assumes that the final success of Copernicus *could somehow have been foreseen* and that we know *in advance* that Orgonomy is completely fanciful, and hopelessly out of touch with reality. But how could we possibly possess such knowledge? Because the existence of Orgon is *inconsistent with contemporary physics*? Copernicus was inconsistent with the physics of his time in a very simple and straightforward manner. The case of Orgon, its relation to the rest of physics, is much more complicated and doubtful. Or shall we reject the Orgon idea or, to take an even better example, the hollow earth theory (the earth is hollow and we live in its inside) because it is *absurd*? Copernicus was regarded as absurd at his time. Read Luther and Sir Francis Bacon. From all this it seems to emerge that from the point of view of their status *before* their success (or failure) there is not much to choose between the hypothesis of the moving earth in the infinite universe on the one hand, and the hollow earth theory, astrology, and Ehrenhaftian physics on the other.[47] This is the difficulty inherent in the optimistic and naive realism which we have described above. It does not allow us to separate the real from the fanciful, the fruitful hypothesis from the products of the crank. How can this difficulty be resolved?

XII

I think it can be resolved only if we are prepared to give up certain very deep-rooted prejudices concerning the nature of empirical support. It is not at all difficult to give up these prejudices once the matter is put in the right light. In order to be able to do this we must first abandon the attitude that the Aristotelian physics was just a heap of rubbish which no careful thinker would ever have supported, and that the inconsistency between it and the heliocentric hypothesis could therefore not at all be counted as an argument against the latter. It must again be emphasized that from the point of view of empirical method the Aristotelian physics was as good as any theory that could have been devised at that early time. It was partly supported by the evidence then available, it was confronted with certain difficulties—and it was in all these respects very similar to the

---

[47] A comparison of degrees of confirmation is completely useless here. If E is the evidence known at the time of Copernicus, A the Aristotelian dynamics, G the geocentric hypothesis, C the Copernican hypothesis, then $P(C/E) = P(G/E)$, but $P(G/AE) = 1$ and $P(C/AE) = 0$, so that the geocentric hypothesis should have been adopted. Altogether I think that a discussion of cases similar to those analyzed in the present paper will reveal the complete inadequacy of the current theories of confirmation.

more detailed theories we possess at the present time.[48] It was therefore completely legitimate to use it in the empirical arguments purporting to show the untenability of the Copernican hypothesis. *At the same time the eventual success of this latter hypothesis shows what empirical arguments of such a kind are worth. It shows that such arguments are not at all final and irrevocable and that it is possible to revise their premises in the light of further research.* And this is not surprising. After all, the laws which are used in such arguments (the laws of the Aristotelian dynamics in the case of the Copernican hypothesis; the second law of thermodynamics in the case of the kinetic theory; the conservation laws, the quantum postulate, and the interference laws in the case of the quantum theory) always go far beyond what could have been shown by experience. They use precise concepts where experience can at most give imprecise information; and they are general statements where experience can at most give rise to a finite number of singular statements of observation.[49] A new point of view which contradicts these laws therefore need not on that account alone be factually inadequate, provided the inconsistency occurs inside the domain of imprecision and outside the class of known facts. Even a singular statement of observation need not be regarded as final and irrevocable. Any such statement will use terms which are part of a fairly comprehensive conceptual system whose postulates have either been explicitly formulated or function implicitly as the "rules of usage" according to which the terms are habitually used. Now a term which is used for describing the result of a (direct) observation will obtain its meaning partly from the impression created by the observed situation, partly from the postulates of the conceptual system to which it belongs. Hence, although it may adequately express what is observed it may still be inadequate if some of those postulates have been found to be incorrect. This being the case, not even a direct observational report is exempt from criticism and reformulation.

Considering all these possibilities of change, the discrepancy between a new point of view on the one hand and accepted theories and observations on the other does not at all decide the fate of the new point of view—and this even if the accepted theories should be as well confirmed and as precisely formulated as is the second law of thermodynamics or the law of conservation of energy. Any such discrepancy creates a problem *that must be further investigated* in order to ascertain whether it is indeed a contradiction between fact and theory that is here being revealed and not, rather, a contradiction between one theory and the as yet untested part of another theory, or between one theory and the as yet untested part of a principle

---

[48] See note 4, *supra*.

[49] For a more detailed account, see §7 of my "Explanation, Reduction, and Empiricism," *Minnesota Studies in the Philosophy of Science*, vol. III, as well as §4 of my paper "How To Be a Good Empiricist," published in vol. I of the *Delaware Studies in the Philosophy of Science* New York, 1963.

contributing to the meaning of a key term in an observational statement. Nobody can say in advance where such further investigation will lead and nobody should therefore allow a controversy between the point of view he likes on one side and facts plus well established theories on the other to deter him from carrying out such an investigation.

XIII

It is here, by the way, that the distinction between "respectable" people and cranks must be drawn. The distinction does not lie in the fact that the former suggest what is plausible and promises success, whereas the latter suggest what is implausible, absurd, and bound to fail. It *cannot* lie in this because we never know in advance which theory will be successful and which theory will fail. It takes a long time to decide this question, and every single step leading to such a decision is again open to revision. Nor can the absurdity of a point of view count as a *general* argument against it. It is a reasonable consideration for the choice of one's own theories to demand that they seem plausible to oneself. This is one's private affair, so to speak. But to declare that only plausible theories should be considered is going too far. No, the distinction between the crank and the respectable thinker lies in the research that is done once a certain point of view is adopted. The crank usually is content with defending the point of view in its original, undeveloped, metaphysical form, and he is not at all prepared to test its usefulness in all those cases which seem to favor the opponent, or even to admit that there exists a problem. It is this further investigation, the details of it, the knowledge of the difficulties, of the general state of knowledge, the recognition of objections, which distinguishes the "respectable thinker" from the crank. The original content of his theory does not. If he thinks that Aristotle should be given a further chance, let him do it and wait for the results. If he rests content with this assertion and does not start elaborating a new dynamics, if he is unfamiliar with the initial difficulties of his position, then the matter is of no further interest. However, if he does not rest content with Aristotelianism in the form in which it exists today but tries to adapt it to the present situation in astronomy, physics, and microphysics, making new suggestions, looking at old problems from a new point of view, then be grateful that there is at last somebody who has unusual ideas and do not try to stop him in advance with irrelevant and misguided arguments.

XIV

I think it is clear now *that there is no harm* in proceeding as Copernicus did and as Professor Bohm does, in introducing unfounded conjectures which are inconsistent with facts and accepted theories and which, more-over, give the impression of absurdity—*provided* the suggestion of such conjectures is followed up by detailed research of the kind outlined in the preceding section. There is no harm in proceeding in this manner. But we have not yet given a single reason why one *should* proceed in this manner. After all, there are many activities which are not harmful but on which nobody would want to waste his time. We can, of course, try to defend such a procedure by reference to its *possible* success. But the accepted theory might also continue to succeed in the future. It might survive all difficulties, and if it does then there is no need of introducing new ideas and getting involved in the laborious investigations connected with such a procedure. Hence, the best method would seem to be to wait until the current theory gets into difficulties and only *then* to start looking for new theories.

This opinion is widely accepted on account of its apparent reasonable-ness. It has a lot to recommend it *provided the difficulties which may beset a theory will always be discovered without any help from other theories*. It is only then that the advice to wait until the accepted point of view has col-lapsed will be feasible. It can be shown, however, that *there exist potential difficulties for any theory that can be detected only with the help of further theories*. Now if this is correct, then the development of such further theories is demanded by the principle of testability, according to which it is the task of the scientist relentlessly to test whatever theory he possesses, and it is also demanded that these further theories *be developed in their strongest possible form, i.e., as descriptions of reality rather than as mere instruments of successful prediction*. Even where it contradicts the accepted views and facts, realism can be justified by methodological considerations. In the next section we shall show how development of additional theories may increase testability of the accepted point of view.

XV

The argument is very simple. Consider a theory T which makes pre-dictions P in a domain D, and assume also that the actual state of affairs P' is different from P but to such a small extent that the difference is far below the experimental possibilities. In this case T is incorrect without our being able to discover this incorrectness. One may now hope that de-velopment of new experimental methods will eventually reveal that P' obtains, and not P. Now when the difference between P' and P is small enough this hope would seem to be as unrealistic (or as realistic) as the hope that invention of a new theory might lead to giving a better account of what is going on in nature. After all, the development of instru-ments of measurement is guided by the ideas and interests of scientists, and it is very unlikely that it will automatically lead to the discovery of all the shortcomings of the theories we possess. Moreover, there are cases where construction of instruments for the direct detection of differences between P and P' *is excluded by laws of nature* and therefore impossible. The behavior of electrons inside an atom can never be investigated by a direct test, i.e., by a test of the kind one would carry out if Newton's celestial mechanics and Coulomb's law were the only theories in existence and the ideas connected with these theories the only ideas we possessed. Nor is it possible by a direct test to discover that Brownian motion is due to a transfer of heat from the embedding fluid to the particle, which means that the difficulties presented for the second law of the phenomenological theory of thermodynamics by the existence of the Brownian particle would in this way never be revealed.[50] Thirdly, it is very unlikely that the discovery of a discrepancy would at once lead to its correct interpretation. Many such discrepancies, if they were small and irregular enough, would be regarded as oddities much in the same fashion as Professor Ehrenhaft's astounding effects are today regarded as oddities rather than as refuting instances for part of contemporary physics. All these circumstances work in the same direction—they tend to hide from us the weaknesses of a theory in which we believe.

Assume now that we introduce alternative theories T', T'', etc. which are inconsistent with T inside D and which predict P' rather than P. Now if we succeed in elaborating one of these theories in such detail that it can be compared with T as regards simplicity and effectiveness, if this theory is confirmed where T was confirmed, if it solves some cases which belonged to the class of unsolved problems of T (see note 4 *supra*), if it furthermore makes predictions which are not made by T, and if these predictions are confirmed, then we shall take T' as our measure of truth and regard T as refuted—and this despite the fact that no *direct* refuting instance has as yet been found for T. This is why invention of new theories which are inconsistent with the accepted point of view is demanded by the principle of testability, and this is also the promised methodological justification for realism.

XVI

To sum up : The issue between realism and instrumentalism has many facets. There are arguments of a philosophical kind which demand that theories be regarded as instruments of prediction only and not be used for inferences concerning the structure of our universe. In the present paper such arguments were only mentioned; they were not discussed. There

---

[50] For details, see §7 of my paper "Explanation, Reduction, and Empiricism."

are other arguments for instrumentalism which concern specific theories such as the quantum theory or the heliocentric hypothesis and which are based upon specific facts and well confirmed theories. It was shown that to demand realism in these cases amounts to demanding support for unplausible conjectures which possess no independent empirical support and which are inconsistent with facts and well confirmed theories. It was also shown that this is a plausible demand which immediately follows from the principle of testability. Hence realism is preferable to instrumentalism even in these most difficult cases.

CHAPTER 20

Observation and the Quantum

BY O. R. FRISCH

IN CLASSICAL PHYSICS, the problem of observation was mainly a practical one : how to construct instruments of sufficient accuracy for observing, say, the planets in their courses or a micro-organism in its nutrient fluid. While there were practical limitations, due, for instance, to the wave length of visible light, there appeared to be no limitation in principle to the accuracy with which any observation might be performed.

The recognition of quantum phenomena during the first quarter of our century has changed that. If we confine ourselves to nonrelativistic quantum theory,* then it is still possible to measure any observable (i.e., any variable that is capable of observation) with arbitrary accuracy and to verify the result by immediate repetition of the measurement. But generally for each observable q there exists a "conjugate" observable p such that the accurate observation of q disturbs the system so that the value of p—which might have been known from a previous measurement—is changed by an amount which is only approximately known. The smaller the uncertainty Δq is made, the bigger Δp becomes, according to the "uncertainty relation"

$$\Delta p \cdot \Delta q \simeq h/2\pi$$

where h is Planck's constant (the quantum of action). A conjugate pair of observables that is often discussed consists of one cartesian coordinate of a particle—i.e., its distance from a given plane—and its momentum component perpendicular to that plane.

Many ingenious attempts were made to devise some measuring pro-

---

* Relativistic quantum theory displays a new difficulty: any attempt to measure with more than a certain accuracy can cause the creation of new particles which confuse the observer.

cedure that would get around the uncertainty relations, and I still remember what hard work it was to discover the subtle flaw in a very convincing *Gedankenexperiment* (thought experiment) sent to me by my friend Karl Popper about 1930. Niels Bohr, in particular, has been indefatigable in refuting such attempts, especially in his memorable argument with Einstein. Today that debate has died down : most physicists are agreed that the framework of quantum theory is consistent and in excellent agreement with experimental facts, and that within that framework the uncertainty relations are valid.

But their interpretation is still under debate. If, after measuring the momentum of a particle, its position is only approximately known to us, does that mean that the particle as such is blurred ? It is hard to say what that would mean. And if we later measure its position, does then the particle suddenly contract? It seems more plausible to say that the blurred wave function of our particle does not describe the real particle but merely our (temporary) knowledge of it; a subsequent measurement of its position does not contract its wave function but simply replaces its wave function by a narrower one when our previous knowledge becomes superseded.

It is then tempting to assume that the particle has an accurate position and momentum all the time, just as in classical physics, although we can never know both at once. The wave function would then be a description, not of the real particle, but merely of the limited information that our observations can give us. Quantum mechanics, although capable of describing all we can know, would not be a complete description of reality.

That was the view presented in the famous paper by Einstein, Podolski, and Rosen, with the following strong argument in its favor. The paper described a *Gedankenexperiment* which showed that it was possible to measure either the position or the momentum of a particle, without interfering with it in any way, purely by making observations on another particle with which it had previously interacted. Surely — the authors argue — any property of an object which you can measure without interfering with that object must be real; so the particle must have both position and momentum even though you can measure accurately only one at a time.

Against this, Bohr has stressed that any prediction of what the particle will do next must be based on what we *know* about it, so that it is a purely academic question whether an observable that has not been measured possesses an unknown but well-defined value. In his view it is indeed meaningless to attribute any value to an observable unless the conditions are such that it can be observed. Yet many of us would like to believe that the particle has real existence, complete with position and momentum, even when it is not observed. Such a "materialistic" belief — though irrelevant to the actual use of quantum-mechanical techniques — is not in visible conflict with the uncertainty principle.

However, such a belief can no longer be upheld in certain variants of the E–P–R experiment. That was shown in a rather mathematical manner by Furry and by Schrödinger, and the specific example which I shall discuss is given in a recent textbook by Dicke and Wittke (as I discovered after this article had been drafted).

The example concerns the correlation between the polarizations of the two photons formed by positron–electron annihilation. Let me first remind you of the relevant properties of linearly polarized light. Such light can be obtained most simply by passing a beam of light through a "polarizer," e.g., a sheet of polaroid. A second piece of polaroid, cut from the same sheet and placed behind the first one without rotating it, will (ideally) transmit all the light that has gone through the first; but if this "analyzer" is turned in its own plane by the angle α the light is weakened by the factor $\cos^2\alpha$, and for $\alpha = 90°$ no light is transmitted.

Polaroid is a plastic which contains minute elongated dyestuff crystals, all parallel to one another and perpendicular to a certain diameter of the polaroid disc; that diameter we shall call its axis. So we can say that the fraction of light transmitted by the analyzer is $\cos^2\alpha$ where α is the angle between the axes of polarizer and analyzer.

This phenomenon, and many others, are in accord with Maxwell's theory, which describes light as a transverse electromagnetic wave. In ordinary (unpolarized) light, not only does the magnitude of the electric vector oscillate rapidly but also its direction varies at random, though it remains perpendicular to the light beam. The dyestuff crystals in the polaroid absorb that component of the electric vector which is parallel to their orientation and hence perpendicular to the axis of the polaroid. So the light transmitted by the polaroid is completely polarized : its electric vector now remains parallel to the axis of the polaroid, irrespective of the polarization of the light before transmission. That explains also why the fraction $\cos^2\alpha$ passes through an analyzer rotated by the angle α relative to the polarization of the beam : it is the square of the factor $\cos\alpha$ by which we have to multiply the electric vector to get its component in the new direction. We need the square because the energy flux in a light wave is proportional to the mean square of the electric vector.

But how is it in the quantum theory? We know that a beam of light is a stream of photons (light quanta) which can be recorded individually by a photocell. Can we then speak of the polarization of individual photons?

We can, within limits. We assume that in a fully polarized beam all the photons are polarized in the same direction; on falling on a polaroid twisted by the angle α with respect to that direction, a photon has a probability $\cos^2\alpha$ of being transmitted; if it is transmitted it emerges with a new polarization parallel to the axis of the polaroid through which it has just passed. With these assumptions we can account for all the phenomena we have described.

But in general the description is not unique. For instance, if half of it gets transmitted, the beam (before striking the polaroid) might have been

1. a mixture of photons with random polarizations,
2. polarized at 45° (relative to the axis of the polaroid),
3. polarized at −45°,
4. a half-and-half mixture of (2) and (3),
5. a half-and-half mixture of photons polarized at 0° and 90°,

or an infinity of other mixtures. Some of those possibilities can be eliminated by turning the polaroid, which will change the transmission in the cases (2) and (3). But if turning has no effect, the beam can still be described by (1), (4), and (5), and there is in fact no way of distinguishing between those descriptions. That is obvious in the framework of classical physics: an unpolarized light wave, with the azimuth of its electric vector varying at random, can be mathematically decomposed into two incoherent* waves of equal average intensity which are linearly polarized in two mutually orthogonal—but otherwise arbitrary—directions, and is indeed identical with a superposition of those two waves.

So the two mixtures (4) and (5) represent different pictures of the same beam, with no physical criterion to tell them apart. It is irrelevant which one we adopt, and there are infinitely many others that will do just as well. On the other hand, if we adopt one of them we can never prove by subsequent analysis of the beam that this is not the true picture.

There is just one case in which the picture is unambiguously given by the experiment: if the beam is completely transmitted by the analyzer it must have been fully polarized parallel to its axis, and all its photons must then have been polarized in that direction. Similarly, if a beam is completely stopped by a polaroid, then all its photons must have been polarized at right angles to its axis.

After this preamble, let us come back to the experiment I want to discuss. It has long been known that the mutual annihilation of a positron and an electron having opposite spins is accompanied by the emission of two short-waved photons in opposite directions. That can be verified by surrounding a source of positrons (e.g., a few micrograms of the radioactive isotope sodium 22) with enough material (about 1 mm. thick) to arrest the positrons, which then quickly annihilate themselves with the electrons present; such a source will emit millions of annihilation photons per second in all directions. By placing two efficient gamma-ray detectors (e.g., scintillation counters) at opposite sides of the source, one can demonstrate that the photons are indeed sent out in pairs, the members of each pair travelling in opposite directions: one observes many simultaneous counts— so-called coincidences—whereas practically no coincidences are found

---

* That is, without persistent phase relations.

when the source is slightly displaced so as to be no longer between the two detectors, although the rate at which each detector counts is hardly affected thereby.

Now Wheeler deduced from quantum theory that each photon and its companion will be polarized at right angles to one another. Yet the radiation in any direction must be unpolarized because the source—positrons of random direction and spin orientation annihilating themselves with electrons of similar random character—is clearly isotropic. It might then be thought that in any one direction we would have an unpolarized beam that can be described by any one of the pictures (1), (4), or (5); but that is not possible, as I shall now try to show.

Let us first try picture (5), that of a random mixture of photons polarized vertically and horizontally (we take photons that travel horizontally, say to the north). The companion photons (going south) will then be a similar mixture, each S-photon being polarized horizontally if its companion N-photon is vertically polarized, and vice versa. We shall now see how such a system would behave if we place "polaroids" in front of our detectors (disregarding, at the moment, the fact that no "polaroids" for those short waves have yet been invented).

Placing a polaroid in front of, say, the N-detector will cut its counting rate to one half; the photons going south are not affected, so the rate of coincidences will fall by the same factor, from its original rate R to $R/2$. If its axis was vertical and we now insert a similar polaroid in front of the S-counter it will stop all the photons whose companions were transmitted, so that the coincidence rate will now be zero; in the "crossed" position, i.e., turned by $90°$, the S-polaroid will transmit all the photons whose companions were recorded by the N-counter, giving a coincidence rate of $R/2$. Those are indeed the rates predicted by the quantum-mechanical calculation; so far, our picture seems to work.

However, if we now turn both polaroids by $45°$—an operation that should have no effect since our source is isotropic—we get a different result. Inserting first one polaroid again cuts the coincidence rate from R to $R/2$ since all the photons—those with vertical and those with horizontal polarizations alike—have a probability $1/2$ of being transmitted; but their companion photons are still a random mixture of the type (4) with respect to the polarizer that we now insert in front of the S-counter, and only half of them will be transmitted. Hence if the polaroids are oriented at $45°$ to the vertical our picture predicts a coincidence rate $R/4$ irrespective of whether their optical axes are parallel or crossed. So our picture cannot be right: those two beams do not represent a phenomenon isotropic about the beam direction, though each beam separately does.

Isotropy can be restored by using picture (1), each beam a random mixture of photons polarized in all directions, and each photon polarized at right angles to its companion photon. With that picture one can compute that crossed polaroids will cut the coincidence rate to $3R/8$, parallel polaroids to $R/8$. But those figures do not agree with $R/2$ and 0, the figures predicted by quantum theory. Only picture (5) gives the predicted figures, and only if the photons are assumed to be polarized parallel and at right angles to the polaroids used, whose orientation is subject to the wishes of the experimenter.

It is, of course, not in itself surprising that by carrying out an observation on a north-going photon one can learn something about the south-going companion : that possibility was stressed in the E–P–R paper. What is surprising is that, by such an observation, the experimenter in the north can force the south-going photon to adopt complete polarization along an axis arbitrarily chosen by himself. If the source is far enough away the experimenter may even set his polaroid after the two photons are emitted : yet when he observes one photon he can be sure that the companion photon —which started a million miles away and is travelling away from him— must be polarized at right angles to the direction in which he had arbitrarily set his polaroid after it had been emitted.

That result is so surprising that one must ask whether it has been experimentally confirmed. Polaroids for those wave lengths do not exist; but such a photon can set an electron in motion (Compton effect) on passing through matter, and the distribution of those electrons about the beam axis is not isotropic if the photons are linearly polarized. Thus a counter, detecting electrons that have been knocked into a particular direction, can be used like a rather inefficient polaroid, whose efficiency can, however, be computed from quantum theory. Using two such counters, Wu and Shaknov have measured the ratio of the coincidence rates in the "parallel" and "crossed" positions and have found it to agree with that computed from the quantum theory. To that extent, the result from the theory has been confirmed experimentally.

But the theory predicts even more. The experimenter could (in principle) use a polarizing device that transmits completely those photons which have a particular *elliptic* polarization, characterized by *two* parameters. In that case the companion photons will again be found to possess a definite elliptic polarization. So the experimenter can compel a far-away photon to adopt a definite polarization—characterized by two independent parameters—by adjusting his own apparatus after the photon has been emitted.

What shall we think about this? Surely one cannot assume that some kind of signal, going faster than light, chases the far-away photon and informs it what the experimenter has done at the other end. Indeed it is not possible to use that experiment to signal faster than light : the man in charge of the N-counter, while he can manipulate his polaroid, cannot let the man by the S-counter know which of the photons he should watch while ignoring the 50 per cent whose companions were absorbed in the N-polaroid. But theories have been mooted in which signals travelling with the negative speed of light ("into the past") inform each photon of the conditions it can expect on arrival. One such theory (Feynman and Wheeler), has achieved its special goal, and perhaps more thought should be given to that idea.

Alternatively we should say that the two photons must not be considered as separate particles but as one system. We know that in quantum mechanics a system of N particles, each of which is described by M coordinates indicating its location, spin state, or polarization, etc., must be described by a wave function in a $M \cdot N$-dimensional space. That function becomes separable (i.e., represented as a product of N functions, each relating to one particle) only if they are independent. But spatial separation does not guarantee independence. Particles can be considered as independent only if what we know about one of them is not affected by any measurements that we may perform on the others. In most practical applications of quantum theory we are concerned with the study of one particle at a time, and we are therefore apt to forget that in all coincidence experiments—where the correlated behavior of two or more particles resulting from one interaction is studied—that interdependence is likely to become apparent. Those wave functions in $M \cdot N$-dimensional space are not merely a mathematical device: they reflect an important feature of the phenomena we observe.

Bibliography

N. Bohr, article in P. A. Schilpp (ed.) *Albert Einstein, Philosopher-Scientist* (New York: Tudor Publ., 1951).

A. Einstein, B. Podolsky, and N. Rosen, *Phys. Rev.*, vol. 47 (1935), 777.

W. H. Furry, *Phys. Rev.*, vol. 49 (1936), 393.

E. Schrödinger, *Proc. Camb. Phil. Soc.*, vol. 31 (1935), 555.

J. A. Wheeler, *Ann. New York Acad. Sci.*, vol. 48 (1946), 219.

C. S. Wu and I. Shaknov, *Phys. Rev.*, vol. 77 (1950), 136.

J. A. Wheeler and R. P. Feynman, *Rev. Mod. Phys.*, vol. 17, (1945) 157; vol. 21 (1949), 425.

CHAPTER 21

Popper on Irreversibility

BY ADOLF GRÜNBAUM

I. Introduction

IN A SERIES OF NOTES published in *Nature* during the years 1956–1958, K. R. Popper[1] has expounded his thesis of the "untenability of the widespread, though surely not universal, belief that the 'arrow of time' is closely connected with, or dependent upon, the law that disorder (entropy) tends to increase." (II) Specifically, he argues in the first three of his four notes that there exist irreversible processes in nature whose irreversibility does not depend on their involvement of an entropy increase. Instead, their irreversibility is *nomologically contingent* in the following sense: the laws of nature governing elementary processes do indeed allow the temporal inverses of these irreversible processes, but these processes are *de facto* irreversible, because the *spontaneous* concatenation of the initial conditions requisite for the occurrence of their temporal inverses is well-nigh physically impossible. Noting that "Although the arrow of time is not implied by the fundamental equations [laws governing elementary processes], it nevertheless characterizes most solutions" (I), Popper therefore rejects the claim that "every non-statistical or 'classical' mechanical process is reversible" (IV). In the fourth of his communications, he maintains against Boltzmann that the statistical behavior of the entropy of physical systems not only fails to be the sole physical basis for the anistropy of time, but does not qualify *at all* as such a basis.[2] For Popper argues that, if it did, the tem-

---

[1] K. R. Popper, *Nature*, vols. 177 (1956), p. 538; 178 (1956), p. 382; 179 (1957), p. 1297; 181 (1958), p. 402. These four publications will be cited hereafter as "I," "II," "III" and "IV," respectively.

[2] In view of the misleading potentialities of Eddington's metaphor "the arrow of time," which is also employed by Popper, I prefer to substitute the nonmetaphorical expression "the anisotropy of time" in my account of Popper's views. A statement of the respects in which I regard "the arrow of time" to be a misleading metaphor is given in A. Grünbaum, *Philosophical Problems of Space and Time*, Alfred A. Knopf, New York, 1963, pp. 209–210; hereafter this book is referred to by the abbreviation "PPST."

poral description of fluctuation phenomena would entail absurdities of several kinds.

In response to the first two of Popper's four notes, E. L. Hill and I published a communication[3] in which we endorsed Popper's contention of the existence of non-entropic, nomologically contingent irreversibility in the form of an existential claim constituting a generalization of Popper's contention.

In view of Popper's criticism (III) of the latter generalization, my aim in the present essay is as follows :

(*i*) To appraise Popper's criticism.

(*ii*) To show that the generalization put forward in the paper by Hill and myself has the important merit of dispensing with the restriction on which Popper predicates his affirmation of nomologically contingent irreversibility : the requirement of the *spontaneity* of the concatenation of the initial conditions requisite to the occurrence of the temporal inverses of the thus conditionally irreversible processes.

(*iii*) To provide a viable version of Boltzmann's conception of a statistically anisotropic time based on the (probabilistic) entropy of systems which is *not* beset by any of the absurdities adduced against that conception by Popper.

It will turn out that the obtaining of certain species of nomologically contingent boundary conditions is a necessary condition for an entropically grounded statistical anisotropy of time. Hence it will become apparent that nomologically contingent properties of the world enter integrally not only into the non-entropic kind of irreversibility affirmed by Popper as the physical basis for the anisotropy of time, but also into the entropic kind of statistical irreversibility to which he denies any such status.

II. Non-entropic, Nomologically Contingent Irreversibility

Independently of O. Costa de Beauregard, who had used the same illustration before him,[4] Popper (I) considers a large surface of water initially at rest into which a stone is dropped, thereby producing an outgoing wave of decreasing amplitude spreading concentrically about the point of the stone's impact. And Popper argues that this process is irreversible in the sense that the "spontaneous" (IV) concatenation on all points

---

[3] E. L. Hill and A. Grünbaum, *Nature*, vol. 179 (1957), p. 1296.

[4] O. Costa de Beauregard, "L'Irréversibilité Quantique, Phénomène Macroscopique," in A. George (ed.), *Louis de Broglie* (Paris, 1953), p. 402.

of a circle of the initial conditions requisite to the occurrence of a corresponding contracting wave is physically impossible, a "spontaneous" concatenation being understood to be one which is *not* brought about by
coordinated influences emanating from a common center. Being predicated on such spontaneity, this nomologically contingent irreversibility is
of a *conditional* kind.

Now, one might object that the attribution of the irreversibility of the
outgoing wave motion to non-thermodynamic causal factors is unsound.
The grounds would be that the statistical entropy law is *not irrelevant* to
this irreversibility, because the diminution in the amplitude of the outgoing
wave is due to the superposition of two independent effects, as follows:
(1) the requirements of the law of conservation of energy (first law of
thermodynamics), and (2) an entropy increase in an essentially closed
system through dissipative viscosity. To be sure, the entropy increase
through dissipative viscosity is a sufficient condition (in the *statistical*
sense of my §III, below!) for the irreversibility of the outgoing wave
motion, i.e., for the absence of a corresponding (spontaneously initiated)
contracting wave motion. But this fact cannot detract from the soundness
of Popper's claim that another, independent *sufficient condition* for the
conditional kind of *de facto* irreversibility affirmed by him is as follows:
the nomologically contingent non-existence of the spontaneous occurrence
of the coordinated initial conditions requisite for a contracting wave
motion. We see that Popper rightly adduces the need for the coherence of
these initial conditions as his basis for denying the possibility of their
spontaneous concatenation, i.e., their concatenation without first having
been coordinated by an influence emanating from a central source. Says
he (III): "Only such conditions can be causally realized as can be organized from one centre. . . . causes which are not centrally correlated are
causally unrelated, and can co-operate [i.e., produce coherence in the
form of isotropic contraction of waves to a precise point] only by accident
. . . . The probability of such an accident will be zero."

In view of the aforementioned conditional character of Popper's nomologically contingent irreversibility, E. L. Hill and I deemed it useful to
point out (cf. note 3, *supra*) the following: there does indeed exist an
important class of processes in infinite space whose irreversibility is (1)
non-entropic and *nomologically contingent,* hence being of the kind correctly envisioned by Popper, yet (2) *not conditional* by not being predicated on Popper's proviso of spontaneity. Without presuming to speak for
Professor Hill, I can say, for my part, that in making that existential
claim I was guided by the following considerations:

(*i*) Popper (II) briefly remarks correctly that the eternal expansion
of a very thin gas from a center into a spatially infinite universe does not
involve an entropy increase, and the *de facto* irreversibility of this process
is therefore non-entropic. For the statistical Maxwell–Boltzmann entropy

is not even defined for a spatially infinite universe: the quasi-ergodic hypothesis, which provides the essential basis for the probability-metric ingredient in the Maxwell–Boltzmann entropy concept, is presumably false for an infinite phase-space, since walls are required to produce the collisions which are essential to its validity. In the absence of some kind of wall, whose very existence would assure the finitude of the system, the rapidly moving particles will soon overtake those moving slowly, leaving them ever further behind for all future eternity instead of mixing with them in a space-filling manner. Moreover, if the number of particles in the infinite universe is only finite, the equilibrium state of maximum entropy cannot be realized, since a finite number of particles cannot be uniformly distributed in a phase space of infinitely many cells. On the other hand, if the number of particles is denumerably infinite, the number W of microscopic complexions in $S = k \log W$ becomes infinite, and no entropy increase or decrease is defined.

(*ii*) Though allowed by the laws of mechanics, there seem to exist no "implosions" at all which would qualify as the temporal inverses of eternally progressing "explosions" of very thin gases from a center into infinite space. In the light of this fact, one can assert the *de facto* irreversibility of an eternal "explosion" *unconditionally*, i.e., without Popper's restrictive proviso of spontaneity with regard to the production of the coherent initial conditions requisite for its inverse. For in an infinite space, there is no possibility at all of even a *non*spontaneous production of the coherent "initial" conditions for an implosion having the following properties: the gas particles converge to a point, after having been moving through infinite space for all past eternity, in a manner constituting the temporal inverse of the expansion of a very thin gas from a point for all future eternity. There can be no question of a nonspontaneous realization of the "initial" conditions required for the latter kind of implosion, since that would involve a self-contradictory condition akin to that in Kant's fallacious First Antinomy: the requirement that a process which has been going on for all infinite past time must have had a finite beginning (production by past initial conditions) after all.

On the other hand, in a spatially *finite* system it is indeed possible to produce nonspontaneously the initial conditions for contracting waves and for implosions of gas particles which converge to a point. Thus, assuming negligible viscosity, there are expanding water waves in finite systems of which the temporal inverses could be produced nonspontaneously by dropping a very large circular object onto the water surface so that all parts of the circular object strike the water surface simultaneously. And hence there are conditions under which contracting waves do exist in finite systems. But Popper's spontaneity proviso is not necessary to assert the *de facto* irreversibility of the eternal expansion of a spherical light wave from a center through infinite space! If space is infinite, the existence of the latter process of expansion is assured by the facts of observation in conjunction with electromagnetic theory; but despite the fact that the laws for a homogeneous and isotropic medium allow the inverse process no less than the actual one,[5] we never encounter the inverse process of spherical waves closing in isotropically to a sharp point of extinction.

In view of the decisive role of the infinitude or "openness" of a physical system (the universe)—as opposed to the finitude of closed systems—in rendering Popper's spontaneity proviso dispensable, Hill and I[6] made the following existential claim concerning processes whose irreversibility is non-entropic and *de facto* in "open" (infinite) systems :

> In classical mechanics the closed systems have quasi-periodic orbits, whereas the open systems have at least some aperiodic orbits which extend to infinity. . . . . there exists a fundamental distinction between the two kinds of system in the following sense. In open systems there always exists a class of allowed elementary processes the inverses of which are unacceptable on physical grounds by requiring a *deus ex machina* for their production. For example, in an open universe, matter or radiation can travel away indefinitely from the "finite" region of space, and so be permanently lost. The inverse process would require matter or radiant energy coming from "infinity," and so would involve a process which is not realizable by physical sources. Einstein's example of an outgoing light wave and Popper's analogous case of a water wave are special finite illustrations of this principle.

It will be noted that Hill and I spoke of there being "at least some aperiodic orbits which extend to infinity" in the classical mechanics of open systems and that we were careful *not* to assert that *every* such allowed process extending to infinity is a *de facto* irreversible one. Instead, we affirmed the existence of a *de facto* irreversibility which is not predicated on Popper's spontaneity proviso by saying : "there always exists a class of allowed elementary processes" that are thus *de facto* irreversible. And, for my part, I conceived of this claim as constituting an extension of Popper's recognition of the essential role of coherence in *de facto* irreversibility to processes of the following kind : processes whose *de facto* irreversibility is not conditional on Popper's finitist requirement of spontaneity because these processes extend to "infinity" in open systems and would hence have inverses in which matter or energy would have to come from *"infinity" coherently* so as to converge upon a point.

I was therefore quite puzzled to find that the communication by Hill and myself prompted the following dissent by Popper (III):

> In this connection, I must express some doubt as to whether the principle proposed by Profs. Hill and Grünbaum is adequate. In formulating their

---

[5] See G. J. Whitrow, *The Natural Philosophy of Time* (London, 1961), pp. 8–10 and 269; also E. Zilsel, *Naturwissenschaften*, vol. 15 (1927), p. 283.

[6] See note 3.

principle, they operate with two ideas : that of the "openness" of a system, and that of a *deus ex machina*. Both seem to me insufficient. For a system consisting of a sun, and a comet coming from infinity and describing a hyperbolic path around the sun, seems to me to satisfy all the criteria stated by them. The system is open; and the reversion of the comet on its track would require a *deus ex machina* for its realization : it would "require matter . . . coming from 'infinity'." Nevertheless, this is an example of just that kind of process which, I take it, we all wish to describe as completely reversible.

Popper's proposed counter-example of the comet coming from "infinity" into the solar system seems to me to fail for the following reasons : (1) neither the actual motion of the comet nor its inverse involve any *coherence*, a feature which I, for my part, had conceived to be essential to the obtaining of non-entropic *de facto* irreversibility in open systems. In my own view, the fact that particles or photons, came from "infinity" in the course of an infinite past does not *per se* require a *deus ex machina*, any more than does their going to "infinity" in the course of an infinite future : in this context, I regard as innocuous the asymmetry involved in the fact that a particle which has come from infinity can be said to have traversed an infinite space by *now*, whereas a particle embarking on an infinite journey will only have traversed a *finite* distance at *any one time* in the future. It is a *coherent* "implosion" from infinity that I believe to require a *deus ex machina*, i.e., to be *de facto* nonexistent, while coherent "explosions" actually do exist. (2) Even ignoring the fact that the motion of Popper's comet does not involve coherence, the issue is *not*, as he seems to think, whether it would require a *deus ex machina* to produce the reversal of any given actual comet in its track; rather, the issue is whether, if no *deus ex machina* would be needed to realize the actual comet motion, a *deus ex machina* would be needed to have another comet execute *instead* a motion inverse to the first one. The answer to this question is an emphatic "no." Unlike the case of outgoing and contracting waves (explosions and implosions), the two comet motions, which are temporal inverses of each other, are *on a par* with respect to the role of a *deus ex machina* in their realization. And even the reversal of the motion of an actual comet at a suitable point in its orbit might in fact be effected by an elastic collision with an oppositely moving other comet of equal mass, and hence would *not* involve, as Popper would have it (III), "a *deus ex machina* who is something like a gigantic tennis player."

It seems to me, therefore, that far from being vulnerable to Popper's proposed counter-example, the existential claim by Hill and myself is fully as viable as Popper's, while having the merit of achieving generality through freedom from Popper's spontaneity proviso. I therefore cannot see any justification at all for the following two assertions which H. Mehlberg made in a very informative recent paper : (1) He states incorrectly that Hill and I have claimed *de facto* irreversibility for "the class of all

conceivable physical processes provided that the latter meet the mild requirement of happening in an 'open' physical system," and (2) he asserts
that "Popper has shown the untenability of the Hill–Grünbaum criterion
by constructing an effective counter example which illustrates the impossibility of their sweeping generalization of his original criterion."[7]

Mehlberg's critical estimate of Popper's own affirmation of non-entropic *de facto* irreversibility likewise seems to me to be unconvincing in
important respects. After asking whether the irreversibility asserted by
Popper is "lawlike or factlike"—a question to which the answer is:
"*avowedly* factlike"—Mehlberg[8] concludes that Popper's temporal asymmetry "seems to be rather interpretable as a local, factlike particularity
of the terrestrial surface than as a universal, lawlike feature . . . which
may be expected to materialize always and everywhere." There are two
points in Mehlberg's conclusion which invite comment: (1) the significance he attaches to the circumstance that the irreversibility of certain
classes of processes is *de facto* or factlike rather than nomological or lawlike, when he assesses the bearing of that irreversibility on the issue of
anisotropy vs. isotropy of time, and (2) the contrast between the epistemological parsimony of Mehlberg's characterization of Popper's irreversibility as a "local . . . particularity of the terrestrial surface" and the inductive boldness of his willingness to affirm a cosmically pervasive nomological isotropy of time on the basis of attributing cosmic relevance, both
spatially and temporally, to the fundamental time-symmetric laws which
have been confirmed in modern man's limited sample of the universe.

As to the first of these two points in Mehlberg's denial of the anisotropy
of time, I note preliminarily that human hopes for an eternal biological
life are no less surely frustrated if all men are indeed *de facto* mortal, i.e.,
mortal on the strength of "boundary conditions" which do obtain permanently, than if man's mortality were assured by some law. By the same
token, I see no escape from the conclusion that if *de facto* irreversibility
does actually obtain everywhere and forever, such irreversibility confers
anisotropy on time. And this anisotropy prevails not one iota less than it
would if its existence were guaranteed by temporally *asymmetrical* fundamentals *laws* of cosmic scope. It is of considerable interest, of course, if
such irreversibility as obtains in nature is *de facto* rather than nomological.
But, in my view, when evaluating the evidence for the anisotropy of time,
Mehlberg commits an error of misplaced emphasis: he wrongly discounts
de facto irreversibility vis-à-vis nomological irreversibility by failing to
show that our warrant for a cosmic extrapolation of time-symmetric *laws*
is actually greater than for a corresponding extrapolation of the factlike

[7] H. Mehlberg, "Physical Laws and Time's Arrow," in: H. Feigl & G. Maxwell
(eds.), *Current Issues in the Philosophy of Science* (New York, 1961), p. 128.
[8] *Ibid.*, p. 126.

conditions making for observed *de facto* irreversibility. For on what grounds can it be maintained that the ubiquitous and permanent existence of the *de facto* probabilities of "boundary conditions" in which Popper rests his affirmation of temporal anisotropy is less well confirmed than that of those laws on whose time-symmetry Mehlberg is willing to base his *denial* of the anisotropy of time? In particular, one wonders how Mehlberg could inductively justify his contention that we are only confirming a "particularity of the terrestrial surface" when we find with Popper (III) that "Only such conditions can be causally realized as can be organized from one centre. . . . causes which are not centrally correlated are causally unrelated, and can co-operate [i.e. produce coherence in the form of *isotropic* contraction of waves to a precise point] only by accident . . . . The probability of such an accident will be zero." If this finding cannot be presumed to hold on all planetlike bodies in the universe, for example, then why are we entitled to assume with Mehlberg that time-symmetric laws of mechanics, for example, are exemplified by the motions of binary stars throughout the universe? Since I see no valid grounds for Mehlberg's double standard of inductive credibility of pervasiveness as between laws and factlike regularities, I consider his negative estimate of Popper's non-entropic *de facto* anisotropy of time as unfounded.

Mehlberg's misplaced emphasis on the significance of lawlike vis-à-vis *de facto* irreversibility likewise seems to me to vitiate the following account which he gives of the import of *de facto* irreversibility in optics, a species of irreversibility which he admits to be of cosmic scope. He writes:

A less speculative example of cosmological irreversibility is provided by the propagation of light *in vacuo,* which several authors have discussed from this point of view. . . . In accordance with Maxwell's theory of light conceived as an electromagnetic phenomenon, they point out that light emitted by a pointlike source, or converging towards a point, can spread on concentric spherical surfaces which either expand or contract monotonically. Yet, independently of Maxwell's theory, the incidence of expanding optical spheres is known to exceed by far the incidence of shrinking spheres. The reason for this statistical superiority of expanding optical spheres is simply the fact that pointlike light-emitting atoms are much more numerous than perfectly spherical, opaque surfaces capable of generating shrinking optical spheres, mainly by the process of reflection. If true, this ratio of the incidences of both types of light waves would provide a cosmological clue to a pervasive irreversibility of a particular class of optical processes.

The bearing of this optical irreversibility upon time's arrow was often discussed. A long time before the asymmetry of expanding and contracting light waves was promoted to the rank of time's arrow, Einstein [A. Einstein, "Über die Entwicklung unserer Anschauungen über die Konstitution und das Wesen der Strahlung," *Physikalische Zeitschrift,* vol. 10 (1910), pp. 817–828] pointed out that the asymmetry of these two types of optical propagation holds only on the undulatory theory of light. Once light is identified instead with a swarm of photons, the asymmetry vanishes. This conclusion holds at least for a spatially finite universe or for optical phenomena confined to a finite spatial region.

Once more, however, the decisive point seems to be that the asymmetry between the two types of light waves depends on factual, initial conditions which prevail in a given momentary cross section of cosmic history or at the "boundaries" of a finite or infinite universe rather than on nomological considerations concerning this history : any other ratio of the incidences of expanding and shrinking light waves would also be in keeping with the relevant laws of nature contained in Maxwell's theory of electromagnetic phenomena. Of course, the aforementioned non-nomological conditions, responsible for the factual ratio of these incidences, are not "local" either, since the whole world is involved — they belong to cosmology. These conditions are nevertheless factlike rather than lawlike, as a comparison with the pertinent laws which can be derived from Maxwell's theory clearly shows.[9]

Contrary to Mehlberg, the decisive point appears *not* to be that "the asymmetry between the two types of light waves depends on factual, initial conditions . . . rather than on nomological considerations." He also asserts that "at least for a spatially finite universe or for optical phenomena confined to a finite spatial region" the corpuscularity of the photon, as conceived by Einstein, invalidates the optical asymmetry which obtains on the undulatory theory of light. I believe, however, that this statement should be amended as follows : The optical asymmetry vanishes, *if at all, only* in a *finite* space. For suppose that one assumes with Einstein that the elementary radiation process is one in which a single emitting particle transfers its energy to only a single absorbing particle. In that case, the fantastically complicated *coherence* needed for the formation of a continuous contracting undulatory spherical shell of light is no longer required. Instead, there is a need for the less complicated coherence among emitting particles located at the walls of a *finite* system and emitting converging photons. But, as Hill and I pointed out,[10] the *de facto* irreversibility of the spatially symmetrical eternal propagation of a pulse of light from a point source into *infinite* space does not depend on whether the light pulse is undulatory instead of being constituted by a swarm of photons.

Neither does *this* irreversibility depend on the acceptance of the steady state theory of cosmology which, in the words of T. Gold,[11] offers the following explanation for the fact that the universe is a non-reflecting sink for radiation :

It is this facility of the universe to soak up any amount of radiation that makes it different from any closed box, and it is just this that enables it to

⁹ *Ibid.*, pp. 123–124.
¹⁰ E. A. Hill and A. Grünbaum, *loc. cit.*

define the arrow of time in any system that is in contact with this sink. But why is it that the universe is a non-reflecting sink for radiation? Different explanations are offered for this in the various cosmological theories and in some schemes, indeed, this would only be a temporary property.[12] In the steady state universe it is entirely attributed to the state of expansion. The red shift operates to diminish the contribution to the radiation field of distant matter; even though the density does not diminish at great distances, the sky is dark because in most directions the material on a line of sight is receding very fast . . .

What Gold appears to have in mind here is that due to a very substantial Doppler shift of the radiation emitted by the receding galaxies, the frequency v becomes very low or goes to zero, and since the energy of that radiation is given by $E = hv$, very little if any is received by us converging from these sources. Gold goes on to say :

> This photon expansion going on around most material is the most striking type of asymmetry, and it appears to give rise to all other time asymmetries that are in evidence. The preferential divergence, rather than convergence, of the world lines of a system ceases when that system has been isolated in a box which prevents the expansion of the photons out into space. Time's arrow is then lost.

We see that Gold's account includes an appreciation of the decisive role played by the infinitude of the space in rendering irreversible the radiation spreading from a point. To be sure, he does emphasize that the Doppler shift due to the expansion makes for the darkness of the sky at night, which would otherwise be lit up by strong radiation. But the crucial point is the following : even if the energy of radiation from receding galaxies were not drastically attenuated by the Doppler shift, such radiation would still *not* be the inverse of a process in which a pulse of photons from a point source forever spreads symmetrically into infinite space from that point source. The inverse of the latter process of outgoing radiation would be a contracting configuration of photons that has been coming from "infinity," i.e., from no sources at all, and has been converging on a point for all infinite past time.

III.   The Statistical Entropy Law and the Anisotropy of Time

Popper has expressed his denial of the relevance of the statistical entropy law to the anisotropy of time as follows (IV):

---

[11] T. Gold, "The Arrow of Time," in *La Structure et L'Évolution de l'Univers*, Proceedings of the 11th Solvay Congress (Brussels, 1958), pp. 86–87.

[12] Presumably Gold is referring here to models of spatially closed or finite universes.

The suggestion has been made (first by Boltzmann himself) that the arrow of time is, either by its very nature, or by definition, connected with the increase in entropy; so that entropy cannot decrease in time because a decrease would mean a reversal of its arrow, and therefore an increase relative to the reversed arrow. Much as I admire the boldness of this idea, I think that it is absurd, especially in view of the undeniable fact that thermodynamic fluctuations do exist. One would have to assert that, within the spatial region of the fluctuation, all clocks run backwards if seen from outside that region.[13] But this assertion would destroy that very system of dynamics on which the statistical theory is founded. (Moreover, most clocks are non-entropic systems, in the sense that their heat production, far from being essential to their function, is inimical to it.)

I do not believe that Boltzmann would have made his suggestion after 1905, when fluctuations, previously considered no more than mathematically calculable near-impossibilities, suddenly became the strongest evidence in favour of the physical reality of molecules. (I am alluding to Einstein's theory of Brownian motion.) As it is, a statistical theory of the arrow of time seems to me unacceptable. The purpose of my first communication was to propose an example of a non-statistical process such that no physicist who finds a film strip of this particular process would need to doubt where the film begins and where it ends : he could determine its arrow of time.

I shall now endeavor to show in detail that *when coupled with specified assumptions as to the boundary conditions,* (1) the statistical behavior of the entropy of physical systems does qualify as a basis for a statistical anisotropy of time, and (2) the modified version of Boltzmann's basic conception to be set forth is *not* beset by any of the absurdities adduced by Popper. We saw that the *non*-entropic irreversibilities countenanced by Popper as a criterion of temporal anisotropy depend on the role played by boundary conditions (in conjunction with the relevant laws). And we shall see that in the context of entropic phenomena, a statistical anisotropy of time is also *not* assured by the laws alone but rather by their conjunction with certain boundary conditions to be specified.

We begin with an account of certain features of the physical world having the character of initial or boundary conditions within the framework of the theory of statistical mechanics. The sought-after basis of the statistical anisotropy of time will then emerge from principles of statistical mechanics relevant to these *de facto* conditions.

---

[13] An attempt to give a *reductio ad absurdum* similar to the one suggested here by Popper has been made by P. W. Bridgman, who writes: ". . . how would one go to work in any concrete case to decide whether time were flowing forward or backward? If it were found that the entropy of the universe were decreasing, would one say that time was flowing backward, or would one say that it was a law of nature that entropy decreases with time?" [P. W. Bridgman, *Reflections of a Physicist*, New York, 1950, p. 165]. For a detailed refutation of Bridgman's version of this attempted *reductio*, see A. Grünbaum, "PPST," pp. 223–232

The universe around us exhibits striking disequilibria of temperatures and other inhomogeneities. In fact, we live by virtue of the nuclear conversion of the sun's reserves of hydrogen into helium, which issues in our reception of solar radiation. As the sun dissipates its reserves of hydrogen via the emission of solar radiation, it may heat a terrestrial rock embedded in snow during the daytime. At night, the rock is no longer exposed to the sun but is left with a considerably higher temperature than the snow surrounding it. Hence, at night, the *warm* rock and the *cold* snow form a quasi-isolated subsystem of either our galactic or solar system. And the relatively low entropy of that subsystem was purchased at the expense of the dissipation of the sun's reserves of hydrogen. Hence, *if* there is some quasi-closed system comprising the sun and the earth, the branching off of our subsystem from this wider system in a state of low entropy at sunset involved an entropy increase in the wider system. During the night, the heat of the rock melts the snow, and thus the entropy of the rock–snow system increases. The next morning at sunrise, the rock–snow subsystem merges again with the wider solar system. Thus, there are subsystems which branch off from the wider solar or galactic system, remain quasi-closed for a limited period of time, and then merge again with the wider system from which they had been separated. Following Reichenbach,[14] we shall use the term "branch system" to designate this kind of subsystem.

Branch systems are formed not only in the natural course of things, but also through human intervention : when an ice cube is placed into a glass of warm ginger-ale by a waiter and then covered for hygienic purposes, a subsystem has been formed. The prior freezing of the ice cube had involved an entropy increase through the dissipation of electrical energy in some larger quasi-closed system of which the electrically run refrigerator is a part. While the ice cube melts in the covered glass subsystem, that quasi-closed system increases its entropy. But it merges again with another system when the then chilled ginger-ale is consumed by a person. Similarly for a cold room that is closed off and then heated by burning logs.

Thus, our environment abounds in branch systems whose initial relatively low entropies are the products of their earlier coupling or interaction with outside agencies of one kind or another. This rather constant and ubiquitous formation of a branch system in a relatively low entropy state resulting from interaction often proceeds at the expense of an entropy increase in some wider quasi-closed system from which it originated. And the *de facto,* nomologically contingent occurrence of these branch systems has the following *fundamental consequence,* at least for our region of the universe and during the current epoch : among the quasi-closed systems whose entropy is relatively low and which behave as if they might remain

---

[14] Cf. H. Reichenbach, *The Direction of Time* (Berkeley, 1956), p. 118.

isolated, the vast majority have not been and will not remain permanently closed systems, being branch systems instead.

Hence, upon encountering a quasi-closed system in a state of fairly low entropy, we know the following to be overwhelmingly probable : the system has *not* been isolated for millions and millions of years and does *not* just *happen* to be in one of the infrequent but ever-recurring low entropy states exhibited by a permanently isolated system. Instead, our system was formed not too long ago by branching off after an interaction with an outside agency. For example, suppose that an American geologist is wandering in an isolated portion of the Sahara desert in search of an oasis and encounters a portion of the sand in the shape of "Coca-Cola." He would then infer that, with overwhelming probability, a kindred person had interacted with the sand in the recent past by tracing "Coca-Cola" in it. The geologist would not suppose that he was in the presence of one of those relatively low entropy configurations which are assumed by the sand particles spontaneously but very rarely, if beaten about by winds for millions upon millions of years in a state of effective isolation from the remainder of the world.

There is a further *de facto* property of branch systems that concerns us. For it will turn out to enter into the temporally asymmetrical statistical regularities which we shall find to be exhibited in the entropic behavior of these systems. This property consists in the following *randomness* obtaining *as a matter of nomologically contingent fact* in the distribution of the W_1 micro-states belonging to the initial macro-states of a *space*-ensemble of branch systems each of which has the same initial entropy $S_1 = k \log W_1$: For each class of *like* branch systems having the *same* initial entropy value S_1, the micro-states constituting the identical initial macro-states of entropy S_1 are *random samples* of the set of all W_1 micro-states yielding a macro-state of entropy S_1.[15] This attribute of randomness of micro-states on the part of the initial states of the members of the space-ensemble will be recognized as the counterpart of the following attribute of the micro-states of one single, permanently closed system : there is equi-probability of occurrence among the W_1 micro-states belonging to the *time*-ensemble of states of equal entropy $S_1 = k \log W_1$ exhibited by one single, permanently closed system.

We can now state the statistical regularities which obtain as a consequence of the *de facto* properties of branch systems just set forth, when coupled with the principles of statistical mechanics. These regularities, which will be seen to yield a temporally asymmetric statistical behavior of the entropy of branch systems, fall into two main groups as follows.

Group 1. In most space-ensembles of quasi-closed branch systems each of which is initially in a state of non-equilibrium or relatively *low* entropy,

---

[15] Cf. R. C. Tolman, *The Principles of Statistical Mechanics* (Oxford, 1938), p. 149.

the majority of branch systems in the ensemble will have *higher* entropies *after* a given time t.[16] But these branch systems simply did not exist as quasi-closed, distinct systems at a time t prior to the occurrence of their initial branching off. Hence, not existing then as branch systems, they did *not* also exhibit the same higher entropy states at the *earlier* times t, which they would indeed have done had they existed as closed systems all along.

The increase after a time t in the entropy of the overwhelming majority of branch systems of initially low entropy—as confirmed abundantly by observation—can be made fully intelligible. To do so, we note the following property of the *time*-ensemble of entropy values belonging to a single, permanently closed system and then affirm that property of the space-ensembles of branch systems : since *large* entropic downgrades or decreases are *far less* probable (frequent) than moderate ones, the *vast majority* of *non*-equilibrium entropy states of a permanently closed system are located either at, or in the immediate temporal vicinity of, the *bottom* of a *dip* of the one-system entropy curve. In short, the vast majority of the *sub*-maximum entropy states are on or temporally very near the *upgrades* of the one-system curve. The application of this result to the space-ensemble of branch systems whose initial states exhibit the aforementioned *de facto* property of randomness then yields the following : Among the initial low entropy states of these systems, the vast majority lie at, or in the immediate temporal vicinity of, the bottoms of the one-system entropy curve at which an upgrade begins.

Group 2. A decisive *temporal asymmetry* in the statistics of the temporal evolution of branch systems arises from the further result that in most space ensembles of branch systems each of which is initially in a state of *equilibrium* or very *high* entropy, the vast majority of the systems will *not* have *lower* entropies after a finite time t, but will still be in equilibrium.[17] For the aforementioned randomness property assures that the vast majority of those branch systems whose initial states are equilibrium states have maximum entropy values lying somewhere *well within* the plateau of the one-system entropy curve, rather than at the extremity of the plateau at which an entropy *decrease* is initiated.[18]

---

[16] Cf. R. Fürth, "Prinzipien der Statistik," *Handbuch der Physik*, vol. 4 (1929), pp. 270 and 192–193.

[17] *Ibid.*, p. 270.

[18] Although the decisive asymmetry just noted was admitted by H. Mehlberg (*op. cit.*, p. 129), he dismisses it as expressing "merely the factual difference between the two relevant values of probability." But here again an asymmetry is no less an asymmetry for depending on *de facto*, nomologically contingent boundary conditions rather than being assured by a *law* alone. Since our verification of laws generally has the same partial and indirect character as that of our confirmation of the existence of complicated *de facto* boundary conditions, the assertion of an asymmetry depending on *de facto* conditions is generally no less reliable than one wholly grounded on a law. Hence when Mehlberg (p. 117, n. 30) urges against Schrödinger's claim of entropic

We see therefore that in the vast majority of branch systems, either one end of their finite entropy curve is a point of low entropy and the other a point of high entropy, or they are in equilibrium states at both ends as well as during the intervening interval. And it is likewise apparent that the statistical distribution of these entropy values on the time axis is such that the vast majority of branch systems have the *same direction of entropy increase* and hence also the same opposite direction of entropy decrease. Thus, the statistics of entropy increase among branch systems assures that in most space ensembles the vast majority of branch systems will increase their entropy in one of the two opposite time directions and decrease it in the other. In this way the entropic behavior of branch systems confers the same statistical anisotropy on the vast majority of all those epochs of time during which the universe exhibits the requisite disequilibrium and contains branch systems satisfying initial conditions of "randomness."[19]

Let us now call the direction of entropy increase of a typical representative of these epochs the direction of "later," as indeed we have done from the outset by the mere assignment of higher time numbers in that direction but without prejudice to our findings concerning the issue of the anisotropy of time. Then our results pertaining to the entropic behavior of branch systems show that the directions of "earlier than" and "later than" are not merely opposite directions bearing decreasing and increasing time coordinates respectively, but are statistically *anisotropic* in an objective physical sense. For I have shown elsewhere in some detail[20] that increasing real numbers can be assigned as time coordinates in a physically meaningful way *without* any commitment to the existence of (*de facto* or nomologically) irreversible kinds of processes. In fact, the use of the real number continuum as a basis for coordinatizing time no more entails the anisotropy of time than the corresponding coordinatization of one of the three dimensions of space entails the anisotropy of that dimension.

It should be noted that I have characterized the positive direction of

---

asymmetry that for every pair of branch systems which change their entropy in one direction, "there is nothing to prevent" another pair of closed subsystems from changing their entropy in the opposite direction, the reply is: Mehlberg's criticism can be upheld only by gratuitously neglecting the statistical asymmetry admitted but then dismissed by him as "merely" factual. For it is the existence of the specified boundary conditions which statistically prevents the existence of entropic time-symmetry in this context.

[19] Readers familiar with Reichenbach's "hypothesis of the branch structure" as set forth in his *The Direction of Time* (p. 136) will note that though I am heavily indebted to Reichenbach, my treatment of the assumptions regarding branch systems departs from his in several *essential* respects. A statement and justification of these departures is given in A. Grünbaum, "PPST," pp. 261–263.

[20] A. Grünbaum, "PPST," pp. 209–216.

time as the direction of entropy increase for a *typical representative* of all those epochs of time during which the universe exhibits the requisite disequilibrium and contains branch systems satisfying initial conditions of "randomness." Accordingly, it is entirely possible to base the customary temporal description of fluctuation phenomena on entropic counter-directedness with respect to the majority of branch systems, a description which is *not* liable to *any* of the absurdities which led Popper to conclude that a statistical (entropic) account of the anisotropy of time is unacceptable.

CHAPTER 22

The Theory of Complex Phenomena

BY F. A. HAYEK

I. Pattern-Recognition and Pattern-Prediction

MAN HAS BEEN IMPELLED to scientific inquiry by wonder and by need. Of these, wonder has been incomparably more fertile. There are good reasons for this. Where we wonder we have already a question to ask. But however urgently we may want to find our way in what appears just chaotic, so long as we do not know what to look for, even the most attentive and persistent observation of the bare facts is not likely to make them more intelligible. Intimate acquaintance with the facts is certainly important; but even systematic observation can start only after questions have arisen. Until we have definite questions to ask we cannot employ our intellect; and questions presuppose that we have formed some provisional hypothesis or theory about the events.[1]

---

[1] See Adam Smith, "The Principles which Lead and Direct Philosophical Inquiries, as Illustrated by the History of Astronomy," in *Essays* (London, 1869), p. 340: "Wonder, therefore, and not any expectation of advantage from its discoveries, is the first principle which prompts mankind to the study of philosophy, of that science which pretends to lay open the concealed connections that unite the various appearances of nature; and they pursue this study for its own sake, as an original pleasure or good in itself, without regarding its tendency to procure them the means of many other pleasures." Is there really any evidence for the now popular contrary view that, e.g., "hunger in the Nile Valley led to the development of geometry" (as Gardner Murphy tells us in the *Handbook of Social Psychology*, ed. by Gardner Linzey, 1954, vol. II, p. 616)? Surely the fact that the discovery of geometry turned out to be useful does not prove that it was developed because of its usefulness.

Questions will arise at first only after our senses have discerned some recurring pattern or order in the events. It is a recognition of some regularity (or recurring pattern, or order), of some similar feature in otherwise different circumstances, which makes us wonder and ask "why?"[2] Our minds are so made that when we notice such regularity in diversity we suspect the presence of the same agent and become curious to detect it. It is to this trait of our minds that we owe whatever understanding and mastery of our environment we have achieved.

Many such regularities of nature are recognized "intuitively" by our senses. We see and hear patterns as much as individual events without having to resort to intellectual operations. In many instances these patterns are of course so much part of the environment we take for granted that they do not cause questions. But where our senses show us new patterns, this causes surprise and questioning. To such curiosity we owe the beginning of science.

Marvellous, however, as the intuitive capacity of our senses for pattern recognition is, it is still limited.[3] Only certain kinds of regular arrangements (not necessarily the simplest) obtrude themselves on our senses. Many of the patterns of nature we can discover only *after* they have been constructed by our mind. The systematic construction of such new patterns is the business of mathematics.[4] The role geometry plays in this respect with regard to visual patterns is merely the most familiar instance. The great strength of mathematics is that it enables us to describe abstract patterns which cannot be perceived by our senses, and to state the common properties of hierarchies of classes of patterns of a highly abstract character. Every algebraic equation or set of such equations defines in this sense

---

[2] See K. R. Popper, *The Poverty of Historicism* (London, 1957), p. 121: "Science . . . cannot start with observations, or with the 'collection of data,' as some students of method believe. Before we can collect data, our interest in *data of a certain kind* must be aroused: the *problem* always comes first." Also in his *The Logic of Scientific Discovery*, London, 1959, p. 59: ". . . observation is always *observation in the light of theories.*"

[3] Although in some respects the capacity of our senses for pattern recognition clearly also exceeds the capacity of our mind for specifying these patterns, the question of the extent to which this capacity of our senses is the result of another kind of (pre-sensory) experience is another matter. See on this, and on the general point that all perception involves a theory or hypothesis, my book *The Sensory Order* (London and Chicago, 1952), esp. para. 6.37. *Cf.* also the remarkable insight expressed by Adam Ferguson (probably derived from George Berkeley) in *The History of Civil Society*, London, 1766, p: 39, that "the inferences of thought are sometimes not to be distinguished from the perception of sense"; as well as H. von Helmholtz's theory of the "unconscious inferences" involved in most perceptions. For a recent revival of these ideas see N. R. Hanson, *Patterns of Discovery* (Cambridge University Press, 1958), esp. p. 19, and the views on the role of "hypotheses" in perception as developed in recent "cognition theory" by J. S. Bruner, L. Postman, and others.

[4] See G. H. Hardy, *A Mathematician's Apology* (Cambridge University Press, 1941), p. 24: "A mathematician, like a painter or poet, is a maker of patterns."

a class of patterns, with the individual manifestation of this kind of pattern being particularized as we substitute definite values for the variables.

It is probably the capacity of our senses spontaneously to recognize certain kinds of patterns that has led to the erroneous belief that if we look only long enough, or at a sufficient number of instances of natural events, a pattern will always reveal itself. That this often is so means merely that in those cases the theorizing has already been done by our senses. Where, however, we have to deal with cases in which there has been no biological reason to develop a pattern, we shall have to invent the pattern before we can discover its presence in the phenomena—or before we shall be able to test its applicability to what we observe. A theory will always define only a kind or class of patterns, and the particular manifestation of the pattern to be expected will depend on the particular circumstances (the "initial and marginal conditions" to which, for the purposes of this article, we shall refer as "data"). How much in fact we shall be able to predict will depend on how many of those data we can ascertain.

It is usual to regard the description of the pattern which the theory provides merely as a tool which will enable us to predict the particular manifestations of the pattern that will appear in specific circumstances. But the prediction that in certain general conditions a pattern of a certain kind will appear is also a significant (and falsifiable) prediction. If I tell somebody that if he goes to my study he will find there a rug with a pattern made up of diamonds and meanders, he will have no difficulty in deciding "whether that prediction was verified or falsified by the result,"[5] even though I have said nothing about the arrangement, size, color, etc., of the elements from which the pattern of the rug is formed.

The distinction between a prediction of the appearance of a pattern of a certain class and a prediction of the appearance of a particular instance of this class is sometimes important also in the physical sciences. The mineralogist who states that the crystals of a certain mineral are hexagonal, or the astronomer who knows that the course of a celestial body in the field of gravity of another will correspond to one of the conic sections, make significant predictions which can be refuted. But in general the physical sciences tend to assume that it will always be possible to specify their predictions to any degree desired.[6] The distinction asssumes, however, much greater importance when we turn from the relatively simple phenomena with which the natural sciences deal to the more complex phenomena of life, mind, and society, where such specification may not always be possible.

---

[5] Charles Dickens, *David Copperfield*, p. 1.

[6] It may be permissible to doubt whether it is in fact possible to predict, for example, the precise pattern which the vibrations of an airplane will at a particular moment produce in the standing waves on the surface of the coffee in my cup.

II. Degrees of Complexity

The distinction between simplicity and complexity raises considerable philosophical difficulties when applied to statements. But there seems to exist a fairly easy and adequate way to measure the degree of complexity of different kinds of abstract patterns. The minimum number of elements of which an instance of the pattern must consist in order to exhibit all the characteristic attributes of the class of patterns in question appears to provide a quite unambiguous criterion.[7]

It has occasionally been questioned whether the phenomena of life, of mind, and of society are really more complex than those of the physical world.[8] This seems to be largely due to a confusion between the degree of complexity characteristic of a peculiar *kind* of phenomenon and the degree of complexity to which, by a combination of elements, any kind of phenomenon can be built up. Of course, in this manner physical phenomena may achieve any degree of complexity. Yet when we consider the question from the angle of the minimum number of distinct variables a

---

[7] See Michael Scriven, "A Possible Distinction between Traditional Scientific Disciplines and the Study of Human Behavior," *Minnesota Studies in the Philosophy of Science*, I (1956) 332: "The difference between the scientific study of behavior and that of physical phenomena is thus partly due to the relatively greater complexity of *the simplest phenomena we are concerned to account for* in a behavioral theory."

[8] Ernest Nagel, *The Structure of Science* (New York, 1961), p. 505: ". . . though social phenomena may indeed be complex, it is by no means certain that they are in general more complex than physical and biological phenomena." See, however, Johann von Neumann, "The General and Logical Theory of Automata," *Cerebral Mechanism in Behavior*, The Hixon Symposium (New York, 1951), p. 24: ". . . we are dealing here with parts of logic with which we have practically no experience. The order of complexity is out of all proportion to anything we have ever known." It may be useful to give here a few illustrations of the orders of magnitude with which biology and neurology have to deal. While the total number of electrons in the Universe has been estimated at 10^{79} and the number of electrons and protons at 10^{100}, there are in chromosomes with 1000 locations [genes] with 10 allelomorphs 10^{1000} possible combinations; and the number of possible proteins is estimated at 10^{2700}. [L. von Bertalanffy, *Problems of Life* (New York, 1952), p. 103]. C. Judson Herrick [*Brains of Rats and Men* (New York, 1926)], suggests that ". . . during a few minutes of intense cortical activity the number of interneuronic connections actually made (counting also those that are actuated more than once in different associational patterns) may well be as great as the total number of atoms in the solar system" (i.e., 10^{56}); and Ralph W. Gerard (*Scientific American*, Sept., 1953, p. 118) has estimated that in the course of seventy years a man may accumulate 15 trillion units of information ("bits") which is more than 1000 times larger than the number of nerve cells. The further complication which social relations superimpose upon this are, of course, relatively insignificant. But the point is that, if we wanted to "reduce" social phenomena to physical events, they would constitute an additional complication superimposed upon that of the physiological processes determining mental events.

formula or model must possess in order to reproduce the characteristic patterns of structures of different fields (or to exhibit the general laws which these structures obey), the increasing complexity as we proceed from the inanimate to the ("more highly organized") animate and social phenomena becomes fairly obvious.

It is, indeed, surprising how simple, in these terms—i.e., in terms of the number of distinct variables—appear all the laws of physics, and particularly of mechanics, when we look through a collection of formulae expressing them. On the other hand, even such relatively simple constituents of biological phenomena as feedback or cybernetic systems, in which a certain combination of physical structures produces an over-all structure possessing characteristic properties, requires for its description something much more elaborate than anything describing the general laws of mechanics. In fact, when we ask ourselves by what criteria we single out certain phenomena as "mechanical" or "physical" we shall probably find that the answer is that these laws are simple in the sense defined. Nonphysical phenomena are more complex because we call physical what can be described by relatively simple formulae.

The "emergence" of "new" patterns, as a result of the increase in the number of elements between which simple relations exist, means that this larger structure as a whole will possess certain general or abstract features which will recur independently of the particular values of the individual data, so long as the general structure (as described, for example, by an algebraic equation) is preserved.[9] Such "wholes," defined in terms of certain general properties of their structure, will constitute distinctive objects of explanation by a theory, even though such a theory may be merely a particular way of fitting together statements about the relations between the individual elements.

It is somewhat misleading to approach this task mainly from the angle of whether such structures are "open" or "closed" systems. There are, strictly speaking, no closed systems within the universe. All we can ask is whether in the particular instance the points of contact through which the rest of the universe can act upon the system we try to single out (and which for the theory become the data) are few or many. These data, or variables, which determine the particular form the pattern described by the theory will assume in the given circumstances, will be more numerous in the case

---

[9] Lloyd Morgan's conception of emergence derives, via G. H. Lewes [*Problems of Life and Mind*, 1st series, vol. II., problem V., Chap. III, section headed "Resultants and Emergents," (American ed.; Boston, 1891), p. 368] from John Stuart Mill's distinction of the "heteropathic" laws of chemistry and other complex phenomena from the ordinary "composition of causes" in mechanics, etc. See his *System of Logic* (London, 1843), Book III, chap. 6 (in vol. I, p. 431 of the first edition), and C. Lloyd Morgan, *The Emergence of Novelty* (London, 1933), p. 12.

of complex wholes, and much more difficult to ascertain and control than in the case of simple phenomena.

What we single out as wholes, or where we draw the "partition boundary,"[10] will be determined by whether we can thus isolate recurrent patterns of coherent structures of a distinct kind which we do in fact encounter. Many complex patterns which are conceivable and might recur we shall not find it worth while to construct. Whether it will be useful to elaborate and study a pattern of a particular kind will depend on whether the structure it describes is persistent or merely accidental. The coherent structures in which we are mainly interested are those in which a complex pattern has produced properties which make the structures self-maintaining.

III.   Pattern Prediction with Incomplete Data

The multiplicity of even the minimum of distinct elements required to produce (and therefore also of the minimum number of data required to explain) a complex phenomenon of a particular kind creates problems which dominate the disciplines concerned with these phenomena, and makes such disciplines appear very different from those concerned with simpler phenomena. The chief problem becomes one of actually ascertaining all the data determining a particular manifestation of the phenomenon in question, a problem with obstacles often difficult to surmount, and sometimes even absolute.[11] Those mainly concerned with simple phenomena are often inclined to think that where this is the case a theory is useless, and that scientific procedure demands that we find a theory of sufficient simplicity to permit deriving from it predictions of particular events. To them the theory, the knowledge of the pattern, is merely a tool whose usefulness depends entirely on its capacity of translation into a representation of the circumstances producing a particular event. Of the theories of simple phenomena this is largely true.[12]

There is, however, no justification for the belief that it must always be possible to discover such simple regularities, and that physics is more advanced because it has succeeded in doing this while other sciences have not yet done so. It is rather the other way around; physics has succeeded be-

---

[10] Lewis White Beck, "The 'Natural Science Ideal' in the Social Sciences," *The Scientific Monthly*, LXVIII (June 1949), 388.

[11] See F. A. Hayek, *The Sensory Order*, para. 8.66–8.86.

[12] See Ernest Nagel, "Problems of Concept and Theory Formation in the Social Sciences," in *Science, Language and Human Rights* (American Philosophical Association, Eastern Division, vol I) (University of Pennsylvania Press, 1952), p. 62: "In many cases we are ignorant of the appropriate initial and boundary conditions, and cannot make precise forecasts even though available theory is adequate for that purpose."

cause it deals with phenomena which, in our sense, are simple. But a *simple* theory of phenomena which are in their nature complex (or, if this expression is preferred, dealing with highly organized phenomena) is probably false merely of necessity—at least, without a specified *ceteris paribus* assumption, after the full statement of which the theory would no longer be simple.

We are, however, not interested only in individual events, and it is also not only predictions of individual events which can be empirically tested. We are equally interested in the recurrence of abstract patterns as such, and the prediction that a pattern of a certain kind will appear in defined circumstances is a falsifiable and therefore empirical prediction. Such knowledge of the conditions in which a pattern of a certain kind will appear, and of what depends on its preservation, may be of great practical importance. The circumstances or conditions in which the pattern described by the theory will appear are defined by the range of values which may be inserted for the variables of the formula. All we need to know in order to make such a theory applicable to a situation is, therefore, that the data possess certain general properties (or belong to the class defined by the scope of the variables). Beyond this we need to know nothing about their individual attributes so long as we are satisfied to know merely the sort of pattern that will appear and not its particular manifestation.

Such a theory, destined to remain "algebraic,"[13] because we are in fact unable to substitute particular values for the variables, ceases then to be a mere tool and becomes the final result of our theoretical efforts. Such a theory will, of course, in Popper's terms,[14] be one of small empirical content because it enables us to predict or explain only certain general features of a situation which may be compatible with a great many particular circumstances. It will perhaps enable us to make only what M. Scriven has called "hypothetical predictions,"[15] i.e., predictions dependent on yet unknown future events, and in consequence the range of phenomena compatible with it will be wide and the possibility of falsifying it correspondingly small. As in many fields, this will be for the present, or perhaps forever, all we can achieve by way of theoretical knowledge, but it will nevertheless extend the range of the possible advance of scientific knowledge.

The advance of science will thus have to proceed in two different directions. While it is certainly desirable to make our theories as falsifiable as possible, we must also push forward into fields where as we advance, the degree of falsifiability necessarily decreases. This is the price we have to pay for an advance into the field of complex phenomena.

---

[13] The useful term "algebraic theories" was suggested to me by J. W. N. Watkins.

[14] K. R. Popper, *Logic* . . . , p. 113; see also pp. 127–145.

[15] M. Scriven, "Explanation and Prediction in Evolutionary Theory," *Science*, August 28, 1959, p. 478.

IV.   Statistics Impotent to Deal with Pattern Complexity

Before we further illustrate the use of those mere "explanations of the principle"[16] provided by "algebraic" theories which describe merely the general character of higher level generalities, and before we consider the important conclusions which follow from the insight into the boundaries of possible knowledge which our distinction provides, it is necessary briefly to turn aside and consider the method which is often but errone-ously believed to give us access to the understanding of complex phe-nomena : statistics. Because statistics is designed to deal with large numbers it is often thought that the difficulty arising from the large number of elements of which complex structures consist can be overcome by recourse to statistical techniques.

Statistics, however, deals with the problem of large numbers essentially by eliminating complexity in the sense of a multiplicity of relations be-tween individual elements and deliberately treating the individual ele-ments which it counts as if they were not systematically connected. It avoids the problem of complexity by substituting for the information on the individual elements information on the frequency with which their different properties occur in classes of such elements, and it deliberately disregards the fact that the relative position of the elements in a structure may matter. In other words, it proceeds on the assumption that informa-tion on the numerical frequencies of the different elements of a collective is enough to explain the phenomena and that no information is required on the manner in which the elements are related. The statistical method is therefore of use only where we either deliberately ignore or are ignorant of the relations between the individual elements with different attri-butes, i.e., where we ignore or are ignorant of any structure into which they are organized. Statistics thereby enables us to regain simplicity and makes the task manageable by substituting a single attribute for the unascertainable individual attributes in the collective. It is for this reason irrelevant to the solution of problems in which it is precisely the relations between individual elements with different attributes which matter.

Statistics may assist us where we have information about many com-plex structures of the same kind, that is, where the complex phenomena and not the elements of which they consist are the elements of the statisti-cal collective. It may provide us, for example, with information on the relative frequency with which particular properties of the complex struc-tures—say, of the members of a species of organisms—occur together; but

---

[16] See F. A. Hayek, "Degrees of Explanation," *The British Journal for the Philosophy of Science*, VI, No. 23 (1955).

it presupposes that we have an independent criterion for identifying structures of the kind in question. Where we have such statistics about the properties of many individuals belonging to a class of animals, or languages, or economic systems, this may indeed be scientifically significant information.[17]

How little statistics can contribute, however, even in such cases to the explanation of complex phenomena is clearly seen if we imagine that computers were natural objects which we found in sufficiently large numbers and whose behavior we wanted to predict. It is clear that we should never succeed in this unless we possessed the mathematical knowledge built into the computers, that is, unless we knew the theory determining their structure. No amount of statistical information on the correlation between input and output would get us any nearer to our aim. Yet the efforts which are currently made on a large scale with regard to the much more complex structures which we call organisms are of the same kind. The belief that it must be possible in this manner to discover by observation regularities in the relations between input and output without the possession of an appropriate theory is in this case even more futile and naive than it would be in the case of computers.[18]

While statistics can successfully deal with complex phenomena where these are the elements of the population on which we have information, it can tell us nothing about the structure of these elements. It treats them, in the fashionable phrase, as "black boxes" which are presumed to be of the same kind but about whose identifying characteristics it has nothing to say. Nobody would probably seriously contend that statistics can elucidate the complex structures of organic molecules, and few would argue that it can help us to explain the functioning of organisms. Yet when it comes to accounting for the functioning of social structures, that belief is widely held. It is largely the product of a misconception about what the aim of a theory of social phenomena is, which is, however, another story.

V. The Theory of Evolution as an Instance of Pattern Prediction

Probably the best illustration of a theory of complex phenomena of great value though it describes merely a general pattern, whose detail we can never fill in, is the Darwinian theory of evolution by natural selection. It is significant that this theory has always been something of a stumbling block for the dominant conception of scientific method. It certainly does

---

[17] See F. A. Hayek, *The Counter-Revolution of Science* (Glencoe, Ill., 1952), pp. 60–63.

[18] Cf. J. G. Taylor, "Experimental Design: A Cloak for Intellectual Sterility," *The British Journal of Psychology*, 49 (1958), esp. pp. 107–108.

not fit the orthodox criteria of "prediction and control" as the hallmarks of scientific method.[19] Yet it cannot be denied that it has become the successful foundation of the whole of modern biology.

Before we examine the character of this theory we must clear up a widely held misconception as to its content. It is often represented as if it consisted of an assertion about the succession of particular species of organisms which gradually changed into each other. This is not the theory but an application of the theory to the particular event which took place on earth during the last two billion years or so.[20] Most of the misapplication of evolutionary theory (particularly in anthropology and the other social sciences) and the various abuses of it (e.g., in ethics) are due to this erroneous interpretation of its content.

The theory of evolution by natural selection describes a kind of process or mechanism which is independent of the particular circumstances in which it has taken place on Earth, which is equally applicable to a course of events in very different circumstances, and which might result in the production of an entirely different set of organisms. The basic conception of the theory is exceedingly simple and it is only in its application to the concrete circumstances that its extraordinary fertility and the range of phenomena for which it can account manifests itself.[21] The basic proposition which has this far-reaching implications is that a mechanism of reduplication, with transmittable variations and competitive selection of those which prove to have a better chance to survive, will in the course of time produce a great variety of structures adapted to continuous adjustment to the environment and to each other. The validity of this general proposition is not dependent on the truth of the particular applications

---

[19] See e.g., Stephen Toulmin, *Foresight and Prediction* (London, 1961), p. 24: "No scientist has ever used this theory to foretell the coming into existence of creatures of a novel species, still less verified his forecast."

[20] Even Professor Popper seems to imply this interpretation when he writes (*Poverty*, . . ., p. 107) that "the evolutionary hypothesis is not a universal law of nature but a particular (or, more precisely, singular) historical statement about the ancestry of a number of terrestrial plants and animals." If this means that the essence of the theory of evolution is the assertion that particular species had common ancestors, or that the similarity of structure always means a common ancestry (which was the hypothesis from which the theory of evolution was derived), this is emphatically not the main content of the present theory of evolution. There is, incidentally, some contradiction between Popper's treatment of the concept of "mammals" as a universal (*Logic* . . ., p. 65) and the denial that the evolutionary hypothesis describes a universal law of nature. The same process might well have produced mammals on other planets.

[21] Charles Darwin himself well knew, as he once wrote to Lyell, that "all the labour consists in the application of the theory" [quoted by C. C. Gillispie, *The Edge of Objectivity* (Princeton, 1960), p. 314]. See also his statement in *The Origin of Species* (New York, 1927), p. 363, that "if we must marvel, let it be at our own presumption in imagining for a moment that we understand the many complex contingencies on which the existence of each species depends."

which were first made of it : if, for example, it should have turned out that, in spite of their structural similarity, man and ape were not joint descendants from a comparatively near ancestor but the result of two convergent strands starting from ancestors which differed much more from each other (such as is true of the externally very similar types of marsupial and placental carnivores), this would not have refuted Darwin's general theory of evolution but only the manner of its application to the particular case.

The theory as such, as is true of all theories, describes merely a range of possibilities. In doing this it excludes other conceivable courses of events and thus can be falsified. Its empirical content, in Popper's terms, consists in what it forbids.[22] If a sequence of events should be observed which cannot be fitted into its pattern, such as, for example, that horses suddenly began to give birth to young with wings, or that the cutting-off of a hind-paw in successive generations of dogs should result in dogs being born without that hind-paw, we should regard the theory as refuted.[23]

The range of what is permitted by the theory is undeniably wide. Yet one could also argue that it is only the limitation of our imagination which prevents us from being more aware of how much greater is the range of the prohibited—how infinite is the variety of conceivable forms of organisms which, thanks to the theory of evolution, we know will not in the foreseeable future appear on Earth. Common sense may have told us before not to expect anything widely different from what we already knew. But exactly what kinds of variations are within the range of possibility and what kinds are not, only the theory of evolution can tell us. Though we could never write down an exhaustive list of the possibilities, any specific question we shall, in principle, be able to answer.

For our present purposes we may disregard the fact that in one respect the theory of evolution is still incomplete because we still know only little about the mechanism of mutation. But let us assume that we knew precisely the circumstances in which (or at least the probability that in given conditions) a particular mutation will appear, and that we similarly knew also the precise advantages which any such mutation would in any particular kind of environment confer upon an individual of a specific constitution. This would not enable us either to explain why the existing species or organisms have the particular structures which they possess, or to predict what new forms will spring from them.

The reason for this is the actual impossibility of ascertaining the particular circumstances which, in the course of two billion years, have decided the emergence of the existing forms, or even those which during the next few hundred years will determine the selection of the types which

----

[22] *Logic* . . ., p. 41.

[23] Cf. Morton Beckner, *The Biological Way of Thought* (Columbia University Press, 1954), p. 241.

will survive. Even if we tried to apply our explanatory scheme to a single species consisting of a known number of individuals each of which we were able to observe, and assuming that we were able to ascertain and record every single relevant fact, their sheer number would be such that we should never be able to manipulate them, i.e., to insert these data into the appropriate blanks of our theoretical formula and then to solve the "statement equations" thus determined.[24]

What we have said about the theory of evolution applies to most of the rest of biology. The theoretical understanding of the growth and functioning of organisms can only in the rarest of instances be turned into specific predictions of what will happen in a particular case, because we can rarely ascertain all the facts which will contribute to determine the outcome. Hence, "prediction and control, usually regarded as essential criteria of science, are less reliable in biology."[25] It deals with pattern-building forces, the knowledge of which is useful for creating conditions favorable to the production of certain kinds of results, while it will in only comparatively few cases be possible to control all the relevant circumstances.

VI.   Theories of Social Structures

It should not be difficult now to recognize the similar limitations applying to theoretical explanations of the phenomena of mind and society. One of the chief results so far achieved by theoretical work in these fields seems to me, indeed, to be a demonstration that here individual events regularly depend on so many concrete circumstances that we shall never be in a position to ascertain them all; that consequently not only must the ideal of prediction and control remain largely beyond our reach but also the hope remain illusory that we can discover by observation regular connections between the individual events. The very insights which theory provides—for example, that almost any event in the course of a man's life may have some effect on almost any of his future actions—make it impossible to translate our theoretical knowledge into specific predictions of particular events. There is no justification for the dogmatic belief that such translation must be possible if a science of these subjects is to be achieved, and what workers in these subjects have merely not yet succeeded in what physics has done, namely, discovering simple relations between a few observables. If the theory which we have yet achieved tells us anything it is that no such simple regularities are to be expected.

I will not consider here the fact that in the case of mind attempting to

---

[24] K. R. Popper, *Logic* . . . , p. 73.

[25] Ralph S. Lillie, "Some Aspects of Theoretical Biology," *Philosophy of Science*, XV, No. 2 (1948), p. 119.

explain the detailed working of another mind of the same order of complexity there seems to exist, besides the "practical" yet nevertheless unsurmountable obstacles, also an absolute impossibility, because the conception of a mind fully explaining itself involves a logical contradiction. This I have discussed elsewhere.[26] It is not relevant here because the practical limits to the possibility of ascertaining all the relevant data lie so far inside the logical limits that the latter have little relevance to what we can do.

In the field of social phenomena only economics and linguistics[27] seem to have succeeded in building up a coherent body of theory. I shall confine myself here to illustrating the general thesis with reference to economic theory, though most of what I have to say would appear to apply equally to linguistic theory.

Schumpeter well described the task of economic theory when he wrote that "the economic life of a non-socialist society consists of millions of relations or flows between individual firms and households. We can establish certain theorems about them, but we can never observe them all."[28] To this must be added that most of the phenomena in which we are interested, such as competition, could not occur at all unless the number of distinct elements involved were fairly large, and that the over-all pattern that will form itself is determined by the significantly different behavior of the different individuals, so that the obstacle of obtaining the relevant data cannot be overcome by treating them as members of a statistical collective.

For this reason economic theory is confined to describing kinds of patterns which will appear if certain general conditions are satisfied; it can rarely if ever derive from this knowledge any predictions of specific phenomena. This is seen most clearly if we consider those systems of simultaneous equations which since Léon Walras have been widely used to represent the general relations between the prices and the quantities of all commodities produced and sold. They are so framed that *if* we were able to fill in all the blanks, i.e., *if* we knew all the parameters of these equations, we could calculate the prices and quantities of all the commodities. But, as at least the founders of this theory clearly understood, its purpose is not "to arrive at a numerical calculation of prices," because it

---

[26] See *The Sensory Order*, para. 8.66–8.86.

[27] See particularly Noam Chomsky, *Syntactic Structures* (s' Gravenhage, 1957), in which he characteristically seems to succeed in building up such a theory, after frankly abandoning the striving after an inductivist "discovery procedure" and substituting for it the search after an "evaluation procedure" which enables him to eliminate false theories or grammars and where these grammars may be arrived at "by intuition, guess-work, all sorts of partial methodological hints, reliance on past experience, etc." (p. 56).

[28] J. A. Schumpeter, *History of Economic Analysis* (Oxford University Press, 1954), p. 241.

would be "absurd" to assume that we can ascertain all the data.[29] Prediction of the formation of this general kind of pattern rests on certain very general factual assumptions (such as that most people engage in trade in order to earn an income, that they prefer a larger income to a smaller one, that they are not prevented from entering whatever trade they wish, etc.) —assumptions which determine the scope of the variables but not their particular values; it is, however, not dependent on the knowledge of the more particular circumstances which we would have to know in order to be able to predict prices or quantities of particular commodities. No economist has yet succeeded in making a fortune by buying or selling commodities on the basis of his scientific prediction of future prices. (Even though some may have done so by selling such predictions!)

To the physicist it often seems puzzling why the economist should bother to formulate such equations, since he admittedly sees no chance of determining the numerical values of the parameters which would enable him to derive the values of the individual magnitudes. Even many economists seem loath to admit that those systems of equations are not a step toward specific predictions of individual events but the final results of their theoretical efforts—a description merely of the general character of the order we shall find under specifiable conditions but can never translate into a prediction of particular manifestations.

Predictions of a pattern are nevertheless both testable and valuable. Since the theory tells us under what general conditions a pattern of this sort will form itself, it enables us to create such conditions and to observe whether the pattern predicted will appear. And since the theory tells us that this pattern assures a maximization of output in a certain sense, it also enables us to create the general conditions which will assure such maximization though we are ignorant of all the particular circumstances determining the pattern that will appear.

It is not really surprising that the explanation of merely a sort of pattern may be highly significant in the field of complex phenomena but of little interest in the field of simple phenomena, such as those of mechanics. The fact is that in studies of complex phenomena general patterns are all that is characteristic of those persistent wholes which are the main object of our interest, because a number of enduring structures will have this general pattern in common and nothing else.[30]

---

[29] V. Pareto, *Manuel d'économie politique* (2nd ed.; Paris, 1927), pp. 223–224.

[30] A characteristic instance of the misunderstanding of this point (quoted by E. Nagel, *The Structure of Science*, p. 61) occurs in Charles A. Beard, *The Nature of the Social Sciences* (New York, 1934), p. 29, where it is contended that if a science of society "were a true science, like that of astronomy, it would enable us to predict the essential movements of human affairs for the immediate and the indefinite future, to give pictures of society in the year 2000 or the year 2500 just as astronomers can map the appearances of the heavens at fixed points of time in the future."

VII.   The Ambiguity of the Claims of Determinism

The insight that we will sometimes be able to say that conditions of a certain class (or of certain classes) will bring about a pattern of a certain kind, but that we will not be able to ascertain the attributes of the individual elements which decide which particular form the pattern will assume, has consequences of considerable importance. In the first instance, when we assert that we know how something is determined, this statement is ambiguous. It may mean that we know what class of circumstances determines a certain kind of phenomena but are unable to specify the particular circumstances that determine which member of the class of patterns predicted will appear; or it may mean that we can also explain the latter. Thus we can reasonably claim that a certain phenomenon is determined by known natural forces and at the same time admit that we do not know precisely how it has been produced. Nor is our explanation of the principle on which a certain mechanism operates invalidated if it is pointed out that we cannot say precisely what it will do at a particular place and time. From the fact that we do know that a phenomenon is determined by certain kinds of circumstances it does not follow that we must be able to know even in one particular instance all the circumstances which have determined all its attributes.

There may well be valid and more grave philosophical objections to the claim that science can demonstrate a universal determinism in human existence, but for all practical purposes the limits created by the impossibility of ascertaining all the data required to derive detailed conclusions from our theories are probably much narrower. Even if the assertion of a universal determinism were justified, scarcely any of the conclusions usually derived from it would therefore follow. In the first of the two senses we have distinguished, for example, we may well be able to establish that every single action of a human being is the necessary result of the inherited structure of his body (particularly of its nervous system) and of all the external influences which have acted upon it since birth. We might even be able to go further and ascertain that if the most important of these factors were in a particular case very much the same as with most other individuals, a particular class of influences would have a certain kind of effect. But this would be an empirical generalization based on a *ceteris paribus* assumption which we could not verify in the particular instance. The chief fact would continue to be, in spite of our knowledge of the principle on which the human mind works, that we cannot state the full set of particular facts which brought it about that the individual did a particular thing at a particular time. The individual personality would remain for us a unique and unaccountable phenomenon, one which we might hope to influence in a desirable direction by such empirically developed practices as praise and blame, but whose specific actions we could not predict or control because we could not obtain the information on all the particular facts which determined it.

VIII. The Ambiguity of Relativism

The same sort of misconception underlies the conclusions derived from the various kinds of cultural "relativism." In most instances relativistic positions on questions of history, culture, or ethics are derived from the erroneous interpretations of the theory of evolution which we have already considered. But the basic conception that the whole of our civilization is the result of a long process of evolution in the course of which values as the aims of human activity appeared and continuously change seems inescapable in the light of our present knowledge. We are probably also entitled to conclude that our present values exist only as the elements of a particular cultural tradition and are significant only for some more or less long phase of evolution, whether extending back to our pre-human ancestors or confined to certain periods of human civilization. We have no more grounds to ascribe eternal existence to them than to the human race itself. There is thus one possible sense in which we may legitimately regard human values as relative and speak of the necessity of their future evolution.

But this general statement is a far cry from the claims of ethical, cultural, or historical relativity or of evolutionary ethics. To put it crudely : we know that all values are relative to something, but we do not know to what. We may be able to indicate the general class of circumstances which have made them what they are, but we do not know the particular conditions to which the values we hold are due, or what our values would be if those circumstances had been different. Most of the illegitimate conclusions are the result of interpreting the theory of evolution as the empirical establishment of a trend. Once we recognize that it gives us no more than a scheme of explanation which might be sufficient to explain particular phenomena *if* we knew all the facts which have operated in the course of history, it becomes evident that the claims of the various kinds of relativism (and of evolutionary ethics) are unfounded. Though we may meaningfully say that our values are determined by a class of circumstances definable in general terms, so long as we cannot state which particular circumstances have produced the existing values, or what our values would be under any specifiable set of different circumstances, no significant conclusions follow from the assertion.

Brief notice is due to how radically different are the practical conclusions derived from the evolutionary approach according as it is assumed that we can in fact know enough about the circumstances to derive specific conclusions from our theory or that we cannot. While the assumption of a sufficient knowledge of the concrete facts generally produces a sort of intellectual *hubris* which deludes itself that reason can judge all values, the insight into the impossibility of such full knowledge induces an attitude of humility and reverence toward that experience of mankind as a whole which has been precipitated in the values and institutions of existing society.

A few observations ought to be added here about the obvious significance of our conclusions for assessing the validity of the various kinds of "reductionism." In the sense of the first part of the distinction we have repeatedly made—in the sense of general description—the assertion that biological or mental phenomena are "nothing but" certain complexes of physical events, or that they are certain classes of structures of such events, is probably defensible. But in the second sense—specific prediction—which alone would permit the more ambitious claims made for reductionism, the assertion is completely unjustified. A full reduction would be achieved only if we were able to substitute for a description of events in biological or mental terms a description in physical terms including an exhaustive enumeration of all the physical circumstances necessary and sufficient to the biological or mental phenomena in question. In fact such attempts always consist—and can only consist—of an illustrative enumeration of classes of events, usually with "etc." added, which might produce the phenomenon in question. Such "etc.-reductions" are not reductions which enable us to dispense with the biological or mental entities, or to substitute for them a statement of physical events, but are mere explanations of the general character of the kind of order or pattern whose specific manifestations we know only through our concrete experience of them.[31]

IX.   The Importance of Our Ignorance

Perhaps it is only natural that in the exuberance generated by the successful advances of science the circumstances which limit our factual knowledge, and the resulting boundaries to the applicability of theoretical knowledge, have been somewhat disregarded. It is high time, however, that we take our ignorance more seriously. As Popper and others have pointed out, ". . . the more we learn about the world, and the deeper our learning, the more conscious, specific, and articulate will be our knowledge of what we do not know, our knowledge of our ignorance."[32] We

---

[31] Cf. my *Counter-Revolution of Science*, pp. 48 *et seq.*, and William Craig, "Replacement of Auxiliary Expressions," *The Philosophical Review*, vol. 65 (1956).

[32] K. R. Popper, "On the Sources of Knowledge and Ignorance," *Proceedings of the British Academy*, vol. 46 (1960), p. 69. See also Warren Weaver, "A Scientist Ponders

have indeed in many fields learned enough to know that we cannot know all that we would have to know for a full explanation of the phenomena.

These boundaries may not be absolute. Though we may never know as much about certain complex phenomena as we can know about simple phenomena, we may partly pierce the boundary by deliberately cultivating a technique which aims at more limited objectives, the explanation not of individual events but "merely" of the appearance of certain patterns or orders. Whether we call these explanations of the principle or pattern predictions or higher level theories does not matter. Once we explicitly recognize that the understanding of the general mechanism which produces patterns of a certain kind is not merely a tool for specific predictions but important in its own right, and may provide important guidance to action (or sometimes indications of the desirability of no action), we may indeed find that this limited knowledge is most valuable.

What we must get rid of is the naive belief that the world must be so organized that it is possible by direct observation to discover simple regularities between all phenomena, and that this is a necessary presupposition for the application of scientific method. What we have by now discovered about the organization of many complex structures should be sufficient to teach us that there is no reason to expect this, and that if we want to get ahead in these fields our aims will have to be somewhat different from aims in the fields of simple phenomena.

Faith," *Saturday Review*, Jan 3, 1959: "Is science really gaining in its assault on the totality of the unsolved? As science learns one answer, it is characteristically true that it also learns several new questions. It is as though science were working in a great forest of ignorance, making an ever larger circular clearing within which, not to insist on the pun, things are clear. . . . But, as that circle becomes larger and larger, the circumference of contact with ignorance also gets longer and longer. Science learns more and more. But there is an ultimate sense in which it does not gain; for the volume of the appreciated but not understood keeps getting larger. We keep, in science, getting a more and more sophisticated view of our ignorance."

CHAPTER 23

The Agreement between Mathematics and Physical Phenomena*

BY JEAN ULLMO

I. Introduction

EINSTEIN'S OFTEN QUOTED REMARK, "The most unintelligible thing about the world is that it is intelligible," conveys the surprise of the scientist at the success of science. Since man has found a method for discovering and mastering the physical world surrounding him, that scientific method —which is the opposite of dogmatism—has taught him modesty, that holding to a preconceived idea will never ensure the attainment of knowledge. The value of science can only be proved by the very development of science; and the old problem of *adequation,* or agreement between mind and nature, is today more mysterious than ever because we have no ready-made solution to offer. However, the elements of an answer begin to appear; this answer is based on the study of the procedures and achievements of science.

Although the problem of adequation is without doubt a philosophical problem *par excellence,* we shall regard it here as just a corollary of the scientific method. We shall also review very cursorily the history of this problem and the various solutions that have so far been proposed.

The scientific tool for knowing the world is mathematical physics. Although dispute continues as to whether mathematics has an empirical origin or is purely rational, it is universally acknowledged that the develop-

---

* Translated by Mario Bunge.

ment of mathematics is the work of the human mind and its peculiar activity. Solution to the problem of adequation, therefore, amounts nowadays to accounting for the success of mathematical physics; it means justifying the agreement of mathematical constructions with experimental data.

Let us limit ourselves to recalling a few modern philosophers for whom the problem had already been posed in the above terms. Their solutions sometimes anticipate what we now think. Descartes saw in Nature and Thought the work of a divine Reason. He regarded them as two different expressions of one and the same higher Reality—this being why the one was adapted to the other. Analytic geometry supplied him with the typical example of this adaptation; geometrical figures, which belong to the external world, are exactly represented by the equations deduced by the mind.

Leibniz introduced a profound conception in the pre-established harmony of his monadology. Each monad or individual consciousness represents a particular view of the world; there is agreement and consistency among all these points of view because they are all perspectives of one and the same object—Reality.

Kant produced the "critical" revolution by subjecting phenomena to the *a priori* frame of thought : we can only perceive of the external world that which satisfies certain requirements of reason. Therefore what we perceive is necessarily found to be adequate to our mathematical constructions, which are the products of that same reason. Outstanding contemporary philosophers, such as Léon Brunschvicg and Gaston Bachelard, have perfected this criticism and rendered it more precise. In his interesting book *La philosophie du non,* Bachelard shows that we experience only what we can understand—that is, what we are able to reproduce or give form to by employing certain definite rules (experimental rules and combinatorial mental rules). Chemical synthesis is a good example of such construction. Every piece of knowledge has a similar character, and the world is known to us insofar as it can be thus constructed.

It will be noticed that Kantianism, by introducing the concept of choice through filtering by reason—which will ignore whatever our minds are unable to handle—replaces the problem of adequation (agreement between a reality that is given and a mind placed in front of it) by the problem of the extent or scope of scientific knowledge. The original question has been reformulated into : What part of nature will satisfy our rational requirements? And won't some things elude us? This is not a question prior to the problem of the value of science; it is a question to be posed once science is finished—that is, never. However, we shall have to understand why and how the choice is made.

The Kantian idea we have just discussed takes on a particular form in relation to the precise question of the success of mathematical physics.

In *La science et l'hypothèse*, Poincaré shows that "mathematics teaches us to combine like with like. Its aim is to guess the result of a combination without having to rebuild the combination piece by piece." This is the essential characteristic of recursive reasoning; it is also peculiar to integration and, in general, to the solution of all differential systems. Now, as Poincaré says, in physics "the observable phenomenon is due to the superposition of similar elementary phenomena." If the elementary phenomena can be put in mathematical form—that is, in symbols—the resulting observation will be symbolizable as well.

By working out this idea of Poincaré's, I have shown[1] how the physicist attains these elementary phenomena by means of *repeatable relations*, which he seeks in experience. The scientific objects are recognized and designated by these repeatable relations, which are translatable into mathematical symbols. They are consequently *defined* in such a way that the outcome is repetition, the "combination of like with like"—which is in turn faithfully pictured by calculation. A parallelism will then result, between the succession in time of the repeatable relations constituting physical laws and the development of the corresponding mental computation.

The above is a first account of the agreement we are discussing; it is concerned with prediction. However, in this paper I propose to work out an alternative and complementary explanation of the agreement, to show how the initial choice of the objects of science is determined by the conditions imposed on any possible experience. These same conditions ensure at the same time the possibility of representing the objects by mathematical objects. If we can then prove that experience can supply to theory only a pre-adapted raw material, we shall have made a decisive step toward the solution of our problem.

II.   Elements of Group Theory

Let us recall some elementary notions of group theory, which we shall need in the discussion to follow. A group is a certain set, finite or infinite, of operations of arbitrary nature on elements of any sort, which has the following properties. Any two operations of the set applied in succession can always be replaced by a single operation belonging to the set. For every operation of the set there is an inverse operation that takes things back to the initial state and belongs to the set as well. Finally, the set includes the identical operation, which changes nothing.

These properties of groups can be summed up by saying that the result of the application of one, two, or several successive operations of a

---

[1] J. Ullmo, *Revue philosophique*, May–June, 1938; *Revue de métaphysique et de morale*, Nov.–Dec., 1936.

group can always be *compensated for* or annulled by a single operation belonging to the group. The group is then a closed set that destroys itself.

The notion of group is intimately linked with the concept of *invariant*. The operations of a group preserve the invariants, i.e., do not change certain relations among the elements to which they are applied; conversely, every group can be defined by its invariants, that is, by a family of relations that subsist upon the operations. In fact, if a set of operations is defined by the peculiarity of leaving certain invariants unchanged, the successive application of two operations of the set will preserve the invariants, hence the resultant effect will be an operation with the same property, an operation that, by definition, still belongs to the set; and this is certainly a group.

Very simple examples will illustrate these ideas. The notion of group is so general that it applies to qualitative operations. With certain limitations, the set of kinship (the operation consisting in going from father to son, from brother to brother, or from nephew to uncle) forms a group the invariant of which is the family; the relation among the elements supporting the operations (that is, the individuals), which subsists during the successive operations, is 'belonging to the same family.'

Take as elements the words of language and as operations the translations from one tongue into another. The effect resulting from two successive translations (what is called their product) from French into English, then from English into Spanish, can be compensated for by a single operation, the translation Spanish→French, which takes us back to the initial word. The invariants of this group of translations are the meanings of words. (As a matter of fact this last example is a groupoid rather than a group; but it helps clarify the basic notions of compensation and invariant.)

The group of musical transpositions has two kinds of invariants: intervals and melodies. This example of the melody shows clearly the relations or structures which subsist among the elements that are individually modified by the operations of a group.

The group of one-one correspondence among collections of arbitrary objects is characterized by the cardinal number of these collections: this number is, then, the invariant of that group. One says then that the cardinal numbers of matching collections are equal; equality is consequently a particular case of an invariance ensured by a group.

Let us finally approach the kind of operations most important for us. The elements will be the points in a space of an arbitrary number of dimensions, and the operations on the points will be *transformations* of this space; a transformation is a transition from every point of this space to another point. If the n-dimensional space E is described with the help of the coordinates x_i $(i = 1, 2, \ldots n)$, a transformation depending upon p parameters a_k $(k = 1, 2, \ldots p)$ will be represented by the system of equations

$$(T_a) \qquad x_i' = f_i(x_1, x_2, \ldots x_n; a_1, a_2, \ldots a_p) \quad (i = 1, 2, \ldots, n) \qquad (1)$$

Given the functions f_i, the parameters a_k determine a (p times infinite) set of such transformations. The application of another transformation of this set, corresponding to the values b_k of the parameters, can be translated by the formula

$$(T_b) \qquad\qquad x_i' = f_i (x_1, x_2, \ldots x_n; b_1, b_2, \ldots b_p) \qquad\qquad (2)$$

If the two transformations T_a and T_b above are applied in succession, every point x_i will pass to the point x_i' defined by formula (1), then to the point x_i'' determined by formula (2) where every x_i in f_i has been replaced by the previously obtained x_i'. If the transformations constitute a group, the product $T_b T_a$ can be replaced by the single transformation T_c of the set :

$$x_i'' = f_i (x_1, x_2, \ldots x_n; c_1, c_2, \ldots c_p) \qquad\qquad (3)$$

The parameters c_k of the product transformation are functions of the parameters a_k, b_k of the component transformations :

$$c_k = g_k (a_1, \ldots a_p; b_1, \ldots b_p) \qquad (k = 1, \ldots p) \qquad (4)$$

In addition, one must have

$$f_i [f_1(x_i, a_k), f_2 (x_i, a_k), \ldots f_n (x_i, a_k); b_1, \ldots b_p] = f_i (x_1, \ldots x_n; c_1, \ldots c_p) \quad (5)$$

Such a set will be called a finite continuous group with p parameters. It involves an infinity of transformations, and particularly the infinitesimal transformations corresponding to parameters which differ very little from those representing the identical transformation.

Simple instances of such groups of transformations are (1) the 6 parameter group of displacements of ordinary space, the invariants of which are the relative distances among the points of this space and consequently the solid volumes; and (2) the projective group in the plane, the operations of which can be regarded as changes of perspective taken from different poles of the three-dimensional space; as is well known, the anharmonic relation is the fundamental invariant of the projective group.

With relation to the group of displacements we find a new definition of equality as a superposition of two solid volumes; we see that it results from the invariance preserved by the group. As to the projective group, the transition from one perspective to another appears like an equality of a more general nature, to which the fundamental axiom applies : if $A = B$ and $B = C$, then $A = C$. Rather than considering this axiom a self-evident truth, it is better to take it as the fundamental property that defines the group; the notion of equality is then generalized for the collections of supporting elements into "possibility of passage by virtue of the operations of a group."

Let us finally remark that equations such as (1), which are simply the correspondence formulas among symbols x_i and x_i', may be interpreted as a change of reference system in space E; one and the same point X of this space will then have the coordinates x_i or x_i' relative to two different systems, instead of the point x_i being displaced to the point x_i' of the same system. Then, instead of formulas (1) to (3) being interpreted as representing a group of transformations they will be interpreted as a group of changes of reference systems. A familiar example is the group of displacements, the corresponding formulas of which can be interpreted as a group of changes of reference trihedrals.

III.   Groups as Criteria of Objectivity

We shall now show that the knowledge of an objective external world is possible only if it satisfies certain *group* conditions. This demonstration will have two stages : first, certain conditions are needed in order that the isolated individual may recognize *objects* independent of his own sensations; second, certain conditions are needed in order that the set of all men may recognize that they observe *the same* external world and may then attribute objectivity to it.

Solids are the type of recognizable objects. However, individual experience does not supply us with an immediate knowledge of solid bodies but with changing perceptions (constituted by sensations or by forms, according as one school of psychology or the other is adopted). Why is every individual driven to externalize his perceptions, to assign them to objects independent of himself? By changing the attitude or the position of his body (the two kinds of changes being probably distinguishable to his consciousness by means of kinesthetic sensations), he may *modify* his perceptions. He must then be led, from this apparent mastering of his perceptions, which he modifies at will, to the belief in an external and independent world. He believes this, obviously, because he must assign to his perceptions a property of permanence exhibited by *repetition* : he changes them at will, but he is always able to find again the same perceptions by another act of his will. The fact that he can find again the same set of perceptions suggests that they have remained virtually present while he experienced other perceptions—hence that they have a permanent origin or support. Each perception that has been found again suggests the inevitable, hence what is independent of ourselves. This is the first stage in the construction of objects.

Now, to be able to find again a given set of perceptions is to be able to annul any change of perceptions produced by our change of attitude or our movement. This amounts to saying that *the operations whereby our perceptions are modified form a group.* If we have produced one, two, or several successive modifications, we will be able to compensate for the resulting modifications by means of a single modification. In short, our early experiences with the external world amount to recognizing the existence of this group. One is reminded here of the classical analyses of Henri Poincaré; but he has not worked out all the consequences.

The solid objects are the invariants of this group. We conclude that the variable perceptions, the variations of which can always be compensated for, are nothing but changing *appearances* covering a permanent reality—the invariant underlying the group. The recognition of the group enables us, then, to *eliminate appearances* and to attain the concept of permanent object. The group points to the stable amid the changing.

This can be expressed in an alternative fashion, namely thus: the deeper process of thought consists of going from the interpretation "group of transformations" over to the interpretation "group of changes of reference systems." In the beginning we know nothing but our body and our perceptions; we are consequently led to regard our body as an absolute reference system, and the modifications in our perceptions as genuine transformations produced by unknown causes. But once we have recognized that these transformations constitute a group (a recognition brought about by the possibility of annulling them at will), we naturally pass over to the alternative interpretation: those unknown causes appear as invariant, we posit that they are permanent objects in a fixed space, and we assign the modifications of our perceptions to variations in our condition of observation, that is, to changements of reference systems; the group supplies *objectification.*

A similar process will take place in the second part of our demonstration. After having permitted the elimination of appearances, the *group criterion* will supply the *conciliation of viewpoints.* A condition for being able to assert the existence of an external world is the consensus of the various observers experiencing it; it is necessary that they reach an agreement, by saying that they observe the same thing. Now they observe it from different points of view, and this time the difference is an essential one, since they are several distinct consciousnesses. How will they attain the consensus?

This agreement among independent consciousnesses, which raises such a difficult philosophical problem, is much simpler from a scientific point of view, because the consensus involves a *necessary condition,* a minimal condition for consistency or overlapping, which will again introduce groups, and these will supply a *sufficient* condition for objectification.

A rough example will show what the minimal condition of consistency is. Let the various observers be individuals speaking different languages and wishing to reach an agreement concerning the objects they observe. If the Frenchman and the Englishman believe that they designate the same object by calling it, respectively *soleil* and *sun,* and if the Englishman and the Italian believe they agree by calling the same object *sun* and *sole,* in order to make sure that there are neither misunderstandings nor phantoms it will be necessary that the Italian and the Frenchman, by consulting their words lists, find the correspondence *sole–soleil,* which will close the circuit. The necessary condition of consistency or overlapping is, then, the closure of all similar circuits for all languages and all words. But this is a group condition (in this case a groupoid condition). Various successive passages from one designation to another must be compensated for by a single passage.

Conversely, once this group property has been ascertained, it will be possible to go much farther. Since there is a group there is an invariant, and *posited* as invariants will be the observed objects, independently of their linguistic designations. It will be affirmed that these objects exist independently of their observers; the group will have supplied the objectification.

The observers of mathematical physics proceed in just this way, though more generalized and refined. Their observations are no longer confined to naming objects: they assign to them, by means of regular measurement procedures, localizing coordinates and observable magnitudes. Two observers must be able to establish correspondence formulas among the quantities they measure and the associated localizations. The condition of consistency must be applicable to the correspondences among three or more observers. It is necessary that the operation consisting of the correspondence between the observations of A and B, combined with the correspondence operation regarding B and C, be compensated for by the single correspondence operation relating the results of C to those of A. This is the condition necessary to convince A, B, and C that they have observed the same phenomena. Now, this is the condition for these operations, i.e., correspondences, to form a group. And we know already that, once this group has been recognized, it will be interpreted as a group of changes of reference systems, or observers' changes; its invariants will represent objective phenomena.[2] *The necessary condition of consistency has become the sufficient condition of objectification*: consensus will have been attained.

IV.   Groups as Products of Rational Activity

That is the clue for the appearance of groups in all the chapters of modern physics: the Galilei group, the Lorentz group, the Einstein group, the group of unitary transformations of the Hilbert space. We do not know

---

[2] As is well known, these invariants are expressed as general vectors and tensors, and all mathematical symbols capable of representing "intrinsic" properties, that is, properties independent of the chosen reference system. They are, indeed, the "mathematical objects" for representing "physical objects."

what group will nature present to us, but we are sure that it will present us one, this being the condition for the existence of an external and objective world that can be recognized as such.

This condition is extremely general, since it occurs at the beginning of the reflection on the character of every possible experience. One could therefore expect that such a general condition would not teach us much. What is remarkable is that it puts in our hands an extremely powerful tool; it is possible to study *a priori* all the sets of operations that form groups, and particularly all transformation groups.

It even seems that the study of groups is peculiar to the activity of reason, and that a very important part of mathematics derives from it. Given one or several operations (e.g., abstracted from experience or imagined by generalization), the mind imagines the effect of the renewal and combination of these operations; every result defines a new operation enabling us to go from the initial datum over to that result, and the new operation is joined to the preceding ones. The process of repetition and combination might be prosecuted continually until a *group* is formed, i.e., until a closed set of operations is attained. It is then, and only then, that the mind comes to a halt, since it has exhausted the possibilities offered by the initial operations.

It should not be surprising, therefore, that most mathematical theories, when they gain in depth, reach a fundamental group that provides the essential infrastructure characterizing them : Galois groups for algebra, groups of Klein's Erlangen program for the geometries, Lie groups and Fuchsian groups for analysis. And the development of mathematics has made it possible to construct *a priori* a very large number of groups, and even *all* those of certain essential species. These are ready-made frames that we propose to nature in order to see whether nature adjusts itself to them. The extraordinarily fertile Lorentz group in quantum physics is but one of the most recent examples.

Let us finally note that Descartes foresaw this structural identity of mathematics and the physical world, an insight that has recently inspired the researches of Dirac, and which is clarified by the fact that the group seems to be at the same time the condition of all experience and the achievement of the constructive activity of reason. Leibniz's proposal was an even more precise anticipation in that he implicitly introduced a group of changes of observers (the group of perspectives) in order to explain his harmony among monads. Finally, Kant rightly saw that reason imposes its frames unto experience, but he was too dogmatic in his belief in the rigidity of these frames and in their permanent *contents* of reason. As a matter of fact, we have seen no content at all in reason—no content such as Euclidean space or absolute time. We have only disclosed an *activity,* namely, reason. This is the capacity for abstracting operations of any kind, for defining, repeating, and combining them. This is how reason builds groups. And if nature adjusts itself to these groups, it is because they are not alien to it. The fact is that nature manifests itself through groups and enables us to abstract them, because an objective world cannot be known, i.e., cannot be represented by symbols, unless these symbols take part in certain groups. The problem of adequation reduces itself, as a consequence, to the question of the existence of such an objective external world, which is supposed to be answered when science begins.

CHAPTER 24

On the Reality of
Elementary Particles

BY WOLFGANG YOURGRAU

I

MY SHOEMAKER AND I often discuss scientific issues. At our last session we decided that man's frantic attempts to conquer outer space can be very simply explained : man originated somewhere in a remote cosmic corner and is now trying hard to return to his primordial home. We elaborated on this conjecture, embellishing some exciting aspects, and what started as a facetious game grew into a pseudo-scientific idea. Of course, I soon felt impelled to sober down our bold speculation and used expressions such as *extrapolative distance, extrapolative leap,* etc., in order to bestow some dignity upon my prosaic hangover. My partner agreed that our galactic venture was totally unrealistic, a fantastic caper, and truly incompatible with our customary mistrust of gigantic hypotheses. After all, were we not both convinced realists?

This conversation suggested the topic for the following haphazard remarks which do not claim any record in the perpetual extrapolative (long) distance race in physical science. I might add that I made a strange observation : whenever Heidegger talks about metaphysics, or Hegel, or even Planck, I throw 'epistemological' tantrums, whereas Popper's frequent references to metaphysical entities or situations elicit from me, at their worst, a grin of approval. Indeed, the same terms, the same clichés, occur in the writing of thinkers with whose views we heartily disagree, as well as in the work of those of whom we approve. And thus, we develop an aversion even against some of the technical expressions we are constrained to employ in our professional vernacular. It is evident that many a worker in the domain of philosophy of science must have undergone a similar experience.

Consistent positivism, materialism, apriorism, and empiricism, all enjoyed their heyday, at one time or another, but they failed in their role as lasting valuable philosophic panaceas. And with the meteoric rise of linguistic analysis it appeared for a while as if any recourse to an 'ism' was foredoomed. But when 'empiristic,' 'analytic,' 'positivistic' were treated as synonyms (though not universally), the triumphant march of the recent fashion to abolish all fashions came to a halt. And the devout believers in the tenet that sense data or sensa (sensations, sensibilia, sensates, sense impressions) can provide us with a plausible sensual theory of knowledge have also lost their appeal—sense impressions are not " 'primary' data, but abstractions."[1] Popper argues convincingly in favor of *rational discussion* whereby the statements of common sense, philosophy, and the natural sciences are *critically* examined; the "one method of philosophy" is equated with "the rational attitude and the critical attitude."[2] I should like to develop this method, this attitude, and plead for a definite *critical, rational realism,* since the adoption of Popper's attitudes induces a commitment to such an 'ism' in physics, provided that one arrives at a plausible, reasonable conception of the ontological status of elementary particles. In other words, a cautious interpretation of the experimental and theoretical findings of atomic physics suggests the tenability of that brand of realism. No doubt, there exists a quagmire of discredited realism theories : Platonic realism, naive realism, historical realism, relative realism, traditional critical realism, etc. Let us confess at the outset that rational, critical realism is somewhat eclectic, inasmuch as it will display materialistic, positivistic, and conventionalistic traits in addition to certain features of the above-mentioned realisms.

To begin with, we may enumerate the prerequisites for a critical, rational realism (CRR) derived from a scrutiny of the physics of elementary particles : intelligibility, communicability, correspondence, consistency, coherence, testability. I contend that any physical theory meant to account for the facts and events pertaining to the external world of material objects *must* invoke those postulates or principles of metaphysical presuppositions—the actually chosen appellation is in this context no more than a question of taste, of personal preference. In a workable physical theory, correspondence and coherence are not mutually exclusive, but complementary; any perusal of a monograph in theoretical physics will support this claim. Admittedly, CRR is marred by paradoxes and no attempt will be made to solve them. Later on we shall have occasion to deal with this 'defect' in some detail.

The obvious query which might be prompted by these remarks is a hackneyed one : what do we mean by 'real'? At the risk of becoming

[1] M. Born, "The Concept of Reality in Physics," *Universitas,* vol. 2, No. 4 (1958–59), p. 355.

[2] K. R. Popper, *The Logic of Scientific Discovery* (London: Hutchinson, 1959), p. 16.

cumbersome let us nevertheless examine this question as if it were still worthy of rational labor. Common-sense language applies this term to visible, tangible, audible things—to houses, trees, lamps, billiard balls, to stars, peoples: in brief, to material objects. Naive realism includes also smells, colors, sounds, tastes, etc., among the qualities pertaining to the external world, existing somewhere. It can be shown that acoustics clearly refutes naive realism.[3] Yet it cannot be denied that vestiges of naive realism are still inherent even in the physicist's jargon, although he doubts the veracity of some cognitive experiences and stands to be corrected. Hence, the naive realist's 'reality' ranges from things that possess solidity, inertia, impenetrability, to data of sensory activities.

However, as soon as he establishes the dichotomy 'real–imaginary,' he will find himself in a labyrinth of epistemological confusion. Centaurs, Pallas Athena, mirror-images, points and curves, miles, shadows, dreams, negative after-images—can these 'entities' claim reality or 'thinghood' because they possess qualities? Bradley uses the subterfuge of 'degrees of reality.' Lewis interprets the statement, 'a thing is known,' as a function of two variables (mind and thing) and thus adds to the myriad of opaque views. Only Moore and Quine successfully try to remedy the fact that 'real' is—as Lewis rightly observes—a "systematically ambiguous" term. They associate 'reality' with the respective attitudes adopted by workers in various domains of inquiry. Mathematicians, philosophers, psychologists, chemists, historians—they all employ the adjective 'real' in a significant manner, discriminate between the 'real' and the 'illusory,' between abstract, theoretical, concrete particular objects and universal objects, and arrive at some intersubjective though elastic criteria for an object to be real; in regard to this issue too, they communicate rationally among the initiated in their respective disciplines.

Lewis recognized the need for a methodological investigation of all these 'reals,' but a thorough and rigorous appraisal of the concept 'real' with its delicate nuances and ramifications was accomplished by Moore[4] and Quine.[5] Moore clarified the interrelation of existence, external objects, and the concept of reality. His assertion, "I can know things which I cannot prove," seems to indicate that he regarded 'real' as a primitive posit, an irreducible property or judgment, capable of a variety of uses. His verdict on this subject is partially supplemented, partially rectified, and

---

[3] B. Bavink, *Ergebnisse und Probleme der Naturwissenschaften* (Zurich: S. Hirzel, 1949), p. 60.

[4] G. E. Moore, *Some Main Problems of Philosophy* (London: Allen & Unwin, 1953); *Philosophical Studies* (London: Routledge & Kegan Paul, 1922); *Philosophical Papers* (London: Allen & Unwin, 1959).

[5] W. V. O. Quine, *From a Logical Point of View* (Cambridge: Harvard University Press, 1953); *Word and Object* (New York: Technology Press of M.I.T. and John Wiley & Sons, 1961).

the whole discussion is transformed to a logico-ontological analysis by Quine. He draws our attention to the not too different "ontic" weights of a term such as 'real'; makes a strong case for hypostatization of abstract entities without succumbing to a preposterous metaphysical dogma; admits an ontic distinction between physical objects and Homeric gods, but regards their epistemological status as hardly different; and, finally, argues reasonably against the claim that "what there is depends on words." However, without intending to corrupt the main trend of his arguments, I find it hard to reconcile statements such as "Existence of the meaning of 'a' must not be confused with existence of a," and "Let . . . terminology not mislead us into viewing geometry too differently from physics." His classification of the conceivable categories of concrete and abstract entities, of physical objects and theoretical concepts, displays almost lexicographical perfection, but the tapestry of the 'real' becomes so copiously patterned that the central ontological issue seems a little blurred.

A much greater sharpness of focus concerning the ontic situation is attained in the pure logical realm. The variables of quantification range indeed over our whole experience so that the quantifiers 'something,' 'everything,' and 'nothing' hold for all cases—provided one accepts Quine's strong reservations against quantifying into modal contexts (possible, impossible, necessary, contingent). I wish to refer here only to a few instances of the type of question which arises if one treats the 'real' in a formal manner. Suppose we wish to state that 'fairies do not exist': '$\sim(\exists x)(x$ is a fairy)' or '$\sim(\exists x)(fx)$' or '$\sim(\exists x)(Fx)$.' Now I want to assert that 'fairies are charitable,' or 'fairies are blue-eyed'; we write '$(\exists x)(fx \cdot gx)$.' Hence we attribute existing qualities to non-existing objects and connect the two by logical constants without any ontological qualms. The implications of this practice can easily lead to incongruous though "well-formed" sentences.

There is an elegant way to deal with quandaries of this kind, a procedural device suggested by Quine. Let '$(\exists x)(Fx \cdot Gx)$' symbolize the statement 'Some children are blond.' The well-formed expression does not tell us anything about the ontological situation: it is free from any abstraction and neutral with respect to class existence. But '$(\exists x)(x \varepsilon y \cdot y \varepsilon z)$' for '$(\exists x)(x \varepsilon$ mankind $\cdot x \varepsilon$ class of blond things)' supposes classes as values of variables. It is easy to prove that

$$(\exists x)\ (Fx \cdot Gx) \equiv (\exists x)\ (x \varepsilon y \cdot x \varepsilon z).$$

Only a "thoughtless thinker" would fail to realize that the effected translation from ontological neutrality to ontological commitment has not settled *the* ontological issue, i.e., the meaning, role, existential import of 'real.' But Quine has clarified some of the most obdurate facets of this problem and thus *reduced a question mark*—to paraphrase Gauss' illustrative aside.

II

We have now reached a stage where we ought to direct our attention to the epistemological fate of fundamental particles in physics. The transition from common-sense knowledge to scientific knowledge has been convincingly described by Popper,[6] and I do not see any need to reproduce here his well-reasoned arguments. In view of the following discussion, however, it might be of some value to pause and mention a few words as to the use of language in physics. We have recently been subjected to a barrage of inquiries into the essential features of ordinary language, its paradoxical blemishes, limitations, immanent vagueness and ambiguity, and so forth. But whereas many philosophers are therefore resigned to philosophizing without any belief in philosophy proper, the theorist interested in the foundations of physics is simply noting the disenchanting findings of the pure analysts or linguistic undertakers. Registering their results, he undauntedly pursues his searches by accepting the inevitable, the irremediable, with the same equanimity with which he accommodates himself to the sad fact that he cannot jump over his own shadow. "Language is a household invention and we must not expect it to be applicable very far beyond the confines of household experience."[7] Once we go down to 10^{-4} sec., we have reached the end of our sensory tether; since our life span is about 10^9 sec., there seems no cause for despair. Further, our visual capacity is confined to 10^{-6} cm. in the domain of micro-phenomena. Strangely enough, supermacro-phenomena do not seem to strain our anthropomorphic potencies. Radiotelescopes inform us that galaxies, undetectable with optical instruments and emitting radio waves, are roughly eight thousand million light-years away. Thus, our analogies with entities of ordinary experience merely break down when the very fast and the very small of the submicroscopic universe enter the stage of atomic physics.

Indeed, we are condemned to hear frequencies covering not more than about 9 octaves, i.e., $n = 50$ vib./sec. to $n = 25,000$ vib./sec., and to see within a range that is less than an octave, namely $n = 4000$ A. to $n = 7500$ A. Psychologically, we do not react to "extra-sensory" stimuli in a uniform, symmetric manner. It is also a truism that many new physical concepts have no counterpart in ordinary common-sense language and that, at the most, we 'stretch' the symbolic conventions adopted in everyday experience to concoct analogies and models of a more or less refined nature. Objections have been raised against the too generous use of models, i.e., artificial constructs, in atomic and theoretical physics. But the model of

---

[6] Popper, *op. cit.*, pp. 18–22.

[7] D. H. Wilkinson, "Towards New Concepts—Elementary Particles," in *Turning Points in Physics* (Amsterdam: North-Holland Publishing Company, 1959), p. 155.

yesterday can sometimes become the 'real' object of today or tomorrow, and our conceptual frustration concerning the limitations of everyday terms and phrases has, after all, never led to a total impasse in the progress of physics.

I cannot agree with D. H. Wilkinson's contention that "we must anticipate contradictions" when we apply our customary linguistic tools of everyday experience to submicroscopic phenomena.[8] The so-called deficiencies of ordinary language have not prevented any worker from performing novel experiments, nor seriously vitiated any theory formation. I think the defect lies somewhere deeper, namely in the circumstance that our ordinary language is not "ordinary" enough. It is permeated with euclidean rigidities at the expense of its former 'haptic innocence'; instead of conceptualizing haptic perception, we overemphasized the visual notions of Greek goemetry. One could perhaps show how our conception of the 'real' has thereby been distorted to an extraordinary degree.

However fascinating it may be to scrutinize many aspects of the dependence of scientific activity upon the wealth and poverty of ordinary language, we shall in this essay be content to concur with the view that physics and language are inextricably interlocked and that our usage of the term 'real' is largely determined by the way linguistic habits have evolved.

Russell maintains that physics is independent of psychological considerations,[9] and he is right if we accept ordinary language at its face value and sharply distinguish between its emotional and its sensorial ingredients. At one time, he investigated the objections to the reality of objects of sense and made a tentative plea for the argument that material objects are a logical construction from sense data. There *are* "actually existing entities" (not an independent metaphysical reality!) such as chairs, rocks, books, etc. Particles, points, and instants do not exist in the real world : they are the results of logical constructions. Yet we are told that the 'thing' is no more than a construction from the series of its aspects. And since particles are the products of the same mental operation, it is not easy to see how things in the real world differ from the "inferred" entities of physics. Are not both the results of logical construction? The "actually existing entities" must, I should say, either belong to an independent, external, objective, real world or be series of aspects, inferences from sense data, constructions. By definition, the artifact and the naturally given cannot be identical—they are mutually exclusive.

Russell's views as to the 'reality' of the external world are a beautiful illustration of the dictum in biology that ontogeny is an epitome of phylogeny. His personal philosophical development in regard to the 'real' can

---

[8] *Ibid.*, pp. 156–157.

[9] B. Russell, *Our Knowledge of the External World* (London: Allen & Unwin, 1926), pp. 106–134.

be likened to the history of ideas, if we make allowance for the absence of strict chronological order. In the final analysis, he seems to have espoused a basically 'realistic' attitude when he admits "that we can know of the existence of realities independent of ourselves."

Unfortunately, he treats instants and particles on a par, and while he has a strong case for regarding 'instants' and 'points' as *constructed* and not to be *found* among the data of experience, it is in no way a matter of epistemological coercion that particles are no more than hypothetical entities or existential constructs. Furthermore, he asserts that the world of physics is *inferred,* but that particles, points, and instants cannot be validly inferred but only constructed. And since he claims that the world of physics is now more familiar to us than the world of sense data, though the physical world is inferred and not given, it is not an easy task to disentangle this cluster of conceptual knots in our search for the 'real.'

In spite of these many logical question marks he possesses the valid insight that common-sense belief in fairly rigid bodies is "a piece of audacious metaphysical theorizing" and that physicists, being quite ignorant of the philosophic implications of their pronouncements, assume the reality of particles, etc., and ironically desist from any claim as to the metaphysical validity of these entities.

Russell then offers the following definition of a 'thing' (one could also dub it the 'real thing') : *"Things are those series of aspects which obey the laws of physics."* This definition is not circular—it merely combines a phenomenalistic approach with the propositions of physics and thus tries to avoid any ontological commitment. Does it succeed? Let us take the statement, 'A proton has mass,' where mass is an aspect or attribute belonging to a characteristic series of data. We now write : $'(x)(px \supset mx)'$. Moreover, since a proton is a member of the class of elementary particles, we may state :

$$'(x) \, [(x\varepsilon\alpha) \supset (x\varepsilon\gamma)]',$$

as 'ε' means 'has the attribute or aspect or quality' as well as 'is a member of the class.'* This point is well presented by Fitch,[10] who appears to me to court here a (implicit) realistic orientation by not treating the expression 'a class of' as a purely mathematical operator, but rather as a concrete general term or category.

My own view is that Russell's definition does not dispose of the ontological issue as long as the question how one should interpret the phrase 'membership in a class' has not been answered, and as long as we have not agreed on the meaning of 'existence,' 'there are,' 'experience,' and so forth. Concepts such a 'real,' 'ideal,' 'virtual' are in physics somehow

---

* 'α' for 'class of protons,' and 'γ' for 'class of elementary particles.'
[10] F. B. Fitch, *Symbolic Logic* (New York: The Ronald Press Co., 1952), p. 89.

related to our knowledge of the world and describe or represent objective, interpersonal states of affairs. Russell does not believe that the physicist is obliged to confer reality upon 'ideal' entities—it is sufficient to see them work. Likewise, Carnap sees no need to encumber the physicist with epistemological inquiries as to the reality of his findings or discoveries : such a task is exclusively reserved for the philosopher.[11]

Granted, the experimentalist is not irritated by questions pertaining to theory of knowledge in general. But can a theoretical physicist or anybody who ponders over the foundations of physics afford to 'change the subject' when the nature of fundamental particles is *sub judice*? The epistemological contributions of a Helmholtz, Duhem, Poincaré, Planck, Einstein, Bohr, Heisenberg, Born, Schrödinger, Bridgman, Margenau, etc., may not always satisfy the severe criteria of current epistemology and methodology of science with its increasing stress on a logico-axiomatic treatment of scientific theories and its inexorable analysis of scientific terms. Nevertheless, modern theory of knowledge has been stimulated and enriched even by their often highly controversial or bizarre conclusions : they were never devoid of cognitive relevance. Whitehead once remarked that Aristotle invented science, but destroyed philosophy; these physicists added to science and spurred philosophic investigation.

III

Let us now spring the snare and try to entrap the 'reality' of elementary particles, since it might prove to be rather difficult to capture a single 'real' one. As far as the reality of our "hunting tools," such as our various technical instruments, is concerned, no experimentalist would ever question that they are real, for the criteria which he customarily applies to his investigations are fully satisfied. In other words, the experimentalist lives in the environment of physical, macro-sized gadgets which are visible, touchable, audible, etc., and even Bohr, Heisenberg, and Jordan, despite their strong bias in favor of positivist ideas, accept a measure of naive realism as inevitable.

Whereas it is an uncontroverted fact that the mechanical apparatus is regarded as real, one encounters dissension of opinion with regard to forces, fields, field oscillators, molecules, atoms, electrons, nucleons, hyperons, energy-parcels, quanta, and so on. Are they idealized concepts, mere models or analogies? Analogies of what—of billiard balls, green fields, bundles of kindling wood, meteors? Rational, critical appraisal might help us to distinguish between fact and theoretical construct, between the

---

[11] R. Carnap, "Über die Aufgaben der Physik," *Kantstudien*, vol. 28 (1923), pp. 90–107.

'real' and the 'fictitious.' First of all, electron microscopes enable us to perceive molecules. Secondly, E. W. Müller, working with a field electron microscope, obtained images of single barium atoms on the screen; his photographs do not allow for any other reasonable interpretation. Further, tracks in the cloud chamber, emulsion phenomena, and discharges in the Geiger-Müller counter can only reasonably be explained as being traces of 'real' things. The arguments submitted by Dingle, who refuses to apply the concept of 'real' to micro-data, have been drastically rebutted by Born.[12]

It is, of course, a question of nomenclature whether or not we choose to call a particle by another name. For instance, we might talk of fundamental 'zoomzooms.' But whatever we call those specific phenomena or results of observation in the micro-domain, they are not ghosts or spirits or dancing demons. Intersubjectivity, reproducibility, testability and many of the previously cited criteria converge upon such indubitable evidence that we deal with 'real' things like particles when we analyze the multitudinous tracks, sounds, photographic records, von Laue's X-ray patterns, and so forth. Born's ingeniously formulated conditions for 'particlehood' are physically and formally sufficient and do not even conflict with positivistic doctrine. The physicist's language was never ambiguous in this respect: Yukawa predicted theoretically the existence of the meson, and Anderson and Neddermeyer 'discovered' such a particle of rest-mass $\approx$ 200 in the cloud chamber. Only recently another elementary particle that had been theoretically predicted, the omega meson (mean lifetime c. 10^{-22} sec., rest-mass 1540, charge $= 0$), has been "discovered." And who has not heard the glamorized saga of our time—the discovery of the positron? Truly, it is not too gigantic an assumption that one *discovers* or *finds* real things, in our case particles; one does not *discover* thought models, hypothetical constructs, or the like.

The characteristics of some particles do still elude our theoretical comprehension, and thus we call them 'strange,' e.g., hyperons. Photons, neutrinos, and antineutrinos possess no mass; the omega meson leaves no tracks in the bubble-chamber; fermions are conserved, but bosons are not conserved—no doubt, the properties of these fundamental particles are not always uniformly shared. But does this "flaw" mean that the term 'particle' is inapplicable? For a thing, an entity, to be 'real' it is not mandatory that it be known down to its last minutiae and that it own all the qualities of its genus. On experimental as well as on theoretical grounds we are justified in using the generic term 'particle' when we talk about electrons, neutrons, muons, etc. It seems to me to be not simply a question of analogy or semantic convenience, but the only correct way of describing a verifi-

---

[12] M. Born, "Physical Reality," *The Philosophical Quarterly*, vol. 3, No. 11 (April 1953), pp. 140–142.

able state of affairs. Expressions such as 'excited states' or 'particle system' or 'wavicle' are, in my view, merely futile attempts to avoid so-called anthropomorphic designations as, for instance, 'real,' 'particle,' etc., because these terms stand for an ontological commitment.

The submicroscopic universe manifests certain traits which affront deeply entrenched habits or conceptions. For instance, elementary particles lack individuality. The fact that we cannot put a blue or pink ribbon around an electron or pion has caused considerable perplexity and consternation among physicists and instilled in Schrödinger a pessimistic attitude vis-à-vis the concept of "elementary particle." In contrast to macro-sized bodies, where we can deal with individual entities, submicroscopic physics operates, on the whole, with ensembles or collectives or aggregates. At the same time, we can follow the track or path of an individual particle; we are able to count photons, but we cannot localize them. We do not experiment with a single electron; although nucleons and electrons are indistinguishable, they are definitely separate individuals but do not possess self-identity. Concepts like 'sameness' or 'individuality' do not hold for a fundamental particle or micro-system. The fact that we observe (macroscopic) tracks produced by a single particle does not, therefore, require that the particle be identifiable as an individual entity. It is irrelevant for our point which type of statistics—Einstein-Bose or Fermi-Dirac—controls the states any particle may assume.

Anomalous as this strange lack of individuality among fundamental particles may appear, it in no way affects the issue whether elementary particles are 'real' or idealized mental models. Nor are any basic logical conditions in jeopardy. '$(\exists x)(x=x)$' is still a valid symbolic expression for the identity nucleon = nucleon. Analyticity, i.e., truth or validity by reason of meanings or definitions, does not depend upon the physicostatistical characteristics of micro-systems! Thus, the universally necessary requirements of consistency and conceptual discreteness are still satisfied, for

$$\text{`}\sim(\exists x)\,[(x = x) \cdot \sim(x = x)]\text{'}$$

And if we wish to stress perceptual discreteness rather than conceptual discreteness, we can write

$$\text{`}\sim(\exists x)\,(fx \cdot \sim fx)\text{'}$$

Nor is our physical statement

$$\text{`}(x)[(x\varepsilon\beta) \supset (x\varepsilon\gamma)]\text{'}$$

i.e., 'An electron is a member of the class of elementary particles,' challenged at all.*

The transition from common-sense beliefs to exact scientific know-

---

* 'β' for 'class of electrons.' Here, as on p. 366, we assume the classes α, β, γ to be non-empty.

ledge is always a hard task, and any theory of elementary particles is hardly more beclouded by paradoxical doctrines or situations. Paradoxes have become permanent features of logic, mathematics, epistemology— and of physics, too. Let us ignore the question whether all these paradoxes can be validly resolved, or if the existence of a single genuine paradox in mathematics would not render all 'true' proofs spurious and thus deprive mathematics of its only absolutely indispensable axiom, namely that of consistency or noncontradiction. Epistemological and logical paradoxes (Epimenides, Zeno, Grelling, König, Richard, Russell) represent a challenge to those who are concerned about any flaws in arguments, logical rigor and coercion. However, physics was not retarded in its advance by Zeno's paradoxes regarding motion: the mathematical paradoxes of Burali-Forti and Russell, the many geometrical paradoxes and antinomies in number theory, did not arrest progress in mathematics proper: Gödel's theorem, Heisenberg's uncertainty principle, the contentious notion of infinities appearing in physical theory have certainly influenced science but not fossilized it. In like vein, the idea of the objectively real, independent, material world is not invalidated by the bankruptcy of traditional causality and determinism, and the demotion of the principle of least action as an all-pervading cosmic property is not tantamount to a death warrant for mathematical physics.

In fact, some paradoxes have proved to be of staggering significance for the evolution of new and correct theories in atomic physics—in the case of the Klein paradox, for example. According to Klein's cogent reasoning, Dirac's theory must lead to the assumption of negative energy states, i.e., of negative mass, and particles of half integer spin would have to obey Fermi statistics, if one wanted to avoid absurd implications in the quantum theory of general fields. Dirac was able to resolve that weighty paradox by the enunciation of his "hole theory." Perhaps we are also on the threshold of deciding satisfactorily the vague and obsolete status of photons. We regard them as particles of a kind, as it were, because they carry energy, possess angular momentum, spin and "stability." Being light quanta, their rest-mass has to be zero, for otherwise it would be infinite. Furthermore, the tendency to treat the photon as a particle gives rise to a highly strained usage of the term 'virtual.' In classical mechanics the distinction between a real and a virtual path is of strong heuristic value, whereas in quantum theory the label 'virtual' becomes recalcitrant and ambiguous.

I think that only a singularly obtuse physicist or philosopher would honestly doubt the cognitive value of the research results in fundamental particle physics for the commonplace reason that this discipline of science, too, is contaminated with a variety of paradoxes. On the other hand, among contemporary creative physicists no less an original thinker than Planck saw in the universal constants "a palpable proof of the existence in nature of something real and independent of every human measurement," "the immutable building blocks of the edifice of theoretic physics."[13] Obviously, we are left with the query : why are molecules, electrons, trees, oceans, galaxies, electromagnetic waves not "really real"? Why do only constants enjoy such an exalted Platonic rank? Again, the reasons for Planck's attitude lies in his anti-positivist warfare and his metaphysical nostalgia to discover in nature something everlasting, something not subject to decay, quantities not doomed to vanish. Einstein placed the constants like *a priori* magnitudes into a structural frame so close to the wishful dream of traditional rationalism that Descartes, Spinoza, and Leibniz in their Elysium must have welcomed him as *frater in spiritu*!

That the passionate belief in the metaphysically privileged role of universal constants has not gone into eclipse is distinctly demonstrated by the almost dogmatic faith in the 'real reality' of invariant quantities in relativistic physics.

The mathematicians Cayley and Sylvester had given the theory of invariants unqualified prominence. In 1872, Felix Klein delivered that memorable inaugural address, the celebrated and dramatically consequential *Erlanger Programm,* in which he emphasized the far-reaching importance of the group concept for the whole of mathematics. The theory of invariants became the most general expression concerning particular transformation groups. Euclidean and non-euclidean geometries, projective geometry, and even topology could be subsumed under the universal category of invariance. Among the many celebrated instances one might cite the fundamental theorem that the cross-ratio of four points is invariant under projection. In short, group and invariance theories invaded, thanks to Klein, almost every mathematical discipline with redoubtable success.

Since Einstein, the search for invariance in the mathematical sense is now also an imperious feature of physics : to wit, the invariant space-time, the fact that "Objectivity becomes equivalent to *invariance* of physical laws, not physical phenomena or observations,"[14] the postulate of invariance. A little reflection will convince us that the mathematician's "invariance under transformations" refers mostly to unchanged relations of properties among given and "deformed" entities; the properties which remain unchanged under transformations are thus relational magnitudes and seldom unrelated quantities. Einstein's invariants can scarcely be compared to universal constants such as h, e, m, c, etc. Even the cosmical constant λ is not a genuine "constant of nature," but rather the result of

[13] M. Planck, *Scientific Autobiography and Other Papers* (London: Williams & Norgate, 1950), pp. 170, 172.

[14] H. Margenau, "Einstein's Conception of Reality," in *Albert Einstein: Philosopher-Scientist* ed. by P. A. Schilpp (Evanston, Ill.: The Library of Living Philosophers, 1949), p. 253.

strictly mathematical reasoning. In view of incontestable evidence I am led to believe that the invariants in physics are 'thought models' or (mathematically) 'inferred entities,' whereas the universal physical constants are found in our study of physical reality.

A close analysis reveals that the non-synonymous terms 'constants' and 'invariants' are often either wrongly interpreted or overinterpreted. Depending upon the mathematical suppositions, this constant of integration, λ, can assume positive, zero, or negative values. On the other hand, space in a so-called model universe with properties of space-time must, according to a postulate of general relativity theory, at any instant possess constant curvature. Or take Hubble's parameter, usually called a 'constant of nature.' Cosmology abounds with such *ad hoc* constants. We might add, for instance, the constant *a* in steady-state theory, which appears when the rate of creation of matter is expressed according to that theory. McVittie's enlightening remarks with regard to the misleading use of the term 'constant' in cosmology[15] could be easily expanded : in mathematics, physics and cosmology the usage of terms such as 'invariant' and 'constant' is under perpetual strain due to logical and methodological vagaries and syntactical confusion. Moreover, the subsequent remarks will indicate how strongly a discussion of invariants reverberates with unwarranted metaphysical overtones.

Now, confessing to apostasy, I refuse to subscribe to the conception of physical invariants recently put forward by Born with powerful eloquence and customary erudition.[16] I believe rest-mass, proper-time, rest-length to be *normative* concepts, computed in analogy to their mathematical counterparts. That is to say, these concepts tell us something about magnitudes or quantities that are independent of clearly defined transformations, i.e., remain invariant for any system of reference. Born's interpretation of invariants in physics is ambiguous. "The idea of invariant is the clue to a rational concept of reality . . . ." Is there not a generic difference between any universal physical constant and the invariant rest-length? To assign to physical invariants the same logical status as to 'things' seems to me a case of stark overinterpretation.

So-called constants of nature may not exactly qualify as ultimate bricks of the universe, but unimpeachable experimental procedures show beyond doubt that many of these constants exist as real quantities (magnitudes). The charge of an electron is a true universal constant no less than the velocity of light. Contrariwise, the notion of proper-time is, strictly speaking, not a property of a 'thing' in the sense in which spin and charge characterize certain elementary particles. Invariants in physics are arti-

---

[15] G. C. McVittie, *General Relativity and Cosmology* (London: Chapman & Hall, 1956). pp. 162, 175–176, 183; *Fact and Theory in Cosmology* (London: Eyre & Spottiswoode, 1961), pp. 78–81, 168.

[16] M. Born, "The Concept of Reality in Physics," *op. cit.*, pp. 356–358.

facts, highly abstract constructs, the results of very sophisticated theoretical computations. Even the expression 'universal constant of nature' is heavily charged with potentially misleading information. To use the mathematical term 'invariant' for certain traits of abstract physical theory—for a 'thought model' so to speak—is admissible when it is clearly understood that these invariant features are not properties of 'things,' of elementary particles, of physical objects pertaining to the external world: they are inferred, derived, idealized entities or relations.

I do not see how one can sustain the contention that "invariants . . . correspond to properties of things . . . ." Invariants, being constituents of transformation theory of physics, are symbolic representations of conceptualized 'super-things' or 'super-relations' in a 'super-theory' which purports to arrive at the most general, the most comprehensive formulation of all physically possible regularities. The 'reality' of the invariants in relativity theory means that in the rational framework of this theory there appear invariant magnitudes (relations) which may, or may not, be testable and thus falsifiable by observation and/or experiment. In high-energy physics we encounter elementary particles when we apply maximal forces to matter. These particles are neither everlasting nor unalterable. Yet they possess certain typical, constant properties; when they are transmuted into another mode of particle existence, then these new units of matter own another set of typical, constant properties.

Constants of nature are corrigible quantities; invariants are, by definition, incorrigible. Hence it makes sense to call the rest-mass of a photon an invariant, since the idea of a photon at rest has no physical meaning, nor does it refer to any actual experimental or observational situation. As soon as we consider an elementary particle whose rest-mass is non-zero, and whose speed v is less than c, we obtain $m = m_0$ if $v/c \ll 1$. This means that the values for mass and rest-mass approximate each other if $v \ll c$. In this case, the above-proposed connotation of the term 'invariant' has undergone some change, because in the low-speed limit the value for the invariant m_0 nearly coincides with the physical magnitude m.

It is very odd that Planck discussed universal constants of nature without juxtaposing them with invariants. It is odder still that Born disregards fundamental physical constants in his appraisal of invariants. Perhaps one is not scholarly just by being aloof enough to enumerate impartially diverse facets of an issue. In a more cautious phrasing than Planck and Born I should like to propose the following layers of constant magnitudes (quantities) in physics:

> parameters
> constants
> fundamental constants
> invariants
> co-variants

It would be wrong to brand this terminological contrivance as an Aristotelian hierarchy. But in the name of CRR (critical, rational realism) one should not confound a universal constant like h with a co-variant describing a field, or the acceleration parameter in cosmology with the invariant proper-time. Symbolic presentation is not the same as a faithful portrait; a mathematical model in physics might evoke esthetic fascination as well as embodying coherence and axiomatic simplicity—and still be devoid of physical significance. The reader is reminded of the mathematical atomism of Boscovich with its abstract *puncta*. L. L. Whyte tries to establish a physically meaningful link between low-energy nucleons and the discrete *point* particles of the *Theoria*.[17] Boscovich took a solitary stand and showed creative imagination, but his *puncta* are neither inferred from a set of phenomena nor extrapolated from physical concepts into (at least indirectly) testable regions; they are the product of sheer speculation, divorced from the body of physics.

It is with some hesitation that I have to cross foils with so brilliant a scholar as Heisenberg in regard to the logic of the system of elementary particles.[18] He avoids a facile epistemological or ontological scepticism, if not nihilism, by deftly changing his ground. Elementary particles, so he tells us, are "really the last units of matter," yet they "are related rather more closely to the Platonic bodies than to the atoms of Democritus." Born regards the *Gestalten* in physics (rest-mass, spin, etc.) not as forms or shapes, but as invariants. Heisenberg declares the elementary particles themselves to be fundamental *forms* which energy must choose to become manifest as matter. It is not easy to see how they can be "the content of the fundamental laws of nature" when they are hardly " 'real' in the true sense of the word." Form, mathematical symmetry—not material objects —are said to provide the basic constituents of current natural science. All we need now to refurbish a Platonic, Eddingtonian Olympus is the logos. And, indeed, Heisenberg obligingly invokes that mystical extra-logical, extra-material, extra-theoretical, but intra-linguistic notion as if it were the magic key to solve all our problems in current physical theory.

It is not easy to understand why Heisenberg nevertheless defines elementary particles as the smallest units of matter yielded by indubitable experimental results. After all, he denies these particles reality and christens them abstractions! In this orgy of physical and epistemological license, any clear, rational discernment is lost and inconsistencies arise, not because the author does not reason with caution, but because he tries too

---

[17] L. L. Whyte (ed.), *Roger Joseph Boscovich* (London: Allen & Unwin, 1961), pp. 106–111.

[18] W. Heisenberg, "The Role of Modern Physics in the Present Development of Human Thinking," *Universitas*, vol. 4, No. 3 (1961), pp. 225–239; "Planck's discovery and the philosophical problems of atomic physics," in *On Modern Physics* (London: The Orion Press Ltd., 1961), pp. 3–19.

hard to satisfy positivistic, realistic, idealistic doctrines all at once. He doubts that matter is "the truly real"; elementary particles are not real because they are ephemeral. The fundamental equation of matter, however, contains at least two constants of nature such as c and h; yet these constants no longer express physical situations. "Natural science stands between nature and man." Well, if this is the case, then I believe that this science must refer to nature, that is, physical reality, to phenomena or 'interphenomena.'

During this meandering account we saw how Born endows only invariants with true reality and how Heisenberg reserves this Platonic 'medal' for fundamental forms alone. Of course, both thinkers do not simply indulge in mere intellectual exercises; no, they are earnestly concerned about the ontological fate of those particles. It is not a small task they set for themselves, but it remains in serious doubt whether or not their conceptions of reality can be generally adopted. Without committing oneself to either of their extreme views, one might be able to make a case for the Klein-Gordon equation's being a structural feature of the sub-microcosm. In a real scalar field, we have

$$(\Box - \kappa^2)\,\psi = 0$$

The Klein-Gordon (Schrödinger-Gordon or Klein-Fock) equation satisfies the quantum-mechanical correspondence between a field and a free spinless particle. In general, when we deal with high-energy particles, we get (relativistically) the Klein-Gordon equation for a particle of rest-mass m_0. The Klein-Gordon equation will retain its validity even if quantum field theory should be drastically reformed or entirely abandoned. For the sake of academic, or better, theoretical purism, one might regard an equation of that type neither as a principle, a law of nature, an invariant, nor as a fundamental form, but as a persistently recurring expression in advanced theory. The equation is of considerable range and has withstood the onslaught of diverse transformations. In spite of occasional extrapolational leaps it always landed safely in a physical region. So far, we have not been able to coin a specific term for that kind of 'invariance.'

Many physicists tend to grant reality to abstract, theoretical constructs. In the light of this comment, let us now examine some essential, and in no way clarified, problems in connection with the here stipulated reality of elementary particles, from the standpoint of our pet 'ism,' CRR.

It is rather bewildering that we cannot quite decide which of the so-called particles are 'elementary' particles. According to current knowledge, a conservative estimate would show that we have approximately forty kinds of elementary particles. Chew, however, considers only pions, nucleons, and kaons to be truly *elementary* particles, for the sole criterion that they are the least massive systems with strangeness as well as with baryon numbers (0,0), (0,1), and (1,0), respectively.[19] Be that as it may, the expression 'elementary particles' is never wrong, though amenable to further more sophisticated classification.

There is increasing evidence of a necessary interplay between observational data gained by the experimentalist and theoretical conjectures, however piecemeal they remain for the time present. To illustrate this assertion, let us tarry and briefly mention dispersion relations that analyze scattering data in high-energy physics. Theorists have recently been deeply stirred by the Mandelstam conjecture (representation) of double dispersion relations.[20] He eliminated the hitherto ever-present Lagrangian and thus removed the distasteful specter of infinite constants. Initially, one had to introduce a postulate of microscopic causality. Now that the intermediary of a Lagrangian is no longer required in a S-matrix theory involving two particles, it is possible to work with a minimum of coupling constants (*empirical* parameters) and a few postulates such as maximal analyticity, unitarity, Lorentz invariance. Mandelstam's representation for scattering amplitudes is a vital part of a theory of long-range strong interactions. Should a generalization of Mandelstam's conjecture (for pion-nucleon scattering) materialize, then such a theory would be complete. It appears as if no spectacular novel concepts have to be adduced, but only ingenious mathematical methods in order to cope with the tractable, though fast increasing, number of degrees of freedom. Progress in double dispersion relations will enable us to explore the ramifications of pion-pion interaction. Pion-pion scattering is experimentally still "elusive." Once Mandelstam and/or his co-workers succeed in developing the generalized theory, some time-honored conceptions in theoretical physics will have to be revised or discarded.

The Mandelstam representation is a further convincing factor in our belief in the reality of elementary particles, because it ensures by theoretical means the value of experimental results in atomic physics. The crisis in quantum field theory has perhaps acquired greater notoriety than the absence of any comprehensive theory that predicts the existence of elementary particles. In quantum field theory we find infinite interactions between fields—this in itself is an immanent imperfection—but none the less, all new results deduced from the theory by adroit manipulation of divergencies agree with experiment. Where lies the defect? The situation is not exactly paradoxical, but intolerably frustrating : the use of wrong theoretical tools produces consistently verifiable results ! There are indi-

---

[19] G. F. Chew, "Double Dispersion Relations and Unitarity as the Basis for a Dynamical Theory of Strong Interactions," in *Dispersion Relations* ed. by G. R. Screaton (Edinburgh: Oliver and Boyd, 1961), p. 172.

[20] S. Mandelstam, *Phys. Rev.*, vol. 112 (1958), p. 1344, and vol. 115 (1959), pp. 1752, 1761; with G. F. Chew, *Phys. Rev.*, vol. 119 (1960), p. 467, and *Nuovo Cimento*, vol. 19 (1961), p. 752.

cations that radical progress in perturbation theory might give us the anxiously expected breakthrough of that field-theoretical barrier.

As we have seen, Born interprets the invariant as the necessary prerequisite for a "rational concept of reality"—a viewpoint not shared by this writer. But whoever concurs in CRR will agree that deductive procedures (not of formal logic!) are indispensable to establish the reality of elementary particles. The interesting case of the *neutral cascade particle* is definitely germane to our discussion.

Powerful accelerators induce in particles huge amounts of kinetic energy and thus enable them to generate new elementary particles. The totality of all the diverse kinds of particles can be subdivided into sharply circumscribed groups or classes. So-called "heavy" particles are collectively listed as baryons. A subgroup of the baryons is known as hyperons. The heaviest among the hyperons has been found to be the negatively charged cascade particle. Now, a combination of experimental data and theoretical arguments adumbrated the existence of a *neutral* cascade particle as the eighth baryon (sixth hyperon) which, together with the negative cascade, should form an isotopic doublet. All the constant physical properties of this particle were determined by Alvarez and his co-workers when they discovered it recently.

We are immediately reminded of the discoveries of the meson and positron. Those particles were "rationally" predicted, but the umbilical cord between theoretical investigator and scientific empiricism was never severed. Newton's law of universal gravitation and Schrödinger's wave equation may have been inspired guesses; nothing so romantic happened in the discovery of the neutral cascade particle. In this case, the extrapolative distance (to use again an expression introduced at the outset) was comparatively very short. Experimental data, definite selection rules, empirical laws, and mathematical analysis sufficed to change a hypothetical entity into a physical, 'observed,' measured particle whose behavior fits smoothly into a theoretical frame. Its expected reality was clearly established on the strength of experimental data.

Our next and obvious question must be: do also anti-particles represent physical reality? Has one good reasons to assume that anti-matter exists? Anti-protons were produced with the aid of Berkeley's bevatron and were detected in cosmic rays too. Likewise, the existence of anti-neutrons has been experimentally confirmed. One might even express a rare optimism by asserting for a moment that all seems to be well in the 'state' of anti-particles. True enough, our investigations are executed with the help of macroscopic instruments and technical gadgets erected in laboratories. The world in which we perform our experiments obeys the laws and rules of classical physics. The expression 'material phenomena' in this context makes good sense. By means of highly involved processes, these macroscopic instruments, consisting of steel, glass, etc., produce elementary particles and anti-particles. However, anti-particles vanish on contact with their "sister" particles. At present, anti-matter cannot be created experimentally in bulk. Matter and anti-matter do not appear together, but it is not too grand a conjecture to assume that some stars and galaxies consist of anti-matter.

Although we are still in the first, naive stages in this border-domain of the physical and unphysical, we may conclude that phenomena can be direct or indirect, occur in different ranges of magnitude, possess 'primary' or 'secondary' reality. (I prefer the latter terminology to 'strong' or 'weak' reality, or to 'layers' of reality.)

In contrast to M. Hesse,[21] I urge a sharp ontological distinction between *virtual* and *real* particles, events, emissions, etc. The virtual path of a moving particle in dynamics does definitely not describe a real path. We have seen that an elementary particle need not be stable in order to qualify for the epithet 'real.' To equate 'virtual' with 'short-time' in the case where we deal with free and bound photons can probably be vindicated. But in any other domain of physical theory—be it electromagnetic, nuclear, or gravitational—the term 'virtual' must not be blurred so that it becomes some kind of boundary or limit of 'real.' In this analysis we have tried to show that the 'real' must not be defined as the 'direct observable.' A hypothetical particle might become a real one, but a virtual particle is understood to be not real. It is just because CRR purports to assign reality to elementary particles that cautious linguistic procedure is absolutely essential.

One requires, of course, a separate analysis if the reality of mirror images, or rather objects within a mirror, i.e., virtual pictures, is scrutinized. Arguments in support of the thesis that virtual images are real, are material objects, were advanced, among others, by Austin. I find them epistemologically wanting, since an examination of images observed first in two parallel mirrors, and then in configurations of three and more suitably arranged mirrors, raises a tortuosity of ever-increasing new logical and ontological problems. At any rate, it does not fall within our purview to treat the reality aspect of virtual representations in this context.

IV

Finally, I wish to refer cursorily to an issue concomitant with a discussion of the reality of elementary particles : natural or physical necessity. Popper[22] defines a statement as naturally or physically necessary if it

---

[21] M. Hesse, *Forces and Fields* (Edinburgh: Thomas Nelson and Sons, 1961), pp. 276–278.

[22] K. R. Popper, *op. cit.*, pp. 425–434.

satisfies the necessary and sufficient condition that it is deducible from a statement function, which holds in all worlds that differ from this world at most with respect to initial conditions. In this spirit, all laws of nature would be naturally or physically necessary. But logical necessity and physical necessity are by definition not identical. Since Schlick we call 'logically possible' any supposition that is not self-contradictory or does not lead to inconsistencies. Empirical possibility is determined by the laws of nature alone. Natural laws are, I believe, 'impatient' generalizations from experience and hence subject to correction in the light of new empirical data or theoretical insights. One might say that *corrigibility* is a necessary condition for any law to be regarded as physical, natural or scientific.

Popper correctly emphasizes that universal (natural) laws transcend experience; a finite number of observable instances can never justify the stipulation of a universal law. But are physical laws strictly necessary in comparison with singular facts and contingent as compared with tautologies? I find this trend to use the term 'necessary' in a formalist, i.e., deduction-theoretic, sense *as well as* in a semantic context highly confusing. Suppose deductive inference applies to formalist, semantic, and axiomatic conceptions of logic. And suppose we can show—as indeed one can—that there is an axiomatic procedure enabling us to subsume *all three* conceptions under one axiomatic schema. In that case, the fundamental distinction between contingent (or empirical) and logically necessary vanishes entirely. It follows that one would be compelled to agree with Kneale's contention that laws of nature are necessary, i.e., tautologies. My resistance to this viewpoint stems from my trust in the tenets of scientific empiricism, which is inextricably linked with CRR. Popper himself opposes Kneale's conception very strongly. Strict implicational inference of the type $U \rightarrow L$ indicates a more rigorous, purely formal deduction, whereas Popper's proposed

$$a \rightarrow b$$
$$N$$

symbolizes natural or physical necessity.

From Popper's postulate of natural necessity follows, of course, that logically necessary statements could become physically necessary too. Hence, the class of physically necessary statements contains logically necessary statements as possible elements or as a subclass. Fitch puts it no less apodictically: ". . . the laws of logic are included among the laws of nature, but not vice versa."[23] But this assertion seems somehow asymmetric with the condition that the expression "logically possible" is wider, less restricted than "physically possible," while "logically necessary" would be narrower, more constrained than "physically necessary." From Fitch's

---

[23] F. B. Fitch, *op. cit.*, p. 70.

claim we must conclude that laws of logic are constituent parts of laws of nature! Again the dichotomy between 'contingent' and 'necessary' is merely removed by decree.

I think that the idea of natural necessity leads to consequences that interfere with the code of logical rigor in formal deduction. I submit the following considerations in order to escape some undesirable 'side effects' of Popper's, Kneale's, and Fitch's approaches to the relation between physical and logical necessity, between laws of nature and tautologies:

1. Closure in the premises is a trait of deductive systems but not of physical laws.
2. Natural laws require heuristic principles, logical identities do not.
3. Formal systems are not reducible to non-formal systems; they can be mathematically interpreted. Physical theories are not simply interpretations of interpretations of formal systems or models.
4. A physical theory differs from a mathematico-deductive system not only with respect to semantic rules that are a necessary condition for any mathematized physical system or natural law, but also in regard to the property of logical necessity that is absent from any even highly mathematized physical theory or natural law.

Popper's discussion of *natural necessity* is not very convincing but it paves the way for any attempt to arrive at a methodical elucidation of this recalcitrant concept. Cleverly, he regards 'necessary' as a mere label or term; any other word would be acceptable as long as it is understood that such a new term is the opposite from 'contingent' or 'accidental,' without being in any way equated with 'logically necessary.'

A severe analysis of this new concept (it is not a mere question of coining a new word) and its implications will, for instance, show us whether the logic of modality can be beneficially applied to physical theories. If this elusive 'natural necessity' becomes nothing but another modality, then I fear that we have not gained much. Carnap and Quine have demonstrated how the specific modal properties can be translated into syntactical or formal properties of extensional logic.

It looks to me as if the question of 'extrapolative distance' and 'physical necessity' are closely related. Strict implication and inference hold in formal schema; physical theories are characterized by extrapolation and a natural "necessity." A dogmatic decision on this issue is not rendered more palatable by our noting that mathematical tableaux do exhibit, on occasion, so-called 'empirical pockets' and physical theories can, on occasion, display some rigorous deductive-inferential islets.

V

All the methodological aspects enumerated in these last paragraphs allude also to elementary particle physics. These are not, as I tried to explain, inferred entities or existential conjectures. To insist upon the reality of elementary particles does not require any *argumenta ex hypothesi,* any abstruse metaphysical hypostatization, or any new epistemological legislation. They are not structural properties of the universe — neither are electromagnetic waves — but real objects, not ghosts or idealized artifacts. In the equations of physics they are variables representing measurable quantities. They are neither invariants nor fundamental constants (like c or h). The fact that they are not directly observable will not alienate a neophyte of CRR. Ontological commitment is perhaps inexpedient; it is certainly more eventful and more promising than epistemological agnosticism — and it is more true.

I wish to thank A. Ambrose, C. Hempel, S. Mandelstam, W. V. O. Quine and J. Swanson for their helpful criticisms and invaluable suggestions.

IV

THE CRITICAL APPROACH TO
SOCIETY AND HISTORY

CHAPTER 25

Social Science and Moral Philosophy:

A Critical Approach to the Value Problem in the Social Sciences

BY HANS ALBERT

THERE IS AN important methodological problem in the social sciences, the solution to which is to a large extent influenced by the philosophical orientation of scientists: the problem of the respective roles of *theoretical* thinking and of *normative* elements in the formation of social knowledge. With respect to this problem, scientists often make a more or less radical distinction between the natural and the social sciences. Thus there are essentialists who hold that theories in social sciences intend to grasp the essence and at the same time the value of social phenomena by asserting empirically untestable propositions. On the other hand there are inductivists who emphasize the absolute value neutrality of their statements and tend to diminish the role of the theoretical element in favor of pure descriptive analysis. Naturally there are other possible combinations of these aspects, for *prima facie* they are not considered to be very closely interconnected. But we will see that this last assumption is hardly justified.

More than half a century ago, Max Weber formulated the so-called *principle of value freedom* (value neutrality) which has not yet been accepted by all social scientists. His thesis was the outcome of a thorough analysis of many aspects of the value problem and took account of the complexity of this problem. Many of his results are still applicable. In certain respects, amendments may be possible.

The confusion of different aspects of this complex problem has often led to apparently simple and plausible but nonetheless questionable solutions. A confusion of language levels, for instance, may suggest that valuations in the range of objects or the value orientation of scientific activity implies that value judgments are necessarily present in scientific statements. It may be useful therefore to split this complex problem into its several parts. First of all, we will enter into some philosophical questions which are connected with the problem of the role of valuation in social science.

I.   The Dilemma of Analytic Moral Philosophy

There is a certain danger of blocking the way to a solution of the problem from the outset by a misleading formulation of the question. That would be the case, for instance, if we began by analyzing the essence of science and the essence of value judgments in order to decide whether value judgments belong to science or not.[1]

The first of these questions is clearly avoidable if we start with a possible objective of scientific activity—for instance, that of exploring the structure of reality to discover the truth about it. But the question whether we need value judgments for that purpose seems to entail immediately the further question: What is a value judgment? Linguistic philosophers try to get around this essentialistic formulation by replacing it with the question of the logical grammar of value judgments, a problem which seems to be soluble by a logical analysis of ordinary language. From this approach we have to expect an interpretation of value judgments that gives us information about the usefulness of statements of this kind for the social sciences, or rather, about their role in the formation of social knowledge. Thus the whole question appears to be reducible to a pure language problem to be solved by the methods of linguistic philosophy. The fact that in Anglo-American moral philosophy the extreme views are gradually disappearing and a mild variety of anti-Platonism is beginning to dominate the scene seems to be at first glance due to the success of these methods, and thus of the analytic program of philosophy, the attempt to reduce philosophy to language analysis.

On closer inspection, however, certain difficulties come to light. It seems to be possible that, at least partially, analytic philosophers have fallen victim of a new kind of essentialism in a linguistic disguise, for *the* logical grammar of value judgments exists just as little as does *the* ordinary language which is logically analyzed to give the desired result.

In modern moral philosophy the distinction between *ethics* and *meta-ethics,* based on the widely accepted difference between object-language and meta-language, has generally gained acceptance. While ethical systems have normative character and refer directly or indirectly to human actions and attitudes, meta-ethical investigations are related to ethical expressions, propositions, and systems, submitting them to a logical analysis. Linguistic

---

[1] Cf. Popper's arguments against essentialism in *The Open Society and Its Enemies* (Princeton, 1950), pp. 34 *et seq.*, which I shall not repeat here.

moral philosophers generally advocate the thesis that this discipline in the main has to be limited to meta-ethical inquiries of this kind. They make a sharp distinction between the propositions of the moral philosophers (meta-ethics), those of the moral scientist (moral psychology, moral sociology, etc.), and finally those of the moralist (ethics). Moreover, they generally assume that between meta-ethical and ethical views there are no necessary relations of logical character, so that two persons can without contradictions adhere to completely different and mutually incompatible moral convictions while they agree in their meta-ethical conceptions, or vice versa. For according to analytic views, philosophy is a *second-order* undertaking and therefore perfectly *neutral* with regard to science, ethics, theology, and so on.[2] It only analyzes the various "language games" and otherwise leaves them untouched.

The various meta-ethical views can be classified, according to whether they assume the existence of non-natural entities in their interpretation of value phenomena or not, into *Platonistic* and *reductionistic* views.[3] As mentioned above, the philosophical discussion in the Anglo-American countries has led in the main to a displacement of Platonism in moral philosophy. In continental Western Europe, however, this kind of meta-ethics still seems to be the dominant view.[4] Within reductionism we can distinguish a *cognitivistic* variant interpreting value judgments as cognitive propositions (naturalism); this view also seems to be retreating. Recently analytic philosophers mostly appear to agree that an adequate interpretation of moral statements has to take into consideration their *normative function*, the *role of logic* and *rational argumentation* in the field of moral discourse, the element of *generality* which particularly is emphasized by the significance of principles, and the *reference* which such statements have *to reality*. On this basis then it is possible to construct interpretative schemes for moral statements or value judgments (normative statements) in general which can make a certain claim to plausibility or adequacy.[5] Ostensibly, language analysis leads to fruitful results in this way.

---

[2] For a critical analysis of this thesis and its implications see my "Ethik und Meta-Ethik. Das Dilemma der analytischen Moralphilosophie," *Archiv für Philosophie*, vol. 11 (1961), pp. 1–2.

[3] I should like to class phenomenology—very successful in continental Europe—Anglo-American intuitionism, and perhaps neo-Kantianism as Platonistic views, while reductionism includes above all emotivism, the value nihilism of the Scandinavian Hägerström school, and American naturalism, which is the only cognitivistic conception of this kind.

[4] Exceptions are the few empirically-minded philosophers like Victor Kraft, [see his *Grundlagen einer wissenschaftlichen Wertlehre* (2, Auflage, Wien, 1951)], and Ernst Topitsch [see his *Ursprung und Ende der Metaphysik* (Wien, 1958)].

[5] For example, see the proposals in the books of R. M. Hare, *The Language of Morals* (Oxford, 1952); Stephen E. Toulmin, *An Examination of the Place of Reason in Ethics* (Cambridge, 1953); Paul Edwards, *The Logic of Moral Discourse* (Glencoe, Ill., 1955); Bernard Mayo, *Ethics and the Moral Life* (London, 1958).

But now the question arises of what is gained by this analysis and what are the foundations of the alleged success of linguistic philosophy. Are the above mentioned results really based on a linguistic analysis compatible with the neutrality thesis? Certain difficulties seem already to arise if the attempt is made to mark off the propositions of morals from normative statements in general. The problem has an unmistakable similarity to the problem of demarcation between science and nonscientific statements with which Karl Popper has dealt in his first book on methodology.[6] As it turned out there, in the realm of the logic of science, such a demarcation in a certain way *cannot be neutral* to the propositions of the object-language since it is based on a criterion for the scientific usefulness of statements and systems. In this case philosophy (or epistemology) may be a second-order enterprise in a certain sense, but certainly not in the sense of linguistic analysis. It will be meta-scientific in a *critical* sense. In this realm of philosophy, most probably the neutrality thesis is not to be rescued.[7] If the linguistic program were strictly carried out, philosophy of science would terminate in an "unprejudiced" description of the actual methodological customs and habits of scientists as far as they are connected with their "language games." Thus philosophy of science would become a part of cultural sociology, the sociological analysis of a subculture of modern industrial society, that is, of science.

The situation in the realm of moral philosophy seems to be entirely analogous. To be sure, the analytic program which is ultimately based on a *resolution* appears to be incontestable for that reason alone. But if it is strictly realized, it yields by no means an autonomous moral philosophy, as the proponents of this program originally believed and intended. Here, too, philosophy must eventually dissolve into empirical science, a science analyzing the actual linguistic usage in certain ranges of culture and closely connected with moral sociology and moral psychology. That may remain concealed to a certain degree as long as in language analysis intuitive arguments and impressionistic allusions take the place of scientific investigations.

The far-reaching agreement between analytic philosophers in their results can possibly be explained by the fact that their explorations have so far taken place within culturally closely circumscribed social groups, perhaps the typical intellectuals of the middle stratum of Anglo-American

---

[6] Cf. Karl Popper *Logik der Forschung, Zur Erkenntnistheorie der modernen Naturwissenschaft* (Wien, 1935); extended English edition: *The Logic of Scientific Discovery* (London, 1959).

[7] For a critique in this direction see J. Agassi, "Epistemology as an Aid to Science: Comments on Dr. Buchdahl's Paper," *The British Journal for the Philosophy of Science*, vol. 10 (1959), p. 135 *et seq.*

culture.[8] An extension of the domain of analysis to groups of a different cultural shaping, for instance, Roman Catholic priests, Prussian officers, or Russian party functionaries, is likely to produce totally different results. Moreover, such an extension would probably show that ethical convictions and moral commitments are rather considerably connected with definite meta-ethical views and with factual interpretations of reality. Surely, the clarification of these connections would be of great significance, especially for cultural sociology. Philosophically, however, this kind of analysis would end in a blind alley. The program of linguistic philosophy is unassailable because it is based on a decision, but it by no means opens up to philosophy an autonomous realm for fruitful investigations. It leads rather to the disappearance of philosophy.

As long as philosophers refuse to realize these consequences of the linguistic program and continue their previous practice of analysis, they are not too far from a linguistically disguised essentialism. All possible elements of traditional moral philosophy reappear, as for instance the Kantian postulate of universality and the utility principle of the utilitarians, but in linguistic reformulation, as necessary elements of the language of morals as it were. It is very difficult to see a significant difference from the older essentialism, with its attempt to solve the problem of the essence of the "good" and so on. The dilemma of analytic moral philosophy presumably consists in the necessity either to dissolve into empirical science or to fall to a linguistic essentialism, unless it is ready to give up its previous orientation — that is, its basic program. As the first alternative entails negating the autonomy of philosophical thinking, a favorite thesis of analytic philosophy, the most convenient escape seems to be to continue analyzing the essence of morals by exhibiting the necessary elements of the language of morals and justifying the results of analysis by allusions to the alleged language behavior of people. The same theoreticians, however, who try to disprove intuitionism as a meta-ethical view are now forced to rely upon their intuition on the meta-ethical level. The result is likely to be a kind of dogmatism tempered by common sense.

II.   A Critical Approach to the Value Problem

The previous considerations suggest the conjecture that the basic mistake in the analytic approach is its emphasis on investigation of the actual functioning of ordinary language. It is difficult to see why the "logical grammar" of ordinary language—with all kinds of superstitions enshrined in it—should be of greater relevance to moral philosophy than to epistemology. He who accepts the critical function of the theory of know-

---

[8] For a critical analysis see Rollo Handy, "Doubts about Ordinary Language in Ethics," *Inquiry*, vol. 3, No. 4.

ledge has no reason to refuse moral philosophy such a function. The dilemma of analytic philosophy is avoidable by passing over from pure language analysis to critical thinking, so that the neutrality thesis need no longer be maintained. The task of philosophy then changes from an adequate interpretation of actual verbal behavior to a critical examination of moral thinking, not only of its formal aspects but also of its content.

The construction of an interpretative scheme for moral statements or normative statements in general may gain a certain, though presumably secondary importance in this context, too. Its purpose may be to show how we *could* interpret normative sentences in a meaningful way, considering the present state of knowledge, *notwithstanding the actual linguistic usage* in existing social groups.

We can sketch a possible proceeding to this purpose as follows: To begin with, the usual distinction will be made between cognitive and noncognitive propositions, cognitive being all propositions which are capable in principle of adopting the truth-values true or false. Within this class are to be distinguished propositions which are true or false on logical grounds (analytic and contradictory propositions), and synthetic propositions[9] to which this does not apply. Within the latter subclass we have to distinguish between empirical and nonempirical statements.[10]

Naturally there are many kinds of *non*cognitive statements, especially those using so-called prescriptive language or language of action.[11] Within this group a species of statements can be distinguished which we shall name *normative* statements. This species is to be characterized not by formal but by functional properties, irrespective of the question of actual usage in existing groups. We propose the following interpretative scheme for this kind of statement.

A sentence is a normative statement if:

1. It marks out in a positive or negative way a possible state of affairs for action or attitude (valuation),

---

[9] The attempt of neo-pragmatists (Quine, Goodman, White) to blur the analytic–synthetic distinction cannot be discussed here. But I think there are fairly good arguments against this view.

[10] This distinction is based on the fact that the positivistic identification of synthetic and empirical propositions has failed. The possibility of a synthetic *a priori* cannot be refuted by an arbitrary convention: see Popper's demarcation of empirical and nonempirical statements in *Logik der Forschung*, and his discussion about synthetic *a priori* propositions in his "Philosophy of Science: A Personal Report," in *British Philosophy in the Mid-Century*, edited by Mace; J. W. H. Watkins, "Between Analytic and Empirical," *Philosophy*, vol. 32, No. 121 (1957); Wolfgang Stegmüller, *Hauptströmungen der Gegenwartsphilosophie* (Stuttgart:Wien, 1952; 2. Aufl. Stuttgart, 1960), discussion about synthetic *a priori* at the end of chap. IX.

[11] For an analysis see Héctor Neri Castañeda, "Outline of a Theory on the General Logical Structure of the Language of Action," *Theoria*, vol. 26 (1960). Castañeda's list of kinds of propositions can possibly be extended.

2. It presupposes a norm (value-standard or action-maxim) as valid which demands a corresponding behavior (action or valuation),

3. It implies a corresponding prescriptive expectation.[12]

The further considerations are partly independent of special properties of this scheme. For the most part, the possibility of a noncognitive interpretation which fulfills certain requirements will suffice, especially to account for the normative function of certain sentences.

Now, we can state that normative statements of this kind are not logically deducible from cognitive propositions, for such a derivation would amount to a naturalistic fallacy. The same holds for all other kinds of prescriptive statements. Starting with this assertion, we can now pass over to a characterization and a criticism of both the cognitivistic forms of meta-ethics: *Platonism* and *naturalism*. If we understand by meta-ethical Platonist the view that value judgments (or moral statements) are a sort of synthetic propositions *a priori,* then the following argument applies: in this view it is not possible to derive prescriptive statements of any kind from value judgments without committing the naturalistic fallacy. If ethics is required to lead to statements with the character of instructions or justification for human action or valuation, then a success in our case is only to be expected if the *descriptive* interpretation of Platonism is combined with a *logical fallacy.* Therefore Platonism is not acceptable.[13]

With regard to meta-ethical naturalism, if it interprets moral sentences as empirical propositions, an analogous argument is possible. If the naturalistic program in meta-ethics is reformulated to the effect that the aim is an explanation of human value-behavior,[14] then meta-ethics dissolves into empirical science in the same manner as analytic moral philosophy is bound

---

[12] See my article: "Ethik und Meta-Ethik," *op. cit.*, pp. 39–47.

[13] This is to be applied, for instance, to the meta-ethics of natural law, which amounts to a ontological identification of *value, essence,* and *reality.* Presumably the essentialistic statements characteristic of this manner of thinking have their source in the tendency to fuse the advantages of three kinds of propositions, the sharp distinction of which is fundamental for modern thinking, namely, *normative, logical,* and *factual* statements. The seeming coincidence of the normative, the necessary, and the real is achieved by the construction of a second "superior" reality, more fundamental than mere empirical reality. For the cognition of this reality, superior cognitive abilities are claimed (intuition of essence, etc.), the common property of which is their lack of intersubjective testability. Values become quasi-facts and their cognition is expressed in propositions which are informative but not controllable by scientific means, necessary but not controllable by logical means, normative but without a personal decision. In short, they afford all possible advantages without the normal "costs." As a meta-ethical conception the natural law doctrine is no longer maintainable; see the critical analyses by Hans Kelsen, Alf Ross, Ernst Topitsch, and others. For a general refutation of essentialism and natural law, see Karl Popper's *The Open Society and Its Enemies.*

[14] For such a reformulation, connected with a critique of analytic moral philosophy, see Paul W. Kurtz, "Naturalistic Ethics and the Open Question," *The Journal of Philosophy,* vol. 52, No. 5 (1955).

to do. Meta-ethical Platonism cannot be saved this way so long as it maintains its aprioristic character and, thus, its identity. The claim of cognitivity for normative sentences has at best the result of dogmatization here, of immunization against critical argument.[15]

This is the outline of an argument against any cognitivistic moral philosophy.[16] Moreover, it shows how it is nevertheless possible to make meaningful normative statements which have a function in certain circumstances, thereby avoiding noncontrollable assumptions concerning non-natural entities of any kind, assumptions without any function in cognition, valuation, and action; the value behavior of men is to be explained without them. It must be conceded that the whole discussion has epistemological foundations. But the demand to keep epistemology and moral philosophy, meta-science and meta-ethics, strictly separate has no interest for us if we are looking for a satisfactory interpretation of all relevant phenomena which is compatible with critical rationalism and which does not operate with superfluous assumptions. To show, however, that a cognitive interpretation of moral statements is impossible *in itself* is an unsoluble task, in the first place because not all possible interpretations are forseeable, and secondly because the respective interpretation schemes can be used in any case for the demarcation of the kind of proposition intended. Even language analysis does not help here, except in the criticism above.

The construction of a possible interpretation scheme for moral statements and normative statements in general has, however, secondary importance to the possible task of moral philosophy of criticizing the prevalent ethical systems, meta-ethical views, moral convictions, and their institutional embodiments in the light of our knowledge—including our empirical knowledge, especially of the social sciences. Moral philosophy and social criticisms are closely connected here. As Popper has shown[17] there

---

[15] Presumably there is the idea of absolute justification in the background, which is also characteristic of epistemological rationalism, as Popper has shown in his "On the Sources of Knowledge and Ignorance," *Proceedings of the British Academy*, vol. 46 (London, 1960). This idea seems to be as important in moral philosophy as in epistemology. Criticism in Popper's sense is based on a refutation of this idea; see note 63, *infra*.

[16] It amounts to a meta-ethical conception which Popper calls "critical dualism" (see his *Open Society* . . . , chap. 5). There may be another kind of cognitivistic meta-ethics, a "degenerate" case so to speak, if we admit the interpretation of moral statements as *analytic*. For such a conception see Felix Kaufmann, *Methodology of the Social Sciences* (London, 1958), §128–138. This meta-ethical logicism is criticizable on the same grounds as Platonism and naturalism. If we make value judgments analytic in the manner proposed by Kaufmann, their normative function will be lost. There is an analogous argument possible with regard to singular causal statements.

[17] K. Popper, "The Nature of Philosophical Problems and their Roots in Science," *The British Journal for the Philosophy of Science*, vol. 3, No. 10 (1952); the quotation is from p. 130.

is no autonomous realm of philosophy : "Genuine philosophical problems are always rooted in urgent problems outside philosophy, and they die if these roots decay."

III. The Problem of Value Neutrality of the Social Sciences

The results of our analysis are immediately applicable to the question of value neutrality of the social sciences. If we interpret this question quite generally as referring to the possible roles of valuations, decisions, and normative statements in the social sciences, then it can be broken down into :

1. The problem of the *value foundation* of the social sciences : the extent to which valuations (decisions, etc.), are the basis of their propositions and systems,
2. The problem of *values in the domain* of the social sciences : the extent to which these sciences have to make valuations (decisions, etc.) the objects of their propositions,
3. The proper problem of *value judgments in the propositions* of the social sciences : the extent to which the statements of these sciences have themselves the character of value judgments.

The first of these problems has an heuristic and a methodological aspect. Frequently, the accusation is made that, because of the alleged dependence of their theories on valuations, the objectivity of social scientists can be doubted. Moreover, sociologists sometimes represent the social determination of social science as an inescapable source of error. Social scientists have occasionally concluded that they are obliged to compensate for these errors by indicating their value-position explicitly, by unveiling their implicit presuppositions through a kind of socioanalysis. This view has been severely criticized by Popper[18] because it implies a serious misinterpretation of scientific method, even its social aspects, which cannot be characterized correctly without taking into account the social institutions rendering free interpersonal criticism possible. The fact that any proposition and any theory in the natural and social sciences (as in ordinary language) is selective and may be influenced by valuations is indisputable, but methodologically without relevance. Neither in the selection of problems nor in the formation of theories must the influence of valuations be eliminated or discounted in order to save the objectivity of science. The opposite idea is founded on confusion of the context of discovery and the context of justification, of the question of origin and the question of validity of

---

[18] See his *Open Society* . . . , chap. 23.

propositions. The intuitions of a scientist are always exposed to the criticisms of others, and to the scientist's own censorship based on logic and experience, and are therefore subject to correction and revision. The situation is the same, in principle, in the social as in the natural sciences, where scientists also are inclined to stick to their favorite ideas until this becomes impossible on account of critical arguments. Also in other respects scientific activity depends on valuations. The methodological rules of science are related to certain aims. In science there are many kinds of decisions to be made, concerning, for instance, the usefulness of hypotheses, the acceptability of methods, the relevance of observations in relation to certain problems. But all these decisions are subject to critical examination. Besides, they are not expressed in the scientific propositions and theories themselves. They do not lead to value judgments within the context of scientific propositions at all. Naturally, the value foundation of science itself can be analyzed in a scientific and "value-free" fashion.

The second of the problems listed above will hardly lead to controversy of any significance. Since social relations and actions are analyzed by the social sciences, valuations of all kinds are fundamental objects of study; they have to be investigated by these sciences. The theory of human behavior can hardly be developed without an analysis of valuations and attitudes.[19] Statements which only inform about them, which describe, explain, and predict the value-behavior and thereby the value judgments of men, however, are statements of fact and not value judgments themselves. They are informative and not normative statements.[20] Here it will be worth while to take into account the difference between object-language and meta-language. Thus also valuations can be handled "value-free."

The third problem listed, which is the proper problem of value judgments, concerns normative statements within the propositions and theories of the social sciences,[21] that is, it concerns the object-language the methodologist must investigate. First of all we must assert that the existence of valuation at the foundation of theory formation and general scientific activity or as an object of research does not make necessary the use of value judgments in the language of scientific statements. It is advisable here to keep separate the different levels—meta-language, object-language, and reality. Very frequently, the rejection of the possibility of a "value-free" social science can be traced back to a confusion of these levels.

The admission of value judgments into the context of the propositions

---

[19] See George C. Homans, *Social Behavior: Its Elementary Forms* (New York: Burlingame, 1961), especially chap. 3.

[20] As is rightly emphasized by Lionel Robbins in *An Essay on the Nature and Significance of Economic Science* (2nd ed.; London, 1952), p. 90.

[21] For an analysis of this problem, see my article, "Das Werturteilsproblem im Lichte der logischen Analyse," *Zeitschrift für die gesamte Staatswissenschaft*, 112. Band 1956.

of a science would make that science normative, a condition recommended for their field by many social scientists today. The methodological question which is to be answered here can be formulated as : What aim would make the use of value judgments necessary in the social sciences or on which conditions can we dispense with such propositions? The answer is that the aim to increase our knowledge of reality makes value judgments within scientific contexts dispensable, the more so since a neutralization of value-laden systems leaves their information content untouched. Now certain reasons for a partial normativization of the social sciences seem to result from an analysis of their practical applicability.[22] It has been noticed for a long while that there is a strong inclination to derive recommendations for practical life, especially for politics, from propositions and theories of the social sciences. The logical derivation of normative sentences from systems of purely informative character would, however, depend, as mentioned above, on committing the naturalistic fallacy. From this fact many social scientists conclude that they are obliged to complete the systems in question by the addition of explicit value judgments appropriate for such derivations.[23] The wish to avoid committing a logical fallacy in connection with the demand for practical applicability of social science can obviously lead to believe in the necessity of a normative social science.

To that the objection can be raised, first of all, that statements in the natural sciences are also applied very successfully without being completed by additional value judgments. Here by practical application is meant the derivation of prognoses relevant to behavior and the employment of such prognoses for the control of occurrences in the realm of objects. For that it is necessary to transform informative theories into technological systems.[24] This transformation is tautological and requires no additional premises, a fortiori no value premises. It is only necessary hypothetically to presuppose certain desiderata in order to be able to decide which of the many possible transformations is relevant to the context in question. Such problems of relevance, which are often to be solved in social as in natural science, belong to the above-mentioned question of the value foundation of science. Here, there is an inclination to confuse different levels and to assume the necessity of value premises in the logical sense. A technological system is, however, value-free in the sense of not containing valuations, recommendations, value judgments, etc. It shows only possibilities, especially possibilities of producing certain effects by

---

[22] For an analysis of this question see my article "Wissenschaft und Politik. Zum Problem der Anwendbarkeit einer wertfreien Sozialwissenschaft," in *Probleme der Wissenschaftstheorie*, Festschrift für Victor Kraft, edited by Ernst Topitsch (Wien, 1960).

[23] For this idea see, for instance, Gerhard Weisser, *Wirtschaftspolitik als Wissenschaft* (Göttingen, 1934).

[24] See Karl Popper, *The Poverty of Historicism* (London, 1957), p. 58 *et seq.*

human action. This is why purely informative theories are very success-fully applicable in practice. Decisive in this connection are their *informa-tion content* and their *relevance* for the practical problems concerned.

The methodological consequence thereof is that the practical, particu-larly the political usefulness of the social sciences depends not on the formulation of value judgments or the explicit introduction of value premises of any kind but on the *choice of their factual problems*. It is not even necessary in every case to build technological systems. Even the elucidation of social reality may sometimes have an extraordinary signifi-cance. In my opinion the opponents of a value-free social science gener-ally underrate the practical significance of positive investigations and over-rate the introduction of value judgments considerably. *Value freedom as a methodological principle is completely compatible with practical rele-vance.*

Summarizing, we can say the following :

1. In a certain sense, *no science* can be value-free : all sciences have a value foundation; they are influenced by valuations.
2. In a certain sense, *social sciences* cannot be value-free : they must analyze valuations in their realm of objects.
3. In a certain sense, *any science* can be value-free : no science is in need of value judgments within the context of its propositions.

There is no value freedom in an absolute sense.[25] But the methodical principle of value freedom can be maintained almost in the sense of Max Weber.

IV.   A Critique of Welfare Economics

One of the areas of investigation which seems unavoidably to involve value questions is the subject of social welfare and, closely connected with it, of the social order. For a long while, these were central problems of political economy, even if they nowadays seem to be partly reserved to a special discipline, i.e., welfare economics. In this discipline the classical tradition of economic thinking, the tradition of natural law and utilitar-ianism,[26] finds its strongest expression, sometimes in a very formalized version. The problems confronting welfare economics are indeed very old. Quite recently a growing scepticism may be observed with regard to these problems.[27] Technical difficulties emerge which seem to be insurmount-

---

[25] As Karl Popper rightly said in his lecture, "Die Logik der Sozialwissenschaften," on the Soziologentag in Tübingen, October, 1961.

[26] As it is analyzed in Gunnar Myrdal's book, *Das politische Element in der national-ökonomischen Doktrinbildung* (Berlin, 1932; Swedish ed., 1930; English ed., 1953).

[27] This seems to be partly under the influence of the famous book by Lionel Rob-bins, *An Essay on the Nature and Significance of Economic Science* (1st ed., London, 1932).

able. Time and again restrictions and reformulations appear to be necessary to save the fundamental ideas. But it is interesting to see that in spite of scepticism the fundamental approach of welfare analysis is often held to be sound and fruitful in principle.

In my opinion, it is precisely in welfare economics that the inherent difficulties of the whole economic approach show themselves in the most outstanding way. They arise from the attempt to view society in a purely economic perspective, to make a theoretical analysis of social life with the aid of a conceptual apparatus which is basically suitable to the analysis of the commercial sector of society, the system of market relations, but is often implicitly understood to show the decisive factors for the determination of social welfare[28]—an apparatus originating with classical political economy and modified and refined by neoclassical theoreticians. This attempt has been exposed to various objections, but the theoreticians, who were not inclined to abandon this tradition, did not react by a fundamental revision of their approach. Rather, they tried to maintain this approach at any price by means which mostly belong to the realm of conventionalist stratagems.[29] As a result, in the domain of welfare economics, above all, there was an increasing isolation of economic thinking from the results of other sciences, permitting the character of political economy as a sociology of commercial relations to be obscured. It was thereby overlooked that the "autonomy" of a province of science is determined exclusively by the extent to which the problems of this field can be solved, not by an aprioristic decision depending on conceptual apparatus transmitted by tradition. We cannot know *a priori* that an analytic apparatus permitting the demarcation of an economic domain of society in a traditional way, and therefore the exclusion of so-called noneconomic factors, suffices for solving the fundamental problems it was created to solve. The development of welfare economics seems to me to be an excellent illustration of the fact that an assumption to the contrary can lead to a blind alley.

Welfare economics in its older, Pigovian form was a reaction to the classical conception,[30] that is, a refutation of certain elements of this view

---

Perhaps it is worth noting that all important difficulties which came into the open in the sometimes very subtle and technical welfare discussion were in principle anticipated by Myrdal's book (see note 26), which, however, was not taken into consideration at all.

[28] See my article "Nationalökonomie als Soziologie," *Kyklos*, vol. 13 (1960). The idea is well represented in Myrdal, *op. cit.*

[29] For a criticism of such methods see Popper's *The Logic of Scientific Discovery*, especially chap. IV, and his "Personal Report" (see note 10).

[30] See Tapas Majumdar, who in *The Measurement of Utility* (London–New York, 1958), pp. 3–4, rightly emphasizes the underlying welfare assumption of classical economics.

by an elaboration of the argument of external effects.[31] But this argument eventually led to a practice remarkably similar to that of the application of the much misused *ceteris-paribus* assumption.[32] Thus this argument finally turned out to be not a vehicle for the critical examination and revision of the whole approach but an instrument for the conservation of an entrenched tradition. We shall later come back to this.

The first serious blunder of Pigovian welfare economics, which was taken over by nearly all later contributors,[33] was the distinction between the *general* and the *economic* welfare of the members of society, the latter defined as that part of general welfare which can be brought directly or indirectly into relation with the measuring rod of money.[34] The import of this distinction appears to be at last to render possible the disregard of the so-called noneconomic factors, that is, to save the autonomy of economic thinking. Thus this distinction was dictated not by the problems but by a questionable theoretical tradition.

In any case, it seemed to facilitate the solution of a problem which must belong to the fundamental questions of any investigation in welfare economics : the problem of the valuation of social aggregate quantities like national income by the application of a universally valid standard. The illusion of the possibility of an "adequate" solution of this problem, however, was destroyed more than thirty years ago by the arguments of Myrdal, which, had they been considered at the time, would have made superfluous a large part of the subsequent discussion.[35] As Myrdal has shown in his analysis, behind the whole argumentation in welfare economics stands the "communist fiction" of a centralized "social housekeeping," a social economy to be conceived not as a mere network of social relations between economic subjects but, so to speak, as the collective subject of social valuation and value imputation. Society appears in the economic perspective as a cooperative unity, endeavoring to overcome the natural scarcity and to bring about the maximal satisfaction of all its members. Leaving out all consideration of conflicts of interests, the ideal of rational

---

[31] By this are meant economies and diseconomies which are not accounted for by the price mechanism of the market.

[32] For a critical analysis of this assumption see T. W. Hutchison, *The Significance and Basic Postulates of Economic Theory* (2nd ed.; New York, 1960), pp. 40–46.

[33] Some theoreticians who analyze the problem of a so-called social welfare function are exceptions, for instance Kenneth J. Arrow. See his famous book, *Social Choice and Individual Values* (New York–London, 1951), which is an investigation of the logical compatibility of certain plausible assumptions about such a function.

[34] This criterion was later justly refuted as insufficient. See J. de V. Graaff, *Theoretical Welfare Economics* (Cambridge, 1957), pp. 4–6. But an adequate substitute seems scarcely to be possible. Graaff questions his own provisional solution.

[35] See Myrdal, *op. cit.*, especially pp. 127–139 of the English edition. To be sure, the discussion has led to a development of the tools of analysis, etc., but in the criticism of fundamental ideas it has hardly led to new insights.

behavior, of acting according to the economic principle, is projected into society as a whole, so that the impression arises of a fundamental harmony of interests. With that a structural fact of social life of the greatest significance has been excluded.[36] That is the *proton pseudos* of economic thinking, even prior to the questionable distinction between general and economic welfare which has contributed to the development of welfare economics as a special province of economics.

In addition, there is to be emphasized the exploitation of the sharp distinction between the *sphere of production and exchange* and the *sphere of the distribution of income,* for the purpose of the determination of welfare.[37] Few economists were convinced that the realization of the ideal of competition in all parts of the economy is bound to lead to an ideal distribution of income. The idea that remuneration according to contribution to production is a just compensation, a normative interpretation of marginal productivity theory, as we find it with J. B. Clark is, to be sure, sometimes found even today, though in a more implicit form. But it was exposed from the outset to heavy doubts. The distinction between the spheres of production and distribution offered the possibility of confining the validity of economic laws to the sphere of production and of allowing in distribution of income the possibility of "ethical" considerations and, with regard to politics, of the exertion of influence by institutional measures. Correspondingly, Pigou distinguished between the formation of national income and its distribution as the determinants of economic welfare. With regard to the formation of national income, the coincidence of the marginal social net product with the marginal private net product and the equality of these magnitudes in all places—that is, the absence of external effects—seemed to him to be the condition for the maximization of welfare; while for the distribution of income the viewpoint of equalization—transfer of income from the rich to the poor, to equalize the marginal utility of income—was held to be decisive. In later developments the tendency was more and more to restrict economic reflections to the sphere of production and to assume as "given" the distribution of income—the determination of which obviously appeared to be subject to ethical viewpoints. Also this qualification seemed necessary for the salvation of the "pure economic" view. It may be that in the background the idea exists of a harmony of interests restricted to the sphere of

---

[36] There is a striking analogy to this in modern sociological thinking; see the criticism of the so-called structural-functional approach by Ralf Dahrendorf in his article "Out of Utopia: Towards a Re-Orientation of Sociological Analysis," *American Journal of Sociology*, vol. 64, No. 2 (1958). For a general criticism of utopian thinking which is highly relevant in this regard see Popper's *Open Society* and his *Poverty of Historicism*.

[37] For a critical analysis see Myrdal, *op. cit.*, pp. 126, 139.

production, while for distribution—the "ethical" aspect—the possibility of conflicts is left open.[38]

Thus welfare analysis centered around the problem of the *ideal output* (or production), identified with the problem of the *optimal allocation* of the means of production and dealt with as a *"purely economic" problem of efficiency*. But the isolation of this problem caused great and ultimately insurmountable difficulties. For welfare economists were forced by their intention, the determination of economic welfare, to relate the ideal output to the satisfaction of the needs of the members of society. This satisfaction, however, must be conceived ultimately as depending on the distribution of income, for whatever needs can be realized is obviously a question of the distribution of purchasing power to the members of society. The determination of the size of the national product—the real counterpart of collective satisfaction of needs—presupposes the valuation of the natural quantities of goods and with it the construction of a system of value coefficients.[39] It is obvious that the given market prices are not sufficient for this purpose.[40] In this regard the index problem causes not only technical difficulties but difficulties of an even more fundamental kind.[41] Behind these difficulties (of finding a practicable standard of evaluation for the national product) stood the problem of *interpersonal comparability* of needs and satisfactions, the solution of which seemed to amount to a normative decision.[42]

The endeavors to avoid this obviously "noneconomic" problem with-

---

[38] The idea of ethics as related preferably to conflicts is to be found elsewhere, too; see, for instance, Bernard Mayo, *Ethics and the Moral Life* (London–New York, 1958), especially pp. 99–114; see also Richard Bevan Braithwaite, *Theory of Games as a Tool for the Moral Philosopher* (Cambridge, 1955). Kenneth E. Boulding, in his interesting article, "Welfare Economics" in *A Survey of Contemporary Economics*, Vol. II, edited by Bernard F. Haley (Homewood, Ill., 1952), pp. 17–19, makes the distinction between "trading" and "conflict" (instead of "allocation" and "distribution") and sees the significance of the so-called Pareto-criterion of the new welfare economics in this distinction. In view of the "strong prejudice" of economists "in favor of trading," as Boulding rightly says, this may be seen as the typical demarcation of economists between the economic and the ethical sphere of investigation.

[39] For a simple representation see Boulding, *op. cit.*, pp. 6–10; see also the interesting article by the same author, "Some Contributions of Economics to the General Theory of Value," *Philosophy of Science*, vol. 23, No. 1 (1956), pp. 8–10.

[40] See my articles "Der Trugschluss in der Lehre vom Gütermaximum," *Zeitschrift für Nationalökonomie*, Band XIV/1, 1953, and "Die Problematik der ökonomischen Perspektive," *Zeitschrift für die gesamte Staatswissenschaft*, Band 117/3, 1961, pp. 453–459.

[41] As Myrdal has rightly emphasized in his book, *op. cit.*, p. 132.

[42] See Lionel Robbins, *op. cit.*, pp. 136–142. This is now contested by I. M. D. Little in his *A Critique of Welfare Economics* (2nd ed.; Oxford, 1957), chap. IV; but, as Streeten emphasizes in his appendix, "Recent Controversies," to the English edition of Myrdal, *op. cit.*, "even if they [i.e., interpersonal comparisons] can be made descriptively, they cannot provide *per se* the 'scientific' basis for social policies" (p. 209). There is a difficult problem of relevance to be solved beforehand.

out giving up the search for an ideal economic situation led eventually, in elaboration of an idea of Pareto's and Barone's, to the formulation of the so-called *Pareto-criterion* and of the rather generally accepted thesis that the *equilibrium of perfect competition* is a *Pareto-optimal* state according to this criterion—though only on the condition of the *absence of external effects.* We shall soon come back to the meaning of this requirement. Actually, a statement that appears to be far less misleading is one by Baumol to the effect that this "ideal output" is only a criterion of the responsiveness of the productive system to consumer demand, and consumer demand is determined by the desires of the individual consumer weighted by the purchasing power at his command.[43] The exclusion of external effects of all kinds is, as we shall see, a typical conventionalist stratagem in Popper's sense, which makes the Pareto-optimality thesis practically useless for the determination of welfare according to the aim of welfare economics.

To show this, it may be appropriate to enter into the particulars of the problem of needs in economics. In economic theory in general, and likewise in welfare economics, the needs of the members of society are mostly considered as given, as "data." That means two different things: first, that the economist regards himself as not entitled to pass *judgment* on individual needs; and second, that he is not inclined to go back *causally* behind the existing needs, that is, to explain them. Causally and normatively the "given" individual needs are conceived as the ultimate points of reference of economic thinking. Their satisfaction seems to be in a completely trivial and obvious sense the ultimate "purpose" of "the economy." It therefore constitutes the "natural" foundation for economic valuations of all kinds. In practice, however, qualifications in economic analysis are made which are remarkably characteristic of the peculiarity of the economic perspective. In it the essentially market-sociological character of the conceptual apparatus of economics finds expression. To begin with, for most purposes all needs are left out of consideration which are not expressed in the sphere of consumption,[44] and consumption is identified with the purchase of consumer goods, that is, with a market transaction. The tools of indifference-curve analysis—stemming from Edgeworth and Pareto—are eminently suitable to freeing the economist from the analysis of real consumption and of the formation of the needs of individuals, and to confining him to the analysis of market decisions. With their aid a watertight separation of economics from social psychological and sociological investigations of all kinds seems possible; sometimes the impression arises that these tools have been contrived precisely for that purpose. It is achieved

---

[43] Cf. William J. Baumol, *Welfare Economics and the Theory of the State* (Cambridge, Mass., 1952), p. 60.

[44] See Myrdal, *op. cit.*, p. 136.

by the identification of satisfaction, consumption, and purchase of consumer goods, and the determination of the last—a market decision—with
the aid of a system of indifference curves, almost devoid of content.

Now, this whole technique of reduction, with its inherent philosophy
of "nothing but," is by no means a harmless one, if we ask how it affects
the solution of the original economic problem: the problem of social
welfare and need-satisfaction. Above all, it must be emphasized that the
interests of men in production are "not confined to its products and their
distribution." They are also undoubtedly strongly directed to "their welfare as producers"[45]; that means, for instance, to their working conditions.
Moreover, the conjecture seems to be justified that these working conditions can have a considerable influence on the consumption needs. That
alone would suffice to render the neglect by economic analysis of the
formation of needs impracticable in view of its own purpose. Besides, it is
no longer contestable today that, in other respects as well, strong effects
on the formation of wants originate in the sphere of production itself, which
make impossible the assumption of the "givenness" of needs. The wants
of individuals often seem to be very nebulous,[46] so that the market process
itself can play a significant role in their determination. The formation of
wants is, after all, a social process, shaped to a large extent by the whole
cultural environment.[47] Without losing ourselves in details we can perhaps
say that the formation of the needs of individuals depends on their group
membership and their reference groups.[48] A change in these relations will
probably very often lead to a modification of individual aspiration levels
and consumption norms. Also the level of an individual's needs, and the
extent to which they are satisfied, will depend on the consumption of
others belonging to the same groups or strata, or to the reference groups of
the individual in question. Thus the relative position of individuals to
each other with regard to income and consumption may be of greater importance than the amount of goods actually consumed. Whoever prefers
to leave such things out of consideration because they transcend the usual
frontiers of economic analysis may occupy himself with the problem of
purchasing decisions, but he certainly does not touch the problem of the
satisfaction of wants and the problem of welfare in the original sense of
this word. Neither will he be able to solve the problem of the market decision of the consumer in this way. Again it seems that the instruments of

---

[45] *Ibid.*, p. 136.

[46] As Baumol rightly states, *op. cit.*, p. 61.

[47] We have to reckon not only with interdependencies of need-structures but also
with interdependencies of satisfaction; see the famous book by James S. Duesenberry,
Income, Saving, and the Theory of Consumer Behavior (Cambridge, Mass., 1959), chap. III,
and Little, *op. cit.*, p. 43.

[48] As this term is used in modern sociology.

analysis and the perspective underlying their use have suggested a procedure which hinders the solution of the original problems and substitutes pseudo-problems for them.

A further limitation consists in the fact that the existence of collective needs and goods cannot be accounted for adequately. Collective needs are to be understood here as those needs of individuals which cannot be satisfied by the individual appropriation of goods to be acquired on the market, but only by facilities provided by the community for public use (collective goods), for instance, public parks, streets, public utilities, etc.[49] Now the preference scales of individuals for individual goods are without doubt strongly influenced by the existence of such collective goods and institutions, so that the provision of such goods by the community causes changes in the structure of individual needs. For instance, the individual demand for cars is probably very strongly influenced by the existence, density, and quality of an adequate network of roads; the demand for books presupposes institutions of education; the wish to possess radio and television sets depends on the existence of corresponding transmitting stations, etc. It is hardly an exaggeration to say that the whole structure of needs expressed in the preference scales for individual consumers goods would be completely incomprehensible without knowledge of the background of the collective "infra-structure," especially in a modern industrial society.

All that is not taken into account if, when reflecting on a social optimum regarding the satisfaction of wants, we only refer, as is usual today, simply to the individual preference scales related to market decisions. Besides, there are many factors reducing the level of individual satisfaction which are to be considered as byproducts of industrialization, urbanization, etc., such as smoke, and water pollution.[50] We could call them negative collective goods. All this belongs to the range of external effects in the Pigovian sense, but there are other effects of this sort in the sphere of production.[51]

The result of our analysis of the problem of needs or wants must be the statement that the treatment of needs as "given," and the manner in which the problem of needs is fitted into the study of welfare, must have destructive consequences for the whole examination of the welfare problem. It is therefore understandable that many authors at present come to a

---

[49] See the interesting extension of the external economies argument to the problem of the state by Baumol, *op. cit.*

[50] See the interesting analysis of these things by K. William Kapp, *The Social Costs of Private Enterprise* (Cambridge, Mass., 1950), which seems to be biased nevertheless by an implicit acceptance of the "communist fiction."

[51] We will content ourselves with the hint at the so-called external effects in technology; for an analysis of this problem, see J. de V. Graaff, *Theoretical Welfare Economics* (Cambridge, 1957), pp. 18–32. Also technology is mostly treated as "given" in welfare economics, which is dangerous with regard to these external effects.

fairly negative judgment with regard to the *relevance* of welfare analyses.[52]

All these manipulations—the distinction between general and economic welfare, the separation of the spheres of production and distribution in order to be able to disregard the latter, the interpretation of the problem of ideal production as a purely economic problem of efficiency, the retreat to the Pareto-criterion as a solution of this problem, the assumption of the individual wants and of technology as "given," and lastly the exclusion of all external effects in the spheres of consumption and production— seem to me to be components of a great attempt to save the conception stemming from classical political economy and elaborated by neo-classical economics, even at the price of keeping virtually nothing of the problems originally treated. This attempt has led to a remarkable refinement of the tools of economics and to fairly sophisticated arguments and discussions, but the formalization of solutions achieved thereby and the elegance of representations only conceals the fact that the result has been a complete sterilization of the problems. All difficulties which have arisen during the discussion have not brought about a *revision* of the fundamental approach but only efforts to *immunize* the respective social optimum theses *against any possible criticism.* Presumably here the application of conventionalist stratagems has been raised to the status of a methodological principle. In the background remains the communist fiction of classical and neo-classical thinking. Indeed, if reality were otherwise, we might be able to formulate certain conditions of a social optimum.

But in addition to that, we have to make a further remark. If one talks about "conditions" of welfare, then it is by no means recommendable to indulge in the illusion of being in the realm of a causal problem, as it seems to be from the often-stated intention of economists to investigate the "causes" of welfare. On the contrary, it is never a question of empirical conditions of a causal or structural kind, as we could suppose *prima facie,* but always a matter of conditions in another sense of this word : of logical or formal conditions. We could even speak of the investigation of the logical grammar of "economic welfare" in a sense analogous to that we find in linguistic moral philosophy. As there the danger emerged of a linguistically disguised essentialism, we see in the case of welfare economics a kind of essentialism in the guise of a formal analysis.[53] While in the first case an

---

[52] See, for instance, E. J. Mishan, "A Survey of Welfare Economics, 1939–1959," *The Economic Journal,* vol. 70, No. 278 (1960), p. 256.

[53] This is not meant exactly in the same sense as in G. C. Archibald's interesting article, "Welfare Economics, Ethics, and Essentialism," *Economica,* vol. 26, No. 104 (1959), though I accept the points made by Archibald. I think, however, that the criticism may go further. Even if we conceive of welfare economics as a part of positive economics, which is certainly possible, we have to criticize the above-mentioned use of conventionalist stratagems and the consequent emptiness of the theorems. It may be that welfare theorems *can be made* refutable, as Archibald says, partly, "if they are

intuitively plausible view is presented as the result of an analysis of ordinary language, in the second case the formalization of an intuitive conception of welfare—in fact the legacy of an old theoretical tradition—appears as a result of a formal analysis of conditions, which strictly speaking was directed to the essence instead of the causes of welfare. A utopian ideal state of affairs, the general equilibrium of perfect competition, the institutional conditions of which remain completely in the dark, thereby receives the dignity of scientific justification. It seems to be possible to have a purely formal insight into the conditions of an abstract common good which is unfortunately difficult to realize.

If we now relate our results of the analysis of the value problem to the problem of welfare as it is treated in welfare economics, we can make the following remarks : The welfare problem, as it has been conceived by classical political economists, may have been of great political significance. It is not astonishing under these circumstances that there was a strong inclination at that time to fuse normative and factual elements (values and facts) to a certain degree in the treatment of this problem. But we know that the practical usefulness of an inquiry does not depend on the formulation of value judgments but on the choice of its factual problems. Consequently, the elimination of the normative elements of classical thinking would not have done any harm to its usefulness, especially if simultaneously the information content would have been increased and its relevance maintained. But in welfare economics the development went in the opposite direction. The result was a highly artificial, fairly *contentless,* and, in relation to the fundamental problem, completely *irrelevant* formal system, continually modified by the application of conventionalist stratagems.

Now there is a tendency to complete the system in question by the addition of value premises,[54] to lay open its implicit value content, in order to protect welfare economics against the reproach of having committed the naturalistic fallacy. Even leaving the justification of this blame undecided, we can say that this modification cannot be of any significance, for it restores to welfare economics neither its relevance nor its content. The same is to be said against a modification in another direction : the complete elimination of normative elements without any other change. This leads to purely formal systems as they are invented by the constructors of the so-called social welfare functions. I see no improvement in such modifications. There is no substitute for content and relevance if we want to have theories which are practically useful. Neither a normative nor a

---

stated in terms of a refutable demand theory" (p. 325). But then the original problem may be lost. "Welfare" may become a "systematically misleading expression." That may be the important insight behind the *prima facie* inconclusive argument of Mishan against Archibald; see Mishan, *op. cit.,* p. 200, note 2.

[54] See Little, *op. cit.,* which is criticized severely by Archibald; see his article "Welfare Economics . . .," *op. cit.*

formal system is so practicable as a technology based on empirical science. Therefore we cannot but agree with the judgments of scientists like Myrdal, who see in the development of welfare economics only a blind alley, a formalization of classical utilitarianism, the philosophical background of classical economics.

V.   The Problem of the Social Order

Perhaps it is no great misinterpretation to see as the core of the classical problem of political economy the question of a social order which solves the problem of poverty,[55] a question which unfortunately has been formulated under the influence of utilitarian ideas as the problem of a social order which guarantees the maximization of want satisfaction of the members of society. In this form the problem lies at the bottom of welfare economics, the results of which again and again have been drawn upon in the discussion about the economic and social order. For a long time the idea of the possibility of economic calculation in a centrally planned economy stood in the center of this discussion. It may easily be shown that nearly all participants in this discussion have used typical welfare arguments — even if they had no connection with welfare economics as a special discipline — in order to demonstrate the possibility or impossibility of a *rational economy* with a central planning body. Thus they have debated the question of the rationality of an economic and social order depending on an aim held to be self-evident : the aim of maximizing the collective satisfaction of wants.[56]

Now with regard to the arguments in this debate, it is striking what significance is generally given to the state of equilibrium of perfect competition — the very state which was characterized as Pareto-optimal and the realization of which involves, in the opinion of the more optimistic participants of the debate, the maximization of social want satisfaction, the sovereignty of the consumer, and perhaps even an ideal distribution of income. Now the welfare discussion itself has, after all, furnished sufficient clues to the fact that the real significance of Pareto-optimality by no means justifies the usual role of this equilibrium-state in the discussion about the problem of the social order. Nevertheless, the whole manner in which the problem of welfare and the connected problem of the social order has been treated so far has scarcely been questioned. Not only do many "self-evi-

---

[55] This formulation, which I owe to Karl Popper—who suggested it to me in a private discussion—seems to be confirmed if we read the brilliant book by Robert L. Heilbroner, *The Worldly Philosophers: The Lives, Times, and Ideas of the Great Economic Thinkers* (rev. ed.; New York, 1961).

[56] Quite recently new problems have emerged, as for instance, the problem of freedom; see Friedrich A. Hayek's important book, *The Road to Serfdom* (Chicago, 1944).

dent" ideas of neo-classical thinking, which ought to be shaken, after all, by the course of the welfare discussion, emerge again and again in their old form in the order-debate. In general it does not even seem possibly conceivable that the whole manner of putting the question must lead astray.

In the hitherto prevailing mode of framing the question of the optimal economic order, however, we see an excellent example of that utopian and holistic thinking[57] which in other systems has led to serious deficiencies and dangerous consequences. A normative model-Platonism, starting from an abstract ideal, an absolute utopia aloof from the course of history, measuring the factual conditions by it and deriving normative consequences from it, is still to be recognized in the most formal welfare constructions. The concentration on the question of maximal satisfaction and on the criteria for its measurement, the formalization and emptiness following therefrom, and the approach to the problem of order from this direction are reliable evidence for the fact that the communist fiction, criticized above, and the utopian approach connected with it, basically speaking, govern the formulation of the question. That also seems to me to be one of the reasons which have induced some social scientists to deem it necessary to normativize social science for practical aims. Only by starting from a supreme abstract principle or system of principles do concrete recommendations for the formation of the social order and the influencing of social life appear to be possible. Only on the basis of a general idea of the common good does a solution of the burning practical problems seem to be possible. On the other hand, only an idea of the common good which takes into consideration the needs of all members of society appears to be justifiable in our time. Therein lies the element of the idea of maximal satisfaction, which is most suggestive for us today. That this conception is completely empty,[58] and the problem of its realization not only practically, but also in principle, insoluble, is overlooked most of the time.

The above-sketched approach has a striking analogy to the theory that truth is manifest, criticized severely by Karl Popper,[59] which prevails in classical rationalism and empiricism and which has an influence not to be underrated on philosophy of science even today. The moral philosophical ideas of natural law and utilitarianism behind this approach bear traits similar to these epistemological conceptions. Indeed, they are the moral philosophical and political versions of the idea that truth is manifest.[60] The

---

[57] The fundamental criticism of this style of thinking we owe to Karl Popper; see his *Poverty of Historicism* and his *Open Society*. . . .

[58] For the problem of such empty formulas, see Ernst Topitsch, "Über Leerformeln," in *Probleme der Wissenschaftstheorie*, Festschrift für Victor Kraft (Wien, 1960).

[59] See, above all, Popper's lecture, "On the Sources of Knowledge and Ignorance."

[60] For an analysis of the connection of epistemological, moral, and political ideas see J. W. N. Watkins, "Epistemology and Politics," *Meeting of the Aristotelian Society*, June 12, 1957.

"rational" element of an intuitively recognizable common good is connected with the "empirical" element of individual need satisfaction. Society is sometimes conceived of as flatly analogous to an induction machine deriving from individual needs adequate decisions in the sphere of production and thereby, in the last analysis, maximum social welfare. That seems to be the economic variant of classical democratic ideology, which is closely connected with the manifestation theory of knowledge.

On the other hand, Popper's approximation theory of knowledge[61] also has an analogue in the realm of social life and of politics. It is, indeed, contained in his ideas about social technology and social engineering. The idea of approximation leads to piecemeal engineering in practical social life. One does not start from the intuition of an absolute optimum, an ideal state of affairs, which is to be realized by all means in social reality, but from concrete evils which are to be eliminated. Here exists a wide domain of application for social technologies as they are characterized above, of technologies which are to be obtained from a value-free theoretical social science with explanatory power, for to *explain* a social phenomenon is *to show how to avoid it* (in principle), as we can say analogous to Quine's famous statement about definitions.[62] The explanation of social evils creates at least one foundation for their elimination.

The idea of *absolute justification*, which is, as Popper has shown,[63] connected with the manifestation doctrine and implicit in all former epistemologies, is, as I see it, also contained in the usual manner of argument with regard to the social and economic order. The notion of maximal social satisfaction is a typical product of a justificatory thinking of this kind. Here too, "truth," that is, the true aim of society, appears to be manifest, and all practical problems soluble if we have grasped it. In the discussions about social welfare and about the optimal social order, the difficulty of such a notion has shown itself, which leaves out of consideration the starting position for all possible political measures : the *institutional a priori* to which all such actions have to refer. Society is no tabula rasa—just as little as the human mind—no vacuum, in which all ideas normatively stated as right can be realized. It can be reformed, but not created anew.

A combination of the critical examination of social life and its institutions with the aid of theoretical thinking and of practical attempts to re-

---

[61] See, for instance, his article, "Über die Zielsetzung der Erfahrungswissenschaft," *Ratio*, vol. I, No. 1 (1957).

[62] See Willard V. O. Quine, *Mathematical Logic* (rev. ed.; Cambridge, Mass., 1955), p. 47.

[63] As Popper kindly wrote to me, William W. Bartley sees the core of Popper's philosophy in his saving of rationalism by showing that ultimate justifications are impossible and that criticism alone is essential and sufficient for rationalism and for questions of validity. This idea is developed in Bartley's book *The Retreat to Commitment*. [See Bartley's contribution to the present volume. Ed.]

form them by application of theoretically founded technological systems—
in short, a combination of social criticism and social technology as the
basis of politics—this would be the realization of the idea of approximation in social life, of approximation to a state for which there is just as
little a criterion as for truth in the realm of knowledge.

CHAPTER 26

Popper and the Critical Philosophy of History

BY W. B. GALLIE

I

THE PHILOSOPHICAL WRITINGS of Karl Popper possess three outstanding merits : they are bold; they are sane; they are thorough—qualities that are particularly significant in the context of contemporary British philosophy. It may seem strange, however, that one who admires *inter alia* the thoroughness of Popper's main contributions to philosophy should choose to write in this volume about a part of his work which is, as I am sure he would agree, tentative and sketchy—an indication of the general direction of his thought, not a fully worked-out and defended thesis.

My excuse is that the fourth section of Popper's *The Poverty of Historicism* seems to me, despite its sketchiness, to succeed precisely where previous critical philosophies of history have failed. It succeeds in getting into connected, graspable, arguable form a number of issues which have previously hung separate, as targets for disconnected *aperçus* or for the kind of cursory survey that belongs to *belles lettres* rather than to philosophy. What I offer here is a few extended glosses on some sentences from the closing pages of Popper's book; glosses which, I hope, will help to suggest further useful lines of research in this field. Possibly my suggestions will not win Popper's approval. I shall be satisfied, however, if I can show how gloss-worthy these passages from his philosophy are.

Popper's *The Poverty of Historicism* is primarily a work of critical demolition. It is directed at a highly influential, though often disguised and allusive body of doctrines found in varying degrees of completeness and coherence in the works of a number of eighteenth, nineteenth, and twentieth century social thinkers. The crucial error of these thinkers, according to Popper, is their belief in the possibility of a *theoretical history*, i.e., an account of past human struggles, achievements, and failures which both explains why these things had to take place in the way they did and, when suitably extrapolated, enables us to predict at least the broad shape and order of the future struggles and achievements of mankind. I agree almost entirely with Popper's diagnosis of this error and with the criticism to which he subjects it : and I know of no significant historicist comeback or counter-argument to Popper's devastating attack. But *The Poverty of Historicism* is inevitably something more than a work of destructive criticism. Periodically Popper has to show his own hand, to outline, and to defend in outline, the positive positions from which he launches his criticisms of historicism.

Insofar as these positions relate to the philosophy of history, their broad character is such as might be expected from Popper's other writings. In order to appreciate the kind of thing that historians have to tell us, or the kind of value which their researches and writings possess, we must ask two questions. First, how should we mark off their peculiar aims and interests from those of theoretical—whether physical or social—scientists? And secondly, when we ask whether a historical thesis or inference is justified, do our methods of testing it differ in principle from those employed in the theoretical sciences? On the first issue, that of the demarcation of historical from theoretical studies, Popper answers by strongly contrasting our aims and interests in the two cases. His position with regard to the peculiar aims and interests of history can be described by his own phrase— introduced, to be sure, for a slightly different purpose—'methodological individualism.' On the second issue, however, Popper answers that there is no difference in principle in our general methods of testing as between history and the theoretical sciences. This is his principle of the unity of method. These basic and closely interwoven theses are clearly expressed in the following passages.

(1)

The thesis of the unity of scientific method . . . can be extended, with certain limitations, even to the field of the historical sciences. And this can be done without giving up the fundamental distinction between theoretical and historical sciences . . . —a distinction which has been so often and emphatically reaffirmed by the best historians. It is the distinction between the interest in universal laws and the interest in particular facts. I wish to defend the view, so often attacked as old-fashioned by historicists, that *history is characterized by its interest in actual, singular, or specific events, rather than in laws or generalizations.* . . . While the theoretical sciences are mainly interested in finding and testing universal laws, the historical sciences take all kinds of universal laws for granted and are mainly interested in finding and testing singular statements.[1]

---

[1] *The Poverty of Historicism*, pp. 143–144.

(2)

In the sense of this analysis, *all* causal explanation of a singular event can be said to be historical in so far as the 'cause' is always described by singular initial conditions. And this agrees entirely with the popular idea that to explain a thing causally is to explain how and why it happened, that is to say, to tell its 'story.' But it is only in history that we are really interested in the causal explanation of a *singular* event.[2]

For Popper, therefore, the main difference between historical and theoretical studies, although of profound importance, is logically very simple. It is indeed a matter of different orientation of interest rather than of logical structure.[3] Essentially the same types of explanation and justification—hypothesis, deduction, and testing of deduced results—gives us the logical structure of both types of study. But in historical study, our interest is in a converse relation to our interest in theoretical studies. In the former we are interested to enlarge our knowledge and understanding (through obtaining acceptable explanations) of particular events; in the latter we are interested to enlarge the range of application (and also the degree of unification) of our generalizations and theories.

So far all is plain sailing. But the following passage begins to reveal complications.

(3)

This leads us to the question of the *uniqueness* of historical events. In so far as we are concerned with the historical explanation of typical events they must necessarily be treated as typical, as belonging to kinds or classes of events. For only then is the deductive method of causal explanation applicable. History, however, is interested not only in the explanation of specific events but also in the description of a specific event as such. One of its most important tasks is undoubtedly to describe interesting happenings in their peculiarity or uniqueness; that is to say, to include aspects which it does not attempt to explain causally, such as the 'accidental' concurrence of causally unrelated events. These two tasks of history, the disentanglement of causal threads and the description of the 'accidental' manner in which these threads are interwoven, are both necessary, and they supplement each other; at one time an event may be considered as typical, i.e. from the standpoint of its causal explanation, and at another time as unique.[4]

At first sight the dual task here ascribed to history—that of describing what is unique or, from a logical point of view, contingent or arbitrary on the one hand, and that of explaining by general laws on the other—might seem quite analogous to the typical situation in the natural sciences. Take Popper's own favorite example of Newton's explanation of Kepler's descriptions (as we may here usefully call them) of the paths of the major

---

[2] *Ibid.*, p. 144.
[3] *Ibid.*, see in particular p. 133, para. 2.
[4] *Ibid.*, p. 146.

planets. According to this explanation, the number of the planets and their respective distances from the sun are, in Popper's terminology, an "accidental concurrence of causally unrelated events"; they just happen to be as they are, and their number and respective positions vis-à-vis the sun had to be established before Newton's explanations could be applied to them. But can the unique, because 'accidental,' interconnections of the events that are the concern of history be assimilated to those of astronomy or physics or any other theoretical science? Popper stresses that history is interested in "the description of a specific event *as such*" (my italics). Can this be said of any theoretical science? Are not the theoretical sciences interested in particular events primarily, if not exclusively, for their being typical events? Of course Newton's laws apply to the planetary motions as described by Kepler, but an essential feature of the explanation is that it is one which would apply equally *to any set of planets,* irrespective of their number, relative positions, etc. Moreover, from the very nature of the mathematical descriptions that we apply to them, the arbitrary or contingent factors or elements in the theoretical explanation are conceived as at least possible candidates for *further explanation* by means of some further no doubt highly complex application of physical laws and theories. But does either of these presuppositions hold in the case of a historian's description and attempted explanation of a particular concatenation of events? The answer is, I think, plainly no; and for two reasons. First, many of the contingencies that play a crucial role in history are of such an extreme kind—they involve the coincidence of such various factors as human decisions, economic trends, lethal epidemics, changes in climate, and so on—that the idea of making a clear deduction from them as a complex premise is quite inconceivable. But second, and perhaps more important, the characteristic force and value of the historian's explanation is to be seen not in the fact that it will apply equally to any properly comparable case, but in the fact that it helps us to make sense of some larger but still *equally particular* theme or period of history.

These differences in the role of explanation as between history and the theoretical sciences seem to suggest that it is the contingent or unexplained facts in history that are the most crucial and most worthy of pursuit by the historian as he composes, and by the reader as he follows, the narrative. It is the unforeseeable incident or discovery or decision, giving the wholly new twist to the seemingly typical flow of events, that fosters historical interest by making us want to hear more of how things actually developed. On the other hand, insofar as a historical event is typical, or insofar as it follows from certain antecedents or gives rise to certain effects as a matter of course, it will usually receive only brief or allusive treatment. What was "only to be expected" can never be of great historical interest.

We may well wonder, however, whether this suggestion is compatible with what Popper affirms in the following passage.

(4)

The historical sciences, it may be remarked, do not stand quite apart in their attitude towards universal laws. Whenever we encounter an actual application of science to a singular or specific problem we find a similar situation. The practical chemist, for example, who wishes to analyse a certain given compound—a piece of rock, say—hardly considers any universal law. Instead, he applies, possibly without much thought, certain routine techniques which, from the logical point of view, are tests of such *singular* hypotheses as "this compound contains sulphur." His interest is mainly a historical one—the description of one set of specific events, or of one individual physical body.[5]

But surely that description is of value, as a piece of applied science, because it succeeds in presenting a particular stage of the rock's history (or a possible particular use of it for certain types of human purpose) as a case of certain combinations of known, even if "hardly considered," physical and chemical laws. The practical chemist's interest is thus in a "case history," i.e., in an instance illustrative of certain general laws or truths. But, although any particular historical event, as treated by a historian, will no doubt *exemplify* certain general laws or truths, this, as Popper admits, is not the exclusive interest of the historian. In this connection we may usefully look back at the closing sentence of quotation (2), above. No doubt Popper is right in alleging that his account of causal explanation agrees with the popular idea that to explain an event causally is "to tell its story." But clearly this is a metaphorical extension of the word 'story' that could very easily prove misleading. The 'story' of how an electric power station failed is (if it is scientifically satisfactory) the story of a *case*. Certain things happened in the sort of way that relevant laws of electricity and mechanics and chemistry would lead us to expect in certain specific circumstances, so that had any sufficiently intelligent engineer been aware of these circumstances he could have predicted the power failure. But this is not true of a story (or at least not of a satisfactory story) in the ordinary sense of the word. If its dénouement were predictable by any sufficiently intelligent person who had studied carefully its *mise en scène,* then it would be a most unsatisfactory story, a story not worth listening to or following, a story that just would not count as a story at all. And what is true of stories in this respect seems to me to be true equally of histories also.

Closely connected with this point is Popper's retort to those historians who declare

(5)

. . . that history has no interest whatever in such [*sc.* universal] laws. To this we may answer that a singular event is the cause of another singular event—which is its effect—only relative to some universal laws. But these

---

[5] *Ibid.,* pp. 145–146.

laws may be so trivial, so much part of our common knowledge, that we need not mention them and rarely notice them. . . .

Among the theories which the political historian presupposes are, of course, certain theories of sociology—the sociology of power, for example. But the historian uses even these theories, as a rule, without being aware of them. He uses them in the main not as universal laws which help him to test his specific hypotheses but as implicit in his terminology.[6]

Popper is certainly right to insist that the historian uses laws and theories, "as a rule, without being aware of them . . . as implicit in his terminology." And in this respect, to be sure, the parallel between the historian's treatment of a particular event and that of the practical chemist can properly and usefully be emphasized. But Popper also sees that it is just as important to keep in mind the following point of contrast between the two kinds of case.

(6)

If we consider the historical sciences in the light of our comparison between them and the theoretical sciences, then we can see that their lack of interest in universal laws puts them in a different position. For in theoretical science laws act, among other things, as centres of interest to which observations are related, or as points of view from which observations are made. In history the universal laws, which for the most part are trivial and used unconsciously, cannot possibly fulfil this function. . . . The only way out of this difficulty is, I believe, consciously to introduce a *preconceived selective point of view* into one's history; that is, to write *that history which interests us*.[7]

In fine, if it is true that in the physical sciences "there is always a theory" to guide separate experimental researches, it is just as true that in history "there is always a story" to unify and guide the historian's successive assessments, interpretations, criticisms, and judgments.

Nevertheless, we must insist that in its capacity as guide to our interest a story differs profoundly from a scientific theory. For example, consider the way in which, from the scientific point of view, we treat contingencies —unexplained or unexplainable intrusions into some carefully designed experimental situation. Insofar as uncontrolled contingent factors enter into any such situation, the scientist believes either that they will "cancel each other out" or else that they will themselves ultimately be explained by means of general laws. By contrast, I have suggested, when we follow either a historical or a fictitious narrative our *dominant* interest is in the surprising or unpredictable events that have happened, or are about to happen : it is upon accidents and other contingencies that the narrative chiefly turns and that our need to follow it through in all its individual details

---

[6] *Ibid.*, p. 145.
[7] *Ibid.*, p. 150.

depends. Further elaboration of this point would carry us too far. All I wish to insist upon is the simple and unquestionable fact that, where we are concerned with descriptions of the actions of human beings, we endeavor to follow their particular histories—sometimes across the most surprising contingencies, coincidences, and so on—*because they interest us* in a way that would be quite unthinkable in the case of any particular events or sequence of events considered primarily as instances or cases of some scientific law or theory.

II

Thus far I have considered what I find incomplete in Popper's demarcation of historical from theoretical scientific interest. Now more briefly I want to discuss how he vindicates his principle of the unity of method with regard to the testing of both historical and theoretical scientific conclusions. This method, we should notice, is described by him chiefly in terms of causal explanations.[8] Now his account of explanation is likely to seem less satisfactory the more stress we place upon history's peculiar interest in what is contingent or unpredictable (as we have just done in the preceding paragraph), as opposed to what can be seen to have been a necessary or probable result of certain prior conditions considered in the light of certain universal laws. We might risk putting the difficulty in the following paradoxical form : Often in reading histories we find ourselves accepting contingencies or unforeseeable developments, and indeed finding them, in a sense, intelligible, although they could not conceivably have been predicted from any previous state of affairs with the fullest imaginable knowledge of general causal laws. How can this situation be explained? Quite easily, I think, if we can agree to accept Popper's general account of causal explanation and his principle of the unity of method (as applied to history) as only a first broad sketch of a position which calls for later modification. And in fact there is every reason for interpreting Popper's position in just this way, as the following interesting passage makes clear.

(7)

. . . there are good reasons . . . for the belief that social science is less complicated than physics . . . for . . . in most, if not in all, social situations there is an element of *rationality*. Admittedly, human beings hardly ever act quite rationally (i.e. as they would if they could make the optimal use of all available information for the attainment of whatever ends they may have), but they act, none the less, more or less rationally; and this makes it possible to construct comparatively simple models of their actions and interactions, and to use these models as approximations.

---

[8] Cf. quotation (1), above, and *The Poverty of Historicism*, p. 133.

. . . I refer to the possibility of adopting, in the social sciences, what may be called the method of logical or rational construction, or perhaps the 'zero method.' By this I mean the method of constructing a model on the assumption of complete rationality (and perhaps also on the assumption of the possession of complete information) on the part of all the individuals concerned, and of estimating the deviation of the actual behaviour of people from the model behaviour, using the latter as a kind of zero co-ordinate. An example of this method is the comparison between actual behaviour (under the influence of, say, traditional prejudice, etc.) and model behaviour to be expected on the basis of the 'pure logic of choice,' as described by the equations of economics.[9]

Now it seems to me that here we have a clue as to how Popper's first broad account of explanation in history can be made consistent with the recognition of the crucial part played by contingencies in historical narratives. In general, if we were to try to understand any sequence of human actions from the point of view of the causal principles which it exemplifies, we would make very little progress toward an intelligible account or narrative of it. But when we work from the supposition of rational action and reaction, and then proceed to note probable deviations from rational norms, we can do a great deal better. This seems to me to be the way in which we learn to follow not only puzzling records of historical events, but the devices of our political opponents, the contrariness of difficult children, and even the freakishness of odd developments in any game of chance and skill. And we may also note in this connection that we can often pick out and appreciate supremely rational or skillful action by referring back to the logical possibilities of the situation, when it would have been quite impossible for us to understand such action in the sense of being able to predict its occurrence from certain factual conditions in the light of general laws.

But now it may be objected that this way of explaining historical events, through a consideration of their 'situational logic,' is explanation of an entirely different type from that which enables us to deduce results from initial conditions by means of general laws. At first sight it might well seem to be so; but the truth is, I believe, that in history the two kinds of explanation are complementary and are found almost inextricably fused in the majority of historical explanations. To explain what I mean here, let me recall Popper's insistence, in quotation (5), above, that the universal laws which historians use "may be so trivial, so much part of our common knowledge, that we need not mention them and rarely notice them"; and this further concession that the historian commonly uses sociological theories "without being aware of them . . . not as universal laws which help him to test his specific hypotheses but as implicit in his terminology." Now in view of these concessions, it is hard to see why Popper assumes that this

---

[9] *Ibid.*, pp. 140–141.

(often implicit) use of general statements always involves or contributes to an *explanation*. Consider, for example, the class of sentences exemplified by :

> Trusting to these reports, the President ordered the attack.
>
> Elated by the good news, the general came in smiling.
>
> Lifting his bat higher than usual, the captain smacked the next ball into the pavilion.

In each of these cases the participial phrase could be regarded as a statement of initial condition from which, in accordance with all manner of (in fact unmentioned) general laws, the main clause could be regarded as a deduced result; i.e., we would in each case have a historical explanation according to Popper's basic account. But wouldn't this be an extremely unnatural claim to make? Would it not be a far better account of the situation in each of the above cases to say that what they give us is a selective description of a train of intelligible actions in which one point (that expressed in the main clause) is particularly emphasized while another part (that expressed by the participial phrase) is somewhat less emphasized? Understanding of the kind of sentence here in question evidently requires a whole background of unspecified generalizations and of possible connections between generalizations—and no doubt of a number of unspecified particular actions as well—which we leave the ordinary intelligent reader to supply for himself.

These considerations lead me to suggest that, in respect of both its 'semantics' and its 'pragmatics,' the meaning of explanation as applied in history differs from its meaning when it is applied in the theoretical sciences. And, briefly and dogmatically, I would outline the characteristic use of explanations in history as follows. The historian is, typically, presenting us with a followable narrative, i.e., one in which successive actions are followed or understood immediately by virtue of certain well-known generalizations and connections between generalizations that are implicit in the terminology that he employs. But suddenly he breaks off from this his usual task; he begins to justify what he is saying or is about to say. Commonly he will contrast his own—no doubt somewhat surprising—new view of the subject with that of previous authors or that which tradition or our natural expectations would have led us to prefer. He then proceeds to recast the relevant part of his narrative in the light of, for example, the logically possible lines of action open in the situation described, the choice of these possible actions which an ideally logical (and/or ideally moral) agent would make, the deviation from this ideal choice that was in fact exemplified in the action or policy chosen, the probable causal explanation of this deviation—an explanation that might be in psychological, economic, sociological, or physical terms—and so on. More simply, I want to suggest that the commonest and most interesting type of explanation that we meet in histories is that in which the historian explains to us, his readers, what he is doing (saying) and why. It should be noted that within an explanation of this kind elements of "situational logic" and elements of straightforward causal explanation are commonly juxtaposed and are often almost inextricably interwoven. Hence both parts of Popper's account of explanation in the historical sciences find a place in this more general account, whose specific feature is its rejection of the assumption that historians are always explaining (in some serious and useful sense of that word) whenever they are applying a well-known or trivial generalization to a particular case.

III

So much with regard to Popper's account of the peculiar orientation of interest and of the methods of explanation and justification that we find in histories. But what has he to say of the terms in which, or of the descriptive level at which, historical narratives are written? On this question his initial position is clear. He propounds his principle of "methodological individualism," viz., "that we must try to understand all collective phenomena as due to the actions, interactions, aims, hopes, and thoughts of individual men, and as due to traditions created and preserved by individual men."[10]

But what of the notorious difficulties that face any attempt to replace statements about nations or armies or economies or legal systems by statements that describe the actual actions and relations of individuals? These difficulties arise on at least two scores. First, there is the fact that every characteristically human action—the sort of individual action that comes to be described in histories—involves a tacit reference to some general institutional fact or norm; i.e., to what any person answering to a given description would do or would be expected to do in the kind of situation in question. Thus, if A is to give an object Z to B, it is not enough that A shall part with Z and B take it up. If A drops Z and B picks it up, A hasn't given Z to B. For an act of giving to take place, it is necessary that A shall regard Z as from now on B's property in all respects, and shall take steps, as and when necessary, to see that other people shall so regard it. Without society and its institutions, without the organization and expectation of certain attitudes within society, we should have no acts of giving; and by a similar argument we could show that we could have no promises, no trade, no insults, no injuries, and even—in characteristically human

---

[10] *Ibid.*, pp. 157–158. Compare also his statement (p. 136) that "the task of social theory is to construct and to analyse our sociological models carefully in descriptive or nominalist terms, that is to say, *in terms of individuals.*" [Popper's italics].

fashion—no fighting, no communication, no love. But secondly, while it may be agreed that institutions have to be manned and traditions sustained and doctrines accepted by individual men, nevertheless in many cases we know little or nothing of who these individual men were, or how many there were or how they stood to each other. Our records are all too often full of gaps, and yet across these gaps we can see or infer lines of development or decline, bifurcations or coalescences of certain institutions, styles, or doctrines. The evidence may afford us a real grasp of the institution's history, but the individuals who made and manned it are not only anonymous, they are totally undescribable and indistinguishable. So why pretend that the history of the institution in question is basically about *them*?

How would Popper deal with these difficulties? Or rather, what indication does he give us of how we should try to meet them? There is one passage at the end of his book, the main logical thread of which is not easy to follow but which nevertheless is alive with exciting pointers and suggestions.

(8)

But is this all? . . . Is there nothing whatever in the historicist idea of 'periods'; of the 'spirit' or 'style' of an age; of irresistible historical tendencies; of movements which captivate the minds of individuals and which surge on like a flood, driving, rather than being driven by, individual men? . . .

I have not the slightest sympathy with these 'spirits'; . . . And yet I feel that they indicate, at least, the existence of a vacuum, of a place which it is the task of sociology to fill with something more sensible, such as an analysis of problems arising within a tradition. There is room for a more detailed analysis of the *logic of situations*. . . . Beyond this logic of situations, or perhaps as a part of it, we need something like an analysis of social movements. We need studies, based on methodological individualism, of the social institutions through which ideas may spread and captivate individuals, of the way in which new traditions may be created, and of the way in which traditions work and break down.[11]

Two antinomies—two pairs of seemingly inconsistent claims—are here curiously combined. First Popper recognizes that it often makes an important difference to our attitude to a question, practical or theoretical, if we recognize that it arises within a tradition. Broadly speaking this will explain why we are forced to approach it in a certain way, for example, with certain presuppositions about the kinds of analyses that are appropriate to it. Hence in recognizing that a certain problem arises within a tradition we must to some extent treat it—and its terms or constituents—as typical, as instances of some general sort or kind. Yet, on the other hand, we have Popper's insistence that any acceptable account of a tradition—or of an institution or of anything else that we are liable to regard as submerging the individuality of individuals—must itself be based on 'methodological

---

[11] *Ibid.*, pp. 147–149.

individualism.' Here is the first antinomy. Secondly we have Popper's reference to the 'logic of situations,' i.e., his reminder that any actual situation or problem admits of a limited number of possible solutions or of moves such as, so far as can be foreseen, could lead toward a satisfactory solution of it. Now insofar as we treat any individual or group of individuals as purely rational or as approximating to rationality in their conduct, we naturally confine our attention to attempted solutions or moves of this limited kind. And this implies surely that our interest in the actions of, for example, the individual man Jones is thus far subordinated to our interest in what any (perfectly or approximately) rational being would or could have done if placed in Jones's shoes. Yet it is quite clear that Popper has no wish to regard historical results as either necessitated by or logically deducible from the logic of the situations from which they arise. We may here cite his dictum that "the human factor is *the* ultimately uncertain and wayward element in social life and in all social institutions,"[12] and we may recall our previous argument to the effect that we can often pick out and appreciate an action as supremely rational when it was one which we (with our feebler powers of creative rationality) could not possibly have foreseen. Experiences of this latter kind must frequently have been enjoyed by the more appreciative disciples of Socrates, for example, or of Galileo.

We are thus brought back, though now in less abstract terms, to something very like the impasse which faced us in section I. Historians, in contrast to theoretical scientists, are particularly concerned with the individual on at least the following scores : (a) in respect of the contingencies, accidents, and coincidences to which his life is inevitably subject; (b) in respect of his personal and idiosyncratic deviations from rational norms of conduct; and (c) in respect of those deviations from traditional ways of thinking and acting which we count as acts of creative intelligence. At the same time the historian's job is not simply to present us with a succession of individual facts. It is to give us a 'followable narrative,' i.e., one in which successions or concurrences of events are accepted because they exemplify familiar generalizations and possible connections between generalizations, or because—as a result of a notable break in the narrative's progress—it is demonstrated by argument that only such and such concurrences or sequences of events are compatible with our evidence and with our current general beliefs regarding man, society, and the world. The question is : how can these two tasks be combined ?

Popper's first—and seemingly official—answer, it seems to me, is that they don't combine, that they cannot be combined, and that the endeavor to combine them is responsible for historicist errors. The simple fact is that genuine historians always describe, but also, whenever possible, explain by reference to either causes or reasons. Now there is, of course, an important

---

[12] *Ibid.*, p. 158.

and basic truth in this clear and simple position. But the paragraph which we have just considered shows that Popper is not wholly satisfied with this first answer; and that he recognizes the need for some qualification or extension of it. My own suggestion with regard to this qualification would here be as follows.

Individuals are no doubt the primary concern of the historian—not individuals as such, however, so much as individuals in respect of the way that they take on, fulfill, fall down on, or transform various social and institutional roles. Every such role is associated with and is expected to conform to a number of traditionally accepted, and in some cases rationally requisite, norms of thought and action. For example, what interests us, and indeed what moves us particularly, about the Three Hundred is that they stood and fought and died in exactly the fashion that Spartans were taught and trained and expected to stand and fight and (if necessary) die. It is only in relation to this role which their state tradition assigned to them that they have a lasting place in history. At the same time, to appreciate their achievement we must bear in mind that they were three hundred individual men, each in his own way affected by his own interests, loves, hopes, and fears, although each succeeded in putting these things into abeyance in the fulfillment of his role as citizen. Nothing, in fine, is more historically crucial, nothing is more revealing of *how individuals make history*, than the way they take on and live out or fail to live up to their institutional roles. Alternatively we could say, the dramatic interest of history lies in the tension between the personal, the wayward, the unpredictable elements in the individuals who make it up, and the necessary or probable consequences of the filling of institutional roles by which men succeed to some degree in molding events to the norms of tradition or reason. And the greatest moments in history are precisely those in which an individual either creates or entirely revolutionizes a social role through his own quite unpredictable new vision of the situation or problem which faces him.

These remarks are the merest sketch of how Popper's discussion of institutional and traditional factors in history might be developed. But perhaps they do something to suggest what I believe to be true, that Popper in a few sentences has here done as much as any of his predecessors to illuminate this crucial issue in the critical philosophy of history.

CHAPTER 27

The Open Society and Its Enemies

BY PIETER GEYL

THIS BOOK, by an Austrian scholar who before the Anschluss sought refuge in England and is now Professor of Logic and Scientific Method in London, was written (in English) during the war, and published in 1945.

The "closed society" for Popper is one which is kept in existence mainly through the influence of petrified forms : the tribal framework, and within it castes and taboos. With the help of reason man can break away from these bonds and direct his fate in the way of security and freedom. The process of liberation started with the Greeks, but it is a lengthy one which again and again has to overcome new obstacles. The "enemies" against whom Popper raises his voice are certain thinkers who managed to clothe prejudice or group interest in attractive forms, and the earliest and greatest of them is Plato. Platonic ideas underlie "the perennial revolt against freedom and reason"; later on, after the French Revolution, Hegel rediscovered them, so to speak, and he may be taken to "represent the 'missing link' between Plato and the modern form of totalitarianism." Marx derived his system from Hegel, and although Popper fully recognizes Marx's true feeling for social justice and spiritual freedom, in Marx's system he finds the old fallacies, bound to lead to disastrous results. Plato, Hegel, Marx—all three are in this work fully treated, analyzed, and opposed. Practically the entire first volume is devoted to Plato. In the second, Hegel is disposed of in relatively few pages, for the author's intellectual contempt for Hegel is too complete for him to deal at length with the man's subtleties and wrong-headed arguments. Marx, on the contrary, is examined thoroughly, and here, as I hinted already, the author carefully discriminates.

As regards Plato, if Popper says that he finds much to admire in him, it is only to add that he feels it to be his task to annihilate what is bad in his philosophy, that is, the totalitarianism of his conception of the state. And devastating indeed is the impression one gathers from this first volume. The totalitarian tendencies of Plato's *Republic* have, of course, been observed before. But all sorts of arguments have been called on to represent them as harmless, as being either no more than the expression of modes of thought current in his time or the dreams of a great mind inclined to carry his thoughts or fancies to their extreme consequences, without intending any connection with humdrum reality. That his ideal state was to be ruled by philosophers, that justice was to be supreme—this seemed to elevate it all into something very exalted, noble, almost unearthly.

Now Popper sets out to demonstrate that there was, on the contrary, a close connection to the bitter party strife between democrats and aristocrats which was then dividing Greece, a struggle between the advocates of an open and of a closed society. That Plato was an adversary of democracy is not a new revelation. But how keenly did he discern its weaknesses, and with what penetrating perception did he, from that particular point of view, describe the social problems of his day? Popper regards Plato as one of the great pioneers of descriptive sociology, but at the same time he examines a little more closely the fine terms under cover of which Plato presents his constructive fancy. Justice? Pericles, the great leader of the democrats, understood by it a system of equality. But Plato misused that patient word to mean that everyone should be content in the class to which he belonged. Sharp division and rigid immobility of the classes—this is the distinctive feature of Plato's state. He distinguishes three classes: the guardians, their armed auxiliaries or warriors, and the working class. "But actually there are only two castes, the armed and educated rulers, and the unarmed and uneducated ruled, the human sheep." "The wise shall lead and rule, and the ignorant shall follow," says Plato. The guardians are indeed to be men of wisdom before everything, but it is a wisdom carefully circumscribed and remote from anything savoring of innovation. Recruited from the warrior class, and admitted to the highest form of philosophic training only at an advanced age, they might rather be called the initiated into the secret lore of the state (or of the privileged caste) than philosophers in the Socratic sense of free seekers after truth. Indeed, Plato constructs an indissoluble tie between philosophy—philosophy of this particular type —and a share in political power: whoever aspires after the one apart from the other shall be forcibly suppressed. Philosophy, in other words, is a matter of state to be kept under strict supervision, and taking this famous *Republic* of Plato's as a whole, it can be read as a textbook of dictatorship. As indeed it *was* read in Hitler's Germany.

But what is so impressive, and at the same time so dangerous, about Plato, is not in these striking practical precepts: It is that they spring naturally from the entirety of a powerful construction. At the foundation of everything there is the conception of the Idea as both primary and perfect, of which the tangible form of appearance as we know it is no more than the faint shadow, ever becoming fainter. History, that is, mankind's life and vicissitudes, is to Plato nothing but that process of decay. All that the philosopher can attempt is to stem, or to restore. A form coming as near as possible to the original, and this guarded by all means from change—a condition of frozen immobility in other words—this is the ideal set forth in *The Republic*. It is indeed the closed society, all issues toward freedom barricaded.

I cannot pretend to any profound knowledge of Plato; I just reproduce Popper's interpretation. But I must say that this interpretation strikes me as completely convincing. And also, that the objections advanced by the Platonists (I have read several such), to the effect that Plato should not be taken down from the regions of heavenly speculation to the everyday world of man and his passions and interests, do not make any impression on me. This kind of argument seems to me entirely unhistorical.

It is not, however, only Plato's practical antidemocratic intentions and his authoritarian, feudal, and racist ideals that Popper detests. The whole of that subjection of the historic process to a preconceived vision he regards as a pernicious game, a game that can fascinate many, but that tends to make them incapable of the free use of reason and of rational methods, required for the maintenance of the open society.

When Popper says that Hegel rediscovered the Platonic ideas he seems to minimize unduly the continuous influence they had exercised all along. The Dutchman cannot help thinking of Groen van Prinsterer,* who, though a younger contemporary of Hegel, came to Plato independently from the German thinker. The conception of an ideal society, ruined by the French Revolution and bound to sink ever more deeply unless it conformed again to an example set once and for all, was obviously Platonic. In a way Hegel was less Platonic than was Groen, because he transformed the process of decline into one of progress : the vision changed from pessimism to optimism.

But Hegel did indeed show the influence of Plato in that he too saw history developing on a set course, a course which man cannot alter, which he can only know, and through that knowledge he is to find his "freedom." Logically connected with this historicism (for this is the term used by Popper, somewhat arbitrarily, to indicate belief in historic laws and the prediction therefrom of future developments) is that other article of Platonic doctrine, the collectivist conception, by which man is made part of a

---

* 1798–1874; one of the small number of intellectuals who, after Bilderdijk (1756–1832), helped to revive orthodox Calvinism. Groen may be regarded as the founder of the Christian-Historical and Anti-Revolutionary parties, both of which, based on orthodox Protestantism, still survive.

whole and said to exist for the sake of the whole, not the other way around. Belief in the organic nature of the state, deification of the state, a low estimate of the individual and of his rights and capacities (among these, especially, reason and free criticism)—all this, some potential good, no doubt, but a great deal of certain evil, flows naturally from this conception.

The "potential good" is a modest insertion of my own. My historic awareness cannot well tolerate an exclusively negative estimate of a thinker who has proved so powerful an influence. I think his treatment of Hegel is a weak spot in Popper's book, even though that same historic awareness of mine would protest a great deal more strongly if the author tried to force me to accept Hegel's immense and impossible system. I understand, also, that what causes Popper's animosity is the fact that the whole of Hegel's philosophy, with its lip service to reason and freedom which seemed to borrow so much from the Enlightenment and the French Revolution, was in fact used by Hegel against these influences, was specially made to serve Prussian royal authority and the Prussian state. Adapted by Marx, the system of course lost those objectionable features, but 'historicism' and collectivism retained their dangerous character.

After his animated, at times impassioned annihilation of Plato, and his partly impatient, partly contemptuous reduction of Hegel, Popper's examination of Marx and his system is far from an anticlimax. I am hardly more of an expert on Marx and Marxism than I am on Plato and Platonism, but I do not think there can be many expositions in so relatively brief a compass that are of this excellence.

Popper assumes that the true motive power of Marx's activities was of an ethical nature. This presents us from the outset with a contradiction to the system. Yet here, Popper believes, we have the explanation of the enormous influence Marx came to exercise, an influence which has, for instance, had a renovating effect on Christianity, much as Luther contributed to the rejuvenation of Roman Catholicism. As presented, *Das Kapital* was, however, nothing but a scientific analysis leading to the demonstration of a law-governed process and a prediction of results; the prediction, as often pointed out, has been disproved by the facts. Marx's mind was governed by the design so completely that he could not very well confess to the ethical urge driving him on. In reality the system left room only for a futuristic ethics, that is, one based on a definition (or what would pass for a definition) of the morality that was to hold sway in the future.

Popper argues (and it is one of the numerous striking disquisitions that enrich his work) that moral futurism does not in principle differ from moral positivism, the system by which Hegel, especially, based morality on existing conditions. Why should man be obliged to subordinate his conscience to the future rather than to the present? He should not, be it on the one pretext or the other, detach himself from his duty to judge individually, independently. The temptation to conceive ethics in a futuristic sense is one of the pernicious effects of a system that pretends to know the future, and that paints it in optimistic colors. And more than that, it damps the zest for action, for practical action, for those piecemeal reforms that are all we can hope to control, at least to some extent, with our reason. The Socialists have managed to free themselves from this Marxist obsession; among the Communists we still see it in full force. And nothing constitutes a greater menace to rational discussion and to the open society that must live thereby.

Oracular Philosophy and the Revolt Against Reason: so runs the title of the next-to-last chapter of the two volumes. The two phenomena are indissolubly connected. It is only natural that Popper in this connection should devote a few critical pages to Toynbee. In his last chapter he discusses the question of whether history has a meaning. The answer is what one might expect after the preceding arguments and speculations: no sense, except for what we impart to it ourselves. And here Popper quotes, very effectively, Karl Barth, who rejects the "doctrine of the revelations of God in history" as presumptuous, as "an encroachment upon the kingly office of Christ."

"My intention in quoting Barth," Popper writes, "is to show that it is not only my 'rationalist' or 'humanist' point of view from which the worship of historical success" appears as idolatrous. And he concludes: "Instead of posing as prophets we must become the makers of our fate."

My review has not given anything like a full impression of the riches contained in this masterly book. Popper is a trained philosopher. In 1934 he entered the difficult ground of the theory of the scientific method as applied in modern physics and mathematics with a book entitled *Logik der Forschung*. Since then he has mastered the social sciences no less thoroughly. The notes added to the chapters of the book under review are a delight for anybody interested in the subject; one cannot but admire the ease with which the author moves in the enormous literature about the widely differing personalities he is discussing. And the text is a model of clarity. The march of the argument is consistently kept up, and what Popper offers is no mere erudite examination of Plato and Aristotle, Kant and Hegel, Marx, Engels and Lenin: it is the exposition of undying problems by which human society is beset, and always the ideas are brought to the test of the author's independent views.

These views may be called essentially modern. Not in the specific sense so often attached to that word and which diminishes or falsifies it. They are modern because the writer is profoundly aware of the life of our times and struggles with its problems. This is what makes the book so fascinating. Plato, Hegel, and Marx are brought very near to us, we learn to what extent they still form part of our mental conceptions. And as regards the spirit in which Popper deals with them, or rather in which he himself engages in our spiritual contest—this polemical spirit may at times result in one-sidedness, but I must say from the bottom of my heart that I should like to see his book in the hands of all those who believe in the value of democracy, freedom, and constructivism. It will help them to strengthen their courage and their powers of resistance.

NOTE

The article here printed in translation, based on a talk for the Dutch radio delivered in March, 1951, was published in *Critisch Bulletin,* The Hague, in April, 1951, after Popper's *Open Society and its Enemies* appeared in a Dutch translation by Justus Meyer. It was to be expected that the Platonists would be aroused by my praise of Popper to give expression to their disapproval of the book. And indeed, "Unworthy Criticism of Plato" was the title of an article contributed by Dr. H. J. M. Broos to the weekly paper of the Society of Secondary School Teachers.

Broos pointed out that Popper had omitted the first lines of the passage printed on the title page of his Part 1, "The Spell of Plato," lines which begin : "As for the armies . . ." In my reply I admitted that it would have been better on Popper's part to give the passage with those opening words. That they do not really affect the sense of what follows becomes clear when one remembers that, as Popper puts it elsewhere, Plato seems to regard his ideal state as being in a permanent state of mobilization. Dr. Broos also pointed to Plato's merciless analysis and total rejection of the τυραννίς and the τύραννος in Books VIII and IX of *The Republic* as sufficient proof that the book could not, as I had done "under the influence of Popper," be described as "a textbook of dictatorship."

To which I replied that this was beside the point, because what Plato wanted was a dictatorship by his carefully chosen and instructed wise men. I also objected to his saying that I had used this description "under the influence of Popper." Popper had made me see its justice much more fully than before, but I had long been familiar with the notion.

One incident sticks in my mind. On May 7, 1945, the day of the liberation of Utrecht, my wife and I went to the former lodging of some S.S. officers to retrieve part of our belongings, taken there a fortnight earlier when the Germans had raided and ransacked our house. (In our cellar arms and munitions of the Dutch underground forces had been hidden; we had just time to 'dive under' before the German military descended upon our stock of hand grenades and Sten guns.) The S.S. officers had left in a hurry. In one of the rooms, on a little table near the disordered bed, I found a book, open as if the owner had been interrupted in his reading. It was *Der Staat,* by Plato. We were there only to recover our own possessions, but I must confess that I took this curious relic of the occupation with me, and I still have it, an edition of 1943, introduction and notes strictly in accordance with the doctrine.

But I also referred Dr. Broos to what Allard Pierson, an outstanding nineteenth century Dutch scholar and thinker, had written in 1883 in an essay devoted to Groen van Prinsterer. On Plato, Pierson said :

The man who contemplated the Idea, did not believe in the power of the Idea. His language, moving and enchanting like music, was used by him to draw up petty precepts. The apologist of Socrates, that most revolutionary thinker, attempted to shut us up in regulations. Plato is the man inspired, to whom one cannot listen without one's soul trembling with the most exalted emotions; but Plato is also the true founder of Monasticism.

He is this, because he is the other. He is a policeman because he is a fanatic; he is anti-liberal because he is an idealist of a certain description.

And after a summary exposition of how Plato builds up his ideal state with minute prescriptions and all-embracing compulsion, Pierson concludes : "The fatal road has been followed to the end. The exalted idealist is a grand inquisitor. The immutable, eternal ideas rest on the gallows. Plato is the jailer of Socrates."

"It is still up to Dr. Broos," I ended my reply, "to prove that the general tendency of Professor Popper's interpretation is wrong." And this my partner in the discussion then set out to do. Not one of the nine points in which he presented his refutation of Popper seemed to me to contain anything of importance; several were completely irrelevant. For instance, that Plato would not have been an admirer of the Third Reich is very likely (the same can be said of Hegel or Nietzsche), and so also that the National Socialists "usurped" him (and the others). These are not arguments that Popper needs for his demonstration of the dangerous tendencies inherent in the Platonic system.

I stated in my article that the Platonists are apt to argue these dangerous tendencies away by insisting that *The Republic* is no more than a fantasy without relation to the everyday reality of political and social life. Dr. Broos does not fail to use this alibi, and at the same time he charges Popper with "having his ears so full of the noise of our present-day alarums that he is lacking in the respectful attention due to the part."

There is no doubt that Popper's book is a tract for the times as much as it is history. Now I believe that the demand for an absolute separation between the two will often produce no more than dull history or, worse, a pretended and false objectivity. I shall not deny that a historian's intense preoccupation with the problems of his own day exposes him to dangers. But Popper's book is, I felt at the time and I still feel, a shining example of the way in which the present can open our eyes for realities of the past which earlier generations constantly overlooked. Notice, for instance, how Pierson, who saw the objectionable features of Plato's political thought clearly enough, did not, writing as he did in the quiet closing decades of the nineteenth century, think for one moment of connecting them with the practical issues as they presented themselves to the Greeks of Plato's own day.

When Broos says that one must not indiscriminately apply to older periods these modern notions of democracy and totalitarianism I am ready to agree—provided I may stress the qualification "indiscriminately." The fact remains that Popper's method has not only produced a book that has clarified our notions about the struggle in which we are now engaged, but has proved historically revealing.

CHAPTER 28

The Tradition of
General Knowledge[*]

BY ERNST H. GOMBRICH

IN ONE of Agatha Christie's detective stories the murderess turns up at a fashionable luncheon party at Claridge's dressed up as her victim, whose death, of course, she wants to conceal as long as possible. At first, all goes well for her wicked plans, for she is an accomplished impersonator; but she has neglected one thing which is apparently needed for success at Claridge's, her classical education. Somebody of that select company mentions the Judgment of Paris, the sadly unsuccessful attempt at arbitration that led to the Trojan War. "The Judgment of Paris?" asks the murderess in her melodious voice. "Why, Paris does not cut any ice nowadays. It is London and New York that count."

"It was an awkward moment," says the narrator. The neighbor on the right drew in his breath sharply, while another guest began to talk violently about Russian opera. But listen to the Duke's reaction. "His lips were drawn tightly together, he had flushed and it seemed he drew slightly away from the speaker . . ." You will not be surprised that this dreadful *gaffe* ultimately leads to the detection of the crime and the undoing of the criminal. She had given herself away.

The episode stuck in my mind because my years at the Warburg Institute have alerted me to the changing context of the classical tradition. If our librarian had sufficient funds left at the end of this quinquennium to buy a Penguin book, he would certainly put *Lord Edgeware Dies* on our shelves. I trust the little story exempts me from the awkward need to define what I mean by general knowledge. It is quite clear, even from Agatha

* Oration delivered at the London School of Economics and Political Science, December 8, 1961.

[431]

Christie's brief account, is it not, that general knowledge is knowledge that is not general. It is only supposed to be general among a certain class of people.

You see, like Mark Antony, I have come to bury Caesar, not to praise him; and like Mark Antony, I have ulterior motives. I believe that this tradition of general knowledge is fading away. I don't think that there are many luncheon parties where the Judgment of Paris is mentioned across the table, and fewer still where admission of ignorance of that old story would cause ladies to talk of Russian opera and Dukes to frown. This is a good riddance. For there were two characteristics of this social convention which nobody would like to call back. One is the obvious temptation to snobbish intolerance toward those who do not share a particular piece of information and, as a corollary, the resentment of those who feel themselves snubbed. The situation at Claridge's would probably be described by an anthropologist as a clash between in-groups and out-groups. Not to know about that classical myth excluded you from a self-appointed élite. Unfortunately, it is likely that even with this dubious touchstone gone, society will keep inventing new signs and symbols with which to maintain the barriers of snobbishness. And the possession of knowledge is not, after all, a worse touchstone than the possession of more tangible status symbols such as cars, yachts, or Impressionist paintings. Indeed, if we must have social hierarchies and a pecking order, the test of general knowledge as a passport to the upper strata is surely preferable to the test of birth, wealth, or an "Oxford accent."

What is more worrying in the idea of general knowledge is its temptation to hypocrisy and superficiality. Those who used to be afraid of frowning dukes were led to countless degrading little lies whenever an item came up which they feared they ought to know. If the admission of ignorance leads to loss of caste and loss of face, people are afraid of asking. Yet how refreshing would it be to hear a person ask "But what is the Judgment of Paris?" It is at such moments of truth, alas, still all too rare on this ancient continent, that one discovers that so-called general knowledge is not only not general but also not really knowledge. Which ancient Greek author tells this famous story? Actually none, at least none that is extant. This fact I did not know before I looked it up. My knowledge about the Judgment of Paris comes mainly from hearsay. I must have read some potted version of Greek legends as a boy, but I would not regard it as general knowledge had I not picked it up again from stray references and from the works of many artists who enjoyed the opportunity of rendering a beauty contest among three Goddesses, as Rubens did in his marvellous canvas in the National Gallery. I think we should admit that most general knowledge is hearsay knowledge of this kind. There would be nothing reprehensible in this if it did not tend to give outsiders a totally different impression. We all know the visitors to our libraries who ask in an awe-struck tone : have you read all these books? and we have to confess that we bought some not to read but to use, and the others in the fond hope that the time would come when we could read them at last. But those who are used to libraries do not have to have this explained, and those who are used to the tradition of general knowledge take it for what it is, a cloud of rumors about a miscellany of things. Nobody who calls an act quixotic or compares a scene with Dante's Inferno wants to imply that he has read Cervantes or studied the *Divine Comedy*. He simply uses a common coin which he knows to be current. I think we can now see why his Grace the Duke suffered so severely when his luncheon companion failed to recognize the reference. I suggest that his sufferings were due less to snobbery than to a breakdown of communications. You do not have to be a Duke to feel embarrassed if your partner responds, let us say, to a reference to "sour grapes" with a lecture on the glucose content of fruit. Aesop's little fable of the fox provides such a convenient shorthand formula to characterize a human situation. What a bore it would be to have to explain it all. I venture to think that unnecessary heat has been generated by the simple fact that all cultures rest on some common stock of knowledge, which those who share it can draw upon at will. If you say "that girl with the Mona Lisa smile" you are scarcely out to impress anyone. You just hope that your partner will know the kind of smile and spot the girl. And if you call somebody a catalyst you equally hope that even non-chemists will understand that you mean a kind of person who brings people together without appearing to be active himself. Clearly all living language is shot through with innumerable such references, hidden or overt, which presuppose some sharing of knowledge, some hearsay in the literal sense of the word.

Perhaps it is only those of us who have had to change from one language to another who fully realize the relevance of such shared knowledge. Who could hope to understand a parliamentary debate in this country who has no inkling of *Alice in Wonderland* or fails to catch references to Jam tomorrow, to Humpty Dumpty's way with words, or to the Red Queen? And how can we fail to feel outsiders when our friends and colleagues talk in cricket terms, of sticky wickets, innings, and hitting for six? If you will allow me to introduce one innocuous technical term, I'd describe such fields of common knowledge as 'sources of metaphor.'

Every culture, of course, has these favored sources of metaphor which facilitate communication among its members and which, by the way, present such a headache to the translator. Common customs, trades, and, of course, the legends and beliefs of the tribe are fused with the language and modes of thought of any civilization. How can we understand the metaphors of Indian poetry and literature without knowing what the cow means to the Indian villager in his life and in his worship? Or how can we understand Islamic culture without sharing some knowledge of the Koran? For religion provides most cultures with a central area of metaphor. The

Olympus or heaven of any nation will offer symbols of power and of com-
passion, of good and of evil, of menace and consolation, and such is the
vitality of these symbols that they will survive the decline of faith. Where
would European poetry be without Venus and Mars, Cupid's dart and
Jove's thunderbolt and even, occasionally, the Judgment of Paris?

> When Priam's sonne in midst of Ida plaine
> Gave one the price, and other two the foile,
> If she for whome I still abide in paine
> Had lived then within the Troyan soile
> No doubt but hers had bene the golden ball,
> Helen had scaped rape, and Troy his fall . . .
>
> Thomas Watson, (*Hekatompathia*)
> London, 1582, XXXIII

Not all our poetry is as studded with such references as that of the
Elizabethan sonneteers. But I still wonder whether we ever can give sufficient credit to those members of other cultures who decide to master our
traditions. It must be as hard for a Chinese student to sort out these van-
ished divinities as it is for us to catch the references to good and bad Em-
perors that are woven into the fabric of Chinese literature. Of course I do
not claim that these obstacles are quite insurmountable. One can try to
understand these allusions as one can try to learn a foreign language. I
may have come intellectually a little closer than I was twenty-five years
ago to understanding what cricket means to my English friends. Thanks to
my son I may even be able to catch an allusion to W. G. Grace and his
beard or to Wisden's Almanack. But I do not deceive myself. I could
never learn to catch a cricket ball, and I could not respond to these allusions as one does to something one has known from childhood. To catch
the ball or the real import of the allusion you must have assimilated the
game into your flesh and blood. You must have grown into it so that it
becomes part of your nature.

As you know, what might be called old-fashioned education emphasized the assimilation of knowledge rather than its acquisition. It is no
accident, I believe, that this tradition stems from classical civilization with
its tremendous emphasis on rhetoric, the mastery of language. Education
was articulation, and the most articulate person was the one who had
assimilated all the sources of metaphor with which to touch the chords of
shared memories.

Our culture sets much less store by oratory. If it did I would certainly
not have been honored by the invitation to deliver an oration. And so,
with this central purpose of assimilated knowledge gone, the question of
what any person ought to learn and ought to know, what should or should
not belong to general knowledge, became in one sense more worrying and
in another less clear. Not that this dispute among disciplines is a new thing
in education. It was in itself a favorite topic of the schools of oratory; you

had to sharpen your wits by proving that medicine was more important than law or law more important than medicine, that the Quadrivium founded on mathematics imparted higher knowledge than the Trivium founded on language, or that language alone made human beings truly human. Not all of those who engage in these conflicts today know of their venerable ancestry. If they did, they would realize also, I believe, how inconclusive are the arguments from relevance and usefulness. Any piece of information, after all, can conceivably be important for something or somebody; but if that were the criterion for including it in the syllabus, the most important textbook would be the telephone directory. It certainly would not be a soft option. But though you could make your students learn it, I doubt if they could assimilate it. It could scarcely become part of their language, a source of metaphor shared by a culture.

It is easy to see and easier to say that language is not everything. A society where everybody could only talk and nobody do would not survive a day. But a society without the assimilation of general knowledge, starting from language and reaching out into the sources of metaphor, would cease to be a society. I shall not waste your time by laboring this point, for nobody thinks that this can ever happen.

Last year, I believe, my very illustrious predecessor Sir Geoffrey Crowther made your flesh creep by warning you of the dire consequences that would overtake us if more know-how in science was not spread sufficiently fast. I have no ambition to compete with him. If the tradition of general knowledge should break, you need not worry that it will hurt. Here ignorance is really bliss, for we certainly cannot miss what we never knew. Our distinctions would become a little cruder and our articulation more coarse-grained, but our language would still serve us, as it has served those millions who were never in touch with these sources of metaphor.

But surely language and metaphor are not only of use, like the telephone-book, to communicate with others. They also help us to articulate and interpret our own world of experience to ourselves and it is here that the shrinking of these sources would ultimately affect us.

If we can believe the psychoanalysts—and here I am sure we can— our first sources of metaphor in this widest sense of the term spring from the earliest experience of our family situation. Mother and father provide for us the emotional models for all the clusters of feeling that go with motherly love or paternal authority. We learn to order and categorize our world of experience along those lines. Soon fresh experiences will provide fresh areas of comparison of a private nature: the toy-cupboard, the world of animals, fairy tales, and of course religion. For the growing child, the teachers at school quite frequently provide an additional range of models and types—and many of them would be surprised if they knew how the child's imagination transforms them into Gods, demons, and devils. Listen to the conversation of schoolchildren and mark how much of it centers on the character, foibles, and charms of their teachers. It is these common references, this shared knowledge, which welds the class together into a little sub-culture. Needless to say, there are other things which groups of adolescents share in our civilization. Television, radio, and cinema provide a whole Olympus of stars and starlets, and those who are excluded by poverty or snobbery are soon made to feel outsiders. It is as embarrassing to those who share this culture to have to spell out an allusion to Cliff Richard (whose name I learned for the purpose of this oration) as it is to others to explain what is meant by the Mona Lisa smile. Everyday speech has become interlaced with phrases and catchwords from popular radio shows which for the habitué evoke a whole chain of associations and are forgotten tomorrow.

I do not think that this kaleidoscopic change of the area of reference is confined to radio and television addicts. Highbrows no less than lowbrows are constantly bombarded with insinuations that they must—or must not—know or read or take up this or that craze if they want to gain entry into some élite. I suspect indeed that some of our Sunday papers and our weeklies cater quite consciously to us busy people who want to have the feeling that though we have no time for reading or even for thinking, we remain in touch with the culture—all of whose members, in theory at least, might be sustained by the same papers in the same comforting illusion.

I am afraid, though, that there is still a difference between being in the swim and being in the culture : it is the same difference as between jargon and language. It is in this context that I can at least advertise my wares with some confidence. I must confess that I find many reasons given for a classical education unconvincing. I do not see, for instance, why the study of Greek and Latin should train the mind more effectively than would an equally intensive study of Chinese. The reason why people in our civilization should always have the possibility of studying the classics is, quite simply, that people in our past cultivated the classics. For in the classical heritage we have an area of metaphor, a common market of symbols and ideas that transcends the boundaries of both nations and periods in a way national literature never can. The Germans, for instance, are proud of having naturalized Shakespeare into their culture and Hamlet or Lear certainly do belong to general knowledge in German-speaking lands. But Jane Austen has not been admitted into that company. Proust, on the other hand, has been naturalized in England, but Goethe, who towers over the Olympus of German *Bildung,* is little known in the Anglo-Saxon world. It was different with the classics, at least up to a generation ago. They had become so profoundly assimilated into the tradition of general knowledge that they were not only less than general and less than knowledge, but also, in a curious way, more than knowledge and also more than general.

To pass from those psychological models of parents and teachers, and from the heroes of the hour to less private and more universal symbols, was to gain contact with the problems that helped to shape our civilization. Athenians and Spartans were to the growing mind not only names of peoples long dead, but also permanent possibilities interpreting the life and duties of man. To learn to hate the Spartans with their beastly ideals was an education in itself—all the more so as this education had to be gained in the teeth of educators who never tired of extolling the virtues of these toughs.

In this way the Lycurgus of Plutarch, the Pericles of Thucydides, the Cato of Cicero have challenged the imagination of many generations to choose between different images of paternal authority. Even today, our living language still carries the traces of these preoccupations. Draconian laws, Stoic endurance, Epicurean living, cynical indifference—these are not only allusions to be used or dropped at will; they are road signs erected at important crossroads. It is true, of course, that when we follow them we may find, for instance, that Epicurus was rather stoical and that many Stoics were rather epicurean. But do not these discoveries alert us to distrust hearsay? The fact that these names have become metaphors, that they stand for whole areas of experience, provides a stimulus to learn more about them. I believe that nearly all worthwhile research in the humanities owes its impulse to these still living forces in our culture. If I may take an example from an eminent member of this School, Karl Popper's criticism of Plato was such a revision, the realization that the political ideas Plato stood for in the pantheon of our culture were less beneficial than we had come to believe. By now it may have become general knowledge among the readers of weeklies that Popper is the anti-Plato. Those who follow this road sign and read his works will also have to revise that stereotype. Anybody deserving the name of a student must learn to mistrust what passes as general knowledge.

Yet our Olympus is not only populated by political father-figures. It merges into Parnassus, where artists survive in our tradition as lasting embodiments of human greatness, its triumphs and temptations. Michelangelo and Raphael, Rubens and Rembrandt, Van Gogh and Cézanne are not only objects of art historical study whose works are investments or status symbols for collectors. They are centers of attraction and repulsion to be loved, admired, criticized, or rejected—living forces with which we get involved. They are culture heroes, Gods of our secular Pantheon, beneficent or baleful, serene or capricious; but like Gods they must be approached with respect and humility, for they can light up for us whole areas of the mind which would have been dark without them.

Take Raphael, now rather a remote deity in our Pantheon but one whose name was once almost a synonym for divine beauty. To study him in the context of our culture is not only to study a historical figure but to examine our own relation to ideal beauty. I am very anxious not to be misunderstood here. I do not want to give my own field a bad name by suggesting that the study of art is necessarily more subjective than a study of scientific problems. There are rigid objective standards which the historian has to respect when he interprets texts and evaluates the authenticity of paintings. What is and must be subjective is the discovery of relevance. The Latin orators had a term for it; it was *tua res agitur,* which the modern advertising agent would probably translate as "this is about you"! The study of Raphael, I suggest, is still about us. The stocks of beauty have fallen considerably during the last 150 years or so, for beauty has become suspected of dishonesty. The very name of the pre-Raphaelite movement was a declaration of war against an element in our culture, and indeed an element in our religion, taking the word in its widest sense. By looking again at Raphael, we are led to examine the justification of this revolt in its turn.

How close must beauty lie to sentimentality? How far was the devotional trash of Victorian oleographs inherent, as a dangerous possibility, in some of Raphael's Madonnas? Personally, I have come to the conclusion that the revolt against Raphael missed its mark. The more I study him, particularly in his drawings, the more I come to admire his genius and the less does his art resemble the stereotype of hollow beauty. I have come to suspect that he is less easy to understand than Michelangelo, whose tremendous images take us by storm. The relationship is somewhat similar to that of Mozart and Beethoven, though Mozart's beauty does not need any defenders now, as it did during the Wagnerian era when many dismissed him as shallow because they had never fathomed his depth.

Thanks, perhaps, to the radio and the long-playing record, music has become the most precious of all shared possessions, of sources of metaphor in our culture. Indeed, if there is anything that belies the modish talk of two cultures, it is the sight of any crowded concert hall. I have never noticed that scientists, engineers, or medical men cared less for this common possession than students of language and literature.

But not everybody is musical, and whether or not you are, you might get a lot out of other elements of our common heritage, a heritage that seems to be receding from our grasp as so many bridges to the past become impassable. It is for this reason that I deplore the fading of the tradition of general knowledge. Granted that it was no more than a cloud of rumors, often false or misleading; even hearsay has its value if it gives you news of something you might enjoy. Life is short and no one could ever follow up a fraction of those cues which the language of our tribe threw in our path. Perhaps, one day, I shall really read *Don Quixote* or learn more about the mystery of catalysts.

How can we make sure that these rumors and these possibilities continue to reach the next generation? Should we, at the university, give courses in General Knowledge?

Many colleagues advocate such courses under the name of General Studies, Surveys of Western Civilization, or the History of Ideas. But though we all follow these experiments in the new universities and abroad with interest, many of us have still to be converted. Among most of my friends 'survey course' is still a dirty word. Having once written a survey of the history of art, I am barred from joining in these denunciations; but I may confess that I am also a bit sceptical, though not quite for the same reasons. What is usually alleged against these attempts is the sin of superficiality. We, it is said, want to teach the student a discipline, we want to build character by making him study a small area in depth. A smattering of this or that may have been good enough for finishing schools preparing for Claridge's; it has no place in the universities.

I hold no particular brief for finishing schools nor, as you know, for Claridge's, but I think this anti-snobbish attitude is tinged with snobbism. Go into any Senior Common Room and hear these same Dons who champion specialization groan about some undergraduate who did not know whether the Greeks came before the Romans or when the Reformation happened.

The truth is that it was always a minority that had the unfair advantage of picking up this hearsay knowledge early in life, by actually hearing these things said during conversations at home, or by freely rummaging among their parents' books. By the time they arrived at the university they were aware of some of the landmarks on the traditional map of knowledge, they knew where they were and where they wanted most to go. It may well have made sense to submit these growing minds mainly to a so-called discipline, to make them explore one area as thoroughly as their time permitted. It is a great comfort to acquire one small allotment in the vastness of the knowable where one feels a little more at home. Yet even this acquisition holds its dangers for the less intelligent. The more they hear about the contrast between a study in depth, which is a good thing, and superficial surveys, which are a bad thing, the more easily they may jump to the conclusion that they, at any rate, are now Masters of Arts. They forget to ask how deep this depth is anyway. Do we measure it in relation to what is knowable or to what is worth knowing? After all, our knowledge is always superficial. Whether we deal with a century in one page or in 5000 pages, the relation to the infinitude of events is scarcely affected.

With all its imperfections, its social dangers, and its temptations, the tradition of general knowledge at least kept alive the feeling that there was much you did not know and ought to learn about; it constantly bombarded you with reminders that there were more classics to be read, more countries to be visited, more scientific theories to be grasped, more languages to be learned than you could hope to assimilate in a lifetime.

We must not snobbishly ignore the fact that increasing numbers of eager and impressionable young people now arrive at the universities who have had no opportunities of hearing these rumors and of forming ideas of these distant landmarks. Yet we propose to let these, too, learn just one square mile of knowledge to some degree of perfection during the three years' work for their B.A. Honours and possibly make them do ten yards of micro-surveying within this square mile for a Ph.D., to see them emerge from five years of successful grind with their character trained, and their minds closed. They may even be asked to teach in their turn without ever having seen a map of the whole country, let alone having crossed the frontiers. It is not all that easy, within our universities, to get permission to do so, as you know. Our students, at least here in London, are not even expected to stray into lectures outside their courses and those who want to change fields, as it is called, after their B.A., encounter further hurdles put up by our Boards of Studies in the form of so-called qualifying examinations to keep them, if possible, on the straight road. I should give such deviants a special bonus; but that is not a proposal likely to go through.

I know that I am neither the first, nor will I be the last, to bemoan the evils of specialization, but I may have explained the reason why I take an even more serious view of this state of affairs than do some of my colleagues. I do not claim to know how this situation affects economics or science; but the humanities certainly draw their strength, their nourishment, and their *reason d'être* from the traditions and general concerns of the culture. To cut them off from these traditions is to kill them. One wonders, of course, how widely their death would be deplored.

But I am afraid their death may even be hastened if we concede the tradition of general knowledge some place in the university curriculum. It could not survive the icy blast of the examination system. It is true that I read in last week's *Observer* that "General studies are *examinable* just as are the subjects of special studies." What a word and what a commendation. The telephone directory may be examinable; our involvement with our heritage is not. Imagine a question : "Compare and contrast Raphael with Mozart and Beethoven with Michelangelo (as metaphors)." These are things you can talk about with like-minded friends late at night after a concert or a visit to a gallery. To make these private reactions the subject of examinations is to kill them for good.

No, I am afraid that as long as the universities are identified in the public mind, not with the passing on of traditions, but with the passing of examinations, those who have these traditions at heart must defend them against bureaucratization. I realize that this increasing bureaucratization of learning and scholarship has come about from the most admirable motives. The desire for examinable knowledge springs from the tradition of fairness to all, from the demand that justice should not only be done but be seen to be done. But unless we look out, this demand, backed as it is by a wonderful system of grants, may turn our academic institutions into a counterpart of the National Egg Marketing Board, places where eggheads are graded and stamped for the convenience of prospective purchasers. Far be it from me to question the reliability of our sister institution, which, I believe, receives the same share of the national income as does higher education. They grade eggs according to size, and though it would be interesting to speculate whether this means height, girth, or weight, I am sure they will know exactly by what criteria they have to sort their product. Who would not envy them their simple task? Having wielded the stamp myself, I confess to being increasingly perplexed as to how to plot such incommensurable qualities as originality and industry along one single coordinate or dimension, and call first the paper and then the total result 53 or 57 out of 100. Most of us who are asked to play this game are agreed that it operates with fictions, particularly along that all-important zone of middling performance that divides the upper from the lower second. And yet it is here that our university places the all-decisive distinction between good eggs and bad eggs—the first to be used for hatching, the others, at best, for cooking.

However, I must not digress. Though I do not want the tradition of general knowledge to be caught in the sausage machine (radically to change the metaphor), I do think that our universities should not necessarily prevent the spread of general knowledge. I am afraid we are on the way to doing precisely this. I do not claim to be original in these strictures either; it has all been said often, and better than I could. It is bad enough that the importance attached to examination marks leaves the conscientious student no choice but to swot and memorize examinable knowledge. It is perhaps worse that our own dissatisfaction with the system of specialization so easily tempts us to overload the syllabus more and more till the undergraduates have to rush from course to course without a moment's time for reflection. I am sure that our first duty to the student is to resist this obvious temptation. We must protect him from all the clamoring bidders for his attention who want to save the idea of general knowledge by making the arts student attend courses on catalysts and the science student courses on the Judgment of Paris. What can be saved of this old tradition can only be saved by leaving the student more time to educate himself, more time to read rather than skim and skip, time to assimilate rather than acquire knowledge, to roam the country round his allotted field, to make his own discoveries and to enjoy looking down on it from a neighboring peak without expecting a diploma in mountaineering to satisfy the bureaucrats that their grant has been well spent.

But though I am sure that we must always try to teach less rather than more, this does not absolve us from our responsibility to help the student to make good use of the time thus gained. On the contrary, we must try very hard to find worthwhile alternatives to yet more courses and examinations, so as to provide the student with the same kind of orientation, at least, which the tradition of general knowledge provided for earlier generations. We need not idealize this tradition, as we have seen. It was neither very coherent nor very accurate. But it cohered at least as much as our language and our culture cohered and it thus counteracted the fragmentation of knowledge into unrelated specialisms. The fact that this unity was subjective speaks in its favor, for what else could it be? Most systems of order are subjective, but it is for this reason that they may help the growing mind to feel at home on the map of culture.

If modesty did not forbit it, I should say that for example the arrangement of the library which Aby Warburg thought out was intended as such a rough map of our cultural universe. Of course, being subjective, any such arrangement has its weaknesses and its idiosyncrasies, and I have sometimes cursed ours when I could not find a book. But this does not detract from the obvious merit of this model : it does not recognize any barriers between the so-called fields of study. The student of artistic traditions is made to realize that he should also read about the history of religion, science, and the economics of patronage; the student of poetry that he should not forget mythology, music, pageants, and social conventions. What is more, he can browse there at his leisure, almost as undisturbed as in an old-fashioned bookshop.

It is not only modesty, though, which prevents me from concluding with the fanfare that all our educated ills could be remedied if all students in these islands came to browse at the Warburg Institute. I surely must end with a more constructive suggestion of how to keep the map of culture before the minds of all these growing specialists.

Perhaps we might take a leaf out of the book of the oldest institution of learning that exists among us, far older and more experienced even than our old universities; I mean the Church. Not that I know anything about the theological syllabus, but I know at least that you cannot specialize in the First Person of the Trinity without having heard of the Second. The Church, it seems, has never had that morbid fear and snobbish contempt of superficiality that has invaded academic life with the rise of professionalism. There are all shades between the most humble and the most learned, but certain essentials of religious knowledge must be common to all. It has invented the Creed and the Catechism as a map of orientation for everyone who wants to be received into its community.

I know that the original purpose of the Creed was to fix the dogma, but the pedagogic wisdom of this simple device can be admired even by heretics who want to know the Creed in order to criticize its tenets. What we can learn from this example is to discard our fear of smattering, our all or nothing attitude to knowledge which is not only unrealistic and inhumane but also, I believe, the greatest obstacle to the survival of our traditions.

The classical tradition itself was only kept alive throughout the Dark Ages because a few learned churchmen such as Isidore of Seville were not ashamed of writing simple compendia to which they committed those few ideas about the universe and about the past which they considered indispensable. It was from writings of this kind that the idea of general knowledge could renew itself.

What I regret most, in the present eclipse of this tradition, is the loss of the historical frame of reference, the amputation of the time dimension from our culture. I think the Church would have known how to remedy this deficiency without imposing a burden. Indeed I have been toying with the idea of secular creeds, as brief and concise, if we can hammer them out, as the Athanasian Creed. It would not need a new course to teach them, for you would read them in three minutes and learn them in an afternoon. We could make such capsuled knowledge available to everybody in our schools irrespective of the type and level of their curricula. A secular creed of this kind should be packed so full with interesting rumors that the many and varied catechisms which should explain it somewhat more fully would always be in demand from the libraries. If we ever convoke a Council of Nicaea to agree on ours, we might also, I suggest, invite other cultures from other continents to tell us their own, which should be available as optional subjects.

It is with some trepidation that I submit for your criticism the first untidy draft of such a creed, biased, subjective, and selective, but containing the kind of iron ration I should like to distribute before we are scattered to the four winds :

I belong to Western Civilization, born in Greece in the first millennium B.C. It was created by poets, philosophers, artists, historians, and scientists who freely examined the earlier myths and traditions of the ancient Orient. It flourished in Athens in the fifth century, was carried East by Macedonian conquests in the fourth century, and in the first by Latin-speaking Romans to large parts of Europe and North Africa.

It was transformed by Christianity, which arose among the Jews of Palestine and spread throughout the Latin- and Greek-speaking world in the second and third centuries A.D. It survived the collapse of the Roman Empire under the pressure of Teutonic tribes in the fifth century, for the Greek and Roman Churches preserved some of its organization, its literature, and its art during the so-called Middle Ages, when most of the barons and their serfs were illiterate. It began to flourish once more in the twelfth and thirteenth centuries when the Gothic style of building spread from France over Europe and when the growing universities of France, Italy, and England gained fresh knowledge of Greek science and learning through translations made by Mohammedan Arabs, who had penetrated through North Africa to Spain. These also brought Arabic numerals from India, and paper, gunpowder, and the marine compass from China, thus assisting the emancipation of the merchant cities of fourteenth and fifteenth century Italy, which encouraged the recovery of Greek and Roman literature, art, and building styles that is called the Renaissance. Its New Learning was disseminated by the printing press which inaugurated the Modern Age and prepared the ground for the Reformation that split Europe in the sixteenth century, while voyages of discovery led to Portuguese, Spanish, and English conquests and settlements across the sea.

It was transformed once more at that time by the renewed faith in the progress of human knowledge, exemplified in the mathematical theories of experimental science created in Italy and developed in the seventeenth century Netherlands and in Protestant England whence, in the eighteenth century, ideals of rationalism and tolerance spread to the Continent. It was thus enabled to survive the rapid increase in population that favored the Industrial Revolution, which led to nineteenth century colonialism, the spread of literacy, and the mass movements of socialism and nationalism. It endangered and transformed, in our century, most other cultures of the globe which has shrunk for us to sputnik size by the invention of flying. I hope there will be a twenty-first century. Amen.

CHAPTER 29

Philosophy of History
Before Historicism

BY GEORGE H. NADEL

THE DOMINANT PHILOSOPHY of history of the nineteenth century is so
familiar to scholars nowadays that a few summary remarks will suffice to
say what it was. Basic to it was the recognition that historical events should
be studied not, as theretofore, as data for a moral or political science, but
as historical phenomena. In practice, this was manifested by the emergence of history as an independent academic discipline, in fact as well as
name. In theory, it found expression in two propositions : that what happened must be explained in terms of when it happened, and that there
exists a science with logical procedures peculiar to itself, the science of
history. Neither proposition was new except in the insistence that was
placed on it, and this insistence is seen in the doctrinal exaggeration of
both. The converse of the first was held to be that the history of anything
constitutes a sufficient explanation of it; and those who imputed a logical
order to the chronological order of events saw the science of history as a
science capable of predicting the future of society. That the term for the
first of these doctrines, historism, has tended to become absorbed, in
English usage, by the term properly applicable to the second alone, historicism, may merely illustrate the difficulty of assimilating to other languages
what were German terms for predominantly German schools of thought.[1]
Or it may attest to a realization that these two doctrines are related, as

---

[1] Dwight E. Lee and Robert N. Beck, "The Meaning of 'Historicism'," *American
Historical Review*, vol. 59 (April 1954), p. 568 n.1, suggest that the English 'historicism,'
which displaced the earlier, correct rendering of *Historismus* as 'historism,' may have
come from the Italian *storicismo* familiar through Croce's writings. That even the
principal German historians of historicism, Troeltsch, Heussi, and Meinecke, seem
to have been greatly mistaken about the history of the term has recently been shown
by Erich Rothacker, "Das Wort 'Historismus'," *Zeitschrift für deutsche Wortforschung*
N.S. I (1960), 4.

indeed they were in Hegel's thought. In any case, it shows, as simplification often does, how widespread interest in historicism in all its aspects has become.

By contrast, we know very little about philosophy of history in the centuries preceding the nineteenth. This is especially true of the sense to which it will be necessary to confine the term here : questions about the purpose of the study of history and the theories advanced to validate the answers given. For it is in this sense that philosophizing about history was principally pursued in the period I shall survey—approximately 1550 to 1750. My concern is primarily descriptive, to say what the dominant conception of history was, where it came from, how it was propagated, and why it declined. To say merely this much is to approach philosophy of history critically, if only by establishing facts which theories about its development can no longer ignore. To attempt more does not seem to me warranted by our present state of knowledge of the subject.

In part the scantiness of dependable knowledge is due to specialization. Historians who have illuminated the Renaissance humanists' ideas on history have generally not gone beyond the fifteenth and sixteenth centuries, and thus the classical humanist influences on the later centuries have been left unexamined except by scholars of art or letters. In part, it is attributable to the general historian of ideas, to whom the most attractive topic in the philosophy of history has been the ideas of Voltaire and other *philosophes* of the later eighteenth century and the alleged sources of their ideas. What came before Voltaire has often been studied by way of projection backward, and topics like providential history or the idea of progress have been made more of than contemporary sources warrant. The concentration on the *philosophes* and their pyrrhonistic predecessors has also led to an overvaluation of their originality. Even scholars like Ernst Cassirer and Friedrich Meinecke have cited as new and revolutionary certain ideas put forward by men like Bayle and Hume which, in fact, were merely paraphrases or quotations from the classics, drearily familiar to any educated person living between the Renaissance and the nineteenth century.[2] In short, the study of philosophy of history in the period between the age of Machiavelli and the age of Voltaire lies thinly cultivated between two well-tilled fields.

I.   The Original of the Exemplar Theory of History

The first thing to bear in mind in an analysis of the historical thought of this period is that it was derived from the classics; and the second is

---

[2] Ernst Cassirer, *The Philosophy of the Enlightenment* (Boston, 1961), pp. 208–209, 226. Friedrich Meinecke, *Die Entstehung des Historismus* (2 vols.) (Munich and Berlin, 1936), vol. I, p. 213.

that these were the writings of the Romans (and their Hellenistic subjects) and not those of the classical Greeks. Latin was the universal language of the educated, and in the perennial sparring between the gentleman of learning and the pedantic or paid scholar, a knowledge of Greek was sometimes reckoned as an unpleasant distinguishing characteristic of the latter. The translators of Greek authors into Latin declared that their purpose was to make these authors accessible to the modern world. Any major work of the sixteenth and seventeenth centuries written in the vernacular was likely to be translated into Latin, as Bacon said of his translation of the *Advancement of Learning*, "to have it read everywhere." The Romans not only provided posterity with a language for scholarship and virtually all genres of literature in prose and verse. They had also, so it seemed, come to grips with the problems of the modern state : legislation, political parties, the function of political offices and the training suitable to their discharge. Political reformers went to historians for information on Roman precedents. A commentary on Tacitus could lose a scholar his academic position or put an historian on trial for treason. In the absence of an academic establishment—the universities were auxiliary rather than principal suppliers of learning—the chief evidence of his scholarly soundness which a writer could offer to his reading public was the massive employment of classical citations and the demonstration that these supported, or authorized, his views. Literature and education were given over to classical didacticism and moralistic purpose, to instilling virtue—not merely Christian virtue, but the Roman concept of virtuous action.

Greek philosophy, however influential later, did not influence historical thought in this period. This was not merely a result of the concern with Roman thinkers rather than Greek, though it was that too. As indicated, and in keeping with contemporary ideas, I employ the term 'philosophy of history' to denote theorizing about the subject of history, about history as *disciplina* and not as *res gestae*. And that was a matter of little interest to the Greek philosopher.[3] To Plato, for example, the subject of history never occurred in his plan for studies in the *Republic*. Even the Stoicism that appears is not the Old Stoicism of the Greek metaphysicians, with its pantheism and periodic world conflagrations, but that of the later schools, of the Romans and their Greek advisers. Similarly, the concept which I discuss below, the "man of action" who makes and writes history, is not Plato's philosopher-king, though it would not be difficult to establish the connection between the two. But such connections were not of particular concern to the historical theorizers of our period, nor did they bother with sociological interpretations. To inflict on them even such plausible notions as the interpretation of Thucydides as a Hippocratic clinician would be

---

[3] Werner Jaeger, *Paideia: The Ideals of Greek Culture* (3 vols.) (New York, 1944), vol. III, pp. 101–102.

anachronistic and would miss the characteristic method of their verbal treatment of sources and authorities. Aristotle's remarks about history, it is true, influenced certain issues; but it was also complained, rightly I think, that he never theorized about history.[4] Speculation about the task and function of the historian, and thus philosophy of history in the more limited sense of the term, belonged to the later, Hellenistic Greeks and to the Romans. It was a development of the lines laid down, not by the Greek philosophers, but by the Greek rhetoricians, especially by Isocrates, himself under the influence of Thucydides. It formed part of the general concern with education, in which the Isocratean antithesis of learning from the experience of the past to learning, as with Plato, from philosophy was elaborated. These elaborations were to determine the conception of history of our period—exemplar history, didactic in purpose, inductive in method, and authorized by the stereotyped dicta of the Roman Stoics, rhetoricians, and historians. I shall sketch these elaborations in turn : political education; the idea of example; and the function of the historian. To know them is to know the later conception of exemplar history in almost all its details, since it was entirely derivative and faithful to the Romans to the letter.

Political education. Since the heyday of classical Greek philosophy the role of the philosopher, of whatever school, had come to be identified as that of a guide to life. To the Romans this meant, in practice, guidance to political life as well as consolation for its not unlikely results, banishment or death. In the two principal Greek influences on Roman thought, rhetoric and Stoic philosophy, practical aims were given more importance than theorizing. The subtleties of Greek rhetoric were, as Quintilian put it, made easy. Stoicism, in its five centuries from Zeno to Marcus Aurelius, was changing from a more or less formally articulated philosophy to a set of attitudes; Seneca and others insisted that philosophy was useless

---

[4] Karl Popper, *The Open Society and Its Enemies* (2 vols.) (3rd ed., London, 1957), vol. II, pp. 7–8, has shown, however, how Aristotle's theory of change, which is not an historical theory but follows from his doctrine of essences, developed into Hegelian philosophy of history. No such developments could be found before the nineteenth century. Aristotle's famous statements about the contrast between poetry and history (*Poet.* 9, 1451a, 36–38; 1451b, 1–10; 23; 1459a, 22–29) affected in the main only discussions on how the two differed. This is understandable, since Aristotle there was only concerned with defending poetry and not with saying anything significant about history. Occasionally, later theorists would quote the opinions of earlier Italian humanists on Plato or Aristotle, or repeat Aristotle's characterization of history as dealing with singulars. If we dig very hard, we might even find an otherwise unknown English philosopher superficially arguing historical knowledge in terms of Aristotle's account (*Analyt. post.* II, xix, 100a 4–6, 16) of how sense perception gives rise to memory and repeated memories to general ideas: [John Petvin] *Letters Concerning the Use and Method of Studying History* (London, 1753), pp. 1–2. But intermittent concern with Aristotelian epistemology, or even metaphysics, led to no philosophy of history.

apart from its practical results. In this process, two Greek Stoics of the second century B.C., active in Roman circles—the philosopher Panaetius and his pupil, the historian Posidonius—were most important. The former became the counselor and friend of Scipio, the latter of Pompey. Their achievement may be summed up by saying that they identified the Stoic ideal with the Roman ideal of public virtue and service to the state, and connected with this, that they imparted to historiography its principal didactic message : moral instruction, which meant training for public service. This message may have been spread indirectly by Posidonius, whose no longer surviving historical work influenced Sallust, Tacitus, Plutarch, and others. Directly, it was spread by the historian Polybius, a member of the Scipionic circle and a Stoic sympathizer. In the opening sentences of his *Histories* he nailed down the didactic purpose of history for centuries to come. Everyone, he writes with deliberate exaggeration, is agreed that the beginning and end of all historical labor is to impress on us "that the soundest education and training for a life of active politics is the study of history, and that the surest and indeed the only method of learning how to bear the vicissitudes of fortune is to recall the calamities of others." (I, 2).

It is no exaggeration to say that, in the period with which I shall be concerned, few, if any, arguments appeared which were not based on one or the other of the two propositions contained in Polybius's sentence. In that period Polybius enjoyed considerable popularity, which was increased by Casaubon's Latin translation (1609), a translation famous for a lengthy preface in which royalty were reminded of the value of history for education in virtue and politics. Moreover, the Polybean propositions were congruent with what all the Roman historians were known to have asserted or implied and with the teachings of the most authoritative of all classical authors, Cicero, who taught that virtue is practical and society its proper sphere. Already in antiquity each of Polybius's propositions was turned into a stereotype. History became the education of rulers, with the corollary, sometimes challenged, that only "men of action," that is, rulers, statesmen, or generals, were capable of writing instructive history; and that history, by presenting the experience of others, was philosophy teaching by examples.

The concept of the man of action began with the image of the statesman being taught by the Stoic philosopher—Sphaerus taught the Spartan king Cleomenes, Blossius taught the Gracchi, and the like—but ended with the image of the statesman self-taught, by reading not philosophy but history. Two of the best-known tales from antiquity were the ascription of Scipio's victory over Carthage to his habit of reading Xenophon, and Lucullus's making of himself into a successful general to reading history.[5]

---

[5] The Scipio model was fixed by Cicero in *Tusc.* II, xxvi: 62 and the Lucullus model in *Acad.* II, i:2. I am not suggesting that the idea of the philosophic adviser dropped out of ancient education or that the idea of the statesman-historian replaced that of the

These models provided, so to speak, the practical political justification for studying history against the claims of those who considered history inferior to other studies. The concept of learning from examples put this justification into a framework of moral philosophy.

The Idea of Example. The ancients' use of 'example' was rich and varied, ranging from the sample of merchandise by which the seller tried to persuade the buyer of the quality of his wares to the precedent in law or history by which the orator tried to do the same thing with his arguments. Metaphorically, the Romans had long used *exemplum* to connote a sign or symptom of states of affairs, or of virtues and vices, or of character, and to denote some specific action or event or person that represented them. The Romans were particularly given to using historical figures to exemplify and define Roman virtues, often as a patriotic assertion of superiority over the Greeks—a habit which even Cicero, hard as he tried to vend Greek philosophy and rhetoric in Rome, did not escape.[6]

In theory, especially in rhetorical theory, the Greek concept of example (*paradeigma*) was more elaborate than the Roman. In practice, in education of the young or in political speech-making, Greeks and Romans seem to have employed examples in much the same way.[7] The young Roman, as Horace and Terence present him, learned not from the ethics of the philosopher, but from his father, who would point to particular examples in the boy's experience and tell him which to avoid and which to imitate.[8] The significance of this pedagogic strategy in relation to the notion of political education is not difficult to see. The Younger Pliny, reminiscing about the good old days, tells us of an established rule among the Romans—the ruling circles, of course—whereby the elders were obliged to transmit to the younger generation the principles which it in turn would have to transmit to the next. This was not to be done merely by what they could hear, that is, by precept, but by what they could see,

---

statesman-philosopher. For example, in Plutarch's essay "That a Philosopher ought to converse especially with Men in Power," the philosophic-adviser model is perpetuated (*Moralia* 777A). Also in his (?) "The Education of Children," the perfect man, the traditional statesman-philosopher, is not advised to school himself specifically in history, though he is told to train above all his memory (*ibid.* 7E, 7–8; 9E).

[6] For the general significance of Cicero's use of examples, see Michel Rambaud, *Cicéron et l'histoire* (Paris, 1953), pp. 36–37. For the point made here, see Karl Alewell, *Über das rhetorishce ΠΑΡΑΛΕΙΓΜΑ* (Leipzig, 1913), 97–8, and Hans Schoenberger, *Beispiele aus der Geschichte, ein Kunstmittel Ciceros Reden* (Augsburg, 1910), 34–5.

[7] It is true that to Plato the *paradeigma* was an ideal pattern, and that he would have regarded the instilling of right conduct by reference to concrete examples mere (necessary) habit-forming and not education in the proper sense. But on Plato's own evidence (*Prot.* 325E–326A), in actual practice the young Athenian of the Periclean age was exposed to much the same method as was the young Roman later.

[8] Horace, *Sat.* I,4:105–128. Terence, *Adel.* III,3:415 and *Heaut.* I,ii:210; II,1:221. The father in Terence's first case was, admittedly, notable for his lack of success.

by example. They were taken to the Senate and there their fathers pointed out the actual procedures to them; they were prepared for politics "by the surest method of instruction, example."[9] The Roman "man of action" thus learned from experience, and since experience was invariably referred to as consisting of "one's own and that of others," history, the experience of others in the past, served the same pedagogic purpose. When Terence has a father give his son the stereotyped advice "it is wise to draw from others' experience the lessons that may profit yourself," he has said almost *verbatim* what Polybius, Livy, Tacitus, and others were to say in justification of the study of history.[10]

From Greek rhetoric, the Romans took over certain technical details in the use of examples, which stayed largely within the rhetorical schools, and certain justifications for using them, which spread more widely. Aristotle had laid down that example was "rhetorical induction" just as the enthymeme was the "rhetorical syllogism" — rather second-rate procedures if compared to dialectic, for rhetoric after all deals with subjects, like human actions, for which we have no systematic rules. Nevertheless, there were rules pertaining to the use of examples, distinctions to be drawn between historical and invented examples, and the greater plausibility of rhetorical induction and the more compelling nature of the rhetorical syllogism, in comparison to their logical counterparts, had to be admitted and examined.[11] Of the three major Roman treatises on rhetoric, Cicero's, the anonymous *Ad Herennium,* and Quintilian's, the first still adhered to Aristotelian technicalities, such as distinguishing example from parallel and similitude as instruments for establishing probability by appeal to precedent; the second modified, and the third simplified these matters.[12] What was happening was that in Rome, where rhetoric in any case soon became a declining art, the role of example in the formal structure of discourse was much less an object of interest than was the general justification for the use of examples which anyone—orator, historian, or poet—could readily employ. In part, this justification was negative, namely, the denial of the efficacy of the alternative to example, precept or principle (*praeceptum*), as a teacher of prudence. Instruction by precept was associated with unpopular, Greek philosophizing; instruction by example offered a superior route to truth and virtue than did the sterile dogmas of the schools. (This was an argument which we shall find again in the seventeenth- and eighteenth-century rejection of scholastic and religious philosophies.) "For if the Greeks are strong on precepts, the Romans are stronger on examples,"

---

[9] Pliny, *Ep.* 8, 14:4 ff.

[10] Terence, *Heaut.*, *loc. cit.*; Polybius, I,35:7–10; Livy, 22,39:10; Tacitus, *Ann.* IV, xxxiii. Cf. also *Rhet. ad Her.* IX,ix:13.

[11] Aristotle, *Rhet.* I,ii:12 (1357a); II,xix:2 (1393a); II,xx:8 (1394a); I,ii:7 (1356b).

[12] Cicero, *De inv.* I,xxx:49. *Rhet. ad Her.* IV,xlix:62. Quintilian, V,xi:1–2.

Quintilian wrote, adding, "which is a far greater thing."[13] If the orator's reasons do not make the facts obvious, he explained, example will lead us to the truth. With Livy the generals learn their tactics from examples; generals inspire the soldier to follow, not their instructions, but their personal example. The Stoic moralist Seneca, who probably used examples more frequently than any other writer of the Roman imperial period, uttered a dictum long recalled by posterity : *longum iter est per praecepta, breve et efficax per exempla,* long is the way if we follow precepts, short and efficacious if we follow examples.

The justification for the use of examples had also a positive side. The rhetoricians held that its vividness made the example persuasive, and soon it was claimed that example was not merely persuasive, but that it irresistibly implanted motives of right conduct. This claim was made for examples of virtue, for in the opposition of virtuous precedent to ethical principle, in terms of their efficacy, something stronger than mere forensic advantage might well have to be claimed for examples. I know of only one ancient who attempted something like a systematic, psychological account of the power of virtuous example. That was Plutarch in his justification for writing his *Lives.* He argued that examples of moral virtue (*to kalon*) create an urge in the observer to be virtuous also; this urge is self-inspired, for it derives not merely from an intuitive attempt, as with a child, at imitation (*mimesis*) of what is perceived but from the investigation (*historia*) into the circumstances of the examples presented to us, which engages our intellectual faculties and in so doing furnishes us with a moral purpose.[14]

Plutarch's argument was a gift to all later theorizers who defended the didactic purpose of history either in itself or in opposition to learning from the precepts of moral philosophy. One might almost say that it was to the persistence of Plutarch's *Lives* as the most popular classical reading from the sixteenth to the nineteenth century that the idea of exemplar history owed much of its persistence.

The Function of the Historian. If we follow the later theorizers, we must not derive ideas about the function of the historian from the general purposes of history-writing among the ancients—record, memorialization, invention of genealogies, patriotic justification, and the like. We have to derive these ideas in a simpler manner, namely, from authoritative pronouncements by specific ancient authors. We shall also have to confine ourselves to those authors whose later influence can be documented, and not adopt more recent judgments, however justifiable in the light of better knowledge, about who is worth citing and who is not.

---

[13] Quintilian, XII,ii:30.
[14] *Pericles* II,3. Plutarch may have had Aristotle in mind here (*Eth. Nich.* VI,viii, 8–9), but that is merely conjecture.

Polybius's *Histories*, of all the major histories of antiquity, was the most method-conscious work. Besides representing older philosophical theories about cyclical recurrence and the role of fortune—theories revived in the Italian Renaissance but of little consequence subsequently—it offered so many explicit arguments with which to sustain the didactic purpose of history that to have read Polybius is to have read most of the advocates of exemplar history, ancient and modern. As mentioned earlier, the "man of action" is Polybius's chief concern, both as a writer and as a reader of history : "Plato tells us that human affairs will then go well when either philosophers become kings or kings study philosophy, and I would say it will be well with history either when men of action (*pragmatikoi ton andron*) undertake to write history . . . or regard training in actual affairs as necessary for writing history." (XII,28 : 1–5). Historians are praised according to whether they took part in politics and thus know what they are talking about or merely wrote for gain or from the seclusion of the study. Mere students of books are unlikely to produce works of utility to readers. And the utility of the historian's work lies in confining his attention to what would interest students of politics only. Polybius praises the great practical utility of such history, especially in his day "when the progress of the arts and sciences has been so rapid that those who study history are, we may almost say, provided with a method for dealing with any contingency that may arise." (IX,2:5). The method proceeds through vicarious experience; of the two experiences that instruct us in practical life, our own misfortunes and those of others, the latter is to be preferred as being safer and not exposing us to harm (I,35:7–10). Thus we are instructed not by precept but by reasoning from experience, especially experience reported in sufficient detail to include the situation and motives and reasons for the failures of past actions (XII,25i:6–9). This, then, is the practical, or as he calls it, pragmatic (*pragmatikos*) function of history. It raises two requirements. First, it requires truth as a necessary criterion of history, history without truth being a body without eyes (XII,12:1–3), incapable of teaching anything. The second requirement is apodictic exposition, which does not mean, however, the philosopher's demonstration or proof by argument. Calling his work apodictic (*tes apodeiktikes historiai*), he merely means by this historical narratives supported by citing of appropriate details and by adducing the causes of events (II,37:3; IV,40:1–3). The apodictic subserves the pragmatic function : "For the mere statement of a fact may interest us but is of no benefit to us : but when we add the cause of it, the study of history becomes fruitful. For it is the mental transference of similar circumstances to our own times that gives us the means of forming presentiments of what is about to happen, and enables us at certain times to take precautions and at others by reproducing former conditions to face with more confidence the difficulties that menace us." (XII,25b:1–3).

and bad forever available. In giving us these sentiments, which are virtually indistinguishable from Seneca's, Diodorus not only proves beyond a doubt the derivation of the notion of "vicarious experience" from Stoic thought. He also gives evidence for how well established the conventional and endlessly repeated catalogue of the virtues of history had become by his time: history gives the young the wisdom of the aged; for the old it multiplies the experience they already possess; it qualifies the private citizen for leadership; it entices to the imitation of good and the avoidance of bad; soldiers become more courageous in the hope of posthumous encomia; the wicked are dissuaded from evil out of fear of everlasting opprobium (I,1:5). If the myth of Hades (in our period, the "fear of hell" was usually substituted for this) can scare people into adherence to piety and justice, much more so can history equip people for righteousness. In short, history was "the mother-city (*metropolis*) of philosophy." (I,2:3).

Among the Roman historians of the later Republican period, Sallust's prefatory reflections to his once enormously popular historical monographs show Stoic influences, but apart from conventional praises of intellect over physical prowess, and some phrases about the importance of the historian to society, he offered little that is relevant here. Livy, in the preface to his history of Rome, insisted strongly on the exemplar function of history. What makes history fruitful is its clear presentation of examples of every kind of conduct; it is from the moral lessons they present that we can select for ourselves, as well as for the state to which we belong, those that are honorable for imitation and those that are dishonorable for avoidance. (I,pref.:10,15–16).

Tacitus, the greatest historian of the imperial period, who unavoidably saw and heard much evil, struggled hard to write *sine ira et studio,* without anger and without preconception, and without affection or hatred —an echo perhaps of Cicero's injunctions, since, as an orator himself, Tacitus freely worked with the stock epigrams accumulated by the rhetorical schools. In his *Annals,* he distinguished between learning right conduct by one's own lights and by the experience of others, and declared the latter more effective (IV,xxxiii). He conceived the first duty of the historian to be not merely the recording of meritorious deeds: firmly convinced that one of the few things that could restrain the villainous Caesars was the fear of what posterity might think of them, he wanted the historian to hold vice in check by threatening to commit it to record (III,lxv). (Later historians, and in particular Tacitus himself, were sometimes referred to as a rod for tyrants—an extreme, but plausible, extension of the historian's role as an educator of princes). In the *Histories* there is a passage on the apodictic function of the historian, reminiscent of what we encountered in Polybius: all the contributory conditions of events must be examined to give us a knowledge of motives and causes (I,4). But Tacitus here says more than Polybius did, and certainly more than Cicero, who defined elaboration of the notion that historical narrative must be true thus came from rhetoric, which need not surprise since discourse on the theory and practice of writing history remained under the aegis of rhetoric for another millenium and a half, and discourse on virtually any academic subject under the aegis of Cicero for even longer.

Dionysius of Halicarnassus's prolix history, the *Roman Antiquities,* would hardly have made him an influence to be reckoned with, despite his deliberately unacknowledged attempt to imitate Polybius's injunctions on the apodictic function of history (V,56:1). But as an outstanding literary critic and essayist on historical, biographical, and rhetorical subjects, he had a great deal to say about the writing of history and the criteria for evaluating good and bad historians. Harking back to classical Greek literature, Dionysius, like Cicero before him, did not find Thucydides a good model for orators to follow, in terms of the best canons of style. In a letter to Pompey he compared Herodotus to Thucydides, dividing history into a subject-matter and a stylistic part, and gave most of the honors to Herodotus. This rapprochement of history and literature is probably Dionysius's chief claim to our attention; he established, though not originated, the notion that the rules of history were literary rules, in their widest sense, and that such things as choice of subject matter, knowledge of where to begin and end narration, decisions about what to include and what to omit, and whether to relate events in logical or chronological order, were governed by esthetic criteria. Moreover, in his *Art of Rhetoric,* a book more properly attributed to Pseudo-Dionysius than Dionysius, he coined the oft-repeated phrase that "history is philosophy teaching by examples."[18]

Diodorus Siculus, another minor Greek figure of the first century B.C., is significant here for other reasons. Having spent thirty years compiling a history of the entire known world from the Creation to his own day, the *Historical Library,* he provided it with a preface in which the moral function of history was elaborated at length. The justification he gives for writing a universal history is that of Stoic philosophy. Rome was realizing the Stoic idea of a world-state or world-city (*cosmopolis*), and Diodorus evidently thought that the historian ought to do the same with his pen. Historians, he says, restore the universal unity of mankind—a unity which time and space had broken up (I,1:3). Our lives are short but history ensures a kind of immortality; everything perishes, but history preserves the experience of all ages and is thus capable of making examples of good

---

[18] XI,2:19–21. The author here ascribes to Thucydides the idea, but not the actual words, that history is philosophy teaching by examples. The reference is to Thucydides' famous remark that his history was meant to be profitable, since the probable recurrence of events makes the records of the past the guide to the future (Thuc., I,22:4). This was also the original pedagogic motif which Isocrates, Polybius, and others elaborated.

Cicero was no historian. But his pronouncements on history were quoted by everyone who wrote about history between the Renaissance and the later eighteenth century, and evidence for their quotation also before and after this period is by no means scant. Since these pronouncements look like *obiter dicta* of no particular profundity, one is tempted to ascribe their longevity primarily to the fact that it was Cicero who had made them. But that is not the whole story. Cicero attempted to weld into a common cultural ideal the systems of thought and instruction of his time—Greek with Roman, philosophy with rhetoric, the ideal of the contemplative with that of the active life. Philosophical rhetoric was to effect this synthesis. The Roman orator, tied to his law-court, was to be turned into a man of wider culture; he was to imitate the sophisticated Greek orators and to study, among other subjects, history. The relationship of history to rhetoric had therefore to be defined for this benefit. It was a peculiar relationship. On the one hand, the orator ought to annex history to oratory; it furnished him with true examples from antiquity which gave authority and credibility to his speech; it was the branch of literature closest to oratory, since it dealt in truths, whereas poetry, whose task it was to provide pleasure, necessarily dealt in falsehoods. On the other hand, rhetoric had no rules for history, nor history for rhetoric ("what Greek rhetorician was ever guided by Thucydides?"); the orator had to raid history at his convenience.[15] The upshot of this was that Cicero declared what was peculiar to history. The first of the two foregoing considerations issued in a recommendation of the particular value of history; the second, in saying what the rules of history were. The value of history was expressed in Cicero's famous definition: history is the witness of the past, the light shed on truth, the life-giving force to memory, the guide to life (*magistra vitae*), the herald of ancient days—and it is the orator's duty to commit her to immortality.[16] The rules of history were laid down in another famous passage. (Under the name of Cicero's law of truth we shall encounter some of them, as well as the indestructible epithet *magistra vitae*, again later.) Every orator, Cicero says, can see that the first law of history is that the historian must not say anything that is false, the second that he must be bold enough to tell the whole truth; and further, there must not be anywhere in his writings a trace of partiality or malice.[17] The most influential

---

[15] *Orator* xxxiv:120. *De leg.* I,ii:5. *De Orat.* II,xii:51 ff. Similarly Quintilian, II,iv:2; X,i:31–34.

[16] *De orat.* II,ix:36.

[17] *Ibid.* II,xiv:62. The strictly rhetorical purport of these remarks should not be forgotten, though subsequently it was. Unlike Thucydides or Polybius, Cicero was here defining the historian's craft and describing neither his importance nor his virtues. He evidently thought these laws applied to the historian's, but not to the rhetorician's treatment of historical narrative (*Brutus* xi,42), and himself did not hesitate to demand of his prospective biographer the maximum of eulogy "and in that respect disregard of the laws of history" (*Ad. Fam.* V,12:3).

contributory conditions of events as "originating in accident, discretion, or foolhardiness."[19] Tacitus's listed contributory conditions were social and psychological factors, and as such left the traditional confines of Roman political historiography. In part, Tacitus fitted the stock requirements of exemplar history; in part, he went beyond it and influenced those who had other conceptions of history. A curious and perhaps not inappropriate comment specifically on this Tacitean passage was made by a German historiographer early in the nineteenth century. The English historians of the second half of the eighteenth century, Hume, Gibbon, and others, he wrote, have become the models for the rest of Europe; their excellence is due to their attempt to live up to Tacitus's demand that the historian make history intelligible by penetrating into the condition of society and into the inner life of man.[20]

Lucian of Samosata, a Syrian Greek in Roman service, shares the distinction with Dionysius and Diodorus of having stated his views on history at length; indeed, he wrote the only full-length treatise on history-writing which survives from antiquity : *How to Write History*. Telling the would-be historians of his day how they should go about their job, he defended the serious, truthful Thucydidean narrative against the Herodotean proclivity to narrate to please. Lucian represents a rather modern point of view, if harking back to Thucydides—who wrote just before the rhetorical invasion of history had begun—can be called modern. In any case, Lucian's insistence that narrative must be truthful owes little to the formulas of the rhetorical school on this particular point; these he undoubtedly knew well, having been a rhetorician once himself. Lucian is also less heavy-handedly didactic than Polybius in his insistence on the utility of history. Polybius's views were, as I have indicated, an elaboration of Thucydides's view, and Lucian went back to the source (41:1).[21] Truth is not merely identified as a characteristic of history for the purpose of teaching rulers or for making clever distinctions between it and oratory and poetry. Lucian has much to say about these distinctions, but to him they arise from practical considerations. The historian, unlike the orator, does not look for what to say, but merely for how to say it; the facts are already there and speak for themselves, and he has to stick to them. If history is to be something serviceable or useful (*to chresimon*), and that is its sole purpose, the incidental function of being amusing or pleasing, and therefore any attempt at making it so, are inimical to history. He condemns all such attempts, be they motivated by a desire to please, to flatter, or to be patriotic, for "history cannot admit a lie, even a tiny one" (7:4). This insistence drives him beyond Tacitus's measured assurance of writing history *sine ira et*

---

[19] *De orat.* II,xv:63.

[20] Ludwig Wachler, *Geschichte der historischen Forschung und Kunst* (2 vols.) (Göttingen, 1818), Vol. II, p. 611.

[21] See note 18 above.

studio. The historian is "in his books a stranger and a man without a country, independent, subject to no sovereign, not reckoning what this or that man will think, but stating the facts." (41:1). We almost feel as if Lucian, by reaching back to Thucydides, and by devaluing those intervening arguments which justified history in pedagogic, rhetorical, or other terms extrinsic to it, is reaching forward to a more modern, perhaps even a nineteenth-century conception of the subject. "The historian's task," Lucian wrote, "is to tell the tale as it happened." (39:1). This, coming from an oft-quoted author, must have sounded like a disappointingly low estimate of the historian's function to those who drew on Polybius rather than Thucydides, on Rome rather than Greece. It is to these exponents of exemplar history that we are now ready to turn.

II.   The Heyday of the Exemplar Theory of History

The conception of history in the centuries preceding the nineteenth was derived principally from reading the authors I have discussed. Other conceptions of history, such as Christian providential history, and other authors, such as Moses, universally acknowledged to have been the first historian, or Josephus, or the early Church Fathers, dealt with history as *res gestae* and not as *disciplina*; even Christian historiographers referred to the classical historians when speculation turned to history as a subject of study. Every major classical historian was available by the fifteenth and sixteenth centuries and translated into the vernaculars by the seventeenth. Translators used their prefaces, understandably, to praise the utility of the study of history, and to hold up as a model the particular author they had translated. Few doubted that in principle the only perfect historians (of secular history) were the classical historians, and that they and the ancient rhetoricians had said all there was to be said about the study of history. There was an awareness that there were also modern historians writing about the history of their own time, and it was generally allowed that some three or four of them—Guicciardini, Davila, de Thou, and Clarendon— could be exempted from the conviction that modern historians must necessarily be inferior to the ancients. This was a very strong conviction. In England, for example, it was reiterated with tiresome frequency—by Bacon, Temple, Bolingbroke, Warburton, and Walpole, among others— that "the history of England has never yet been written" (as Hume put it), and that what passed for history was ridiculous compared to what there was from antiquity. Antiquity provided authoritative historians. A few moderns might be read because they had something interesting to say, exposing the wickedness of this or that religious or political party; but no modern historian could be taken as a guide to the science of history as such.[22]

But if Livy and Tacitus had no peers, Dionysius and Lucian did. A branch of scholarship had sprung up, practiced by the writers of historical manuals, who undertook both to define history and to explain the method of studying and writing it. It is to them that we must look for the primary evidence of historical theorizing. There were a great many of these books, and the complaint that there were more people designing systems for history than there were good modern historians was neither uncommon nor without foundation. These books originally were referred to as *Artes Historicae* and bore such titles as *How to Write History, The Conception of History, The Art of History,* or *The Study of History.* There was also another genre of manuals, books dealing with more technical problems, like source and textual criticism; although sometimes written by the same people who wrote the former type of book, they need not concern us here.[23]

The *Artes Historicae* as a recognizable species originated with the Italian humanists in the sixteenth century, but soon spread to other countries. At first much occupied with the literary or rhetorical approach to history, they developed into justifications of the pragmatic, Polybean approach and the problems connected with it, though as late as the eighteenth century many of them still showed considerable concern with such rhetorical problems as order of argument, arrangement of subject matter, and style. There are national differences between them too : the early Italian ones, for example, were sometimes written by Platonists and cast in the form of Socratic dialogues; the German specimens showed more concern with scriptural history than the Italian, the French, or the English, since they were caught up in Reformation theology. The inspiration for these historical manuals is quite plain. An early collection of these works in two large volumes, the *Treasury of Historical Art* (1579), contains in one volume seven contemporary specimens and Dionysius's essay on Thucydides, and in the other, nine contemporary works and Lucian's essay on how to write history.[24] Studded with citations from the ancients,

---

[22] This last point was a very durable one. The opponents of the establishment of a school of modern history at Oxford in the 1850's asked: "Is the subject suitable for Education? Is it an exercise of the mind? . . . Where is the standard author like Thucydides, etc?" Quoted in R. W. Southern, *The Shape and Substance of Academic History* (Oxford, 1961), p. 10.

[23] Arnaldo Momigliano, *Contributo alla storia degli studi classici* (Rome, 1955), p. 81, describes the latter type.

[24] *Artis Historicae Penus* (2 vols.), Johann Wolff (ed.) (Basel, 1579). The collection is surveyed in P. C. F. Daunou, *Cours d'études historiques* (20 vols.) (Paris, 1842), vol. VII, pp. 34–69; in John L. Brown, *The Methodus ad Facilem Historiarum Cognitionem of John Bodin* (Washington, 1939), pp. 46–48; and most recently by Beatrice Reynolds, "Shifting Currents in Historical Criticism," *Journal of the History of Ideas,* Vol. 14 (Oct. 1953), pp. 417–492.

imitative of them as well as of one another, often platitudinous, and always didactic, the *Artes Historicae* were the principal disseminators of exemplar history. When they were written for more formally educational purposes—or when their arguments were retailed for these purposes—there was much emphasis on the Christian as well as the classical conception of virtue, as befitted books written for the young; in point of time, that line goes back to the uses of exemplar history by medieval hagiographers and chroniclers. I wish to illustrate the contents of the *Artes* in our period by briefly citing the three major works of the genre as well as by reference to more specifically pedagogic works.

The most important history manual in the sixteenth century was that of the French legal theorist Jean Bodin, who is better known for his later work on political theory. His manual, the *Method for the Easy Comprehension of History* (1566), influenced the English ones written in the seventeenth century, and was still cited in the eighteenth by admirers who acclaimed Bodin as the discoverer of the geometric method before Descartes and by critics who thought him old-fashioned and unable to distinguish fact from fiction. The seventeenth century produced the most detailed manual, that by the Dutch classicist Vossius (Gerard Johann Vos), *The Art of History, or the Nature of History and Historics* (1623)—"historics" or "historic," a term in use until the nineteenth century, denoting the theory of historiography. The last considerable manual was Lord Bolinbroke's *Letters on the Study and Use of History* (1738), which is perhaps the only one still occasionally read today. The tradition of writing these manuals, in imitation of the earlier models, continued for some time, though usually as formal lecture courses for students; Joseph Priestley's *Lectures on History and General Policy* (1788) was among the more prominent later patchworks of that sort. There are of course great differences between the three major *Artes Historicae* I have cited. Bodin represented the Renaissance humanist viewpoint in his subdivision of history into human, natural, and divine (which, since usually nothing much further was said about divine history, served as a device for excluding it from consideration), and, rather atypically, attempted to establish naturalistic principles, such as climatic theories, for connecting the first two. Vossius, a classical philologist to the bone, developed the notion of the art of historics by a series of definitions, and offered a catalogue of all the possible characteristics of history, substantive and formal, using Cicero's injunctions as the basis. Bolingbroke, the only one of the three who did not write in Latin, reflected the aristocratic, belletristic mannerisms of the English Augustan age, and several contemporary themes—such as anti-clericalism, pyrrhonism, and the theory of self-love—are mixed in, often inconsistently, with the conventional topics discussed. But granted these differences, all three uphold exemplar history after the manner of their Roman predecessors; in large part, they are little else but paraphrases and citations, with suitable elaboration, from the authors we surveyed in the preceding section; the "man of action," the function of examples to inspire or deter, history as *magistra vitae*, history as the prolongation of individual experience, the historian's concern with truth, the persuasiveness of example—these and all the rest occur with varying emphasis in each one of them.

A representative sample from more specifically pedagogic literature in France and England shows the same picture, but with greater stress on moral elevation conceived in Christian terms. Among the Jesuit authors, the poet Le Moyne (1670) has much on history as an "exemplary philosophy" and its role as a teacher of virtue, especially for the young. Bishop Bossuet (1681), addressing his history to the Dauphin, stressed the use of examples of the past as a means of extending experience for princes in particular. Locke's liberal and pious friend Jean Le Clerc, who used the maxims of the ancient historians to fight the religious prejudices of the day, devoted much of his inaugural address (1712) as professor of ecclesiastical history at Amsterdam to the notion of example and its irresistible effect on virtue. The great French Catholic historiographer Lenglet du Fresnoy, in a massive survey of the entire known historical theory and practice of his day (1713), declared that the purpose of history is to instill prudential rules and extol the practice of virtue through representing persons possessed of it in an eminent degree. Rollin, the principal of the University of Paris, whose textbook on *Belles Lettres* (1726–28) went through innumerable editions in English and French and was still read in the nineteenth century, devoted much of one volume to urging that history be studied before all other subjects on the grounds that it affords exemplar lessons of virtue. (Among men of letters more interested in wit than pedagogy, Saint-Réal [1671] introduced what was perhaps the only novel idea on the subject, namely, that since men are invariably motivated by prejudices and passions, it is more useful to study examples of vice than of virtue.)[25] With the English writers, the picture is the same. Degory Whear, the incumbent of the first endowed history professorship at Oxford (1622), proposed two ways in which historical examples can be used : for imitation of the deeds of good men and avoidance of those of the wicked, and for the extraction of morally elevating maxims from particular historical sequences. Hearne (1703), the Bodleian librarian famous as an antiquary, discoursed in his

[25] Pierre Le Moyne, *De l'histoire* (Paris, 1670), Diss. I, chap. vii, ix. Jacques Benigne Bossuet, *Discours sur l'histoire universelle* (9th Fr. ed.; London, 1707), pp. 5 ff. *An Oration Concerning the Usefulness and Excellence of Ecclesiastical History . . . by Mr. Le Clerc* in Jabez Hughes, *Miscellanies in Verse and Prose* (London, 1737), pp. 235–238. N. Lenglet Dufresnoy, *New Method of Studying History . . .* translated by R. Rawlinson (2 vols.) (London, 1728), vol. I, p. 24. [Charles] Rollin, *The Method of Teaching and Studying the Belles Lettres* (10th ed.; London, 1804), vol. II, esp. pp. 232 ff. "De l'usage de l'histoire," in *Oeuvres de M. L'Abbé [César Vichard] Saint-Réal* (The Hague, 1722), vol. I, pp. 68–73.

guide to history on the irresistible nature of historical examples even to those who are wicked themselves. And in a later textbook, Manwaring (1737), giving thirty-three rules for the study of history, limits the subject matter of history to matter novel and weighty, full of virtuous examples.[26]

It is true that most of the exponents of exemplar history raised issues which post-date antiquity, like divine providence, scriptural chronology, or periodization of history. But these formed mostly a small proportion of their discourse, the bulk of which dwelt on the problems with which the ancients had dealt. Polybius's "man of action" was a favorite not only with the Jesuit and other pedagogues, but even with the principal historians of the period, themselves often statesmen or office-holders. Polybius's apodictic requirement is included in virtually every manual, as is the Stoic concept of vicarious experience and the prolongation of life by history—not only by the neo-Stoics of the seventeenth century but by influential churchmen like Bossuet and Le Clerc. Tacitus's advice to explore motives held great appeal for those forever advising historians to expose the inner springs and resorts of historical characters; from Bacon to Saint-Réal to Gibbon, the fascination with the springs of character attests to both preoccupation with psychological ethics and fascination with an as yet not understood mechanical problem, the spring being an object that mysteriously stored energy. Lucian's analysis of impartiality was to protect the reader of histories written by partisans of religious and political factions. Dionysius of Halicarnassus came in for his share when one of his translators, deploring the fact that Aristotle had left rules for rhetoric but not for history, proved at some length that Dionysius's rules provided what Aristotle had left undone; and, indeed, surpassed those that Lucian and even Cicero had laid down.[27] Nothing, of course, could be written without reference to Cicero. Cicero's definition of history as *magistra vitae* occurs in every manual for a period of over two centuries; either it or his laws of truth even provided the standard motto or emblem for the frontispiece of major historical works, like Raleigh's or Clarendon's, or graced the dedicatory preface, like de Thou's.

In all these discourses, new ideas consisted of adding yet another virtue which an example could instill, or yet another reason why examples were irresistible. Differences of opinion revolved around whether or not the study of philosophy was a prerequisite to the study of history (denied by the majority); differences of conception of the subject matter were re-

---

[26] Degory Whear, *The Method and Order of Reading both Civil and Ecclesiastical Histories*, translated by E. Bohun (3rd ed.; London, 1698), pp. 343–344. [Thomas Hearne], *Ductor Historicus* (2 vols.) (2nd ed.; London, 1705), vol. I, p. 115. Edward Manwaring, *A Historical and Critical Account of the Most Eminent Classic Authors . . .* (London, 1737), pp. 357 ff.

[27] Edward Spelman, *The Roman Antiquities of Dionysius Halicarnassensis* (4 vols.) (London, 1758), vol. I, pp. xv ff.

flected in whether the rhetorical rules of arrangement, order, and style were included (as in Vossius) or omitted altogether (as in Bolingbroke), and in advising which classical historian offered the best model.

The practicing historian had little hope of distilling any meaning from all this. The Frenchman Rapin, the author of the best-known history of England before Hume's, wrote: "When realizing that I had now undertaken to write a formal history, I wanted to thoroughly instruct myself in the rules which the masters had given for succeeding in this enterprise; but believe me, I did not find there the aids I had expected. Some of these rules are so vague that one can only regard them as useless; not that they are not excellent in themselves, but because they teach nothing in particular or nothing new, nothing that does not come naturally to any person with common sense. Of this kind, for example, are the following: That a historian must speak the truth. That he be on no account biased. . . . One advises to take Livy as a model. Another prescribes Caesar. A third would like all histories written in the manner of Tacitus . . . . there is no less diversity in all the advice about the form of history, the arrangement of materials, the style, and similar matter. All this advice is absolutely unnecessary . . . the most distinguished historians never tie themselves down to any rules except those dictated by reason and common sense."[28]

What, then, is the significance of this extensive literature? It does not suffice to say, though it is perfectly true, that the authors of the *Artes Historicae* and related literature simply copied the ancients (and one another). We have to ask what they thought they were doing when they did this. It is safe to interpret the irrelevance of this literature to the practicing historian as *prima facie* evidence that we are dealing with philosophy of history, with propositions intelligible in the context of ethics or other philosophic concerns but void of anything either sufficient or necessary to the study of history. And in this context the clue, here as with the ancients, is the juxtaposition of example and precept.

III. The Significance and Decline of the Exemplar Theory of History

The many virtues imputed to learning from example—the quickness and irresistibility, the greater safety of having vicarious rather than personal experiences—were particularly evident in moral instruction. "In moral philosophy," a Christian humanist had declared long ago, "examples are of more avail than precepts; for everyone more promptly and more willingly imitates what he admires." We follow Christ "rather by the

---

[28] Paul de Rapin Thoyras, *Histoire d'Angleterre* (10 vols.) (2nd ed.; The Hague, 1727), vol. I, pp. ii–iii.

example of martyrs, than by the admonition of theologians." Le Clerc and other theologians engaged in opposing narrow dogmas, said the same thing.[29] But not only theologians. Vossius, engaged in a polemic against sixteenth-century German pyrrhonists who had asserted that history is no more than an imperfect record of singular events, not a discipline subject to a definite method, enlarged on the relation of singular to universal, and on the necessary confirmation of universal precepts by particular examples. Vossius did not maintain that we arrive at precepts by induction from historical particulars. History was merely "the knowledge of singular events whose memory it is useful to conserve in order to ensure a good and happy life"; all this knowledge does is to confirm the precepts (of morality) established by philosophy. It was a way of teaching philosophy. History "is philosophy teaching by examples. From this affinity and conjunction of philosophy with history it follows that the philosopher may illustrate his precepts with historical examples, and that the historian, in turn, may test (*expendere*) certain facts in the light of the philosopher's precepts."[30] Already Bodin had advised the construction of a commonplace book, in which the examples of history, suitably graded as base or honorable, useful or useless, could be systematically arranged to yield something like a training manual in virtue.[31] Among Bodin's English followers, Whear had gone further and explained how, in addition to guiding us to the good, examples might be used in the actual construction of all kinds of precepts. Formally what he proposed was typical of the new scientific method; substantively it was rather unenlightening. We are told to take the story of those indestructible cast-out brothers Romulus and Remus from Livy. This story would yield two or three "axioms": the foreknowledge of the deity preserves those destined for great work later; or, the foundation of empires is sometimes laid in shameful beginnings but attended by miraculous events; or, the foundations of empires appear to us contemptible, but actually show the operation of divine providence. Then, to confirm the axiom, we have to look for other examples—like Cyrus's foundation of the Medo-Persian empire—to see whether other founders were similarly cast out and miraculously preserved.[32]

With Bacon and Hobbes, who wrote about the relation of historical example to precept at the same time as Vossius and Whear, these tendencies are carried further. Example begins to play a decisive role in the

---

[29] Juan Luis Vives, *De tradendis disciplinis* (1531), in *Vives: On Education*, translated by Foster Watson (Cambridge, 1913), p. 234. Le Clerc, *op. cit.* in Hughes, *Miscellanies* (1737), p. 235.

[30] *Gerardi Ioannis Vosii Ars historica, sive de historiae & historices natura . . . commentatio* (Leyden, 1623), pp. 16, 30.

[31] John Bodin, *Method for the Easy Comprehension of History*, translated by Beatrice Reynolds (New York, 1945), pp. 35–36.

[32] Whear, *op. cit.*, p. 344.

construction of a prudential ethic. Hobbes, it has been argued, used history in his rejection of traditional ethical norms just as he later used political philosophy to the same end.[33] Bacon, as might be expected, not only used historical examples but wrote at length on the validity of doing so. He held that moral philosophers had discoursed on virtue and vice but had found no convincing method for making their teachings effective. Their failure to relate their teachings to the emotional life and to the forces which composed it made them inferior to the historian as observers, and therefore teachers, of morals. "Aristotle's and Plato's moral doctrines are admired by many; but Tacitus utters observations on morals that are much truer to life."[34] The philosophers have had little to say about the emotions which, like factions in a state, are at war within man. "But the poets and writers of histories are the best doctors of this knowledge." History, poesy, experience, and not the precepts of traditional philosophy, teach those things that are useful to life. Tacitus's astute case histories provided one part of the argument; the method of that great consumer of classical examples, Machiavelli, provided the other. The best method "is that which Machiavel chose wisely and aptly for government; namely, *discourse upon histories or examples.* For knowledge drawn freshly and in our view out of particulars, knoweth the way best to particulars again. And it hath much greater life for practice when the discourse attendeth upon the example, than when the example attendeth upon the discourse."[35]

Bolingbroke, a professed Baconian, and in his epistemology more radical than Locke but less radical than Hume, examined the notion of example in great detail. He did this both as a philosopher striving for something new to say and as a humanist anxious to show that he could quote whatever the ancients had said—a somewhat difficult undertaking. The

---

[33] Leo Strauss, *The Political Philosophy of Hobbes* (Chicago, 1952), chap. vi, where Hobbes's preface to his translation of Thucydides (whose narrative instructs the reader "more effectually than can possibly be done by precept") is convincingly shown to be the key to the argument. Hobbes, incidentally, admitted to following Cicero and Lucian, among others.

[34] *Temporis Partus Masculus*, in *The Works of Francis Bacon*, J. Spedding, R. L. Ellis, and D. D. Heath (eds.) (12 vols.) (Boston, 1863), vol. VII, p. 31.

[35] *Advancement of Learning, op. cit.*, vol. VI, p. 359. What I have quoted from Bacon does little more than hint at his elaborate arguments. For Bacon's defense of examples see VI, 97 ff; on history, VIII, 385; on Tacitus, VI, 235. It might be noted that admiration for Tacitus usually implied admiration, even if qualified, for Machiavelli. The orthodox opposed both as anti-Christian and subversive and sometimes averred that the teaching of Tacitus produced Machiavelli's. Bolingbroke, a freethinker, chose Tacitus as his model for history and Machiavelli for political theory; others chose the other way about. Hume put a motto from Tacitus's *Histories* on the title page of the *Treatise* (which, according to its first learned reviewer, in 1739, "sufficiently betrayed the author's evil intentions"); he classed Machiavelli among the best historians. All this points to a connection of exemplar history with political theory which is in need of study.

two ways of experience, personal and historical, were not only quoted from Tacitus and Polybius, but examined and compared. The inefficacy of precepts, however true they might be, was considered in the light of Seneca's pronouncements, and also connected with the then current psychological theory in which the passions—amenable to example, but impervious to precepts—received much attention. Indeed, Bolingbroke gave examples of examples to make his points, but seemed to have found only one argument that had not yet been "insisted on by those who have writ concerning the method to be followed." He noticed that ancient examples were singularly inapplicable to modern times. If a general were to imitate the Roman generals who committed suicide in order to fulfill an oracle's prediction of victory for their troops, "he might pass for a hero, but I am sure he would pass for a madman." The correct use of examples was not to imitate the particular modes of action, which are the subject of custom and therefore the product of different periods and nations, but to study them philosophically, as exemplifications of general principles and rules of conduct. (This fertile notion was taken further by Bolingbroke's early follower, Voltaire, and is sometimes identified as Voltaire's philosophy of history.) Such principles, we are told, must always be true because they are conformable to the invariable nature of things; they are discoverable by induction from historical example. "He who studies History as he would study Philosophy will soon distinguish and collect them, and by so doing will soon form to himself a general System of Ethicks and Politicks on the surest foundations, on the trial of these principles and Rules in all ages, and on the confirmation of them by universal experience."[36]

The drift of the arguments advanced by those who wrote about history should now be clear. History is, or ought to be, what came to be called in the eighteenth century "the empirical part of moral philosophy." Throughout the period surveyed the term "moral" referred to both civic and private morality, to politics and ethics. (To the exponents of exemplar history, the Roman conception of virtue as public virtue was paramount. That specifically Christian virtues could be gained from the study of historical examples, was, for the most part, the stock-in-trade of Christian educators and authors of history schoolbooks, who retailed exemplar history for pedagogic purposes, as I have noted.) But the term 'philosophy' by the end of this period, as is well known, had become detached from its scholastic moorings and was defined in terms of the characteristics which first Bacon and then Newton had assigned to natural philosophy.

This stage in the development of history as moral philosophy heralded the end of exemplar history as a serious intellectual endeavor. For history now had to be regarded as a science, or as part of a science, in the sense in which natural philosophy implied that term. The systematization

---

[36] Henry Saint-John, Lord Viscount Bolingbroke, *Letters on the Study and Use of History* (London, 1738), Let. III, pp. 12–13.

which all along had been hinted at, feebly by Bodin and Whear, dogmatically by Bacon, and optimistically by Bolingbroke, was now demanded. It was one thing to oppose, just as the Romans had, precept to example, and to define history as "philosophy purified from the pedantry of the schools and free from the encumbrances of division and arguments, and reduced to action and example."[37] It was another thing to show by what kind of systematic study of historical facts moral philosophy could be produced. Few thinkers, even among Englishmen who had never heard of Descartes's, Malebranche's, or Bayle's scepticism of history, could find a basis for doing this. After Locke and Shaftesbury—Hobbes was still not respectable—any significant attempts at constructing scientific systems of ethics would have to be based on psychology and not on history; to put it epigramatically, on man, not men. Whoever wanted to convert the aspirations of exemplar history into rigorous scientific propositions would soon find either that the propositions he advanced were scientific only in name or that he had in fact abandoned history for social science. This did indeed happen. I shall mention, without further elaboration, an instance of each position.

The Scottish philosopher and pedagogue George Turnbull, who wrote about the same time as Bolingbroke, found himself inspired by a remark in Newton's *Optics* to the effect that the methods of the natural philosopher were applicable to those of the moral philosopher. He wrote first one book designed to prove that this was so, and then another in which he assigned to the study of history that role in moral science which observation and reasoning from observed facts played in the physical sciences. He thought he had established that history was the empirical part of moral philosophy. By the latter term he seems to have meant as often as not political science; for the (unstated) moral laws established inductively from historical examples are said to make prediction of political changes and the establishment of good government possible, and Aristotle as well as Polybius are mentioned as his predecessors.[38] What in fact he had done was to write an *Ars Historica* of the traditional kind on which the fashionable vocabulary of axiom, law, and hypothesis was superimposed. His principal original suggestion was that the study of history become the paradigm for the study of the physical sciences, since proceeding from "facts or examples" was better than experimental proof of preconceived hypotheses. Turnbull, evidently mindful of Newton's *Queries*, thought that such hypotheses, though used in the natural sciences, were ultimately "fatal" to the development of them as well as of moral science.[39] His work was uninfluential and little read.

---

[37] Le Moyne, *op, cit.*, Diss. I, chap. viii.

[38] Turnbull, *The Principles of Moral Philosophy* (London, 1740), pp. 200–201.

[39] Turnbull, *Observations Upon Liberal Education in All Its Branches* (London, 1742), p. 391.

A different approach was taken in a slightly later attempt to theorize about history in a scientific—in contemporary usage, "philosophic"—style. It emanated from a school of theorists, chiefly Scottish philosophers, of the second half of the eighteenth century. Their approach was described by one of their contemporaries as *"Theoretical* or *Conjectural History,* an expression which coincides pretty nearly in its meaning with that of *Natural History,* as employed by Mr. Hume and with what some French writers have called *Histoire Raisonnée.*[40] This school was principally devoted to historical sociology or anthropology. Its theory was to concentrate on the reconstruction of unknown stages in the development of society from known stages, by means of certain hypotheses about social, cultural, physiological, or economic phenomena. As one of its practitioners said, Montesquieu was the Bacon and Adam Smith the Newton of this new science of society. Contemporaries, for a while at least, saw in it an unexpected fulfillment of two centuries of speculation about history, an advance "which even the prophetic genius of Bacon could not foresee." But neither it nor the similar philosophy of history espoused by the *philosophes* of the French Enlightenment was destined to succeed the idea of exemplar history as the dominant speculation about history, though they exercised a profound influence on the practices of social scientists and historians, respectively.

The exemplar conception of history survived; what had failed was the attempt to integrate it with current scientific assumptions, except in a superficial way.[41] By the second half of the eighteenth century politics and ethics had been put, individually and jointly, on what seemed to be a secure scientific base, and exemplar history had little to offer the majority of the sophisticated. It no longer had a philosophic function, since history was no longer the principal antischolastic source of moral enquiry. It was no longer plausible as a philosophy of history, though it could still purport to be a theory of its study—at least so long as history was a lay pursuit

---

[40] Dugald Stewart, *Account of the Life and Writings of Adam Smith* (1793] in *The Collected Works of Dugald Stewart,* Sir William Hamilton (ed.) (10 vols.) (Edinburgh, 1858), vol. X, p. 34. The practitioners of this history have been surveyed by Gladys Bryson, *Man and Society: The Scottish Enquiry of the Eighteenth Century* (Princeton, 1945), who includes Hume among them (pp. 102–109); and by W. C. Lehmann, *Adam Ferguson and the Beginning of Modern Sociology* (New York, 1930).

[41] A popularized notion like the principle of the uniformity of human nature was used to explain to the ordinary reader the exemplar value of the deeds of kings and generals, which he is said to have often complained had not the slightest bearing on his own problems. (With the *Artes* often defining history as "an exemplary philosophy invented for the instruction of the great," nothing much lower than kings and generals were admitted as examples for imitation.) The typical reply to this complaint was that, morally, high and low were constituted alike, "that the frame of men's minds is much the same, let their condition be what it will. The passions, virtues and vices, operate in all alike, though upon different objects." John Oldmixon, *The Critical History of England* . . . (London, 1726), p. 2.

whose chief interest, then as now, was in individuals and their behavior. As a moral philosophy it could survive only in its Christian-pedagogic version, which was least likely to be affected by the decline of classical conceptions of moral and political enquiry. After about 1800, exemplar history was largely confined to schoolbooks—some of them reprints of early eighteenth-century manuals—since it was only for the education of children that the justification of the study of history in terms of its moral utility could now be plausibly upheld. (That was also Hegel's view, who, like Burke before him, denied that exemplar history was for statesmen.[42])

The authoritative position ascribed to the ancients, from which exemplar history derived, also survived, but again with significant changes. Among the classical historians the philosophizing Polybius, the father of pragmatic history, was eclipsed by Herodotus and Thucydides, who were more consonant, not merely with the intensified interest in Greek antiquity, but with the new view that history should be studied for its own sake.

This new view—new in its acceptance only, but long underlying the work of scholars and antiquarians—succeeded in deposing exemplar history even as a theory of study, because history as a subject of study had ceased to be what it had been in the heydey of the *Artes Historicae*. It was rapidly becoming a professional enterprise written by professors for students. That metamorphosis hailed from Germany, especially from Göttingen, where scholars in the late eighteenth and early nineteenth centuries were able to appropriate quietly the fruits of English, Dutch, and French historical scholarship without concern for a lay public (to which, unlike their French and English counterparts, political history had in any case not been a *Bildungswissenschaft*), and where essentially different intellectual and social conditions obtained. From there, and later from universities elsewhere, the tendency to make historical studies strictly professional was spread. The perennial struggle between the antiquarian and the philosophic historian, the pedantic scholar and the educated gentlemen, fought in antiquity and again since the Renaissance, was coming to an end. The

---

[42] "It may be allowed that examples of virtue elevate the soul and are applicable in the moral instruction of children for impressing excellence upon their minds. But the destinies of peoples and states . . . present quite another field. Rulers, statesmen, nations, are wont to be emphatically commended to the teaching which experience offers in history. But what experience and history teach is this—that peoples and governments never learned anything from history, or acted on principles deduced from it." G. W. F. Hegel, *The Philosophy of History*, translated by J. Sibree (New York, 1956), p. 6. Burke, reacting against the attempts at systematization which I have described, declared that if history were studied "as a repertory of cases and precedents . . . a thousand times better that a statesman had never learned to read." *The Works of the Right Honourable Edmund Burke*, John C. Nimmo (ed.) (12 vols.) (London, 1887), vol. IV, p. 468. Nietzsche, going back to Polybius, was to take a different view, and in his *Use and Abuse of History* resurrected the "man of action" (*Thätige*)as the chief beneficiary of exemplar history.

érudit was triumphing over the *philosophe,* and only a few transition figures—Gibbon, it has been suggested, was one of them—succeeded in being both.[43]

The view that history should be studied for its own sake was not a philosophic view, though it received that gloss from the romantics; it sprang from practice, not from theory. It was a fact of professional life which, so to speak, falsified the exemplar theory, though it was soon called upon to support other and perhaps less harmless delusions. By the early nineteenth century the moralist who had equated the study of history with the highminded art of instructing the public in drawing lessons from the past would be confronted with the identification of the study of history as the lowly pursuit of past facts. The reality and the significance of this confrontation have become forgotten. But the sentence which was passed on the old view and which heralded the triumph of the new is still remembered : "To history has been attributed the office to judge the past and to instruct the present to make its future useful," Ranke wrote in his first work in 1824, "at such high functions this present attempt does not aim—it merely wants to show how things really were."[44]

---

[43] For Gibbon see Momigliano, *op. cit.,* p. 199. A review of the late eighteenthcentury German translation of Bolingbroke's *Letters* declared that it was preferable to study history "as mere dates and dry narration" than as philosophy teaching by examples, since emphasis on individuals and their behavior had proved a handicap to technical, historical subjects like administrative history; Bolingbroke was for beginners, not for professionals. *Göttingische Anzeigen von gelehrten Sachen,* vol. 199 (1794), pp. 1991–1993.

[44] Leopold von Ranke, *Geschichten der romanischen und germanischen Völker,* W. Andreas (ed.) (Wiesbaden, 1957), p. 4.

WRITINGS OF KARL R. POPPER

Writings of Karl R. Popper

[References to later re-editions are in *square brackets*. Most of the re-editions are revised and enlarged.]

1925   "Über die Stellung des Lehrers zu Schule und Schüler," *Schulreform* (Vienna), **4**, pp. 204 ff.

1927   "Zur Philosophie des Heimatgedankens," *Die Quelle* (Vienna), **77**, pp. 899 ff.

1931   "Die Gedächtnispflege unter dem Gesichtspunkt der Selbsttätigkeit," *Die Quelle* (Vienna), **81**, pp. 607 ff.

1932   "Pädagogische Zeitschriftenschau," *Die Quelle* (Vienna), **82**, pp. 301, 580, 646, 712, 778, 846, and 930.

1933   "Ein Kriterium des empirischen Charakters theoretischer Systeme," *Erkenntnis*, **3**, pp. 426 ff. [English translation contained in 1959(*a*).]

1934(*a*) "Zur Kritik der Ungenauigkeitsrelationen," *Die Naturwissenschaften*, **22**, p. 807.

(*b*) *Logik der Forschung*, Julius Springer, Vienna (with the imprint '1935'). [English translation contained in 1959(*a*).]

1935   "Induktionslogik und Hypothesenwahrscheinlichkeit," *Erkenntnis*, **5**, pp. 170 ff. [English translation contained in 1959(*a*).]

1938(*a*) "A Set of Independent Axioms for Probability," *Mind*, **47**, N.S., pp. 275–277. (See also pp. 415, 552). [Also contained in 1959(*a*).]

(*b*) "Isaac Newton," *The Press* (Christchurch, New Zealand), October 15, 1938, p. 20. (Review of J. W. N. Sullivan's biography of Newton.)

1940(*a*) "What is Dialectic?" *Mind*, **49**, N.S., pp. 403–426. [Also contained in 1963(*a*).]

(*b*) "Interpretation of Nebular Red Shifts," *Nature*, **145**, pp. 69 ff.

1943   "Are Contradictions Embracing?" *Mind*, **52**, N.S., pp. 47–50.

1944(*a*) "The Poverty of Historicism I" (Parts I & II), *Economica*, **11**, No. 42, pp. 86–103.

(*b*) "The Poverty of Historicism II" (Part III), *Economica*, **11**, No. 43, pp. 119–137.

1945(a) "The Poverty of Historicism III" (Part IV), *Economica*, **12**, No. 46, pp. 69–89. [Book edition of this and the two foregoing articles : 1957(g).]
   (b) *The Open Society and Its Enemies:* vol. I, *The Spell of Plato*. London : George Routledge & Sons.
   (c) *The Open Society and Its Enemies:* vol. II, *The High Tide of Prophecy*. London : George Routledge & Sons.
   (d) "The Growth of German Historicism" (review article), *Economica*, **12**, pp. 259–261.

1946   "Why are the Calculi of Logic and Arithmetic Applicable to Reality?" *Aristotelian Society, Supplementary Volume* XX, pp. 40–60. [Also in 1963(a).]

1947(a) "New Foundations for Logic," *Mind*, **56**, N.S., pp. 193–235. See also Corrections in **57**, pp. 69 ff.
   (b) "Logic Without Assumptions," *Proceedings of the Aristotelian Society*, **47**, pp. 251–292.
   (c) "Functional Logic without Axioms or Primitive Rules of Inference," *Proceedings of the Kon. Nederlandsche Akademie van Wetenschappen*, **50**, No. 9, pp. 1214 ff, and *Indagationes Mathematicae*, **9**, Fasc. 5, pp. 561 ff.
   (d) "L'Utopie et la Violence," *Les rencontres philosophiques des Bruxelles*. (French translation of 1948(a).)
   (e) Second printing of *The Open Society*. (London : Routledge.)

1948(a) "Utopia and Violence," *The Hibbert Journal*, **46**, January 1948, pp. 109 ff. [Also in 1963(a).]
   (b) "On the Theory of Deduction : Part I, Derivation and its Generalizations," *Proceedings of the Kon. Ned. Akademie van Wetenschappen*, **51**, No. 2, pp. 173 ff.
   (c) "On the Theory of Deduction : Part II, The Definitions of Classical and Intuitionist Negation," *ibid.*, No. 3, pp. 322 ff.
   (d) "The Trivialization of Mathematical Logic," *Proceedings of the Xth International Congress of Philosophy, Amsterdam*, 1948, vol. I, pp. 722 ff.
   (e) "Prediction and Prophecy, and their Significance for Social Theory," *ibid.*, pp. 82 ff. [Also in 1963(a).]
   (f) "What can Logic do for Philosophy?" *Aristotelian Society, Supplementary Volume* XXII, pp. 141–154.

1949(a) "A Note on Natural Laws and so-called 'Contrary-to-Fact Conditionals'," *Mind*, **58**, N.S., No. 229, pp. 62–66.
   (b) "Towards a Rational Theory of Tradition," *The Rationalist Annual*, London 1949, pp. 36 ff. [Also in 1963(a).]
   (c) Third printing of *The Open Society*. (London : Routledge.)
   (d) "Naturgesetze und Theoretische Systeme," in *Gesetz und Wirklichkeit*, edited by S. Moser. Innsbruck. Pp. 43 ff.

1950(a) *The Open Society and Its Enemies*, revised and enlarged edition in one volume. Princeton, N.J. : Princeton University Press.

(b) Indeterminism in Quantum Physics and in Classical Physics I,"
The British Journal for the Philosophy of Science, 1, No. 2, pp. 117–
133.

(c) "Indeterminism in Quantum Physics and in Classical Physics II,"
ibid., No. 3, pp. 173–195.

(d) *De Vrije Samenleving en Haar Vijanden* (Dutch translation of
1947(e), by Justus Meyer), 2 vols. Bussum, Holland : F. G. Kroonder.

1952(a) *The Open Society and Its Enemies;* 2nd ed., further revised and
enlarged. London : Routledge and Kegan Paul.

(b) "Humanism and Reason," *The Philosophical Quarterly,* 2, pp. 166 ff.
[Also in 1963(a).]

(c) "The Nature of Philosophical Problems and Their Roots in Science,"
The British Journal for the Philosophy of Science, 3, No. 10, pp.
124–156. [Also in 1963(a).]

(d) "La poverta dello storicismo," *L'Industria* (Milano), 1952, No. 3,
pp. 285–307. (Italian translation of 1944(a).)

1953(a) "Language and the Body-Mind Problem," *Proceedings of the XIth
International Congress of Philosophy,* vol. VII, pp. 101 ff. [Also in
1963(a).]

(b) "The Principle of Individuation," *Aristotelian Society, Supplementary Volume* XXVII, pp. 97–120.

(c) "The Sociology of Knowledge," in *Readings in Philosophy of
Science,* edited by P. P. Wiener. New York : Scribners. Pp. 357 ff.
(Part of chapter 23 of 1950(a).)

(d) "A Note on Berkeley as Precursor of Mach," *The British Journal for
the Philosophy of Science,* 4, No. 13, pp. 26–36. [Also in 1963(a).]

(e) "La poverta dello storicismo," *L'Industria,* 1953, No. 1, pp. 69–93;
No. 2, pp. 283–307; No. 3, pp. 451–491. (Italian translation of
1944(b) and 1945(a).)

1954(a) "Immanuel Kant : Philosopher of the Enlightenment," *The Listener,*
51, pp. 291 ff. [Also in 1963(a).]

(b) "Immanuel Kant. Zu seinem 150. Todestag," *Engl. Rundschau,*
Koeln, 19.3.54. (German translation of 1954(a).)

(c) "Self-Reference and Meaning in Ordinary Language," *Mind,* 63,
N.S., No. 250, pp. 162–169. [Also in 1963(a).]

(d) "Degree of Confirmation," *The British Journal for the Philosophy
of Science,* 5, pp. 143–149. (See also 5, No. 20, pp. 334, 359). [Also
in 1959(a).]

(e) *Miseria dello storicismo.* Milano : Editrice L'Industria. (Book
edition with revised text of 1952(d) and 1953(e).)

(f) "The Advocate of Democratic Criticism," in *The State versus
Socrates, A Case Study in Civic Freedom,* edited by J. D. Mont-
gomery. Boston : The Beacon Press. Pp. 160 ff. (Part of chapter
10 of 1945(b).)

(g) "Has History any Meaning?" in *Contemporary Philosophy,* edited
by J. L. Jarrett & S. N. McMurrin. New York. (From chapter 25
of 1945(c).)

1955(a) "On a Proposed Solution of the Paradox of the Liar," *The Journal of Symbolic Logic*, **20**, No. 1, pp. 93 ff.

   (b) "Two Autonomous Axiom Systems for the Calculus of Probabilities," *The British Journal for the Philosophy of Science*, **6**, No. 21, pp. 51–57. (See also **6**, No. 22, p. 176, and No. 24, p. 351, where some misprints are noted.)

   (c) "A Note on the Body-Mind Problem," *Analysis*, **15**, N.S., No. 6, pp. 131–135. [Also in 1963(a).]

   (d) "A Note on Tarski's Definition of Truth," *Mind*, **64**, pp. 388–391.

   (e) "'Content' and 'Degree of Confirmation': A Reply to Dr. Bar-Hillel," *The British Journal for the Philosophy of Science*, **6**, No. 22, pp. 157–163.

   (f) "Verso Una Teoria Liberale Dell'Opinione Pubblica," *Il Politico* (Milano), **20**, No. 2, pp. 181–189. (Italian translation of 1956(e).)

   (g) "The Demarcation between Science and Metaphysics," a contribution to *The Philosophy of Rudolf Carnap*, The Library of Living Philosophers, edited by P. A. Schilpp. Mimeographed. [Also in 1963(a) and (j).]

1956(a) *Misère de l'historicisme*. Paris : Librairie Plon. (French translation of 1954(e).)

   (b) "The Arrow of Time," *Nature*, **177**, No. 4507, p. 538.

   (c) Second printing of the one-volume edition of *The Open Society and Its Enemies*. Princeton University Press.

   (d) Special book-club edition of *The Open Society and Its Enemies*. New York : The Basic Book Club.

   (e) "Die oeffentliche Meinung im Lichte der Grundsaetze des Liberalismus," *Ordo*, **8**, pp. 7–17. Duesseldorf und Muenchen : Helmut Kuepper Verlag. [English translation in 1963(a).]

   (f) "Three Views Concerning Human Knowledge," in *Contemporary British Philosophy*, edited by H. D. Lewis. London : Allen & Unwin. Pp. 355–388. [Also in 1963(a).]

   (g) "Irreversibility and Mechanics," *Nature*, **178**, No. 4529, p. 382. (Continuation of 1956(b).)

   (h) "Reply to Professor Carnap," *The British Journal for the Philosophy of Science*, **7**, No. 27, pp. 244–245.

   (i) "Adequacy and Consistency," *The British Journal for the Philosophy of Science*, **7**, No. 27, pp. 249–256.

1957(a) "Philosophy of Science : A Personal Report," in *British Philosophy in the Mid-Century*, edited by C. A. Mace. London : Allen & Unwin. Pp. 155–191. [Also in 1963(a).]

   (b) "A Second Note on Degree of Confirmation," *The British Journal for the Philosophy of Science*, **7**, No. 28, pp. 350–353. [Also in 1959(a).]

   (c) *La Sociedad Abierta Y Sus Enemigos*, Biblioteca de Psicologia Social y Sociologia, Buenos Aires : Editorial Paidos. (Spanish translation of 1950(a).)

   (d) "Irreversible Processes in Physical Theory," *Nature*, **179**, No. 4573, p. 1297. (Continuation of 1956(b) and (g).)

(e) "The Propensity Interpretation of the Calculus of Probability, and the Quantum Theory," in *Observation and Interpretation, Proceedings of the Ninth Symposium of the Colston Research Society,* University of Bristol. (Vol. IX of the Colston Papers.)

(f) "Irreversibility; or Entropy since 1905," *The British Journal for the Philosophy of Science,* **8**, pp. 151–155.

(g) *The Poverty of Historicism.* London : Routledge & Kegan Paul; Boston, Mass. : The Beacon Press. (Revised English Book edition of 1954(e).)

(h) *The Open Society and Its Enemies;* 3rd rev. ed., with a new *Addendum* (Plato and Geometry) to vol. I. London : Routledge & Kegan Paul.

(i) "The Aim of Science," *Ratio* (Oxford), **1**, No. 1, pp. 24–35. [Italian translation contained in 1962(l).]

(j) *Die offene Gesellschaft und ihre Feinde,* 1. Bd., *Der Zauber Platons.* Bern, Switzerland : Francke Verlag. (German translation by P. K. Feyerabend of Vol. I of 1952(a).)

(k) "Probability Magic or Knowledge out of Ignorance," *Dialectica,* **11**, No. 43/44, pp. 354–374.

(l) "Ueber die Zielsetzung der Erfahrungswissenschaft," *Ratio* (Frankfurt am Main), **1**, No. 1, pp. 21–31. (German translation of 1957(i).)

1958(a) "A Third Note on Degree of Corroboration," *The British Journal for the Philosophy of Science,* **8**, pp. 294–302. [Also in 1959(a).]

(b) "Irreversible Processes in Physical Theory," *Nature,* **181**, pp. 402–403. (Continuation of 1957(d).)

(c) "Das Problem der Nichtwiderlegbarkeit von Philosphien," *Deutsche Universitätszeitung,* **13**, January 1958. (German translation of part of 1958(f).)

(d) *Tre concezioni sulla conoscenza umana.* Milano : Editrice L'Industria. (Italian translation of 1956(f).)

(e) "Back to the Presocratics" (The Presidential Address), *Proceedings of the Aristotelian Society,* N.S., **59**, pp. 1–24. [Also in 1963(a).]

(f) "On the Status of Science and of Metaphysics," *Ratio* (Oxford), **1**, No. 2, pp. 97–115. [Also in 1963(a).]

(g) "Ueber die Moeglichkeit der Erfahrungswissenschaft und der Metaphysik," *Ratio* (Frankfurt), **1**, No. 2. (German translation of 1958(f).)

(h) "On Mr. Roy Harrod's New Argument for Induction," *The British Journal for the Philosophy of Science,* **9**, pp. 221–224.

(i) *Die offene Gesellschaft und ihre Feinde,* 2. Bd., *Falsche Propheten, Hegel, Marx & die Folgen,* Bern, Switzerland : Francke Verlag. (German translation by P. K. Feyerabend of Vol. II of 1952(a).)

1959(a) *The Logic of Scientific Discovery.* London : Hutchinson (January); New York : Basic Books (March). (Contains English translations of 1934(b), 1933, 1935; also 1938(a), and other material.)

(b) *The Poverty of Historicism* (Arabic translation of 1957(g) by Abdulhamid Sabra). Alexandria : Galal.

(c) *A sociedade democràtica e seus inimigos.* Belo Horizonte, Brazil : Editora Itatiata. (Portuguese translation of 1950(a).)

(d) "Ueber die Unwiderlegbarkeit philosophischer Theorien," *Forum,* **6**, Heft 61, January 1959, Vienna. (A revised version of 1958(c).)

(e) "The Propensity Interpretation of Probability," *The British Journal for the Philosophy of Science,* **10**, pp. 25–42.

(f) "Testability and 'Ad Hocness' of the Contraction Hypothesis," *The British Journal for the Philosophy of Science,* **10**, p. 50.

(g) "Woran glaubt der Westen," in *Erziehung zur Freiheit,* edited by A. Hunold. Zurich und Stuttgart : Eugen Rentsch Verlag.

(h) "On the Sources of Our Knowledge," *The Indian Journal of Philosophy,* **1**, No. 1, pp. 3–7. (Forms part of 1960(d).)

(i) "On Subjunctive Conditionals with Impossible Antecedents," *Mind,* **68**, N.S., No. 272, pp. 518–520.

(j) "Has History Any Meaning?" in *The Philosophy of History in Our Time,* edited by Hans Meyerhoff. New York : Doubleday Anchor Books. (Part of chapter 25 of 1950(a).)

(k) "Critical Rationalism," in *Philosophy for a Time of Crisis, An Interpretation with Key Writings by Fifteen Great Modern Thinkers,* edited by Adrienne Koch. New York : Dutton & Co. (Part of chapter 24 of 1950(a).)

(l) "Prediction and Prophecy in the Social Sciences," in *Theories of History,* edited by Patrick Gardiner. Glencoe, Illinois : The Free Press. (A revised version of 1948(e).) [Also in 1963(a).]

1960(a) *The Logic of Scientific Discovery,* revised. London : Hutchinson.

(b) "Probabilistic Independence and Corroboration by Empirical Tests," *The British Journal for the Philosophy of Science,* **10**, pp. 315–318.

(c) *The Poverty of Historicism,* revised 2nd ed. London : Routledge & Kegan Paul.

(d) "On the Sources of Knowledge and of Ignorance," *Proceedings of the British Academy,* **46**, pp. 39–71. [Also in 1963(a).]

1961(a) "The Moral Theory of Historicism," in *Society, Law, and Morality,* edited by Frederick A. Olafson. Englewood Cliffs, N.J. : Prentice-Hall Inc. (Part of chapter 22 of 1950(a).)

(b) *The Poverty of Historicism.* A Routledge Paperback (revised 2nd edition).

(c) *The Poverty of Historicism,* 1st and 2nd Japanese editions (translated by Saburo Ichii and Osamu Kuno). Tokyo : Chuo-Koron Sha, Inc.

(d) "Selbstbefreiung durch das Wissen," in *Der Sinn der Geschichte,* edited by Leonhard Reinisch, Muenchen : Beck'sche Verlagsbuchhandlung.

(e) *The Logic of Scientific Discovery,* 3rd printing, further revised, paperback. New York : Science Editions Inc.

(f) *On the Sources of Knowledge and of Ignorance,* Annual Philosophical Lecture, Henriette Hertz Trust, British Academy. London : Oxford University Press. (Separate edition of 1960(d).)

(g) *La Miseria del Historicismo.* Madrid : Taurus. (Spanish translation of 1957(g).)

(h) "Philosophy and Physics," *Atti del XII Congresso Internazionale di Filosofia* (Venezia 1958), vol. II, Firenze 1960, pp. 367–374.

1962(a) "On Carnap's Version of Laplace's Rule of Succession" *Mind,* **71,** N.S., No. 281, pp. 69–73.

(b) "Historical Explanation : An Interview." (Replies to questions by Michael Tanner, John Dunn, and Alistair Young.) *Cambridge Opinion,* **28,** pp. 20–25.

(c) *The Open Society and Its Enemies:* vol. I, *The Spell of Plato,* 4th ed., further revised with two new *Addenda* : "The Dating of the Theaetetus," and "Reply to a Critic." London : Routeledge & Kegan Paul.

(d) *Ibid.:* vol. II, *The High Tide of Prophecy,* 4th ed. further revised, with a new *Addendum* : "Facts, Standards and Truth : A Further Criticism of Relativism." London : Routledge & Kegan Paul.

(e) *Ibid.,* volume I, a Routledge Paperback.

(f) *Ibid.,* volume II, a Routledge Paperback.

(g) "Selbstbefreiung durch das Wissen" (second edition of 1961(d), Muenchen 1962).

(h) "Julius Kraft 1898–1960," *Ratio* (Oxford), **4,** No. 1, pp. 2–12.

(i) "Julius Kraft 1898–1960," *Ratio* (Frankfurt), **4,** No. 1, pp. 2 ff. (German translation of 1962(h).).

(j) "Some Comments on Truth and the Growth of Knowledge," in *Proceedings of the 1960 International Congress for Logic, Methodology, and Philosophy of Science.* Edited by E. Nagel, P. Suppes, and A. Tarski. Stanford : Stanford University Press.

(k) "Ueber Geschichtsschreibung und über den Sinn der Geschichte," in *Geist und Gesicht der Gegenwart,* edited by Otto Molden. Zürich : Europa Verlag. (Revised version of part of chapter 25 of 1958(i).)

(l) "Lo Scopo della Scienza," *La Scuola in Azione,* **13,** Milano 1962, pp. 5–22. (Italian translation of 1957(i).)

(m)"Die Logik der Sozialwissenschaften," *Kölner Zeitschrift für Soziologie und Sozialpsychologie,* Köln, **14,** No. 2, pp. 233–248.

(n) "Critical Rationalism," in *Philosophy for a Time of Crisis.* Essays by Albert Einstein, E. M. Foster, Karl R. Popper, Bertrand Russell, edited by Kozo Tada. Tokyo : Kinseido Ltd.

(o) "A Comment on the New Prediction Paradox," *The British Journal for the Philosophy of Science,* **13,** p. 51.

(p) "On the Sources of Knowledge and of Ignorance," *Encounter,* **108,** Sept., pp. 42–57. (Abbreviated version of 1960(d) and 1961(f); also in 1963(a).)

(q) "The Propensity Interpretation of the Calculus of Probability and the Quantum Theory," in *Observation and Interpretation,* edited by S. Körner. New York : Dover Publications. (Paperback edition of 1957(e).)

1963(a) *Conjectures and Refutations: The Growth of Scientific Knowledge.* London : Routledge and Kegan Paul, and New York : Basic Books.

(b) "Kindai Ninshikiron no Shinwa : Chishiki to Muchi no Gensen ni tsuite" (The Myth of Modern Epistemology : On the Sources of Knowledge and of Ignorance), *Jiyu* (Tokyo), **5**, No. 1. (Japanese translation of 1960(d) by Tooru Yoshimura.)

(c) *La Lógica de la Investigación Científica*, Madrid : Editorial Tecnos. (Spanish translation of 1960(a) by Sanchez de Zavala.)

(d) "Kirk on Heraclitus, and on Fire as the Cause of Balance," *Mind*, **72**, N.S., No. 287, pp. 386–392.

(e) "The Erewhonians and the Open Society," *ETC: A Review of General Semantics* (San Francisco), **20**, No. 1, pp. 5–22.

(f) "La Opinion Publica a la Luz de los Principios del Liberalismo," in *La Economia Mercado*. Madrid : Sociedad de Estudios y Publicaciones. (Spanish translation of 1956(e).)

(g) "Creative and Non-Creative Definitions in the Calculus of Probability," *Synthèse*, **15**, No. 2, pp. 167–186.

(h) "Science : Problems, Aims, Responsibilities," *Federation Proceedings*, Federation of American Societies for Experimental Biology, **22**, No. 4, Part I, pp. 961–972.

(i) *Jiyu Shakai No Tetsugaku to sono Ronteki* (The Philosophy of the Free Society and its Enemies). Osaka City : The Izumiya Co., Inc. (Japanese translation of 1962(c) and (d) by Hiromichi Takeda.)

(j) "The Demarcation between Science and Metaphysics," in *The Philosophy of Rudolf Carnap, The Library of Living Philosophers,* edited by Paul A. Schilpp. Chicago : The Open Court Publishing Co. (Also in 1963(a).)

(k) "Über die Unwiderlegbarkeit philosophischer Theorien," in *Club Voltaire*, I, edited by Gerhard Szczesny. Munich : Szczesny Verlag. Pp. 271–279.

(l) *The Open Society and Its Enemies,* 4th Ed., further rev., with two new *Addenda* ("The Dating of the Theaetetus" and "Reply to a Critic") to vol. I, and a new *Addendum* ("Facts, Standards, and Truth : A Further Criticism of Relativism") to vol. II. Princeton, N.J. : Princeton University Press.

(m)*Ibid.*, in two volumes, Harper Torchbooks, The Academy Library. New York and Evanston : Harper & Row.

(n) "Plato as Enemy of the Open Society," in *Plato: Totalitarian or Democrat?* edited by Thomas L. Thorson, a Spectrum book. Englewood Cliffs, N.J. : Prentice-Hall Inc. (Part of 1950(a).)

(o) "Problems of Scientific Knowledge," *Bulletin of the International House of Japan* (Tokyo), No. 12, October, pp. 23–28.

1964(a) *The Poverty of Historicism* (rev. 2nd ed.), Harper Torchbooks, The Academy Library. New York and Evanston : Harper & Row.

Science and Technology Studies

This series includes monographs in the philosophy, sociology, and history of science and technology. "Science" and "technology" are taken in the broad sense: the former as including mathematics and the natural and social sciences, and the latter the social technologies or policy sciences. It is hoped that this series will help raise the level of the current debate on science and technology, by sticking to the standards of rationality and the concern for empirical tests that are being challenged by the current Counter-Enlightenment wave.